HYDRODYNAMICS

M.M. Rahman

ATLANTIC

PUBLISHERS & DISTRIBUTORS (P) LTD

Published by
ATLANTIC
PUBLISHERS & DISTRIBUTORS (P) LTD
7/22, Ansari Road, Darya Ganj,
New Delhi-110002
Phones : +91-11-23273880, 23275880, 23280451
Fax : +91-11-23285873
Web : www.atlanticbooks.com
E-mail : info@atlanticbooks.com

Branch Office
5, Nallathambi Street, Wallajah Road,
Chennai-600002
Phones : +91-44-64611085, 32413319
E-mail : chennai@atlanticbooks.com

Printed in India at Nice Printing Press, A-33/3A, Site-IV, Industrial Area, Sahibabad, Ghaziabad, U.P.

PREFACE

This book *Hydrodynamics* is intended to cover the subject syllabi of M.A. and M.Sc. of all Indian universities as well as competitive examinations of equivalent standard.

After teaching for more than twenty years to honours and postgraduate students, I have experienced that average students find most of the existing books either too difficult to comprehend or too inadequate to meet their requirements. This book is complete in the sense that it is rich in subject matter and coverage, and yet easily accessible. The stress has been given on fundamental ideas, so that students may grasp the topic easily. Each topic has been introduced and developed in a simple and straightforward manner as much as possible and made well within the comprehension of students of all categories. Chapters are arranged in a systematic order and the fundamental principles involved have been amply illustrated by diagrams and worked-out examples given correspondingly, so that the students may understand the methods clearly and easily. Most of the questions have been taken from the question papers of various universities in the hope that these will facilitate thorough examination preparation. I am hopeful that the book will prove extremely beneficial to the postgraduate students of Mathematics as well as aspirants of UPSC, State Public Services and other competitive examinations.

The present book is not an outcome of my solo efforts. I have been fortunate to receive help from many people in accomplishing this task. I am thankful to all those persons who have helped directly or indirectly in bringing out this book.

More particularly, I am grateful to my students whose constant inspiration and help encouraged me to complete the present book. I also express my sincere thanks to my wife N. Jahan and sons Tauseef, Tauqeer and Ayan for their constant cooperation without which this book would not have seen the light of day.

Last but not least, I am thankful to Dr. K.R. Gupta, Chief Editor of Atlantic Publishers and Distributors (P) Ltd., for taking keen interest in publishing this book.

M.M. Rahman

CONTENTS

1. **Kinematics** 1-74

1.1. Hydrodynamics, Fluid, Shearing Stress, Perfect Fluid, Viscous or Real Fluid, Difference between Perfect Fluid and Real Fluid, Viscosity, Velocity of a Fluid Particle, Methods of Describing Fluid Motion, Lagrangian Method, Eulerian Method, Relation between the Lagrangian and Eulerian Methods, Local and Individual Time Rates of Change, Relation between the Local and Individual Time Rates, Acceleration, Kinds of Motion, Uniform and Non-Uniform Flows, One Dimensional, Two Dimensional and Three Dimensional Flow, Lines of Flow or Streamlines, Path Lines or Path of a Particle, Difference between Streamlines and Path Lines, Stream Tube or Tube of Flow, Stream Surface, Stream Filament, Streak Lines or Filament Lines, Velocity Potential 1

2. **Equations of Motion** 75-175

2.1. Euler's Equation of Motion 75

2.2. Euler's Dynamical Equation (Cartesian Coordinates) 78

2.3. Conservative Field of Force 80

2.4. Pressure Equation 80

2.5. Bernoulli's Equation for Steady Motion 82

2.6. If the motion of an ideal fluid, for which density is a function of pleasure p only is steady and the

external forces are conservative, then there exists a family of surfaces which contain the stream-lines and vortex lines 84
2.7. Integration of the Equation of Motion, Cartesian Form 86
2.8. Lagrange's Equation of Motion 88
2.9. Helmboltz's Vorticity Equation 89
2.10. Cauchy's Integrals 91
2.11. General Equation of Impulsive Motion 97
2.12. Equation of Motion under Impulsive Forces (Cartesian Form) 100
2.13. Energy Equation 102

3. Motion in Two Dimensions 176-256
3.1. Motion in Two Dimensions 176
3.2. Lagrange's Stream Function or Current Function 177
3.3. Physical Significance of Stream Function 178
3.4. Irrotational Motion in Two Dimensions 179
3.5. Complex Potential and Velocity 180
3.6. Two Dimensional Sources, Sinks, Strength 181
3.7. Complex Potential Due to a Source 181
3.8. Two Dimensional Doublets 183
3.9. Complex Potential for a Doublet 183
3.10. Image 184
3.11. To Find the Image of a Source with Respect to a Straight Line 184
3.12. Image of a Doublet in a Straight Line 185
3.13. Image of a Source with Respect to a Circle in Two Dimensions 185
3.14. Image of a Doublet with Regard to a Circle 187
3.15. The Milni Thomson Circle Theorem or Circle Theorem 188
3.16. To Find Image of a Source with Respect to a Circle of Radius *a* with the help of Circle Theorem 189
3.17. To Find the Image of a Doublet with Respect to a Circle by Circle Theorem 191

3.18. Theorem of Blosius 193
3.19. Conformal Representation or Transformation of a Mapping 196
3.20. Application of Conformal Representation to Hydrodynamics 199
3.21. A Fundamental Theorem 200

4. Motion of Cylinders 258-343

Part I—Motion of Circular Cylinder

4.1. Introduction 258
4.2. Boundary Conditions for the Stream Function 258
4.3. General Motion of a Cylinder in Two Dimensions 259
4.4. Motion of a Circular Cylinder in a Uniform Stream 260
4.5. To Discuss the Motion in the Case of a Liquid Streaming Past a Fixed Circular Cylinder of Radius *a* with Velocity *U* 262
4.6. Application of Circle Theorem 263
4.7. Initial Motion Between Two Coaxial Circular Cylinders 264
4.8. Kinetic Energy of a Finite Liquid 267
4.9. Equation of Motion of a Circular Cylinder without Circulation 267
4.10. Equation of Motion of the Circular Cylinder with Circulations 270
4.11. Circulation about a Circular Cylinder 273
4.12. Streaming and Circulation about a Fixed Circular Cylinder 274

Part II—Elliptic Cylinder

4.13. Elliptic Coordinates 304
4.14. Solution of Laplace's Equation 305
4.15. The Joukowski Transformation 305
4.16. Streaming Past a Fixed Elliptic Cylinder 306
4.17. Motion of a Moving Elliptic Cylinder 308
4.18. Motion of a Fixed Elliptic Cylinder 310
4.19. Motion of a Rotating Elliptic Cylinder 312

4.20. Kinetic Energy of the Liquid Contained in Rotating Elliptic Cylinder 315
4.21. Path of the Particles 317
4.22. Circulation about an Elliptic Cylinder 319

5. Irrotational Motion in Three Dimensions (Motion of a Sphere) 344-410
5.1. Introduction 344
5.2. Equation of Continuity 344
5.3. To Find the Velocity Potential when a Sphere of Radius *a* is Moving with Velocity *U* in a Liquid at Rest at Infinity 345
5.4. Liquid Streaming Past a Fixed Sphere 347
5.5. Equations of Motion of a Sphere 347
5.6. Motion of a Sphere Under Gravity 351
5.7. Concentric Spheres (Problem of Initial Motion) ... 351
5.8. Stoke's Stream Function 354
5.9. A Property of Stoke's Stream Function 355
5.10. Irrotational Motion 356
5.11. Polar Coordinates 357
5.12. Solution for ψ 358
5.13. Motion of a Solid of Revolution along the Axis ... 359
5.14. Three-Dimensional Source and Sink 362
5.15. To Determine ϕ due to a Three-dimensional Source μ at the Origin 362
5.16. Three-Dimensional Doublet 362
5.17. Value of Stoke's Stream Function in Simple Cases 363
5.18. Image of a Source with Regard to a Sphere 364
5.19. Image of a Doublet Relative to a Sphere or Image of a Radial Doublet in a Sphere 366

6. General Theory of Irrotational Motion 411-453
6.1. Some Elementary Definitions 411
6.2. Irrotational Motion 412
6.3. Kinetic Energy of a Finite Liquid 414
6.4. Kinetic Energy of Infinite Liquid 415
6.5. Kelvin's Minimum Energy Theorem 416

6.6. Mean Value of Velocity Potential over a Spherical Surface 418
6.7. Uniqueness Theorem 419
6.8. Green's Theorem 420
6.9. Kelvin's Circulation Theorem 421

7. Vortex Motion 454-551

7.1. Introduction 454
7.2. Some Definitions 454
7.3. Properties of Vortex Filament 455
7.4. Strength of Vortex 457
7.5. Rectilinear Vortices 457
7.6. Complex Potential of a Two-Dimensional Vortex Motion 458
7.7. Motion due to *m* Vortices 460
7.8. Single Vortex in the Field of Several Vortices 461
7.9. Centre of Vortices 462
7.10. System of Two Vortex Filaments 463
7.11. Vortex Pair 465
7.12. Vortex Doublet or Dipole 467
7.13. Spiral Vortex 469
7.14. Image of a Vortex Filament in a Plane 470
7.15. Image of a Vortex in a Quadrant 471
7.16. Vortex Inside an Infinite Circular Cylinder 472
7.17. Vortex Outside a Circular Cylinder 473
7.18. Image of Vortex Outside a Circular Cylinder 474
7.19. Image of a Vortex Inside a Circular Cylinder 476
7.20. Vortex Rows 477
7.21. Infinite Single Row of Vortices of Equal Length ... 477
7.22. Two Infinite Rows of Vortices 480
7.23. Karman's Vortex Street 482
7.24. Rectilinear Vortex with Circular Section 483
7.25. Rankine's Combined Vortex 486
7.26. Rectilinear Vortex with Elliptic Section 488
7.27. Vortex between Parallel Walls 491
7.28. Conformal Transformation, Routh's Theorem 493
7.29. Vortex Sheet 496

8. **Waves** 552-622
8.1. Introduction 552
8.2. Wave Motion 552
8.3. Some Definitions 554
8.4. Standing or Stationary Waves 555
8.5. Kinds of Wave Motion 556
8.6. Kinematical Surface Condition for Wave Profiles of Small Elevation 557
8.7. Velocity of Propagation of Liquid Waves 559
8.8. Paths of Particles in a Canal of Finite Depth 563
8.9. Progressive Wave on Deep Water 565
8.10. Path of Particles in a Deep Canal 566
8.11. Standing or Stationary Waves on the Surface of a Canal of Finite Depth 567
8.12. Standing or Stationary Wave on the Surface of a Deep Canal 571
8.13. Kinetic Energy and Potential Energy of a Wave per Unit Thickness 573
8.14. Energy of a Progressive Wave 574
8.15. Energy of Stationary Waves 576
8.16. Steady Motion of Progressive Waves 577
8.17. Steady Motion of Progressive Waves on Deep Water 580
8.18. Waves at the Common Surface of Two Liquids ... 582
8.19. Waves at the Common Surface of Two Liquids with Upper Surface Free 586
8.20. Motion of a Surface Particle 590
8.21. Energy of a Long Wave 591
8.22. Long Waves at the Common Surface of Two Liquids Bounded Above and Below by Two Fixed Horizontal Planes 592
8.23. Group Velocity 594
8.24. Transmission of Energy 596
8.25. Surface Tension and Capillary Waves 597
8.26. Capillary Waves at the Common Surface of Two Liquids 598

List of Important Formulae 623

1 KINEMATICS

1.1. Definition and Basic Concepts

1. Hydrodynamics. Hydrodynamics is that branch of Mathematics which deals with the motion of fluids or that of bodies in contact with fluids.

2. Fluid. By fluid we mean a substance which is capable of flowing.

Actual fluids are divided into two categories: (i) liquids and (ii) gases. We regard liquids as incompressible fluids for all practical purposes and gases as compressible fluids. Actual fluids have five physical properties: density, volume, temperature, pressure and viscosity.

3. Shearing stress. Two types of forces act on a fluid element. One of them is body force and the other is surface force. The body force is proportional to the mass of the body on which it acts. The body force is proportional to the mass of the body on which it acts while the surface force acts on the boundary of the body and so it is proportional to the surface area.

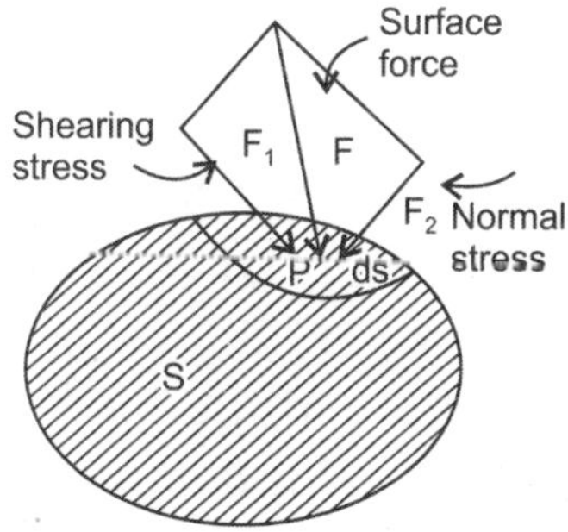

Fig. 1.1

Let F be a surface force acting on an elementary surface area ds at the point P of the surface S.

Let F_1 and F_2 be the resolved parts of F in the direction of tangent and normal at P. The normal force per unit area is called normal stress and is also called pressure. The tangential force per unit area is called shearing stress. Hence, F_1 is a shearing stress and F_2 is a normal stress.

4. Perfect fluid. A fluid is said to be perfect if it does not exert any shearing stress however small. From the definition of shearing stress and body force it is clear that body force per unit area at every point of surface of a perfect fluid acts along the normal to the surface at that point.

The following have the same meaning: perfect, frictionless, non-viscous, inviscous and ideal.

5. Viscous or Real fluid. A viscous fluid is an aggregate of particles (a fluid) which exert friction on another particles, when in motion i.e. offers resistance to the motion of a body in it. The following have the same meaning: compressible and real.

6. Difference between Perfect fluid and Real fluid. The main difference between real fluid and perfect fluid is that stress across any plane surface of perfect fluid is always normal to the surface, while in case of real fluid both shearing stress and normal stress exist.

7. Viscosity. The property by which the particles of a fluid exert friction on one another is called viscosity. Viscosity is also known as internal friction of fluid. It may be noted that viscosity comes into play when the fluid is in motion. Hence, for fluids at rest, there is no tangential action or shearing stress. All known fluids have this property in varying degrees. Viscosity of glycerine and oil is large in comparison to viscosity of water or gases.

8. Velocity of a fluid particle. Let a fluid particle be at P and Q at time t and $t + \delta t$ such that $\overrightarrow{OP} = \vec{r}$ and $\overrightarrow{OQ} = \vec{r} + \delta\vec{r}$.

The velocity of the particle at P is then defined by the vector

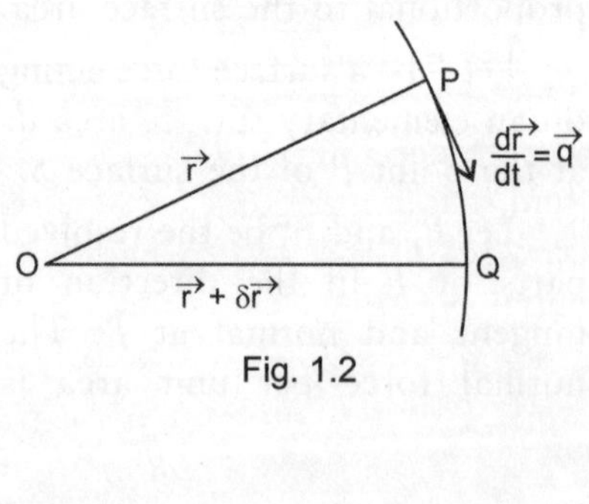

Fig. 1.2

$$\vec{q} = \underset{\delta t \to 0}{Lt} \frac{(\vec{r} + \delta\vec{r}) - \vec{r}}{(t + \delta t) - t}$$

$$= \underset{\delta t \to 0}{Lt} \frac{\delta\vec{r}}{\delta t} = \frac{d\vec{r}}{dt}$$

Thus, the velocity $\vec{q}$ is a function of $\vec{r}$ and t can be expressed mathematically as

$$\vec{q} = \frac{d\vec{r}}{dt} = \vec{f}\,(\vec{r}, t)$$

If u, v, w are the components of $\vec{q}$ along the Cartesian coordinate axes and $\vec{i}, \vec{j}, \vec{k}$ the unit vectors along the axes, then

$$\vec{q} = u\vec{i} + v\vec{j} + w\vec{k}$$

9. Methods of describing Fluid motion. There are two methods for describing motion in a fluid system. The first in which the history of individual fluid particles is described i.e. to consider a particle point of view, is called the *Lagrangian Method* while the second, which concentrates on fixed points of space filled by the fluid i.e. to adopt a field point of view is called the *Eulerian Method.*

10. Lagrangian method. A description of fluid motion based on studying the history of an individual fluid particles is called Lagrangian. The flow quantities are represented as functions of time and of the choice of a material element of fluid. In this method, any particle of the fluid is selected and pursue it on its onward, course observing the changes in path, velocity, density or any other property associated with it at each point and each instant. The selected fluid particle is identified in such a way that its linear extension is not involved.

Let a fluid particle be initially at the point P_0 (a, b, c). After a lapse of time t, let the same fluid particle be at P (x, y, z). Since the particle initially chosen, is any particle in the fluid, t is not the only independent variable. The coordinates (x, y, z) of the particle will be functions of t and also of its initial position (a, b, c) so that

$$x = f_1\,(a, b, c, t),\; y = f_2\,(a, b, c, t),\; z = f_3\,(a, b, c, t)$$

If the motion is everywhere continuous, then f_1, f_2, f_3 are continuous functions of a, b, c and t such that their first and second partial derivatives with respect to a, b, c and t exists. Then the components of the velocity and acceleration of the fluid particle are given by

$$\ddot{x} = \frac{\partial^2 f_1}{\partial t^2}, \ddot{y} = \frac{\partial^2 f_2}{\partial t^2}, \ddot{z} = \frac{\partial^2 f_3}{\partial t^2}$$

$$\dot{x} = \frac{\partial f_1}{\partial t}, \dot{y} = \frac{\partial f_2}{\partial t}, \dot{z} = \frac{\partial f_3}{\partial t}$$

Remark. While in ordinary dynamics, the coordinates (x, y, z) of a particular particle are dependent on time t, it is not so in the case of a fluid particle whose coordinates (x, y, z) are functions of four independent variables because three other variables are necessary to specify which fluid particle has been chosen.

11. Eulerian method. In this method, we select a general point in the space occupied by the fluid and study whatever changes take place in velocity pressure and density at this point rather than the variations of velocities and accelerations of particle as they follow their own paths. For example, if u, v, w be the components of velocity at the point (x, y, z) at time t, then u, v, w are functions of x, y, z and t. For a particular value of t, u, v, w define the motion at all points in the fluid at that time. As t changes, u, v, w tells us what happens at the point (x, y, z) and thus give a history of the variations of velocity at that point. So unless some further meanings are assigned to the symbols $\frac{dx}{dt}, \frac{d^2x}{dt^2}$ etc. are meaningless.

Remark. In Eulerian method, we study the motion of a fluid particle at a fixed point, whereas in Lagrangian method we study the motion of every fluid particle at every point. Hence, Euler's method corresponds to local time rate of change and Lagrangian method corresponds to individual time rate of change.

12. Relation between the Lagrangian and Eulerian methods. In order to establish relationship between these two methods, we establish a relation between the particle parameters and points in the space.

(i) ***Lagrangian to Eulerian.*** Let Q be some quantity defined in terms of Lagrangian description

$$Q = Q\,(a, b, c, t) \qquad \dots (1)$$

We shall express a, b, c in terms of the coordinates x, y, z of a point in space. In Lagrangian method, it is defined as

$$x = f_1(a, b, c, t),\ y = f_2(a, b, c, t),\ z = f_3(a, b, c, t) \qquad \dots (2)$$

By solving these relations to obtain a, b, c in terms of Eulerian variables, x, y, z and t, we have

$$a = g_1(x, y, z, t),\ b = g_2(x, y, z, t),\ c = g_3(x, y, z, t) \qquad \dots (3)$$

From (1) and (3), we get

$$Q = Q[g_1(x, y, z, t), g_2(x, y, z, t), g_3(x, y, z, t)] \qquad \dots (4)$$

which represents the Eulerian description.

(ii) ***Eulerian to Lagrangian.*** Let Q be some quantities defined in terms of Eulerian description

$$Q = Q(x, y, z, t) \qquad \dots (5)$$

We shall express x, y, z in terms of the particle parameters a, b, c. Let u, v, w be the velocity components at the point x, y, z at any instant t, which is defined as

$$u = F_1(x, y, z, t),\ v = F_2(x, y, z, t),\ w = F_3(x, y, z, t) \qquad \dots (6)$$

Again, from the Lagrangian description, we have

$$u = \frac{\partial x}{\partial t},\ v = \frac{\partial y}{\partial t},\ w = \frac{\partial z}{\partial t} \qquad \dots (7)$$

where x, y, z are functions of the variables a, b, c and t.

From (6) and (7), we have

$$\left.\begin{aligned} &\frac{\partial x}{\partial t} = F_1(x, y, z, t),\ \frac{\partial y}{\partial t} = F_2(x, y, z, t), \\ &\frac{\partial z}{\partial t} = F_3(x, y, z, t) \end{aligned}\right\} \qquad \dots (8)$$

which represents the first order differential equation. By integrating, we have

$$\left.\begin{aligned} x = f_1(x_0, y_0, z_0, t), y = f_2(x_0, y_0, z_0, t), \\ z = f_3(x_0, y_0, z_0, t) \end{aligned}\right\} \quad \dots (9)$$

where x_0, y_0, z_0 are the initial values of x, y, z at initial instant assumed to be constants of integration. Taking $a = x_0$, $b = y_0$ and $c = z_0$ respectively, we have

$$Q = Q[f_1(a, b, c, t), f_2(a, b, c, t), f_3(a, b, c, t)]$$

which represents the Lagrangian description.

13. Local and individual time rates of change. We consider a fluid motion associated with scalar point of function $\phi(x, y, z, t)$ or $\phi(\vec{r}, t)$. Then keeping the point $P(\vec{r})$ fixed, the change in ϕ during an interval of time δt is,

$$\phi(x, y, z, t + \delta t) - \phi(x, y, z, t)$$

or $\phi(\vec{r}, t + \delta t) - \phi(\vec{r}, t)$

Hence, the local time rate of change is given by

$$\frac{\partial \phi}{\partial t} = \underset{\delta t \to 0}{Lt} \frac{\phi(\vec{r}, t + \delta t) - \phi(\vec{r}, t)}{\delta t}$$

Now keeping the particle fixed, change in ϕ is

$$\phi(\vec{r} + \delta\vec{r}, t + \delta t) - \phi(\vec{r}, t)$$

where $\delta\vec{r}$ is the change in the position of the fixed particle during the short time δt. Therefore,

$$\frac{d\phi}{dt} = \underset{\delta t \to 0}{Lt} \frac{\phi(\vec{r} + \delta\vec{r}, t + \delta t) - \phi(\vec{r}, t)}{\delta t}$$

gives the individual time rate of change.

14. Relation between the Local and Individual time rates. Let u, v, w be the components of velocity $\vec{q}$ along the coordinates axes, so that

$$\vec{q} = u\vec{i} + v\vec{j} + w\vec{k}$$

where $\dfrac{dx}{dt} = u$ etc.

Now $\phi = \phi(x, y, z, t)$

$$\therefore \quad \frac{d\phi}{dt} = \frac{\partial\phi}{\partial x}\frac{dx}{dt} + \frac{\partial\phi}{\partial y}\frac{dy}{dt} + \frac{\partial\phi}{\partial z}\frac{dz}{dt} + \frac{\partial\phi}{\partial t}$$

$$= u\frac{\partial\phi}{\partial x} + v\frac{\partial\phi}{\partial y} + w\frac{\partial\phi}{\partial z} + \frac{\partial\phi}{\partial t}$$

$$= \left(u\vec{i} + v\vec{j} + w\vec{k}\right)\cdot\left(\vec{i}\frac{\partial\phi}{\partial x} + \vec{j}\frac{\partial\phi}{\partial y} + \vec{k}\frac{\partial\phi}{\partial z} + \frac{\partial\phi}{\partial t}\right)$$

$$= \left(\vec{q}\cdot\vec{\nabla}\right)\phi + \frac{\partial\phi}{\partial t}$$

Thus,

$$\frac{d\phi}{dt} = \left(\vec{q}\cdot\vec{\nabla}\right)\phi + \frac{\partial\phi}{\partial t}$$

This is the relation between two time rates.

Note. A similar expression for a vector point function $\vec{f}$ can be derived in the form

$$\frac{d\vec{f}}{dt} = \frac{\partial\vec{f}}{\partial t} + \left(\vec{q}\cdot\vec{\nabla}\right)\vec{f}$$

15. Acceleration. Acceleration is defined as the total derivative of velocity w.r.t. time

$$\vec{a} = \frac{d\vec{q}}{dt} = \frac{\partial\vec{q}}{\partial t} + \left(\vec{q}\cdot\vec{\nabla}\right)\vec{q} \qquad \text{... (1)}$$

Let $\vec{a} = a_1\vec{i} + a_2\vec{j} + a_3\vec{k}$

$\vec{q} = u\vec{i} + v\vec{j} + w\vec{k}$

then from (1), we have

$$\vec{a} = \frac{d\vec{q}}{dt} = \left(\frac{\partial}{\partial t} + u\frac{\partial}{\partial x} + v\frac{\partial}{\partial y} + w\frac{\partial}{\partial z}\right)\vec{q}$$

$$\Rightarrow \quad a_1\vec{i} + a_2\vec{j} + a_3\vec{k}$$

$$= \left(\frac{\partial}{\partial t} + u\frac{\partial}{\partial x} + v\frac{\partial}{\partial y} + w\frac{\partial}{\partial z}\right)\left(u\vec{i} + v\vec{j} + w\vec{k}\right)$$

Equating the coefficients of $\vec{i}, \vec{j}, \vec{k}$ from both the sides, we get

$$a_1 = \left(\frac{\partial}{\partial t} + u\frac{\partial}{\partial x} + v\frac{\partial}{\partial y} + w\frac{\partial}{\partial z}\right)u$$

$$a_2 = \left(\frac{\partial}{\partial t} + u\frac{\partial}{\partial x} + v\frac{\partial}{\partial y} + w\frac{\partial}{\partial z}\right)v$$

$$a_3 = \left(\frac{\partial}{\partial t} + u\frac{\partial}{\partial x} + v\frac{\partial}{\partial y} + w\frac{\partial}{\partial z}\right)w$$

16. Kinds of motion

1. Streamline motion. A fluid motion is said to be streamline motion, if the path traced out by it from parts of some regular curves.

2. Turbulent motion. A fluid motion is said to be turbulent if the paths are widely irregular.

3. Steady motion. A fluid motion is said to be steady if the condition at any point in the fluid at any time remains the same for all time. Thus a fluid motion is said to be steady if $\frac{\partial f}{\partial t} = 0, \frac{\partial p}{\partial t} = 0, \frac{\partial \vec{q}}{\partial t} = 0$, where $f, p, \vec{q}$ denote density, pressure and velocity of the fluid respectively.

4. Rotational motion. A fluid motion is said to be rotational if $\overrightarrow{W} = \text{curl}\, \vec{q} \neq 0$ at every time and at every point.

5. Irrotational motion. A fluid motion is said to be irrotational if $\overrightarrow{W} = \text{curl}\, \vec{q} = 0$ at every point and at every time.

17. Some definitions

1. Uniform and non-uniform flows. If at every point the velocity vector is identical in magnitude and direction at any given instant, or, the conditions and properties are independent of the coordinate of the direction in which the fluid is moving then the motion is said to be *uniform*. If the flow characteristics, at any given time, change with distance, it is said to be *non-uniform*.

For example, a liquid being pumped through a long straight pipe has uniform flow, whereas a liquid flowing through a curved pipe has non-uniform flow.

2. ***One dimensional, two dimensional and three dimensional flow.*** *One dimensional flow* neglects variations or changes in velocity, pressure etc., transverse to the main flow direction. The flow characteristics vary only in the direction of flow.

The flow is said to be dimensional if the velocity vector $\vec{q}$ is everywhere at right angles to a certain direction and independent of displacements parallel to that direction.

The components of $\vec{q}$ (u, v, w) in a two dimensional flow are (u, v, o) where u and v are independent of z. The flow conditions are identical in planes perpendicular to the z-axis.

A flow is said to be three dimensional of the velocity components (u, v, w) in mutually perpendicular directions are the functions of space coordinates and time x, y, z and t.

3. ***Lines of flow or streamlines.*** A streamline or line of flow is a curve such that the tangent at any point of it at any instant of time is in the direction of motion of the fluid at that point.

Thus velocity at any point is along the tangent to the streamline through that point.

Let $\vec{r} = x\vec{i} + y\vec{j} + z\vec{k}$ be the position vector of a point P on a straight line and let $\vec{q} = u\vec{i} + v\vec{j} + w\vec{k}$ be fluid velocity at P. Then $\vec{q}$ is parallel to $d\vec{r}$ at P on the streamline and so

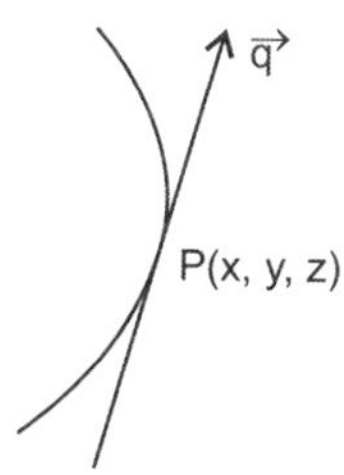

Fig. 1.3

$$\vec{q} \times d\vec{r} = 0$$

$$\Rightarrow \begin{vmatrix} \vec{i} & \vec{j} & \vec{k} \\ u & v & w \\ dx & dy & dz \end{vmatrix} = 0$$

$$\Rightarrow \vec{i}\,(vdz - wdy) - \vec{j}\,(udz - wdx) + \vec{k}\,(udy - vdx) = 0$$

Hence $vdz - wdy = 0, wdx - udz = 0, udy - vdx = 0$

So that $\frac{dx}{u} = \frac{dy}{v} = \frac{dz}{w}$

4. ***Path lines or path of a particle.*** A path line is a curve, which a particular fluid particle describes during its motion.

The differential equation of path lines are

$$\frac{d\vec{r}}{dt} = \vec{q} = u\vec{i} + v\vec{j} + w\vec{k}$$

i.e. $\frac{dx}{dt} = u, \frac{dy}{dt} = v, \frac{dz}{dt} = w$

where $\vec{r} = x\vec{i} + y\vec{j} + z\vec{k}$

5. ***Difference between streamlines and path lines.*** The tangents to the streamlines give the directions of velocities of fluid particles at various points at a given time, while tangents to the path lines give the directions of the velocities of a given fluid particle at various times. That is to say *streamlines show how each fluid particle is moving at a given instant whereas the path line show how a given particle is moving at each instant.*

In steady flow, streamlines do not vary with time and coincide with path lines.

6. ***Stream tube or tube of flow.*** The streamlines drawn through each point of a closed curve in the fluid called a stream tube or tube of flow.

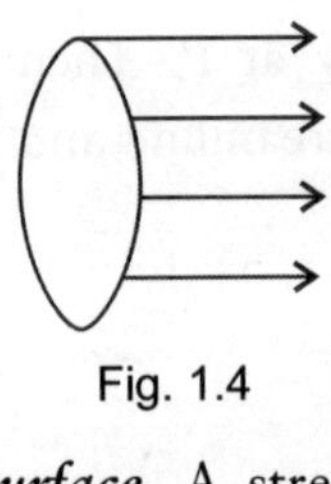

Fig. 1.4

7. ***Steam surface.*** A stream surface is a surface made by the streamlines passing through an arbitrary line in the fluid region at any instant of time.

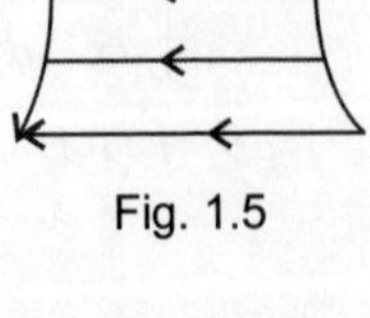

Fig. 1.5

*8. **Stream filament.*** A stream tube of infinitesimal cross-section is called a stream filament.

*9. **Streak lines or filament lines.*** A streak line is a line on which lie all those fluid elements that at some earlier instant passed through a certain point in space. A streak line connects the locations at one instant of particles moving with the fluid which passed through a certain point. The coloured dye injected a fluid stream or the smoke particles rising from the cigarette cnd exhibits streak lines. When the flow is steady streak line coincide with path lines and streamlines.

18. Velocity potential. If the expression $udx + vdy + wdz$ is an exact differential $-\,d\phi$ so that

$$udx + vdy + wdz = -\,d\phi = -\frac{\partial\phi}{\partial x}dx - \frac{\partial\phi}{\partial y}dy - \frac{\partial\phi}{\partial z}dz$$

$$\therefore \quad u = -\frac{\partial\phi}{dx},\ v = -\frac{\partial\phi}{\partial y},\ w = -\frac{\partial\phi}{\partial z}$$

then ϕ is called the velocity potential or velocity function.

If the velocity is denoted by the vector $\vec{q}$, then

$$\vec{q} = u\vec{i} + v\vec{j} + w\vec{k}$$

$$= -\left(\vec{i}\frac{\partial\phi}{\partial x} + \vec{j}\frac{\partial\phi}{\partial y} + \vec{k}\frac{\partial\phi}{\partial z}\right)$$

$$\vec{q} = -\vec{\nabla}\phi \qquad \text{... (1)}$$

The negative sign in the equation $\vec{q} = -\vec{\nabla}\phi$ is a convention. It ensures that flow takes from higher to lower potentials. The velocity potential is a scalar function of space and time.

The necessary and sufficient condition for (1) to hold is

$$\vec{\nabla} \times \vec{q} = 0$$

$$\text{or} \quad \begin{vmatrix} \vec{i} & \vec{j} & \vec{k} \\ \dfrac{\partial}{\partial x} & \dfrac{\partial}{\partial y} & \dfrac{\partial}{\partial z} \\ u & v & w \end{vmatrix} = 0$$

$$\Rightarrow \vec{i}\left(\frac{\partial w}{\partial y}-\frac{\partial v}{\partial z}\right)+\vec{j}\left(\frac{\partial u}{\partial z}-\frac{\partial w}{\partial x}\right)+\vec{k}\left(\frac{\partial v}{\partial x}-\frac{\partial u}{\partial y}\right)=0$$

$$\Rightarrow \xi\vec{i}+\eta\vec{j}+\zeta\vec{k}=0$$

where $\xi=\frac{\partial w}{\partial y}-\frac{\partial v}{\partial z}, \eta=\frac{\partial u}{\partial z}-\frac{\partial w}{\partial x}, \vec{k}=\frac{\partial v}{\partial x}-\frac{\partial u}{\partial y}$ are known as spin components.

The necessary condition of the existence of a velocity potential in a fluid is irrotational motion. Thus potential flow is the irrotational flow of an inviscid or perfect fluid.

19. To show that the surfaces exist which cut the streamlines orthogonally whenever velocity potential exists.

The differential equation of streamlines are given by

$$\frac{dx}{u}=\frac{dy}{v}=\frac{dz}{w} \quad \text{... (1)}$$

The surfaces which cut (1) orthogonally are given by

$$udx + vdy + wdz = 0 \quad \text{... (2)}$$

The necessary and sufficient condition for the existence of (2) is that (2) must have a solution of the type

$$\phi\ (x, y, z) = c \quad \text{... (3)}$$

where c is the constant of integration.

The necessary and sufficient condition for the existence of (3) is that

$$u\left(\frac{\partial v}{\partial z}-\frac{\partial w}{\partial y}\right)+v\left(\frac{\partial w}{\partial x}-\frac{\partial u}{\partial z}\right)+w\left(\frac{\partial u}{\partial y}-\frac{\partial v}{\partial x}\right)=0 \quad \text{... (4)}$$

Since velocity potenntial exists,

$$\therefore \quad u=-\frac{\partial\phi}{\partial x}, v=-\frac{\partial\phi}{\partial y}, w=-\frac{\partial\phi}{\partial z}$$

Now L.H.S. of (4)

$$=u\left(-\frac{\partial^2\phi}{\partial z\partial y}+\frac{\partial^2\phi}{\partial y\partial z}\right)+v\left(-\frac{\partial^2\phi}{\partial x\partial z}+\frac{\partial^2\phi}{\partial z\partial x}\right)+w\left(-\frac{\partial^2\phi}{\partial x\partial y}+\frac{\partial^2\phi}{\partial x\partial y}\right)$$

$$= 0 = \text{R.H.S. of (4)}$$

Hence surface exists which cut the streamlines orthogonally when velocity potential exists.

The surfaces $\phi\,(x, y, z, t)$ = constant are called equipotential surfaces.

20. Vorticity vector. If $\vec{q}$ be the velocity vector of a fluid particle then the quantity $\vec{w} = \vec{\nabla} \times \vec{q} = \text{curl}\,\vec{q}$ is called vorticity vector or simply the vorticity. In fact the angular velocity (the velocity of rotation) of an infinitesimal element is the vorticity.

Let $\vec{w} = \xi\vec{i} + \eta\vec{j} + \zeta k$

$$\text{Then } \vec{w} = \begin{vmatrix} \vec{i} & \vec{j} & \vec{k} \\ \dfrac{\partial}{\partial x} & \dfrac{\partial}{\partial y} & \dfrac{\partial}{\partial z} \\ u & v & w \end{vmatrix}$$

$$\xi\vec{i} + \eta\vec{j} + \zeta\vec{k} = \vec{i}\left(\frac{\partial w}{\partial y} - \frac{\partial v}{\partial z}\right) + \vec{j}\left(\frac{\partial u}{\partial z} - \frac{\partial w}{\partial x}\right) + \vec{k}\left(\frac{\partial v}{\partial x} - \frac{\partial u}{\partial y}\right)$$

Equating the coefficients of $\vec{i}, \vec{j}, \vec{k}$ on both the sides, we get

$$\xi = \frac{\partial w}{\partial y} - \frac{\partial v}{\partial z},\ \eta = \frac{\partial u}{\partial z} - \frac{\partial w}{\partial x},\ \zeta = \frac{\partial v}{\partial x} - \frac{\partial u}{\partial y}$$

A fluid motion is said to be irrotational if $\xi = 0$, $\eta = 0$, $\zeta = 0$, otherwise rotational.

21. Vortex line. A vortex line is a curve drawn in the fluid such that the tangent to it at each point at any instant of time t is in the direction of vorticity vector at that point. It means that $\vec{w}$ is paralle to $d\vec{r}$.

Hence $\vec{w} \times d\vec{r} = 0$

$$\Rightarrow \begin{vmatrix} \vec{i} & \vec{j} & \vec{k} \\ \xi & \eta & \zeta \\ dx & dy & dz \end{vmatrix} = 0$$

$$\Rightarrow \vec{i}\,(\eta dz - \zeta dy) - \vec{j}\,(\xi dz - \zeta dx) + \vec{k}\,(\xi dy - \eta dx) = 0$$

$$\Rightarrow \eta dz - \zeta dy = 0,\ \zeta dx - \xi dz = 0,\ \xi dy - \eta dx = 0$$

$$\Rightarrow \quad \frac{dx}{\xi} = \frac{dy}{\eta} = \frac{dz}{\zeta}$$

where $\vec{w} = \xi\vec{i} + \eta\vec{j} + \zeta\vec{k}$

$$\vec{r} = x\vec{i} + y\vec{j} + z\vec{k}$$

Note. In an irrotational motion of fluid, there are no vertex lines.

22. Vortex tube. The vortex lines drawn through each point of a closed curve enclose a tubular space in the fluid called a vortex tube. A vortex tube of infinitesimal cross section is called a vortex filament or simply a vortex.

23. Beltranic flow. A fluid motion is said to be Beltranic flow if $\vec{q}$ is parallel to $\vec{w}$ i.e. $\vec{q} \times \vec{w} = 0$. In this case $\vec{q}$ is called Beltranic vector.

24. Flux of a fluid. We consider a fixed geometrical surface *S*. Let us take a point *P* on it and let *df* be an element of *S* enclosing *P*. Let $q(u, v, w)$ be the velocity of the fluid at *P* at the instant *t* and (l, m, n) the direction cosines of the normal *PN* at *P* to the surface. The component of *q* along *PN* will be $lu + mv + nw$. If θ be the angle between the direction of the normal *PN* and the velocity *q* then $q \cos\theta = lu + mv + nw$.

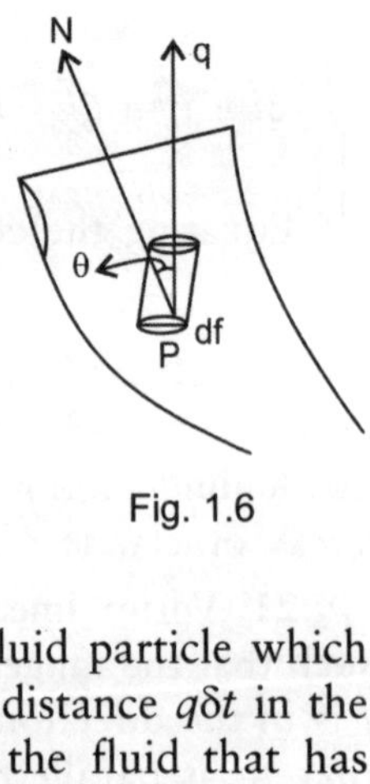

Fig. 1.6

After a small interval of time δt, the fluid particle which was upon *df* at time *t*, will have moved a distance $q\delta t$ in the direction of *q*. Therefore, the volume of the fluid that has streamed past *df* in time δt is equal to the volume of the fluid contained inside the cylinder whose base is *df* and whose length is $q\delta t$.

$$\begin{aligned}\text{Volume of the cylinder} &= q\delta t df \cos\theta \\ &= q \cos\theta df\delta t \\ &= (lu + mv + nw)\, df\delta t\end{aligned}$$

Hence mass of the fluid that has streamed past df in time δt

$= \rho\,(lu + mv + nw)\,df\delta t$

= (density) × (normal velocity) × (area of the surface) × δt

This is defined as the flux of the fluid across df in time δt. If $\vec{q}$ be the velocity vector and $d\vec{f}$ be the vectorial area, the flux of the fluid in unit time in vector notation is $\rho\vec{q}\cdot d\vec{f}$.

25. Equation of continuity. The equation of continuity simply expresses the law of conservation of mass in mathematical form. The law of conservation of mass states that fluid mass can be neither created nor destroyed.

Thus equation of continuity expresses that the rate of generation of mass within a given volume must be balanced by an equal net outflow of mass from the volume. This amount to saying that matter is neither created or destroyed.

26. Equation of continuity by Euler's method (Vector form). Let us consider a fixed closed surface S, enclosing a volume v in the region occupied by moving fluid. Let $\vec{n}$ be a unit outward normal vector drawn on the surface element ds, where fluid velocity is $\vec{q}$ and fluid density is ρ. Then the inward normal velocity is $(-\vec{q}\cdot\vec{n})$.

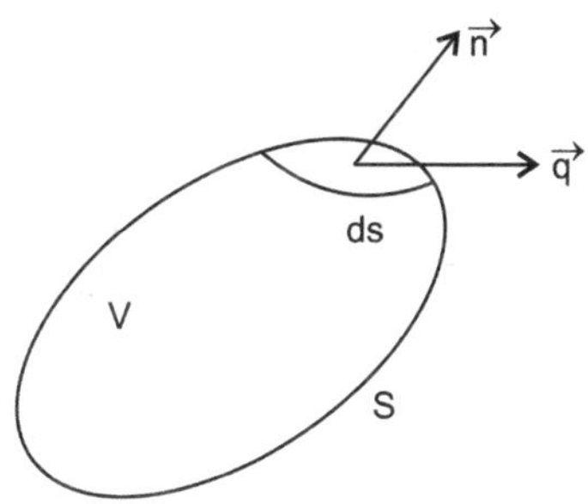

Fig. 1.7

The mass of the fluid entering per unit time through the element ds is $(-\rho\vec{q}\cdot\vec{n}\,ds)$.

Therefore, the mass of fluid entering per unit time across the whole surface S

$$= -\int_S \rho(\vec{q}\cdot\vec{n})\,ds$$

$$= -\int_S \vec{n}\cdot\rho\vec{q}\,ds$$

$$= \int_V \vec{\nabla} \cdot (\rho \vec{q})\, dv \qquad \text{... (1)}$$

[By Gauss divergence theorem]

The mass of the fluid within volume V, enclosed by S is

$$= \int_V \rho\, dv$$

Rate of generation of the fluid within the volume

$$= \frac{\partial}{\partial t} \int_V \rho\, dv$$

$$= \int_V \frac{\partial \rho}{\partial t}\, dv \qquad \text{... (2)}$$

$$\left[\because \rho \frac{\partial}{\partial t}(dv) = \rho \left(\frac{dv}{dt} \right) = \rho \cdot 0 = 0 \text{ as volume is constant w.r.t. time} \right]$$

By the principle of continuity, we have

$$\int_V \frac{\partial \rho}{\partial t}\, dv = -\int_V \vec{\nabla} \cdot (\rho \vec{q})\, dv$$

$$\Rightarrow \quad \int_V \left[\frac{\partial \rho}{\partial t} + \vec{\nabla} \cdot (\rho \vec{q}) \right] dv = 0$$

Since S is arbitrary and so V is arbitrary. Hence integrand of the last integral vanishes.

$$\therefore \quad \frac{\partial \rho}{\partial t} + \vec{\nabla} \cdot (\rho \vec{q}) = 0 \qquad \text{... (3)}$$

This is Eulerian equation of continuity.

Deductions (1) From (3),

$$\frac{\partial \rho}{\partial t} + \left(\vec{q} \cdot \vec{\nabla} \right) \rho + \rho \left(\vec{\nabla} \cdot \vec{q} \right) = 0$$

$$\Rightarrow \quad \left[\frac{\partial}{\partial t} + \vec{q} \cdot \vec{\nabla} \right] \rho + \rho \left(\vec{\nabla} \cdot \vec{q} \right) = 0$$

$$\Rightarrow \quad \frac{d\rho}{dt} + \rho\left(\vec{\nabla}\cdot\vec{q}\right) = 0 \qquad \ldots (4)$$

This is alternate form of (3).

(2) Since $\vec{q} = u\vec{i} + v\vec{j} + w\vec{k}$

$$\therefore \quad \vec{\nabla}\cdot(\rho\vec{q}) = \frac{\partial}{\partial x}(u\rho) + \frac{\partial}{\partial y}(v\rho) + \frac{\partial}{\partial z}(w\rho)$$

Hence from (3), we have

$$\frac{\partial\rho}{\partial t} + \frac{\partial}{\partial x}(u\rho) + \frac{\partial}{\partial y}(v\rho) + \frac{\partial}{\partial z}(w\rho) = 0$$

$$\text{or} \quad \frac{\partial\rho}{\partial t} + u\frac{\partial\rho}{\partial x} + v\frac{\partial\rho}{\partial y} + w\frac{\partial\rho}{\partial z} + \rho\left(\frac{\partial u}{\partial x} + \frac{\partial v}{\partial y} + \frac{\partial w}{\partial z}\right) = 0 \quad \ldots (5)$$

This is the Cartesian form of equation of continuity.

(3) If the fluid be incompressible, then $\frac{d\rho}{dt} = 0$

Hence from (4), we have

$$\rho\left(\vec{\nabla}\cdot\vec{q}\right) = 0$$

$$\Rightarrow \quad \vec{\nabla}\cdot\vec{q} = 0$$

$$\Rightarrow \quad \frac{\partial u}{\partial x} + \frac{\partial v}{\partial y} + \frac{\partial w}{\partial z} = 0$$

Thus the velocity vector $\vec{q}$ is solenoidal clearly $\vec{\nabla}\cdot\vec{q}$ gives the rate of volume expansion of a fluid element. For this reason it may be called "dilatation" or "expansion".

(4) Let the motion be irrotational and incompressible. Then there exists velocity potential ϕ such that $\vec{q} = -\vec{\nabla}\phi$.

Also in this case $\frac{d\rho}{dt} = 0$

Hence equation (4) becomes

$$\rho\left[\vec{\nabla}\cdot\left(-\vec{\nabla}\phi\right)\right] = 0$$

$$\Rightarrow \quad -\vec{\nabla}^2\phi = 0$$

$$\Rightarrow \quad \frac{\partial^2\phi}{\partial x^2} + \frac{\partial^2\phi}{\partial y^2} + \frac{\partial^2\phi}{\partial z^2} = 0$$

This is the equation of continuity in this case. Thus equation of continuity reduces to Laplace's equation, when the liquid is incompressible and irrotational.

(5) If the motion be symmetrical. In this case velocity has only one component, say u.

Then from equation (5), we have

$$\frac{\partial \rho}{\partial t} + u\frac{\partial \rho}{\partial x} + \rho\frac{\partial u}{\partial x} = 0$$

or $$\left(\frac{\partial}{\partial t} + u\frac{\partial}{\partial x}\right)\rho + \rho\frac{\partial u}{\partial x} = 0$$

27. Equation of continuity by Lagrange's method. Let initially a fluid particle be at (a, b, c) at time $t = t_0$, when its volume is dv_0 and density is ρ_0. After a lapse of time t, let the same fluid particle be at (x, y, z) when its volume is dv and density ρ. Since the mass of the fluid particle remains invariant during its motion.

$$\therefore \quad \frac{d}{dt}(\rho dv) = 0$$

$\Rightarrow$ ρdv = constant = $\rho_0 \; dv_0$ (say)

In Cartesian rectangular coordinates

$$dv = dxdydz$$

and $dv_0 = dadbdc$

$\therefore$ $\rho dxdydz = \rho_0 \; dadbdc$

or $$\rho_0 dadbdc = \rho\frac{\partial(x,y,z)}{\partial(a,b,c)} dadbdc$$

or $$\rho_0 = \rho J \text{ where } J = \frac{\partial(x,y,z)}{\partial(a,b,c)}$$

28. Equivalence of the Eulerian and Lagrangian forms of the equation of continuity. Let initially a fluid particle be at (a, b, c) at time $t = t_0$ when its volume is dv_0 and density ρ_0. After a lapse of time t, let the same fluid particle be at (x, y, z) when its volume is dv and density is ρ. The components of velocity in the two systems are connected by the equations

$$u = \frac{dx}{dt}, v = \frac{dy}{dt}, w = \frac{dz}{dt}$$

Also $x = x\ (a, b, c, t)$, $y = y\ (a, b, c, t)$, $z = z\ (a, b, c, t)$

Now $\dfrac{\partial u}{\partial a} = \dfrac{\partial}{\partial a}\left(\dfrac{\partial x}{\partial t}\right) = \dfrac{d}{dt}\left(\dfrac{\partial x}{\partial a}\right)$ and so on.

Firstly we shall determine $\dfrac{dJ}{dt}$.

Now $J = \dfrac{\partial(x,y,z)}{\partial(a,b,c)}$

$$= \begin{vmatrix} \frac{\partial x}{\partial a} & \frac{\partial y}{\partial a} & \frac{\partial z}{\partial a} \\ \frac{\partial x}{\partial b} & \frac{\partial y}{\partial b} & \frac{\partial z}{\partial b} \\ \frac{\partial x}{\partial c} & \frac{\partial y}{\partial c} & \frac{\partial z}{\partial c} \end{vmatrix}$$

$$\therefore \frac{dJ}{dt} = \begin{vmatrix} \frac{\partial u}{\partial a} & \frac{\partial y}{\partial a} & \frac{\partial z}{\partial a} \\ \frac{\partial u}{\partial b} & \frac{\partial y}{\partial b} & \frac{\partial z}{\partial b} \\ \frac{\partial u}{\partial c} & \frac{\partial y}{\partial c} & \frac{\partial z}{\partial c} \end{vmatrix} + \begin{vmatrix} \frac{\partial x}{\partial a} & \frac{\partial v}{\partial a} & \frac{\partial z}{\partial a} \\ \frac{\partial x}{\partial b} & \frac{\partial v}{\partial b} & \frac{\partial z}{\partial b} \\ \frac{\partial x}{\partial c} & \frac{\partial v}{\partial c} & \frac{\partial z}{\partial c} \end{vmatrix} + \begin{vmatrix} \frac{\partial x}{\partial a} & \frac{\partial y}{\partial a} & \frac{\partial w}{\partial a} \\ \frac{\partial x}{\partial b} & \frac{\partial y}{\partial b} & \frac{\partial w}{\partial b} \\ \frac{\partial x}{\partial c} & \frac{\partial y}{\partial c} & \frac{\partial w}{\partial c} \end{vmatrix}$$

$$= \frac{\partial(x,y,z)}{\partial(a,b,c)} + \frac{\partial(x,v,z)}{\partial(a,b,c)} + \frac{\partial(x,y,w)}{\partial(a,b,c)}$$

$$= J_1 + J_2 + J_3 \text{ (say)} \qquad \dots (1)$$

Now $J_1 = \begin{vmatrix} \frac{\partial u}{\partial a} & \frac{\partial y}{\partial a} & \frac{\partial z}{\partial a} \\ \frac{\partial u}{\partial b} & \frac{\partial y}{\partial b} & \frac{\partial z}{\partial b} \\ \frac{\partial u}{\partial c} & \frac{\partial y}{\partial c} & \frac{\partial z}{\partial c} \end{vmatrix}$

$$= \begin{vmatrix} \frac{\partial u}{\partial x}\frac{\partial x}{\partial a} + \frac{\partial u}{\partial y}\frac{\partial y}{\partial a} + \frac{\partial u}{\partial z}\frac{\partial z}{\partial a} & \frac{\partial y}{\partial a} & \frac{\partial z}{\partial a} \\ \frac{\partial u}{\partial x}\frac{\partial x}{\partial b} + \frac{\partial u}{\partial y}\frac{\partial y}{\partial b} + \frac{\partial u}{\partial z}\frac{\partial z}{\partial b} & \frac{\partial y}{\partial b} & \frac{\partial z}{\partial b} \\ \frac{\partial u}{\partial x}\frac{\partial x}{\partial c} + \frac{\partial u}{\partial y}\frac{\partial y}{\partial c} + \frac{\partial u}{\partial z}\frac{\partial z}{\partial c} & \frac{\partial y}{\partial c} & \frac{\partial z}{\partial c} \end{vmatrix}$$

$$= \begin{vmatrix} \frac{\partial u}{\partial x} & \frac{\partial x}{\partial a} & \frac{\partial y}{\partial a} & \frac{\partial z}{\partial a} \\ \frac{\partial u}{\partial x} & \frac{\partial x}{\partial b} & \frac{\partial y}{\partial b} & \frac{\partial z}{\partial b} \\ \frac{\partial u}{\partial x} & \frac{\partial x}{\partial c} & \frac{\partial y}{\partial c} & \frac{\partial z}{\partial c} \end{vmatrix} + \begin{vmatrix} \frac{\partial u}{\partial y} & \frac{\partial y}{\partial a} & \frac{\partial y}{\partial a} & \frac{\partial z}{\partial a} \\ \frac{\partial u}{\partial y} & \frac{\partial y}{\partial b} & \frac{\partial y}{\partial b} & \frac{\partial z}{\partial b} \\ \frac{\partial u}{\partial y} & \frac{\partial y}{\partial c} & \frac{\partial y}{\partial c} & \frac{\partial z}{\partial c} \end{vmatrix}$$

$$+ \begin{vmatrix} \frac{\partial u}{\partial z} & \frac{\partial z}{\partial a} & \frac{\partial y}{\partial a} & \frac{\partial z}{\partial a} \\ \frac{\partial u}{\partial z} & \frac{\partial z}{\partial b} & \frac{\partial y}{\partial b} & \frac{\partial z}{\partial b} \\ \frac{\partial u}{\partial z} & \frac{\partial z}{\partial c} & \frac{\partial y}{\partial c} & \frac{\partial z}{\partial c} \end{vmatrix}$$

$$= \frac{\partial u}{\partial x}\begin{vmatrix} \frac{\partial x}{\partial a} & \frac{\partial y}{\partial a} & \frac{\partial z}{\partial a} \\ \frac{\partial x}{\partial b} & \frac{\partial y}{\partial b} & \frac{\partial z}{\partial b} \\ \frac{\partial x}{\partial c} & \frac{\partial y}{\partial c} & \frac{\partial z}{\partial c} \end{vmatrix} + \frac{\partial u}{\partial y}\begin{vmatrix} \frac{\partial y}{\partial a} & \frac{\partial y}{\partial a} & \frac{\partial z}{\partial a} \\ \frac{\partial y}{\partial b} & \frac{\partial y}{\partial b} & \frac{\partial z}{\partial b} \\ \frac{\partial y}{\partial c} & \frac{\partial y}{\partial c} & \frac{\partial z}{\partial c} \end{vmatrix}$$

$$+ \frac{\partial u}{\partial z}\begin{vmatrix} \frac{\partial z}{\partial a} & \frac{\partial y}{\partial a} & \frac{\partial z}{\partial a} \\ \frac{\partial z}{\partial b} & \frac{\partial y}{\partial b} & \frac{\partial z}{\partial b} \\ \frac{\partial z}{\partial c} & \frac{\partial y}{\partial c} & \frac{\partial z}{\partial c} \end{vmatrix}$$

$$= \frac{\partial u}{\partial x}\frac{\partial(x,y,z)}{\partial(a,b,c)}$$

$$= J\frac{\partial u}{\partial x}$$

$$\therefore \quad J_1 = J\frac{\partial u}{\partial x}$$

Similarly $J_2 = J\dfrac{\partial v}{\partial y}, J_3 = J\dfrac{\partial w}{\partial z}$

Now (1) becomes

$$\frac{dJ}{dt} = J\left(\frac{\partial u}{\partial x} + \frac{\partial v}{\partial y} + \frac{\partial w}{\partial z}\right) = J\vec{\nabla}\cdot\vec{q} \qquad \text{... (2)}$$

In the Lagrangian form the equation of continuity is

$$\rho J = \text{constant} = \rho_0$$

$$\therefore \quad \rho\frac{dJ}{dt} + J\frac{d\rho}{\partial t} = 0$$

$$\Rightarrow \quad \rho J\left(\vec{\nabla}\cdot\vec{q}\right) + J\frac{d\rho}{dt} = 0 \qquad \text{[From (2)]}$$

$$\Rightarrow \quad J\left(\frac{d\rho}{dt} + \rho\,\vec{\nabla}\cdot\vec{q}\right) = 0$$

$$\Rightarrow \quad \frac{d\rho}{dt} + \rho\vec{\nabla}\vec{q} = 0\;(\text{Since } J \neq 0)$$

which is the Eulerian form of equation of continuity.

Again Eulerian form of equation of continuity is

$$\frac{d\rho}{dt} + \rho\vec{\nabla}\cdot\vec{q} = 0$$

$$\Rightarrow \quad \frac{d\rho}{dt} + \rho\frac{1}{J}\frac{dJ}{dt} = 0 \qquad \text{[From (2)]}$$

$$\Rightarrow \quad J\frac{d\rho}{dt} + \rho\frac{dJ}{dt} = 0$$

$$\Rightarrow \quad \frac{d}{dt}(\rho J) = 0$$

Integrating we get $\rho J = \rho_0$

where ρ_0 is a constant

which is the Lagrangian equation of continuity.

29. Generalised orthogonal curvilinear coordinates. Let $f_1(x, y, z) = a_1$, $f_2(x, y, z) = a_2$, $f_3(x, y, z) = a_3$ be three independent orthogonal families of surfaces, where (x, y, z) are Cartesian coordinates of a point. The surfaces a_1 = const, a_2 = const, a_3 = const form an orthogonal system in which (a_1, a_2, a_3) may be used as the orthogonal curvilinear coordinates of a point in the space. The relation between the two systems of coordinates (x, y, z) and (a_1, a_2, a_3) can also be expressed by the equations $x = x(a_1, a_2, a_3)$, $y = y(a_1, a_2, a_3)$, $z = z(a_1, a_2, a_3)$.

Now
$$dx = \frac{\partial x}{\partial a_1} da_1 + \frac{\partial x}{\partial a_2} da_2 + \frac{\partial x}{\partial a_3} da_3$$
$$dy = \frac{\partial y}{\partial a_1} da_1 + \frac{\partial y}{\partial a_2} da_2 + \frac{\partial y}{\partial a_3} da_3$$
$$dz = \frac{\partial z}{\partial a_1} da_1 + \frac{\partial z}{\partial a_2} da_2 + \frac{\partial z}{\partial a_3} da_3$$

Since the element of length ds in both the systems must be the same, we have

$$\begin{aligned}(ds)^2 &= (dx)^2 + (dy)^2 + (dz)^2 \\
&= \left\{\left(\frac{\partial x}{\partial a_1}\right)^2 + \left(\frac{\partial y}{\partial a_1}\right)^2 + \left(\frac{\partial z}{\partial a_1}\right)^2\right\}(da_1)^2 \\
&\quad + \left\{\left(\frac{\partial x}{\partial a_2}\right)^2 + \left(\frac{\partial y}{\partial a_2}\right)^2 + \left(\frac{\partial z}{\partial a_2}\right)^2\right\}(da_2)^2 \\
&\quad + \left\{\left(\frac{\partial x}{\partial a_3}\right)^2 + \left(\frac{\partial y}{\partial a_3}\right)^2 + \left(\frac{\partial z}{\partial a_3}\right)^2\right\}(da_3)^2 \\
&\quad + 2da_1da_2\left\{\frac{\partial x}{\partial a_1}\frac{\partial x}{\partial a_2} + \frac{\partial y}{\partial a_1}\frac{\partial y}{\partial a_2} + \frac{\partial z}{\partial a_1}\frac{\partial z}{\partial a_2}\right\} \\
&\quad + 2da_2da_3\left\{\frac{\partial x}{\partial a_2}\frac{\partial x}{\partial a_3} + \frac{\partial y}{\partial a_2}\frac{\partial y}{\partial a_3} + \frac{\partial z}{\partial a_2}\frac{\partial z}{\partial a_3}\right\} \\
&\quad + 2da_3da_1\left\{\frac{\partial x}{\partial a_3}\frac{\partial x}{\partial a_1} + \frac{\partial y}{\partial a_3}\frac{\partial y}{\partial a_1} + \frac{\partial z}{\partial a_3}\frac{\partial z}{\partial a_1}\right\}\end{aligned}$$

By orthogonal property, the terms containing da_1da_2, da_2da_3, da_3da_1 vanish.

Hence

$$(ds)^2 = (dx)^2 + (dy)^2 + (dz)^2$$
$$= (h_1da_1)^2 + (h_2da_2)^2 + (h_3da_3)^2$$

where $h_1^2 = \left(\frac{\partial x}{\partial a_1}\right)^2 + \left(\frac{\partial y}{\partial a_1}\right)^2 + \left(\frac{\partial z}{\partial a_1}\right)^2$ etc.

30. Equation of continuity in generalised orthogonal curvilinear coordinates. Let ρ be the fluid density at a curvilinear point $P(a_1, a_2, a_3)$ enclosed by a small parallelopiped with edges of lengths h_1da_1, h_2da_2, h_3da_3.

Let q_1, q_2, q_3 be the velocity components along OA, OB, OC respectively.

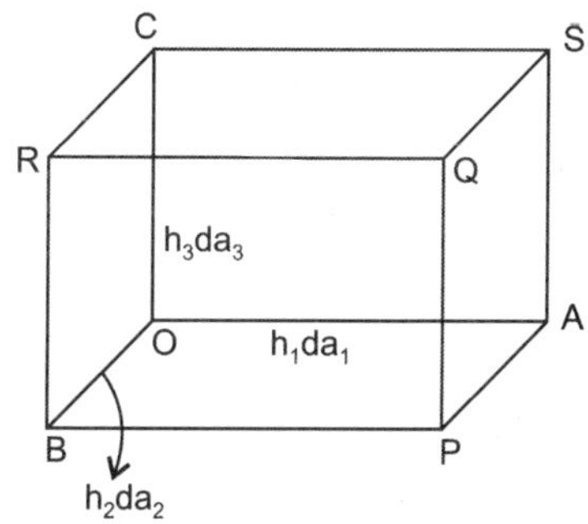

Fig. 1.8

Mass of the fluid that passes in unit time across the face $OBRC$

$$= \text{density} \cdot \text{area} \cdot \text{normal velocity}$$
$$= \rho\,(h_2da_2h_3da_3)\,q_1$$
$$= \rho q_1h_2h_3da_2da_3$$
$$= f\,(a_1, a_2, a_3)\ \text{(say)}$$

Mass of the fluid that passes in unit time across the face $PACQ$

$$= f\,(a_1 + \delta a_1, a_2, a_3)$$
$$= f(a_1, a_2, a_3) + \delta a_1 \frac{\partial f}{\partial a_1}$$

Now the excess of flow in over flow out from the face $OBRC$ and $PACQ$ in unit time

$$= f - \left(f + \delta a_1 \frac{\partial f}{\partial a_1}\right)$$

$$= -\delta a_1 \frac{\partial f}{\partial a_1}$$

$$= -\delta a_1 \frac{\partial}{\partial a_1}(a_1, a_2, a_3)$$

$$= -\frac{\partial}{\partial a_1}(\rho q_1 h_2 h_3)\, da_2 da_3 da_1$$

Similarly the excess of flow in overflow out from the faces $QRSC$ and $BPAO$, $OASC$ and $PBRQ$ are respectively

$$-\frac{\partial}{\partial a_3}(\rho q_3 h_1 h_2)\, da_1 da_2 da_3$$

and $-\dfrac{\partial}{\partial a_2}(\rho q_2 h_1 h_3)\, da_1 da_2 da_3$

Also rate of increase in the mass inside the parallelopiped

$$= \frac{\partial}{\partial t}(\rho h_1 da_1 h_2 da_2 h_3 da_3)$$

$$= \frac{\partial \rho}{\partial t} h_1 h_2 h_3 da_1 da_2 da_3$$

By the principle of continuity

Increase in mass = total excess of flow in over flow out

$$\therefore \quad \frac{\partial \rho}{\partial t} h_1 h_2 h_3 da_1 da_2 da_3$$

$$= -\left[\frac{\partial}{\partial a_1}(\rho q_1 h_2 h_3) + \frac{\partial}{\partial a_2}(\rho q_2 h_1 h_3) + \frac{\partial}{\partial a_3}(\rho q_3 h_1 h_2)\right] da_1 da_2 da_3$$

$$\Rightarrow \quad \frac{\partial \rho}{\partial t} + \frac{1}{h_1 h_2 h_3}\left[\frac{\partial}{\partial a_1}(\rho q_1 h_2 h_3) + \frac{\partial}{\partial a_2}(\rho q_2 h_1 h_3) + \frac{\partial}{\partial a_3}(\rho q_3 h_1 h_2)\right] = 0$$

This is the required equation of continuity.

Particular Cases

(1) Rectangular Cartesian Coordinates

$ds^2 = dx^2 + dy^2 + dz^2 = (h_1 da_1)^2 + (h_2 da_2)^2 + (h_3 da_3)^2$

Hence $h_1 = 1$, $h_2 = 1$, $h_3 = 1$, $a_1 = x$, $a_2 = y$, $a_3 = z$

Hence, the equation of continuity becomes

$$\frac{\partial \rho}{\partial t} + \frac{\partial}{\partial x}(\rho q_1) + \frac{\partial}{\partial y}(\rho q_2) + \frac{\partial}{\partial z}(\rho q_3) = 0$$

(2) Spherical Polar Coordinates

In this case

$$(ds)^2 = (dr)^2 + (rd\theta)^2 + (r \sin \theta d\phi)^2$$
$$= (h_1 da_1)^2 + (h_2 da_2)^2 + (h_3 da_3)^2$$

Hence $h_1 = 1$, $h_2 = r$, $h_3 = r \sin \theta$, $a_1 = r$, $a_2 = \theta$, $a_3 = w$

In this case equation of continuity becomes

$$\frac{\partial \rho}{\partial t} + \frac{1}{r^2 \sin \theta}\left[\frac{\partial}{\partial r}(\rho q_1 \cdot r \cdot r \sin\theta)\right.$$
$$\left. + \frac{\partial}{\partial \theta}(\rho q_2 \cdot 1 \cdot r \sin\theta) + \frac{\partial}{\partial w}(\rho q_3 \cdot 1 \cdot r)\right] = 0$$

$$\Rightarrow \quad \frac{\partial \rho}{\partial t} + \frac{1}{r^2}\frac{\partial}{\partial r}(\rho q_1 r^2) + \frac{1}{r \sin\theta}\frac{\partial}{\partial \theta}(\rho q_2 \sin\theta)$$
$$+ \frac{1}{r \sin \theta}\frac{\partial}{\partial w}(\rho q_3) = 0$$

(3) Cylindrical Coordinates

Here we have $(ds)^2 = (dr)^2 + (rd\theta)^2 + (dz)^2$

and so, $h_1 = 1$, $h_2 = r$, $h_3 = 1$, $a_1 = r$, $a_2 = \theta$, $a_3 = z$

Then, the equation of continuity takes the form

$$\frac{\partial \rho}{\partial t} + \frac{1}{r}\left[\frac{\partial}{\partial r}(\rho q_1 r) + \frac{\partial}{\partial \theta}(\rho q_2) + \frac{\partial}{\partial z}(\rho q_3 r)\right] = 0$$

$$\Rightarrow \quad \frac{\partial \rho}{\partial t} + \frac{1}{r}\frac{\partial}{\partial r}(\rho q_1 r) + \frac{1}{r}\frac{\partial}{\partial \theta}(\rho q_2) + \frac{\partial}{\partial z}(\rho q_3) = 0$$

31. Equation of continuity in Cartesian coordinates. Let ρ denote fluid density $P(x, y, z)$ enclosed by a small parallelopiped with edges of lengths δx, δy, δz. Let u, v, w be the components of velocity at P parallel to the rectangular coordinate axes.

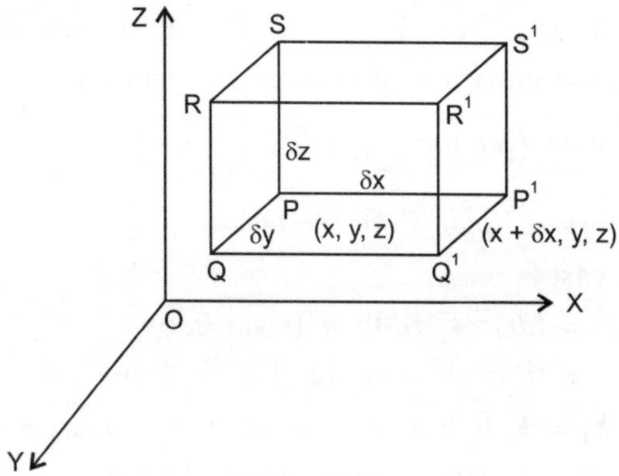

Fig. 1.9

Now mass of the fluid that passes in unit time through the plane face $PQRS$ in unit time

$$= \text{density} \cdot \text{area} \cdot \text{normal velocity}$$

$$= \rho \delta y \delta z u$$

$$= f(x, y, z) \text{ (say)}$$

Now mass of the fluid that passes through the opposite face $P^1Q^1R^1S^1$ in unit time

$$= f(x + \delta x, y, z)$$

$$= f + \delta x \frac{\partial f}{\partial x}$$

Therefore the excess of flow in over flow out from the face $PQRS$ and $P^1Q^1R^1S^1$ in unit time

$$= f - \left(f + \delta x \frac{\partial f}{\partial x} \right)$$

$$= -\delta x \frac{\partial f}{\partial x}$$

$$= -\delta x \frac{\partial}{\partial x} (\rho u \delta y \delta z)$$

$$= -\frac{\partial}{\partial x} (\rho u)\, \delta x\, \delta y\, \delta z$$

Similarly the excess of flow in over flow out from the faces QQ^1RR, PP^1S^1S and PP^1Q^1Q and SS^1R^1R

$$= -\frac{\partial}{\partial y}(\rho v)\,\delta x\,\delta y\,\delta z$$

and $-\dfrac{\partial}{\partial z}(\rho w)\,\delta x\,\delta y\,\delta z$

Therefore total excess of flow in over flow out from all the faces

$$= -\left[\frac{\partial}{\partial x}(\rho u) + \frac{\partial}{\partial y}(\rho v) + \frac{\partial}{\partial z}(\rho w)\right]\delta x\,\delta y\,\delta z$$

Also the total mass in the parallelopiped $= \rho\delta x\delta y\delta z$

Hence rate of increase in the mass inside the parallelopiped

$$= \frac{\partial}{\partial t}(\rho\delta x\,\delta y\,\delta z) = \frac{\partial \rho}{\partial t}\,\delta x\,\delta y\,\delta z$$

By the principle of continuity, we have

Increase in mass = total excess of flow in over flow out

$$\frac{\partial \rho}{\partial t}\delta x\,\delta y\,\delta z = -\left[\frac{\partial}{\partial x}(\rho u) + \frac{\partial}{\partial y}(\rho v) + \frac{\partial}{\partial z}(\rho w)\right]\delta x\,\delta y\,\delta z$$

$$\Rightarrow \quad \frac{\partial \rho}{\partial t} + \frac{\partial}{\partial x}(\rho u) + \frac{\partial}{\partial y}(\rho v) + \frac{\partial}{\partial z}(\rho w) = 0 \qquad \text{... (1)}$$

This is the equation of continuity in Cartesian coordinates.

Deductions

(1) If the fluid is incompressible, then (1) becomes

$$0 + \rho\left(\frac{\partial u}{\partial x} + \frac{\partial v}{\partial y} + \frac{\partial w}{\partial z}\right) = 0$$

$$\Rightarrow \quad \frac{\partial u}{\partial x} + \frac{\partial v}{\partial y} + \frac{\partial w}{\partial z} = 0$$

(2) The equation (1) can also be written as

$$\frac{\partial \rho}{\partial t} + u\frac{\partial \rho}{\partial x} + v\frac{\partial \rho}{\partial y} + w\frac{\partial \rho}{\partial z} + \rho\left(\frac{\partial u}{\partial x} + \frac{\partial v}{\partial y} + \frac{\partial w}{\partial z}\right) = 0$$

$$\Rightarrow \quad \left(\frac{\partial}{\partial t} + u\frac{\partial}{\partial x} + v\frac{\partial}{\partial y} + w\frac{\partial}{\partial z}\right)\rho + \rho\left(\frac{\partial u}{\partial x} + \frac{\partial v}{\partial y} + \frac{\partial w}{\partial z}\right) = 0$$

(3) If velocity has one component u say then eqn. (1) becomes

$$\frac{\partial \rho}{\partial t} + \frac{\partial}{\partial x}(\rho u) = 0$$

32. Equation of continuity in spherical polar coordinates. Let ρ be the fluid density at a point $P(r, \theta, \phi)$ enclosed by a small parallelopiped with edges δr, $r\delta\theta$, $r \sin \theta\delta\phi$.

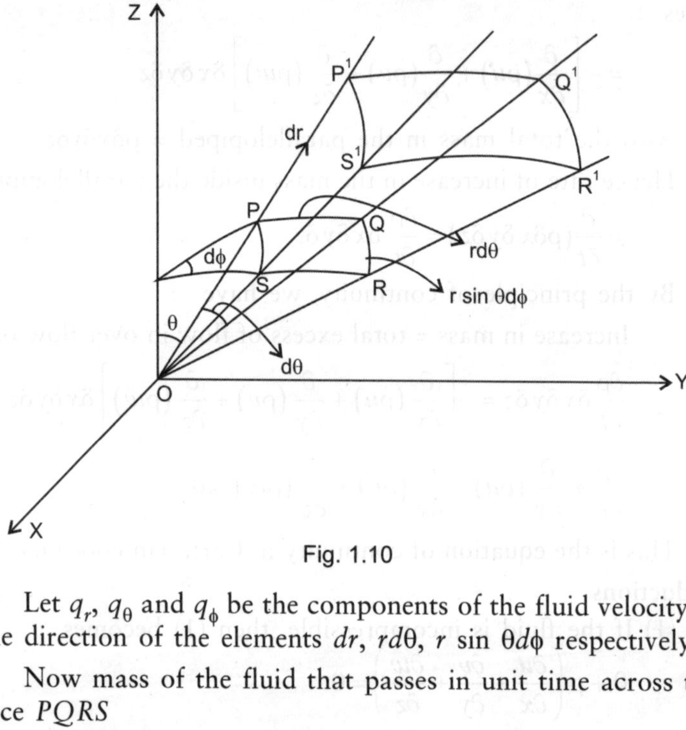

Fig. 1.10

Let q_r, q_θ and q_ϕ be the components of the fluid velocity in the direction of the elements dr, $rd\theta$, $r \sin \theta d\phi$ respectively.

Now mass of the fluid that passes in unit time across the face $PQRS$

$$= \text{density} \cdot \text{area} \cdot \text{normal velocity}$$

$$= \rho\ (r\delta\theta\ r \sin \theta\delta\phi)\ q_r$$

$$= f\ (r, \theta, \phi) \quad \text{(say)}$$

Also mass of the fluid that passes in unit time across the face $P^1Q^1R^1S^1$

$$= f\ (r + \delta r, \theta, \phi)$$

$$= f + \delta r \frac{\partial f}{\partial r}$$

Therefore excess of flow in over flow out from the face $PQRS$ and $P^1Q^1R^1S^1$ in unit time

$$= f - f - \delta r \frac{\partial f}{\partial r}$$

$$= -\delta r \frac{\partial f}{\partial r}$$

$$= -\delta r \frac{\partial}{\partial r}(\rho q_r r\delta\theta\, r \sin\theta\delta\phi)$$

Similarly the excess of flow in over flow out from the faces PSS^1P^1, QRR^1Q^1 and PQQ^1P^1, SRR^1S^1 are respectively

$$= -r\sin\theta\,\delta\phi \frac{\partial}{r\sin\theta\partial\phi}\left(\rho\, q_\phi r\delta\theta\,\delta r\right)$$

and $-r\,\delta\theta \dfrac{\partial}{r\partial\theta}(\rho q_\theta \delta r\, r\sin\theta\,\delta\phi)$

Thus total excess of flow in over flow out from all faces

$$= -\frac{\partial}{\partial r}\left(\rho r^2 q_r\right)\sin\theta\,\delta r\,\delta\theta\,\delta\phi - \frac{\partial}{r\partial\theta}(\rho\sin\theta\, q_\theta)r^2\delta r\,\delta\theta\,\delta\phi$$
$$- \frac{\partial}{r\sin\theta\,\partial\phi}\left(\rho q_\phi\right)r^2\sin\theta\,\delta r\,\delta\theta\,\delta\phi$$

Also total mass inside the parallelopiped

$$= \rho\,\delta r\; r\delta\theta\; r\sin\theta\delta\phi$$

Hence rate of increase in mass of the fluid inside the parallelopiped

$$= \frac{\partial}{\partial t}(\rho\delta r r\,\delta\theta r\sin\theta\delta\phi)$$

$$= \frac{\partial\rho}{\partial t} r^2 \sin\theta\delta r\,\delta\theta\,\delta\phi$$

By the principle of continuity, we have

Increase in mass = total excess of flow in over flow out

$$\frac{\partial\rho}{\partial t} r^2\sin\theta\,\delta r\,\delta\theta\,\delta\phi = -\left[\frac{\partial}{\partial r}\left(\rho r^2 q_r\right)\sin\theta\right.$$
$$\left. + \frac{\partial}{\partial\theta}(\rho q_\theta\sin\theta)\,r + r\frac{\partial}{\partial\phi}\left(\rho q_\phi\right)\right]\delta r\,\delta\theta\,\delta\phi$$

$$\Rightarrow \quad r^2 \sin\theta \frac{\partial \rho}{\partial t} + \sin\theta \frac{\partial}{\partial r}\left(\rho r^2 q_r\right) + r\frac{\partial}{\partial \theta}(q_\theta \sin\theta) + r\frac{\partial}{\partial \phi}\left(\rho q_\phi\right) = 0$$

$$\Rightarrow \quad \frac{\partial \rho}{\partial t} + \frac{1}{r^2}\frac{\partial}{\partial r}\left(\rho r^2 q_r\right) + \frac{1}{r\sin\theta}\frac{\partial}{\partial \theta}(q_\theta \sin\theta) + \frac{1}{r\sin\theta}\frac{\partial}{\partial \phi}\left(\rho q_\phi\right) = 0$$

This is the required equation of continuity.

33. Equation of continuity in cylindrical coordinates. Let ρ be the fluid density at the point $P\ (r, \theta, z)$ enclosed by a small parallelopiped with edges of lengths δr, $r\delta\theta$, δz. Let q_r, q_θ and q_z be the velocity of components in the direction of the elements respectively.

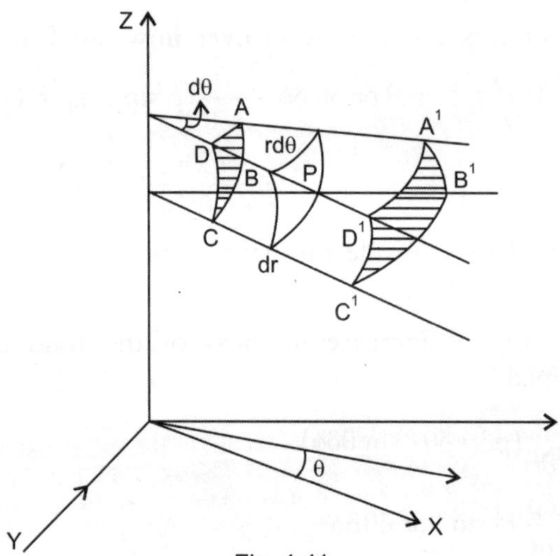

Fig. 1.11

Mass of the fluid that passes through the face $ABCD$ in unit time

$= $ density · area · normal velocity

$= \rho\ r\delta\theta\delta z\ q_r$

$= f\ (r, \theta, z)$ (say)

Mass of the fluid that passes in unit time from the face $A^1B^1C^1D^1$

$$= f(r + \delta r, \theta, z)$$

$$= f + \delta r \frac{\partial f}{\partial r}$$

Now excess of flow in over flow out from the faces $ABCD$ and $A^1B^1C^1D^1$

$$= f - \left(f + \delta r \frac{\partial f}{\partial r}\right)$$

$$= -\delta r \frac{\partial f}{\partial r}$$

$$= -\delta r \frac{\partial}{\partial r}(\rho q_r r\delta\theta\delta z)$$

$$= -\frac{\partial}{\partial r}(\rho q_r r)\,\delta r\,\delta\theta\,\delta z$$

Similarly the excess flow in over flow out from the other two faces are respectively

$$-\frac{\partial}{r\partial\theta}(\rho q_\theta) r\delta r\,\delta\theta\,\delta z$$

and $-\dfrac{\partial}{\partial z}(\rho q_z) r\delta r\,\delta\theta\,\delta z$

Hence total excess of flow in over flow out

$$= -\left[\frac{\partial}{\partial r}(\rho q_r) + \frac{\partial}{r\partial\theta}(\rho q_\theta) r + \frac{\partial}{\partial z}(\rho q_z) r\right] \delta x\,\delta\theta\,\delta z$$

Also total mass inside the parallelopiped = $\rho\delta r r\delta\theta\delta z$

Hence rate of increases in mass of the fluid within the parallelopiped

$$= \frac{\partial}{\partial t}(\rho\delta r r\delta\theta\delta z)$$

$$= \frac{\partial\rho}{\partial t} r\delta r\delta\theta\delta z$$

Therefore, by the principle of continuity

$$\frac{\partial \rho}{\partial t} r\delta r\delta\theta\delta z = -\left[\frac{\partial}{\partial r}(\rho r q_r) + \frac{\partial}{\partial \theta}(\rho q_\theta) + \frac{\partial}{\partial z}(\rho q_z) r\right] \delta r\,\delta\theta\,\delta z$$

$$\Rightarrow \quad \frac{\partial \rho}{\partial t} + \frac{1}{r}\frac{\partial}{\partial r}(\rho r q_r) + \frac{1}{r}\frac{\partial}{\partial \theta}(\rho q_\theta) + \frac{\partial}{\partial z}(\rho q_z) = 0$$

This is the required equation of continuity.

34. Certain symmetrical forms of equation of continuity.

1. Spherical symmetry. In this case, the motion is symmetrical about the centre and the velocity q along any radius r is a function of r and t.

We consider two consecutive concentric spheres of radii r and $r + \delta r$.

Mass of the fluid which passes in unit time across the inner sphere

$$= \text{density} \cdot \text{area} \cdot \text{normal velocity}$$

$$= \rho \cdot 4\pi r^2 \cdot q = f(r, t) \text{ (say)}$$

Mass of the fluid that passes across the outer sphere in unit time

$$= f(r + \delta r, t)$$

$$= f + \delta r \frac{\partial f}{\partial r}$$

Hence the excess of the flow in over flow out from these two faces

$$= f - \left(f + \delta r \frac{\partial f}{\partial r}\right)$$

$$= -\delta r \frac{\partial f}{\partial r}$$

$$= -\delta r \frac{\partial}{\partial r}\left(\rho 4 \cdot \pi r^2 q\right)$$

$$= -\frac{\partial}{\partial r}\left(\rho r^2 q\right) 4\pi\delta r$$

Also mass between the two spheres at time $t = 4\pi r^2 \delta r \rho$

Therefore, the rate of increase in the mass of the fluid within the spheres

$$= \frac{\partial}{\partial t}\left(4\pi r^2 \delta r \rho\right)$$

$$= \frac{\partial \rho}{\partial t} 4\pi r^2 \delta r$$

Hence by the principle of continuity

$$\frac{\partial \rho}{\partial t} 4\pi r^2 \delta r = -\, 4\pi \delta r \frac{\partial}{\partial r}\left(\rho r^2 q\right)$$

$$\Rightarrow \quad \frac{\partial \rho}{\partial t} + \frac{1}{\partial^2} \frac{\partial}{\partial r}\left(\rho r^2 q\right) = 0$$

Note. If the fluid is incompressible, then ρ = constant and the above equation reduces to

$$0 + \frac{\rho}{r^2} \frac{\partial}{\partial r}\left(r^2 q\right) = 0$$

$$\Rightarrow \quad \frac{\partial}{\partial r}\left(r^2 q\right) = 0$$

Integrating, we get $r^2 q$ = constant = $f(t)$.

2. *Cylindrical symmetry.* In this case the velocity q at any point is perpendicular to a fixed axis and is a function of r and t only, where r is perpendicular distance of the point from the axis and t is the time. We consider two consecutive cylinders of radii r and $r + \delta r$ bounded by planes at unit distance apart.

Then flow across the inner surface

$$= \rho q \; 2\pi r$$

$$= f(r, t) \quad \text{(say)}$$

Flow acros the outer surface

$$= f(r + \delta r, t) = f + \delta r \frac{\partial f}{\partial r}$$

Hence the excess of flow in over flow out

$$= f - \left(f + \delta r \frac{\partial f}{\partial r}\right) = -\, \delta r \frac{\partial f}{\partial r}$$

$$= -\,\delta r \frac{\partial}{\partial r}(\rho q\, 2\pi r)$$

$$= -\,2\pi\delta r \frac{\partial}{\partial r}(\rho r q)$$

Also rate of increase in the mass of the fluid contained in the cylinders

$$= \frac{\partial}{\partial t}(\rho \cdot 2\pi r \cdot \delta r)$$

$$= \frac{\partial \rho}{\partial t} 2\pi\, r\delta r$$

Hence by the principle of continuity, we have

$$\frac{\partial \rho}{\partial t} 2\pi r\delta r = -\,2\pi\,\delta r \frac{\partial}{\partial r}(\rho r q)$$

$$\Rightarrow \quad \frac{\partial \rho}{\partial t} + \frac{1}{r}\frac{\partial}{\partial r}(\rho r q) = 0$$

Note. If the fluid be incompressible then ρ = constant and the last equation becomes

$$\frac{\partial}{\partial r}(rq) = 0$$

Integrating, we get rq = cons = $f(t)$.

35. Boundary Surface. Physical conditions that should be satisfied on given boundaries of the fluid are called as boundary conditions. At the boundary of the fluid the equation of continuity is replaced by a special surface condition. When the fluid is in contact with an impermeable bounding surface, the

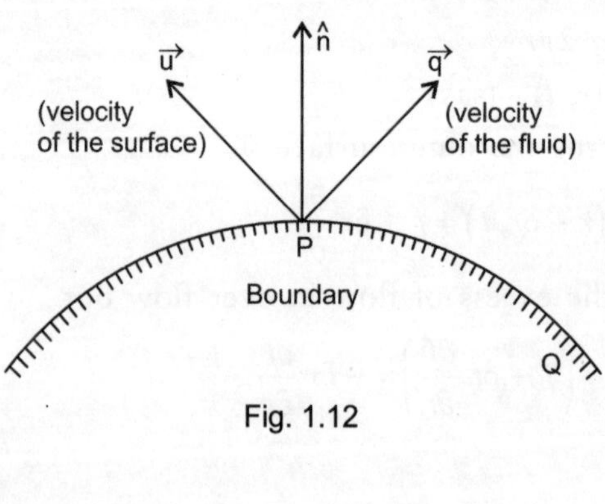

Fig. 1.12

velocity of a fluid particle at any point of the boundary relative to the surface must be tangential to the boundary. Thus at a fixed boundary, the velocity of the fluid perpendicular to the surface must vanish and the normal component of the velocity of the fluid must be equal to the normal component of the velocity of the surface.

Let P be an arbitrary point on the boundary surface $F(\vec{r}, t) = 0$, where the fluid velocity is $\vec{q}$ and velocity of the surface is $\vec{u}$. Let $\hat{n}$ be the unit normal vector drawn at P on the boundary surface.

Since normal component of the velocity of the fluid = normal component of the velocity of the surface.

Hence $\vec{q} \cdot \hat{n} = \vec{u} \cdot \hat{n}$ ... (1)

Since $\vec{\nabla} F$ is normal to the surface $F(\vec{r}, t) = 0$. Hence $\hat{n}$ and $\vec{\nabla} F$ both are parallel vectors.

Hence (1) becomes

$$\vec{q} \cdot \vec{\nabla} F = \vec{u} \cdot \vec{\nabla} F \qquad \text{... (2)}$$

Let $P(\vec{r}, t)$ move to a point $Q(\vec{r} + \delta\vec{r}, t + \delta t)$ in time δt. Since Q lies on $F(\vec{r}, t) = 0$.

Hence $F(\vec{r} + \delta\vec{r}, t + \delta t) = 0$

$$\Rightarrow \quad F(\vec{r}, t) + \delta\vec{r} \cdot \vec{\nabla} F + \delta t \frac{\partial F}{\partial t} = 0$$

Also $F(\vec{r}, t) = 0$

Hence $\dfrac{\partial \vec{r}}{\partial t} \cdot \vec{\nabla} F + \dfrac{\partial F}{\partial t} = 0$

$$\Rightarrow \quad \frac{\partial F}{\partial t} = -\frac{d\vec{r}}{dt} \vec{\nabla} F$$

$$= -\vec{u} \cdot \vec{\nabla} F$$

As $\vec{u}$ is the velocity of the surface.

Now (2) becomes

$$\vec{q} \cdot \vec{\nabla} F = -\frac{\partial F}{\partial t}$$

$$\Rightarrow \quad \frac{\partial F}{\partial t} + \vec{q} \cdot \vec{\nabla} F = 0$$

$$\Rightarrow \quad \left[\frac{\partial}{\partial t} + \vec{q} \cdot \vec{\nabla}\right] F = 0$$

$$\Rightarrow \quad \frac{dF}{dt} = 0$$

$$\Rightarrow \quad \frac{\partial F}{\partial t} + u\frac{\partial F}{\partial x} + v\frac{\partial F}{\partial y} + w\frac{\partial F}{\partial z} = 0$$

This is the required condition for the surface to be a possible form of boundary surface. If the surface is a rigid surface, then the condition becomes

$$u\frac{\partial F}{\partial x} + v\frac{\partial F}{\partial y} + w\frac{\partial F}{\partial z} = 0$$

Also the normal component of velocity for the boundary

$$= \vec{q} \cdot \hat{n} = \frac{q \cdot \vec{\nabla} F}{|\vec{\nabla} F|}$$

$$= \frac{-\partial F/\partial t}{|\vec{\nabla} F|} \left(\because \frac{\partial F}{\partial t} + \vec{q} \cdot \vec{\nabla} F = 0\right)$$

$$= \frac{-\partial F/\partial t}{\sqrt{\left(\frac{\partial F}{\partial x}\right)^2 + \left(\frac{\partial F}{\partial y}\right)^2 + \left(\frac{\partial F}{\partial z}\right)^2}}$$

Alternative Method

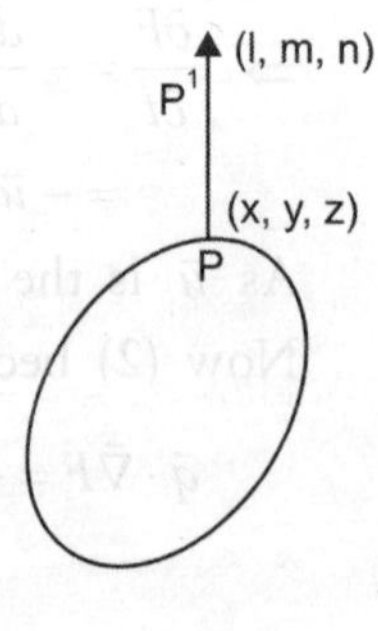

Let (l, m, n) be the direction cosines of the normal at the point $P(x, y, z)$ on the boundary surface $F(x, y, z, t) = 6$ and let the normal velocity of the boundary at P be $\dot{v}$. After time δt, P will move to P^1. Where $PP^1 = \dot{v}\delta t$ and the projections of PP^1 on the axes are $l\dot{v}\delta t$, $m\dot{v}\delta t$, $n\dot{v}\delta t$, therefore the coordinates of P^1 are $(x + l\dot{v}\delta t, y + m\dot{v}\delta t, z + n\dot{v}\delta t)$. But at time $t + \delta t$, P^1 is on the bounding surface therefore we must have

$$F(x + l\dot{v}\delta t,\ y + m\dot{v}\delta t,\ z + n\dot{v}\delta t) = 0$$

$$\Rightarrow\ F(x, y, z, t) + \frac{\partial F}{\partial x} l\dot{v}\delta t + \frac{\partial F}{\partial y} m\dot{v}\delta t + \frac{\partial F}{\partial z} n\dot{v}\delta t + \frac{\partial F}{\partial t}\delta t = 0$$

Neglecting higher order than $\frac{\partial F}{\partial x}, \frac{\partial F}{\partial y}, \frac{\partial F}{\partial z}$,

$$\dot{v}\left(l\frac{\partial F}{\partial x} + m\frac{\partial F}{\partial y} + n\frac{\partial F}{\partial z}\right) + \frac{\partial F}{\partial t} = 0$$

$$\Rightarrow\ \dot{v} = \frac{-\frac{\partial F}{\partial t}}{l\frac{\partial F}{\partial x} + m\frac{\partial F}{\partial y} + n\frac{\partial F}{\partial z}}$$

But $\dfrac{l}{\frac{\partial F}{\partial x}} = \dfrac{m}{\frac{\partial F}{\partial y}} = \dfrac{n}{\frac{\partial F}{\partial z}} = \dfrac{\sqrt{l^2 + m^2 + n^2}}{\sqrt{\left(\frac{\partial F}{\partial x}\right)^2 + \left(\frac{\partial F}{\partial y}\right)^2 + \left(\frac{\partial F}{\partial z}\right)^2}}$

$$= \frac{1}{\sqrt{\left(\frac{\partial F}{\partial x}\right)^2 + \left(\frac{\partial F}{\partial y}\right)^2 + \left(\frac{\partial F}{\partial z}\right)^2}}$$

$$\therefore\ l\frac{\partial F}{\partial x} + m\frac{\partial F}{\partial y} + n\frac{\partial F}{\partial z} = \frac{\left(\frac{\partial F}{\partial x}\right)^2 + \left(\frac{\partial F}{\partial y}\right)^2 + \left(\frac{\partial F}{\partial z}\right)^2}{\sqrt{\left(\frac{\partial F}{\partial x}\right)^2 + \left(\frac{\partial F}{\partial y}\right)^2 + \left(\frac{\partial F}{\partial z}\right)^2}}$$

$$= \sqrt{\left(\frac{\partial F}{\partial x}\right)^2 + \left(\frac{\partial F}{\partial y}\right)^2 + \left(\frac{\partial F}{\partial z}\right)^2}$$

Hence $\dot{v} = \dfrac{-\frac{\partial F}{\partial t}}{\sqrt{\left(\frac{\partial F}{\partial x}\right)^2 + \left(\frac{\partial F}{\partial y}\right)^2 + \left(\frac{\partial F}{\partial z}\right)^2}}$ ... (1)

But $\dot{v}$ must also be equal to the normal velocity of the fluid at P which is $lu + mv + nw$, (u, v, w) be the components of the velocity of the fluid at P.

$$\therefore \quad \dot{v} = lu + mv + nw$$

$$= \frac{u\dfrac{\partial F}{\partial x} + v\dfrac{\partial F}{\partial y} + w\dfrac{\partial F}{\partial z}}{\sqrt{\left(\dfrac{\partial F}{\partial x}\right)^2 + \left(\dfrac{\partial F}{\partial y}\right)^2 + \left(\dfrac{\partial F}{\partial z}\right)^2}} \qquad \text{... (2)}$$

Hence from (1) and (2), we get

$$u\frac{\partial F}{\partial x} + v\frac{\partial F}{\partial y} + w\frac{\partial F}{\partial z} = -\frac{\partial F}{\partial t}$$

$$\Rightarrow \quad u\frac{\partial F}{\partial x} + v\frac{\partial F}{\partial y} + w\frac{\partial F}{\partial z} + \frac{\partial F}{\partial t} = 0$$

WORKED OUT EXAMPLES

I. Problems related to streamlines and possible liquid motion

1. Determine the streamlines and the path lines of the particle when the velocity field is given by $\left(\dfrac{x}{1+t}, \dfrac{y}{1+t}, \dfrac{z}{1+t}\right)$.

[I.A.S. 94]

Solution: Here $u = \dfrac{x}{1+t}, v = \dfrac{y}{1+t}, w = \dfrac{z}{1+t}$

The equations of streamlines are given by

$$\frac{dx}{u} = \frac{dy}{v} = \frac{dz}{w}$$

or $$\frac{dx}{\dfrac{x}{1+t}} = \frac{dy}{\dfrac{y}{1+t}} = \frac{dz}{\dfrac{z}{1+t}}$$

or $$\frac{dx}{x} = \frac{dy}{y} = \frac{dz}{z} \qquad \text{... (1)}$$

Taking first two equations of (1), we get

$$\frac{dx}{x} = \frac{dy}{y}$$

Integrating, we have $\log x = \log y + \log A$

$$\Rightarrow \log x = \log (yA)$$

$$\Rightarrow x = Ay \qquad \text{... (2)}$$

Taking first and last equation of (1), we get

$$\frac{dx}{x} = \frac{dz}{z}$$

Integrating, we get

$$\log x = \log z + \log B$$

$$\Rightarrow \log x = \log (zB)$$

$$\Rightarrow x = Bz \qquad \text{... (3)}$$

Hence streamlines are given by the intersection of (2) and (3)

The differential equations of path lines are given by

$$\vec{q} = \frac{d\vec{r}}{dt}$$

$$\Rightarrow \frac{dx}{dt} = \frac{x}{1+t}, \frac{dy}{dt} = \frac{y}{1+y}, \frac{dz}{dt} = \frac{z}{1+t}$$

$$\Rightarrow \frac{dx}{x} = \frac{dt}{1+t}, \frac{dy}{y} = \frac{dt}{1+t}, \frac{dz}{z} = \frac{dt}{1+t}$$

Integrating, we get

$$\log x = \log A + \log (1 + t)$$

$$\log y = \log B + \log (1 + t)$$

$$\log z = \log C + \log (1 + t)$$

$$x = A (1 + t), y = B (1 + t), z = C (1 + t)$$

These are the required equations of path of the particles.

2. Show that

$$u = \frac{-2xyz}{\left(x^2 + y^2\right)^2}, v = \frac{\left(x^2 - y^2\right)z}{\left(x^2 + y^2\right)^2}, w = \frac{y}{x^2 + y^2}$$

are the velocity components of a possible liquid motion. Is this motion irrotational? [I.A.S. 79, 93]

Solution: To show that the motion is possible, we have to show that the equation of continuity $\frac{\partial u}{\partial x}+\frac{\partial v}{\partial y}+\frac{\partial w}{\partial z}=0$ is satisfied.

Here $\frac{\partial u}{\partial x}=\frac{-2yz}{\left(x^2+y^2\right)^4}\left[\left(x^2+y^2\right)^2-2\left(x^2+y^2\right)2x^2\right]$

$$=\frac{-2yz\left(y^2-3x^2\right)}{\left(x^2+y^2\right)^3}$$

$$\frac{\partial v}{\partial y}=\frac{z\left[-2y\left(x^2+y^2\right)^2-2\left(x^2+y^2\right)2y\left(x^2-y^2\right)\right]}{\left(x^2+y^2\right)^4}$$

$$=\frac{-2yz\left[x^2+y^2+2x^2-2y^2\right]}{\left(x^2+y^2\right)^3}$$

$$=\frac{-2yz\left(3x^2-y^2\right)}{\left(x^2+y^2\right)^3}$$

$$\frac{\partial w}{\partial z}=0$$

Now $\frac{\partial u}{\partial x}+\frac{\partial v}{\partial y}+\frac{\partial w}{\partial z}=\frac{2yz\left[-y^2+3x^2-3x^2+y^2\right]}{\left(x^2+y^2\right)^3}=0$

Hence u, v, w are the velocity components of a possible liquid motion.

Again condition for irrotational motion is

$$\frac{\partial u}{\partial y}-\frac{\partial v}{\partial x}=0$$

$$\frac{\partial v}{\partial z}-\frac{\partial w}{\partial y}=0$$

and $\dfrac{\partial w}{\partial x} - \dfrac{\partial u}{\partial z} = 0$

Now $\dfrac{\partial u}{\partial y} - \dfrac{\partial v}{\partial x} = \dfrac{-2xz\left(x^2 - 3y^2\right)}{\left(x^2 + y^2\right)^2} - \dfrac{2xz\left(3y^2 - x^2\right)}{\left(x^2 + y^2\right)^2}$

$$\frac{\partial v}{\partial z} - \frac{\partial w}{\partial y} = \frac{x^2 - y^2}{\left(x^2 + y^2\right)^2} - \frac{x^2 - y^2}{\left(x^2 + y^2\right)^2} = 0$$

$$\frac{\partial w}{\partial x} - \frac{\partial u}{\partial z} = \frac{-2xy}{\left(x^2 + y^2\right)^3} + \frac{2xy}{\left(x^2 + y^2\right)^3} - 0$$

Hence the motion is irrotational.

3. If the velocity of an incompressible fluid at the point (x, y, z) is given by

$$\left(\frac{3xz}{r^5}, \frac{3yz}{r^5}, \frac{3z^2 - r^2}{r^5}\right)$$

Prove that the liquid motion is possible and the velocity potential is $\cos\theta/r^2$. Also determine the streamlines.

Solution: Given $u = \dfrac{3xz}{r^5}, v = \dfrac{3yz}{r^5}, w = \dfrac{3z^2 - r^2}{r^5}$

The condition for the possible liquid motion is given by $\dfrac{\partial u}{\partial x} + \dfrac{\partial v}{\partial y} + \dfrac{\partial w}{\partial z} = 0$

Now $r^2 = x^2 + y^2 + z^2$

$$\frac{\partial r}{\partial x} = \frac{x}{r}$$

Now $u = \dfrac{3xz}{r^5}$

$$\frac{\partial u}{\partial x} = \frac{3z}{r^{10}}\left(r^5 - 5r^4\frac{x^2}{r}\right)$$

$$= \frac{3z}{r^{10}}\left(r^5 - 5r^3x^2\right)$$

$$\frac{\partial v}{\partial y} = \frac{3z}{r^{10}}\left(r^5 - 5r^3y^2\right)$$

$$\frac{\partial w}{\partial z} = \frac{1}{r^{10}}\left[\left(6z - 2r\frac{z}{r}\right)r^5 - 5r^4\frac{\partial r}{\partial z}\left(3z^2 - r^2\right)\right]$$

$$= \frac{1}{r^{10}}\left[4zr^5 - 5r^4\frac{z}{r}\left(3z^2 - r^2\right)\right]$$

$$= \frac{1}{r^{10}}\left[4zr^5 - 5r^3z\left(3z^2 - r^2\right)\right]$$

Now $\frac{\partial u}{\partial x} + \frac{\partial v}{\partial y} + \frac{\partial w}{\partial z}$

$$= \frac{3z}{r^{10}}\left[2r^5 - 5r^3\left(r^2 - z^2\right)\right] + \frac{1}{r^{10}}\left[4zr^5 - 5zr^3\left(3z^2 - r^2\right)\right]$$

$$= \frac{1}{r^{10}}\left[6zr^5 - 15r^3z\left(r^2 - z^2\right) + 4zr^5 - 5zr^3\left(3z^2 - r^2\right)\right]$$

$$= \frac{zr^3}{r^{10}}\left[6r^2 - 15r^2 + 15z^2 + 4r^2 - 15z^2 + 5r^2\right]$$

$$= 0$$

If ϕ be the velocity poential then

$$d\phi = \frac{\partial\phi}{\partial x}dx + \frac{\partial\phi}{\partial y}dy + \frac{\partial\phi}{\partial z}dz$$

$$= -(udx + vdy + wdz)$$

$$= -\frac{1}{r^5}\left[3xzdx + 3yzdy + \left(3z^2 - r^2\right)dz\right]$$

$$= -\frac{1}{r^5}\left[3z\left(xdx + ydy + zdz\right) - r^2dz\right]$$

$$= -\frac{1}{r^5}\left[\frac{3z}{2}d\left(r^2\right) - r^2dz\right]$$

$$= -\frac{3z}{2}\frac{d\left(r^2\right)}{r^5} + \frac{dz}{r^3}$$

$$= -\frac{3zdr}{r^4} + \frac{dz}{r^3}$$

$$= d\left(\frac{z}{r^3}\right)$$

Integrating, we get

$$\phi = \frac{z}{r^3} = \frac{r\cos\theta}{r^3} = \frac{\cos\theta}{r^2}$$

neglecting constant of integration.

The equation of streamlines are given by

$$\frac{\partial x}{u} = \frac{dy}{v} = \frac{dz}{w}$$

or $$\frac{dx}{\frac{3xz}{r^5}} = \frac{dy}{\frac{3yz}{r^5}} = \frac{dz}{\frac{\left(3z^2 - r^2\right)}{r^5}}$$

or $$\frac{dx}{3xz} = \frac{dy}{3yz} = \frac{dz}{3z^2 - r^2} = \frac{xdx + ydy + zdz}{3z\left(x^2 + y^2 + z^2\right) - r^2 z} \qquad \text{... (1)}$$

Taking the first two members of (1), we get

$$\frac{dx}{x} = \frac{dy}{y}$$

Integrating, we get

$$\log x = \log y + \log c$$

$$\log x = \log (cy)$$

or $x = cy$ ... (2)

Taking the first and last members of (1), we get

$$\frac{dx}{3xz} = \frac{xdx + ydy + zdz}{3zr^2 - r^2 z}$$

$$\Rightarrow \quad \frac{dx}{3x} = \frac{xdx + ydy + zdz}{2r^2}$$

$$\Rightarrow \quad \frac{4dx}{x} = \frac{(2xdx + 2ydy + 2zdz)}{\left(x^2 + y^2 + z^2\right)}$$

Integrating, we get

$$4 \log x = 3 \log (x^2 + y^2 + z^2) + \log b$$

$$\Rightarrow x^4 = b\,(x^2 + y^2 + z^2)^3 \qquad \ldots (3)$$

The equations (2) and (3) represent the streamlines.

4. Show that the velocity potential of an irrotational fluid motion is equal to $A\left(x^2 + y^2 + z^2\right)^{-\frac{3}{2}} z \tan^{-1}\left(\frac{y}{x}\right)$, the lines of flow lie on the series of the surfaces $x^2 + y^2 + z^2 = K^{\frac{2}{3}}\left(x^2 + y^2\right)^{\frac{2}{3}}$.

Solution: Given

$$\phi = A\left(x^2 + y^2 + z^2\right)^{-\frac{3}{2}} z \tan^{-1}\frac{y}{x}$$

$$\Rightarrow \quad \phi = Ar^{-3} z \tan^{-1}\frac{y}{x}$$

where $r^2 = x^2 + y^2 + z^2$

$$\therefore \quad \frac{\partial r}{\partial x} = \frac{x}{r}, \frac{\partial r}{\partial y} = \frac{y}{r}, \frac{\partial r}{\partial z} = \frac{z}{r}$$

$$u = \frac{-\partial\phi}{\partial x} = Az\left[-3r^{-4}\frac{\partial r}{\partial x}\tan^{-1}\frac{y}{x} + \frac{x^2}{x^2 + y^2}\frac{y}{x^2}r^{-3}\right]$$

$$= Az\left[-3r^{-4}\frac{x}{r}\tan^{-1}\frac{y}{x} + \frac{yr^{-3}}{x^2 + y^2}\right]$$

$$= 3Azxr^{-5}\tan^{-1}\frac{y}{x} + \frac{Azyr^{-3}}{x^2 + y^2}$$

$$V = \frac{-\partial\phi}{\partial y} = 3Azyr^{-5}\tan^{-1}\frac{y}{x} - \frac{Azxr^{-3}}{x^2 + y^2}$$

$$W = \frac{-\partial\phi}{\partial z} = 3Az^2r^{-5}\tan^{-1}\frac{y}{x} - Ar^{-3}\tan^{-1}\frac{y}{x}$$

The equations of lines of flow are given by

$$\frac{dx}{u}=\frac{dy}{v}=\frac{dz}{w}$$

$$\Rightarrow \frac{dx}{3Azxr^{-5}\tan^{-1}\frac{y}{x}+\frac{Azyr^{-3}}{x^2+y^2}}=\frac{dy}{3Azyr^{-5}\tan^{-1}\frac{y}{x}-\frac{Azxr^{-3}}{x^2+y^2}}$$

$$=\frac{dz}{3Az^2r^{-5}\tan^{-1}\frac{y}{x}-Ar^{-3}\tan^{-1}\frac{y}{x}} \qquad \text{... (1)}$$

Each member of (1)

$$=\frac{xdx+ydy+zdz}{A\tan^{-1}\frac{y}{x}\left[3x^2zr^{-5}+3y^2zr^{-5}+3z^3r^{-5}-zr^{-3}\right]}$$

$$=\frac{xdx+ydy}{3Ar^{-5}\tan^{-1}\frac{y}{x}z\left(x^2+y^2\right)}$$

$$\Rightarrow \frac{xdx+ydy+zdz}{A\tan^{-1}\frac{y}{x}2zr^{-3}}=\frac{xdx+ydy}{3Ar^{-5}\tan^{-1}\frac{y}{x}z\left(x^2+y^2\right)}$$

$$\Rightarrow \frac{xdx+ydy+zdz}{2}=\frac{xdx+ydy}{3r^{-2}\left(x^2+y^2\right)}$$

$$\Rightarrow \frac{2xdx+2ydy+2zdz}{2}=\frac{r^2\left(2xdx+2ydy\right)}{3\left(x^2+y^2\right)}$$

$$\Rightarrow \frac{2xdx+2ydy+2zdz}{x^2+y^2+z^2}=\frac{2}{3}\frac{\left(2xdx+2ydy\right)}{x^2+y^2}$$

Integrating, we get

$$\log\left(x^2+y^2+z^2\right)=\frac{2}{3}\log\left(x^2+y^2\right)+\frac{2}{3}\log K$$

or $x^2+y^2+z^2=K^{\frac{2}{3}}\left(x^2+y^2\right)^{\frac{2}{3}}$

5. Given $u = -wy$, $v = wx$, $w = 0$, show that the surfaces intersecting the streamlines orthogonally exist and are the planes through z axis, although the velocity potential does not exist.

Solution: The motion will be possible if it satisfies the equation of continuity $\frac{\partial u}{\partial x}+\frac{\partial v}{\partial y}+\frac{\partial w}{\partial z}=0$.

Here $\frac{\partial u}{\partial x}+\frac{\partial v}{\partial y}+\frac{\partial w}{\partial z}=0+0+0=0$

Hence the motion is a possible liquid motion.

The surfaces which cut the streamlines orthogonally are

$$udx + vdy + wdz = 0$$

$$\Rightarrow \quad -wydx + wxdy = 0$$

$$\Rightarrow \quad \frac{dx}{x}-\frac{dy}{y}=0$$

Integrating, we have

$$\log\frac{x}{y}=\log a$$

$$\Rightarrow \quad \frac{x}{y}=a$$

$$\Rightarrow \quad x = ay$$

which represents a plane through z-axis and cuts the streamline orthogonally.

The velocity potential will exist if $udx + vdy + wdz$ is a perfect differential.

Now

$$udx + vdy + wdz = -wydx + wxdy + odz$$
$$= w\,(xdy - ydx)$$
$$= Mdx + Ndy \quad \text{(say)}$$

Here $\frac{\partial M}{\partial y}=w, \frac{\partial N}{\partial x}=w$

Hence $\frac{\partial M}{\partial y}\neq\frac{\partial N}{\partial x}$

Therefore the equation is not exact so that $d\phi$ is not a perfect differential.

Hence the surface intersecting streamlines orthogonally exists and are the planes through z-axis, although the velocity potential does not exist.

6. In the steady motion of homogeneous liquid if the surfaces $f_1 = a_1$, $f_2 = a_2$ define the streamlines, prove that the most general values of the velocity components u, v, w are

$$F(f_1,f_2)\frac{\partial(f_1,f_2)}{\partial(y,z)},\ F(f_1,f_2)\frac{\partial(f_1,f_2)}{\partial(z,x)},\ F(f_1,f_2)\frac{\partial(f_1,f_2)}{\partial(x,y)}$$

Solution: Since the motion is steady and hence streamlines are independent of t. Therefore f_1 and f_2 are functions of x, y, z only.

Now $f_1 = a_1$, $f_2 = a_2$

$\Rightarrow \quad df_1 = 0,\ df_2 = 0$

$$\Rightarrow \quad \frac{\partial f_1}{\partial x}dx + \frac{\partial f_1}{\partial y}dy + \frac{\partial f_1}{\partial z}dz = 0$$

$$\Rightarrow \quad \frac{\partial f_2}{\partial x}dx + \frac{\partial f_2}{\partial y}dy + \frac{\partial f_2}{\partial z}dz = 0$$

$$\Rightarrow \quad \frac{dx}{\frac{\partial f_1}{\partial y}\frac{\partial f_2}{\partial z} - \frac{\partial f_2}{\partial y}\frac{\partial f_1}{\partial z}} = \frac{dy}{\frac{\partial f_1}{\partial z}\frac{\partial f_2}{\partial x} - \frac{\partial f_1}{\partial x}\frac{\partial f_2}{\partial z}} = \frac{dz}{\frac{\partial f_1}{\partial x}\frac{\partial f_2}{\partial y} - \frac{\partial f_2}{\partial x}\frac{\partial f_1}{\partial y}}$$

$$\Rightarrow \quad \frac{dx}{\frac{\partial(f_1,f_2)}{\partial(y,z)}} = \frac{dy}{\frac{\partial(f_1,f_2)}{\partial(z,x)}} = \frac{dz}{\frac{\partial(f_1,f_2)}{\partial(x,y)}}$$

$$\Rightarrow \quad \frac{dx}{J_1} = \frac{dy}{J_2} = \frac{dz}{J_3} \qquad \text{... (1)}$$

where $J_1 = \dfrac{\partial(f_1,f_2)}{\partial(y,z)}$, $J_2 = \dfrac{\partial(f_1,f_2)}{\partial(z,x)}$, $J_3 = \dfrac{\partial(f_1,f_2)}{\partial(x,y)}$

But the differential equations of streamlines are

$$\frac{dx}{u}=\frac{dy}{v}=\frac{dz}{w} \qquad \dots (2)$$

Comparing (1) and (2), we get

$$\frac{u}{J_1}=\frac{v}{J_2}=\frac{w}{J_3}=F \quad \text{(say)}$$

$$u = J_1F,\ v = J_2F,\ w = J_3F$$

Here F is an arbitrary function. We shall now determine the nature of F by applying the condition

$$\frac{\partial u}{\partial x}+\frac{\partial v}{\partial y}+\frac{\partial w}{\partial z}=0$$

which u, v, w must satisfy for motion.

This condition gives

$$F\left(\frac{\partial J_1}{\partial x}+\frac{\partial J_2}{\partial y}+\frac{\partial J_3}{\partial z}\right)+\left(J_1\frac{\partial F}{\partial x}+J_2\frac{\partial F}{\partial y}+J_3\frac{\partial F}{\partial z}\right)=0$$

By the property of Jacobian $\frac{\partial J_1}{\partial x}+\frac{\partial J_2}{\partial y}+\frac{\partial J_3}{\partial z}=0$

Hence

$$\frac{\partial(f_1,f_2)}{\partial(y,z)}\frac{\partial F}{\partial x}+\frac{\partial(f_1,f_2)}{\partial(z,x)}\frac{\partial F}{\partial y}+\frac{\partial(f_1,f_2)}{\partial(x,y)}\frac{\partial F}{\partial z}=0$$

$$\Rightarrow \quad \frac{\partial(F,f_1,f_2)}{\partial(x,y,z)}=0$$

This proves that F, f_1, f_2 are not independent.

Therefore $F = F(f_1, f_2)$

Hence

$$u=\frac{\partial(f_1,f_2)}{\partial(y,z)}F(f_1,f_2)$$

$$v=\frac{\partial(f_1,f_2)}{\partial(z,x)}F(f_1,f_2)$$

$$w=\frac{\partial(f_1,f_2)}{\partial(x,y)}F(f_1,f_2)$$

7. Determine whether the motion specified by

$$\vec{q} = \frac{A\left(x\vec{j} - y\vec{i}\right)}{x^2 + y^2}, (A = \text{const})$$

is a possible motion for an incompressible fluid. If so, determine the equations of the streamlines. Also show that the motion is of potential kind. Find the velocity potential.

Solution: We know that condition for possible motion of an incompressible fluid is $\vec{\nabla} \cdot \vec{q} = 0$.

Now

$$\begin{aligned}
\vec{\nabla} \cdot \vec{q} &= A\left(\vec{i}\frac{\partial}{\partial x} + \vec{j}\frac{\partial}{\partial y} + \vec{k}\frac{\partial}{\partial z}\right) \cdot \left[\frac{x\vec{j} - y\vec{i}}{x^2 + y^2}\right] \\
&= A\left[-\frac{\partial}{\partial x}\left(\frac{y}{x^2 + y^2}\right) + \frac{\partial}{\partial y}\left(\frac{x}{x^2 + y^2}\right)\right] \\
&= A\left[\frac{2xy}{\left(x^2 + y^2\right)^2} - \frac{2yx}{\left(x^2 + y^2\right)^2}\right] \\
&= 0
\end{aligned}$$

Hence the motion is possible.

The equations of the streamlines are

$$\frac{dx}{u} = \frac{dy}{v} = \frac{dz}{w}$$

$$\Rightarrow \quad \frac{dx}{\dfrac{-Ay}{x^2 + y^2}} = \frac{dy}{\dfrac{Ax}{x^2 + y^2}} = \frac{dz}{0}$$

$$\Rightarrow \; xdx + ydy = 0, \; dz = 0$$

Integrating, we get

$$x^2 + y^2 = \text{constant}, \; z = \text{constant}$$

Thus the streamlines are circles whose centres are on z-axis, their planes being perpendicular to this axis.

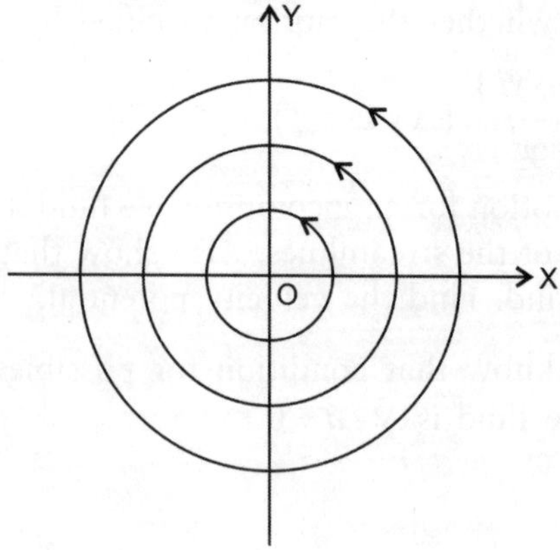

Fig. 1.13

Again

$$\vec{\nabla} \cdot \vec{q} = \begin{vmatrix} \vec{i} & \vec{j} & \vec{k} \\ \dfrac{\partial}{\partial x} & \dfrac{\partial}{\partial y} & \dfrac{\partial}{\partial z} \\ \dfrac{-Ay}{x^2+y^2} & \dfrac{Ax}{x^2+y^2} & 0 \end{vmatrix}$$

$$= \vec{k}\left[\frac{\partial}{\partial x}\left(\frac{Ax}{x^2+y^2}\right) + \frac{\partial}{\partial y}\left(\frac{Ay}{x^2+y^2}\right)\right]$$

$$= \vec{k}A\left[\frac{y^2-x^2}{\left(x^2+y^2\right)^2} + \frac{x^2-y^2}{\left(x^2+y^2\right)^2}\right] = 0$$

Thus the flow is of potential kind, so we can determine $\phi(x, y, z)$ such that $\vec{q} = -\vec{\nabla}\phi$

$$\Rightarrow \quad \frac{\partial \phi}{\partial x} = -u = \frac{Ay}{x^2+y^2} \qquad \text{... (1)}$$

$$\frac{\partial \phi}{\partial y} = -v = -\frac{Ax}{x^2+y^2} \qquad \text{... (2)}$$

$$\frac{\partial \phi}{\partial z} = -w = 0$$

which shows that ϕ is independent of z, hence $\phi = \phi(x, y)$

Integrating (1), we get

$$\phi(x, y) = f(y) + A \tan^{-1}\left(\frac{x}{y}\right)$$

$$\Rightarrow \quad \frac{\partial \phi}{\partial y} = f^1(y) - \frac{Ax}{x^2 + y^2}$$

$\Rightarrow \; f^1(y) = 0 \qquad$ From (2)

Integrating $f(y)$ = constant

Since the constant can be omitted while writing velocity potential. Therefore $\phi(x, y) = A \tan^{-1}\left(\frac{x}{y}\right)$.

8. The velocity vector in the flow field is given by $\vec{q} = \vec{i}\,(Az - By) + \vec{j}\,(Bx - Cz) + \vec{k}\,(Cy - Ax)$. Where A, B, C are non zero constants. Determine the equations of vortex lines.

Solution: Let $\vec{q} = u\vec{i} + v\vec{j} + w\vec{k}$

Then $u = Az - By$, $v = Bx - Cz$, $w = Cy - Ax$

Let Ωx, Ωy, Ωz be the velocity components, then

$$\Omega x = \frac{\partial w}{\partial y} - \frac{\partial v}{\partial z} = C + C = 2C$$

$$\Omega y = \frac{\partial u}{\partial z} - \frac{\partial w}{\partial x} = A + A = 2A$$

$$\Omega z = \frac{\partial v}{\partial x} - \frac{\partial u}{\partial y} = B + B = 2B$$

The equation of the vortex lines are

$$\frac{\partial x}{\Omega x} = \frac{dy}{\Omega y} = \frac{dz}{\Omega z}$$

$$\Rightarrow \quad \frac{dx}{2C} = \frac{dy}{2A} = \frac{dz}{2B} \qquad \ldots (1)$$

Taking the first two members of (1), we have

$$\frac{dx}{C}=\frac{dy}{A}$$

Integrating, we get

$$Ax - Cy = C_1 \qquad \text{... (2)}$$

Taking the last two members of (1), we have

$$\frac{dy}{A}=\frac{dz}{B}$$

Integrating, we get

$$By - Az = C_2 \qquad \text{... (3)}$$

The required vortex lines are the straight lines of the intersection of (2) and (3).

9. At a point in an incompressible fluid having spherical coordinates (r, θ, ϕ) the velocity components are $[2Mr^{-3}\cos\theta,\ Mr^{-3}\sin\theta,\ 0]$, where M is a constant. Show that the velocity is of the potential kind. Find the velocity potential and the equations of the streamlines.

Solution: Here

$$q_r = 2Mr^{-3}\cos\theta,\ q_\theta = Mr^{-3}\sin\theta,\ q_\phi = 0$$

Thus

$$\vec{q} = 2Mr^{-3}\cos\theta\,\hat{r} + Mr^{-3}\sin\theta\,\hat{\theta} + 0\left(\hat{\phi}\right)$$

and

$$\operatorname{curl}\vec{q} = \frac{1}{r^2\sin^2\theta}\begin{vmatrix} \hat{r} & r\hat{\theta} & r\sin\theta\,\hat{\phi} \\ \dfrac{\partial}{\partial r} & \dfrac{\partial}{\partial \theta} & \dfrac{\partial}{\partial \phi} \\ q_r & q_\theta & q_\phi \end{vmatrix}$$

$$= \frac{1}{r^2\sin^2\theta}\begin{vmatrix} \hat{r} & r\hat{\theta} & r\sin\theta\,\hat{\phi} \\ \dfrac{\partial}{\partial r} & \dfrac{\partial}{\partial \theta} & \dfrac{\partial}{\partial \phi} \\ 2Mr^{-3}\cos\theta & Mr^{-3}\sin\theta & 0 \end{vmatrix}$$

$$= 0$$

Hence the flow is of potential kind.

Let $\phi^1(r, \theta, \phi)$ be the required velocity potential

Then

$$-\frac{\partial\phi^1}{\partial r} = q_r = 2Mr^{-3}\cos\theta$$

$$-\frac{\partial\phi^1}{r\partial\theta} = q_\theta = Mr^{-3}\sin\theta$$

and $$-\frac{\partial\phi^1}{r\sin\theta\,\partial\phi} = q_\phi = 0$$

$$\therefore \quad d\phi^1 = \frac{\partial F}{\partial r}dr + \frac{\partial F}{\partial\theta}d\theta + \frac{\partial F}{\partial\phi}d\phi$$

$$= -\left(2Mr^{-3}\cos\theta\right)dr - \left(Mr^{-2}\sin\theta\right)d\theta + 0d\phi$$

$$= d\,(Mr^{-2}\cos\theta)$$

Integrating, we get

$$\phi^1 = Mr^{-2}\cos\theta$$

Finally the streamlines are given by

$$\frac{dr}{q_r} = \frac{rd\theta}{q_\theta} = \frac{r\sin\theta\,d\phi}{q_\phi}$$

$$\Rightarrow \quad \frac{dr}{2\,Mr^{-3}\cos\theta} = \frac{rd\theta}{Mr^{-3}\sin\theta} = \frac{r\sin\theta\,d\phi}{0}$$

which gives $d\phi = 0$ and $2\cot\theta\,d\theta = \dfrac{dr}{r}$.

Integrating, we get the equation of streamlines as

$$\phi = C_1 \text{ and } r = C_2\sin^2\theta$$

The equation ϕ = constant shows that the required streamline lies in a plane which pass through the axis of symmetry $\theta = 0$.

10. Show that $\phi = (x - t)\,(y - t)$ represent the velocity potential of an incompressible two dimensional fluid. Show that the streamlines at time 't' are the curves

$$(x - t)^2 - (y - t)^2 = \text{constant}$$

and that the paths of the fluid particles have the equations

$$\log(x-y)=\frac{1}{2}\left[(x+y)-a(x-y)^{-1}\right]+b$$

where a, b are constants.

Solution: Given $\phi = (x - t)(y - t)$

$$u=\frac{-\partial\phi}{\partial x}=-(y-t)$$

$$v=\frac{-\partial\phi}{\partial y}=-(x-t)$$

and so $\dfrac{\partial u}{\partial x}=0$ and $\dfrac{\partial v}{\partial y}=0.$

Thus the equation of continuity $\dfrac{\partial u}{\partial x}+\dfrac{\partial v}{\partial y}=0$ is satisfied. Hence ϕ represents the velocity potential of an incompressible two dimensional flow.

Also, the equations of streamlines are

$$\frac{dx}{u}=\frac{dy}{v}$$

$$\Rightarrow \quad \frac{dx}{-(y-t)}=\frac{dy}{-(x-t)}$$

$$\Rightarrow (x-t)\,dt-(y-t)\,dy=0$$

Integrating, we get

$$(x-t)^2-(y-t)^2=\text{constant}$$

Also equation of paths of particles are given by

$$u=\frac{dx}{dt}=-(y-t)$$

and $v=\dfrac{dy}{dt}=-(x-t)$

$$\therefore \quad \frac{dx}{dt}=t-y$$

and $\dfrac{dy}{dt}=t-x$ ... (1)

Hence

$$\frac{dx}{dt}+\frac{dy}{dt}=2t-(x+y)$$

$$\Rightarrow \quad \frac{d}{dt}(x+y)+(x+y)=2t$$

which is a linear differential equation in $(x + y)$

Integrating factor $=e^{\int dt}=e^t$

Its solution is

$$(x + y)\ e^t = 2\int t\ e^t dt$$

$$= 2\left[t\int e^t dt - \int e^t dt\right]$$

$$(x + y)\ e^t = 2e^t\ (t - 1) + C_1$$

$$(x + y) = 2\ (t - 1) + C_1e^{-t}$$

Again from (1)

$$\frac{dx}{dt}-\frac{dy}{dt}=x-y$$

$$\Rightarrow \quad \frac{d}{dt}(x-y)=x-y$$

$$\Rightarrow \quad \frac{d(x-y)}{x-y}=dt$$

Integrating, we get

$$\log\ (x - y) - \log\ C_2 = t$$

or $x - y = C_2e^t$ ... (2)

Now $(x + y) - a\ (x - y)^{-1}$

$$= 2t-2+C_1e^{-t}-\frac{a}{C_2}e^{-t}$$

$$= 2t-2, \text{ where } C_1=\frac{a}{C_2}$$

$$\Rightarrow \quad \frac{1}{2}\left[(x+y)-a\,(x-y)^{-1}\right]=t-1$$

From (2)

$$\frac{1}{2}\left[(x+y)-a(x-y)^{-1}\right]=\log(x-y)-\log C_2-1$$

$$\Rightarrow \frac{1}{2}\left[(x+y)-a(x-y)^{-1}\right]=\log(x-y)+b$$

where $b = -\log C_2 - 1$

II. Solved Problems related to boundary surface

11. Show that the variable ellipsoid

$$\frac{x^2}{a^2k^2t^4}+kt^2\left[\left(\frac{y}{b}\right)^2+\left(\frac{z}{c}\right)^2\right]=1$$

is a possible form for the boundary surface of a liquid at any time t. [I.A.S. 92]

Solution: The surface

$$F(x, y, z, t) = 0 \qquad \text{... (1)}$$

can be a possible boundary surface, if it satisfies the condition

$$\frac{dF}{dt}=\frac{\partial F}{\partial t}+u\frac{\partial F}{\partial x}+v\frac{\partial F}{\partial y}+w\frac{\partial F}{\partial z}=0 \qquad \text{... (2)}$$

provided that u, v, w satisfied the equation of continuity

$$\frac{\partial u}{\partial x}+\frac{\partial v}{\partial y}+\frac{\partial w}{\partial z}=0 \qquad \text{... (3)}$$

Here

$$F(x,y,z,t)=\frac{x^2}{a^2k^2t^4}+kt^2\left[\left(\frac{y}{b}\right)^2+\left(\frac{z}{c}\right)^2\right]$$

Now

$$\frac{\partial F}{\partial t}=\frac{-4x^2}{a^2k^2t^5}+2kt\left(\frac{y^2}{b^2}+\frac{z^2}{c^2}\right)$$

$$\frac{\partial F}{\partial x}=\frac{2x}{a^2k^2t^4},\ \frac{\partial F}{\partial y}=\frac{2kt^2y}{b^2},\ \frac{\partial F}{\partial z}=\frac{2kt^2z}{c^2}$$

Putting these values in (2), we get

$$\frac{-4x^2}{a^2k^2t^5} + 2kt\left(\frac{y^2}{b^2} + \frac{z^2}{c^2}\right) + u\,\frac{2x}{a^2k^2t^4} + v\,\frac{2kt^2y}{b^2} + w\frac{2kt^2z}{c^2} = 0$$

$$\Rightarrow \quad \frac{2x}{a^2k^2t^4}\left(u - \frac{2x}{t}\right) + \frac{2kyt}{b^2}(y + vt) + \frac{2zkt}{c^2}(z + wt) = 0$$

which will hold if

$$u - \frac{2x}{t} = 0,\ y + vt = 0 \text{ and } z + wt = 0$$

$$\Rightarrow \quad u = \frac{2x}{t},\ v = \frac{-y}{t},\ w = \frac{-z}{t}$$

Now $\dfrac{\partial u}{\partial x} = \dfrac{2}{t},\ \dfrac{\partial v}{\partial y} = -\dfrac{1}{t},\ \dfrac{\partial w}{\partial z} = -\dfrac{1}{t}$

Now $\dfrac{\partial u}{\partial x} + \dfrac{\partial v}{\partial y} + \dfrac{\partial w}{\partial z} = \dfrac{2}{t} - \dfrac{1}{t} - \dfrac{1}{t} = 0$

Hence (1) is a possible form of boundary surface.

12. Show that $\dfrac{x^2}{a^2}\tan^2 t + \dfrac{y^2}{b^2}\cot^2 t - 1 = 0$ is a possible form of boundary surface and find an expression for normal velocity.

Solution: We know that the surface

$$F(x, y, z, t) = 0 \qquad \text{... (1)}$$

is a possible boundary surface, if it satisfies the boundary condition

$$\frac{\partial F}{\partial t} + u\,\frac{\partial F}{\partial x} + v\,\frac{\partial F}{\partial y} + w\,\frac{\partial F}{\partial z} = 0 \qquad \text{... (2)}$$

where u, v, w satisfy the equation of continuity

$$\frac{\partial u}{\partial x} + \frac{\partial v}{\partial y} + \frac{\partial w}{\partial z} = 0$$

Here $F(x, y, t) = \dfrac{x^2}{a^2}\tan^2 t + \dfrac{y^2}{b^2}\cot^2 t - 1 = 0$

so that

$$\frac{\partial F}{\partial t} = \frac{2x^2}{a^2}\tan t \sec^2 t - \frac{2y^2}{b^2}\cot t \operatorname{cosec}^2 t$$

$$\frac{\partial F}{\partial x} = \frac{2x}{a^2}\tan^2 t, \frac{\partial F}{\partial y} = \frac{2y}{b^2}\cot^2 t$$

Putting these values, (2) becomes

$$\frac{2x^2}{a^2}\tan t \sec^2 t - \frac{2y^2}{b}\cot t \operatorname{cosec}^2 t + u\frac{2x}{a^2}\tan^2 t + v\frac{2y}{b^2}\cot^2 t = 0$$

$$\Rightarrow \quad \frac{x \tan t}{a^2}\left(x \sec^2 t + u \tan t\right) + \frac{y \cot t}{b^2}\left(-y \operatorname{cosec}^2 t + v \cot t\right) = 0$$

which will hold if

$$x \sec^2 t + u \tan t = 0$$

and $-y \operatorname{cosec}^2 t + v \cot t = 0$

$$\Rightarrow \quad u = -x \sec^2 t \cot t = -\frac{x}{\sin t \cdot \cot t}$$

and $-y \operatorname{cosec}^2 t + v \cot t = 0$

$$\Rightarrow v = y \operatorname{cosec}^2 t \tan t$$

$$= \frac{y}{\sin t \cot t}$$

Now $\dfrac{\partial u}{\partial x} = \dfrac{-1}{\sin t \cot t}, \dfrac{\partial v}{\partial y} = \dfrac{1}{\sin t \cot t}, \dfrac{\partial u}{\partial x} + \dfrac{\partial v}{\partial y} = 0$

Hence (1) is a possible form of boundary surface.

Also normal velocity

$$= \frac{-\dfrac{\partial F}{\partial t}}{\sqrt{\left(\dfrac{\partial F}{\partial x}\right)^2 + \left(\dfrac{\partial F}{\partial y}\right)^2}}$$

$$= \frac{\frac{2y^2}{b^2}\cot t \operatorname{cosec}^2 t - \frac{2x^2}{a^2}\tan t \sec^2 t}{\sqrt{\frac{4x^2}{a^4}\tan^4 t + \frac{4y^2}{b^4}\cot^4 t}}$$

$$= \frac{a^2y^2 \cot t \operatorname{cosec}^2 t - x^2b^2 \tan t \sec^2 t}{\sqrt{x^2b^4 \tan^4 t + y^2a^4 \cot^4 t}}$$

13. Determine the restrictions on f_1, f_2, f_3 if

$$f_1(t)\frac{x^2}{a^2} + f_2(t)\frac{y^2}{b^2} + f_3(t)\frac{z^2}{c^2} = 1$$

is a possible boundary surface of a liquid. [I.A.S. 95]

Solution: Let

$$F(x,y,z,t) = \frac{x^2}{a^2}f_1(t) + \frac{y^2}{b^2}f_2(t) + \frac{z^2}{c^2}f_3(t) - 1 = 0 \qquad \dots (1)$$

We know that the surface $F(x, y, z, t) = 0$ can be a possible boundary surface, if it satisfies the boundary condition

$$\frac{\partial F}{\partial t} + u\frac{\partial F}{\partial x} + v\frac{\partial F}{\partial y} + w\frac{\partial F}{\partial z} = 0 \qquad \dots (2)$$

where the velocity components u, v, w satisfy the equation of continuity

$$\frac{\partial u}{\partial x} + \frac{\partial v}{\partial y} + \frac{\partial w}{\partial z} = 0 \qquad \dots (3)$$

Now

$$\frac{\partial F}{\partial t} = \frac{x^2}{a^2}f_1^1(t) + \frac{y^2}{b^2}f_2^1(t) + \frac{z^2}{c^2}f_3^1(t)$$

$$\frac{\partial F}{\partial x} = \frac{2x}{a^2}f_1(t), \frac{\partial F}{\partial y} = \frac{2y}{b^2}f_2(t), \frac{\partial F}{\partial z} = \frac{2z}{c^2}f_3(t)$$

Putting these values (2) becomes

$$\frac{x^2}{a^2}f_1^1(t) + \frac{y^2}{b^2}f_2^1(t) + \frac{z^2}{c^2}f_3^1(t) + u\frac{2x}{a^2}f_1(t)$$

$$+ v\frac{2y}{b^2}f_2(t) + w\frac{2z}{c^2}f_3(t) = 0$$

$$\Rightarrow \frac{2x}{a^2} f_1(t)\left[u + \frac{x f_1^1(t)}{2f_1(t)}\right] + \frac{2y}{b^2} f_2(t)\left[v + \frac{yf_2^1(t)}{2f_2(t)}\right] + \frac{2z}{c^2} f_3(t)\left[w + \frac{zf_3^1(t)}{2f_3(t)}\right] = 0$$

which will hold if

$$u + \frac{xf_1^1(t)}{2f_1(t)} = 0$$

$$v + \frac{yf_2^1(t)}{2f_2(t)} = 0$$

$$w + \frac{2f_3^1(t)}{2f_3(t)} = 0$$

$$\Rightarrow \quad u = \frac{-xf_1^1(t)}{2f_1(t)}, v = \frac{-yf_2^1(t)}{2f_2(t)}, w = \frac{-zf_3^1(t)}{2f_3(t)}$$

Also

$$\frac{\partial u}{\partial x} = \frac{-f_1^1(t)}{2f_1(t)}, \frac{\partial v}{\partial y} = \frac{-f_2^1(t)}{2f_2(t)}, \frac{\partial w}{\partial z} = \frac{-f_3^1(t)}{2f_3(t)}$$

Hence from (3), we have

$$-\frac{y}{2}\left[\frac{f_1^1(t)}{f_1(t)} + \frac{f_2^1(t)}{f_2(t)} + \frac{f_3^1(t)}{f_3(t)}\right] = 0$$

$$\Rightarrow \quad \frac{d}{dt}\left[\log(f_1 f_2 f_3)\right] = 0$$

$$\Rightarrow \quad \log f_1 f_2 f_3 = \text{constant}$$

$$\Rightarrow \quad f_1 f_2 f_3 = \text{constant}$$

which is the required restriction on f_1, f_2 and f_3.

14. Show that all necessary conditions can be satisfied by a velocity potential of the form $\phi = \alpha x^2 + \beta y^2 + \gamma z^2$ and a bounding surface of the form $F = ax^4 + by^4 + cz^4 - x(t) = 0$, where $x(t)$ is a given function of time and α, β, γ, a, b, c are suitable functions of time.

Solution: The necessary conditions are:

(1) that ϕ satisfies the Laplace's equation i.e. $\vec{\nabla}^2\phi = 0$ for incompressible fluid flow.

(2) that ϕ satisfies the condition for boundary surface

$$\frac{\partial F}{\partial t} + u\frac{\partial F}{\partial x} + v\frac{\partial F}{\partial y} + w\frac{\partial F}{\partial z} = 0 \qquad \text{... (1)}$$

We have

$$\phi = \alpha x^2 + \beta y^2 + \gamma z^2$$

$$\frac{\partial \phi}{\partial x} = 2\alpha x$$

$$\frac{\partial^2 \phi}{\partial x^2} = 2\alpha, \frac{\partial^2 \phi}{\partial y^2} = 2\beta, \frac{\partial^2 \phi}{\partial z^2} = 2\gamma$$

Hence the Laplace's equation $\frac{\partial^2 \phi}{\partial x^2} + \frac{\partial^2 \phi}{\partial y^2} + \frac{\partial^2 \phi}{\partial z^2} = 0$ becomes

$$2\ (\alpha + \beta + \gamma) = 0$$

$$\text{or} \quad \alpha + \beta + \gamma = 0 \qquad \text{... (2)}$$

Also $F\ (x, y, z) = ax^4 + by^4 + cz^4 - x\ (t) = 0$ ... (3)

$$\therefore \quad \frac{\partial F}{\partial t} = \frac{\partial a}{\partial t}x^4 + \frac{\partial b}{\partial t}y^4 + \frac{\partial c}{\partial t}z^4 - x^1(t)$$

$$\frac{\partial F}{\partial x} = 4ax^3, \frac{\partial F}{\partial y} = 4by^3, \frac{\partial F}{\partial z} = 4cz^3$$

$$u = \frac{-\partial \phi}{\partial x} = -2\alpha x, v = \frac{-\partial \phi}{\partial y} = -2\beta y, w = \frac{-\partial \phi}{\partial z} = -2\gamma z$$

Hence (1) becomes

$$x^4\frac{\partial a}{\partial t} + y^4\frac{\partial b}{\partial t} + z^4\frac{\partial c}{\partial t} - x^1(t) - 2\alpha x\left(4ax^3\right)$$

$$- 2\beta y\left(4by^3\right) - 2\gamma z\left(4cz^3\right) = 0$$

$$\Rightarrow \quad x^4\left(\frac{\partial a}{\partial t} - 8\alpha a\right) + y^4\left(\frac{\partial b}{\partial t} - 8\beta b\right)$$

$$+ z^4\left(\frac{\partial c}{\partial t} - 8\gamma c\right) - x^1(t) = 0 \qquad \text{... (4)}$$

Since (3) and (4) both have to hold for all points (x, y, z) on the surface, they should be identical and so, on comparison, we get

$$\frac{\dfrac{\partial a}{\partial t} - 8a\alpha}{a} = \frac{\dfrac{\partial b}{\partial t} - 8b\beta}{b} = \frac{\dfrac{\partial c}{\partial t} - 8c\gamma}{c} = \frac{x^1(t)}{x(t)}$$

Hence

$$\frac{da}{dt} - 8a\alpha = \frac{a\,dx}{x\,dt}$$

$$\Rightarrow \quad \frac{da}{a} = 8\alpha\,dt + \frac{dx}{x}$$

Integrating, we get

$$\log a = 8\int \alpha\,dt + \log x(t)$$

Similarly

$$\log b = 8\int \beta\,dt + \log x(t)$$

$$\log c = 8\int \gamma\,dt + \log x(t)$$

When α, β, γ and x are known, a, b, c can be found out.

III. Solved problems related to equation of continuity

15. If every particle moves on the surface of a sphere, prove that the equation of continuity is

$$\frac{\partial \rho}{\partial t}\cos\theta + \frac{\partial}{\partial \theta}(\rho w \cos\theta) + \frac{\partial}{\partial \phi}\left(\rho w^1 \cos\theta\right) = 0$$

ρ being the density, θ, ϕ the latitude and longitude respectively of an element and w, w^1 the angular velocities of any element in the latitude and longitude respectively. [I.A.S. 91]

Solution: We consider an elementary parallelopiped $PQRS$ and $P^1Q^1R^1S^1$ on the surface of the sphere, whose length of edges are $PP^1 = \delta r$, $PQ = r\delta\theta$, $PS = r\cos\theta\,\delta\phi$.

Since w and w^1 are the angular velocities of the elements along latitude and longitude respectively, so rw and $r\cos\theta w^1$ are the velocities along PQ and PS. Since the particles moves

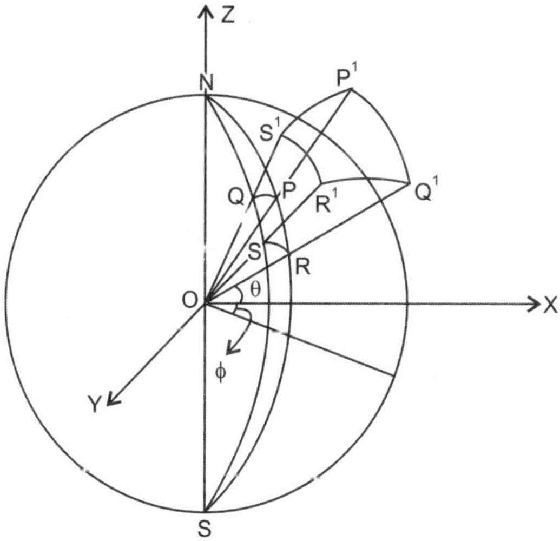

Fig. 1.14

on the surface of the sphere, it follows that there will be no velocity normal to the surface of the sphere i.e. the velocity along PP^1 vanishes.

Now excess of flow in over flow out along the faces perpendicular to PQ i.e. from the face QQ^1R^1R and PP^1S^1S is

$$= - r\delta\theta \frac{\partial}{\partial\theta}[\rho\ \delta r\ r \cos\theta\ \delta\phi\ rw] \text{ per unit time}$$

Similarly excess of flow in over flow out along the faces perpendicular to PS is

$$= -r \cos\theta\ \delta\phi \frac{\partial}{r\cos\theta\ \partial\phi}\left(\rho\ \delta r\ r\ \delta\theta\ r\cos\theta\ w^1\right)$$

per unit time

Hence total excess of flow in over flow out through all the faces

$$= \left[-r\delta\theta \frac{\partial}{r\ \partial\theta}(\rho\ \delta r\ r\cos\theta\ \delta\phi\ rw) - r\cos\theta\delta\phi \frac{\partial}{r\cos\theta\partial\phi}\left(\rho\delta r \cdot r\delta\theta\ r\cos\theta w^1\right)\right]$$

Also mass of the parallelopiped

$$= \rho \cdot \delta r\ r\delta\theta \cdot r\cos\theta\delta\phi$$

Thus rate of increase in mass per unit time

$$= \frac{\partial}{\partial t}(\rho\delta r\ r\ \delta\theta\ r\cos\theta\ \delta\phi)$$

Hence by the principle of continuity

$$\frac{\partial}{\partial t}(\rho\delta r\ r\ \delta\theta\ r\cos\theta\delta\phi) = -r\delta\theta\frac{\partial}{r\partial\theta}(\rho\delta r\ r\cos\theta\delta\phi\ rw)$$

$$-r\cos\theta\delta\phi\frac{\partial}{r\cos\theta\delta\phi}\left(\rho\delta r\ r\delta\theta\ r\cos\theta w^1\right)$$

$$\Rightarrow \quad \frac{\partial\rho}{\partial t}\left(r^2\cos\theta\ \delta r\ \delta\theta\ \delta\phi\right) + r\delta\theta\frac{\partial}{r\partial\theta}(\rho w\cos\theta)r^2\delta r\delta\phi$$

$$+ r\cos\theta\ \delta\phi\frac{\partial}{r\cos\theta\ \partial\phi}\left(\rho w^1\cos\theta\right)r^2\delta r\ \delta\theta = 0$$

$$\Rightarrow \quad \frac{\partial\rho}{\partial t} + \frac{1}{\cos\theta}\frac{\partial}{\partial\theta}(\rho w\cos\theta) + \frac{1}{\cos\phi}\frac{\partial}{\partial\phi}\left(\rho w^1\cos\theta\right) = 0$$

$$\Rightarrow \quad \cos\theta\frac{\partial\rho}{\partial t} + \frac{\partial}{\partial\theta}(\rho w\cos\theta) + \frac{\partial}{\partial\phi}\left(\rho w^1\cos\theta\right) = 0.$$

16. If the lines of motion are curves on the surface of cones having their vertices at the origin and the axis of z for common axis, prove that the equation of continuity is

$$\frac{\partial\rho}{\partial t} + \frac{\partial}{\partial r}(\rho q_r) + \frac{2\rho q_r}{r} + \frac{\operatorname{cosec}\theta}{r}\frac{\partial}{\partial w}(\rho q_w) = 0.$$

Solution: Let O, the common vertex, be the origin and OZ the common axis, the axis of Z. We consider a cone OPQ of semi vertical angle θ. Let $A(r, \theta, w)$ be a point on the surface of the cone and q_r, q_θ, q_w be the components of velocity along the edges

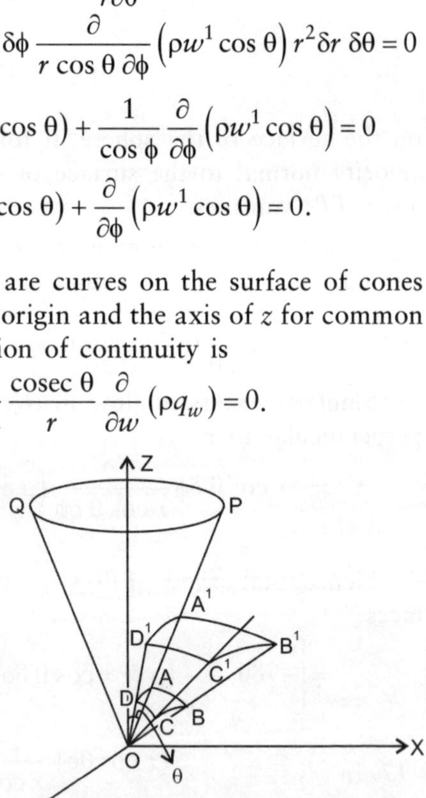

Fig. 1.15

of the parallelopiped AA^1, AB and AD respectively. Also $AA^1 = \delta r$, $AB = r\delta\theta$, $AD = r \sin \theta\delta w$.

Since the lines of motion are curves on the surfaces of cones, so there will be no motion perpendicular to the surface of cone.

Now excess of flow in over flow out in the direction AA^1 i.e. from the face $ABCD$ and $A^1B^1C^1D^1$

$$= -\frac{\partial}{\partial r}(\rho q_r \, r\delta\theta \, r \sin \theta\delta w)\delta r \text{ per unit time}$$

and the excess of flow in over the flow out in the direction AD i.e. from the face ABB^1A^1 and the opposite face DCC^1D^1

$$= -\frac{\partial}{r \sin \theta \, \partial \phi}(\rho q_w r \, \delta\theta \, \delta r) \, r \sin \theta \, \delta w \quad \text{per unit time.}$$

Hence total excess of mass of flow in over flow out through all the faces

$$= -\frac{\partial}{\partial r}(\rho q_r r\delta\theta \, r \sin \theta\delta w)\,\delta r$$
$$- \frac{\partial}{r \sin \theta \, \partial w}(\rho q_w r \, \delta\theta\delta r) \, r \sin \theta\delta w$$

Also mass of the fluid element $= \rho r\delta\theta \cdot r \sin \theta\delta w \, \delta r$

Hence rate of increase in the mass of the element

$$= \frac{\partial}{\partial t}(\rho r \, \delta\theta \, r \sin \theta \, \delta w \, \delta r)$$

Hence by the principle of continuity, we have

$$\frac{\partial \rho}{\partial t}\left(r^2 \sin \theta \, \delta r \, \delta\theta \, \delta w\right) = -\frac{\partial}{\partial r}\left(\rho r^2 q_r\right) \delta r \, \delta\theta \, \delta w \sin \theta$$
$$- \frac{\partial}{r \sin \theta \, \partial w}(\rho q_w) r^2 \sin \theta \, \delta\theta \, \delta r \, \delta w$$

$$\Rightarrow \quad \frac{\partial \rho}{\partial t} + \frac{1}{r^2}\frac{\partial}{\partial r}\left(\rho r^2 q_r\right) + \frac{1}{r \sin \theta}\frac{\partial}{\partial w}(\rho q_w) = 0$$

$$\Rightarrow \quad \frac{\partial \rho}{\partial t} + \frac{\partial}{\partial r}(\rho q_r) + \frac{2\rho q_r}{r} + \frac{\operatorname{cosec} \theta}{r}\frac{\partial}{\partial w}(\rho q_w) = 0$$

17. In the lines of motion are curves on the surfaces of spheres all touching the plane of xy at the origin 0, the equation of continuity is

$$r \sin\theta \frac{\partial \rho}{\partial t} + \frac{\partial}{\partial \phi}(\rho v) + \sin\theta \frac{\partial}{\partial \theta}(\rho u) + \rho u\,(1 + 2\cos\theta) = 0$$

where r is the radius CP of one of the spheres, θ the angle PCO, u the velocity in the plane PCO, v the perpendicular velocity, and ϕ the inclination of the plane PCO to a fixed plane through z-axis.

Solution: We consider any two consecutive spheres with centre C and C^1.

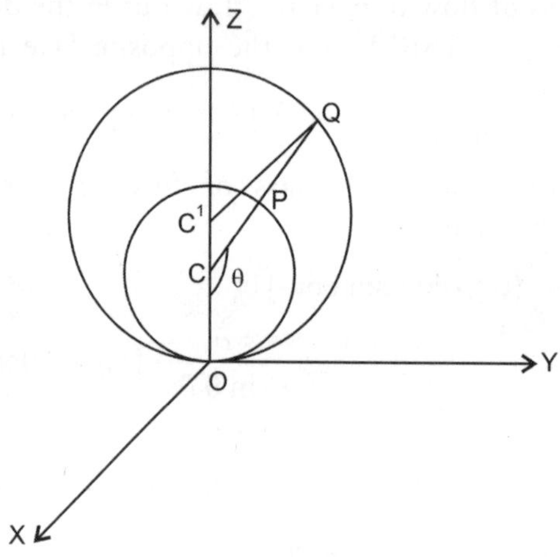

Fig. 1.16

Let $CP = r$, $C^1Q = r + \delta r$, $\angle PCO = \theta$.

Then $CC^1 = \delta r$

$CQ = CP + PQ = r + PQ$

From the ΔCC^1Q

$$C^1Q^2 = C^1C^2 + CQ^2 - 2C^1CCQ \cos(\pi - \theta)$$

$$\Rightarrow (r + \delta r)^2 = (\delta r)^2 + (r + PQ)^2 + 2\delta r\,(r + PQ)\cos\theta$$

$$\Rightarrow 2r\delta r = 2rPQ + (PQ)^2 + 2r\delta r\cos\theta + 2PQ\delta r\cos\theta$$

$$\Rightarrow r\delta r\,(1 - \cos\theta) = PQ\,(r + \delta r\cos\theta) \text{ neglecting } (PQ)^2$$

$$\Rightarrow PQ = r\delta r\,(1 - \cos\theta)\,(r + \delta r\cos\theta)^{-1}$$

$$= \delta r\,(1 - \cos\theta)\left(1 - \frac{\delta r}{r}\cos\theta\right)$$

$= \delta r\ (1 - \cos\ \theta)$

neglecting δr^2 and its highest power.

The angle between the plane PCO and a fixed plane XOZ is ϕ. Since the lines of motion are curves on surfaces of spheres touching the plane $z = 0$, there would no flow along the element PQ. Now we consider a parallelopiped with edges of lengths $(1 - \cos\ \theta)\ \delta r$, $r\delta\theta$, $r \sin\ \theta\delta\phi$, the velocities along these elements are u, v, w respectively.

Hence the excess of flow in over flow out in the direction PR per unit time

$$= -\frac{\partial}{r\partial\theta}\left[\rho ur \sin\ \theta\delta\phi\delta r\left(1-\cos\ \theta\right)\right]r\delta\theta$$

Similarly the excess of flow in over flow out in the direction PS per unit time

$$= -\frac{\partial}{r \sin\ \theta\partial\phi}\left[\rho vr\delta\theta\delta r\left(1-\cos\ \theta\right)\right]r \sin\ \theta\delta\phi$$

Thus the total excess of flow in over flow out through all the faces

$$= -\frac{\partial}{r\partial\theta}\left[\rho ur \sin\ \theta\phi\delta r\left(1-\cos\ \theta\right)\right]r\delta\theta$$
$$-\frac{\partial}{r \sin\ \theta\partial\phi}\left[\rho vr\delta\theta\delta r\left(1-\cos\ \theta\right)\right]r \sin\ \theta\delta\phi$$

Also mass of the elementary parallelopiped

$$= \rho\delta r\left(1-\cos\ \theta\right)r\delta\theta r \sin\ \theta\ \delta\phi$$

Hence the rate of increase in the mass of the parallelopiped

$$= \frac{\partial}{\partial t}\left[\rho\left(1-\cos\ \theta\right)\delta r\ r\delta\theta\ r \sin\ \theta\delta\phi\right]$$

Thus by the principle of continuity, we have

$$\frac{\partial}{\partial t}\left[\rho\left(1-\cos\ \theta\right)\delta r\ r\delta\theta\ r \sin\ \theta\delta\phi\right]$$
$$= -\frac{\partial}{r\partial\theta}\left[\rho ur \sin\ \theta\ \delta\phi\ \delta r\left(1-\cos\ \theta\right)\right]r\delta\theta$$
$$-\frac{\partial}{\partial \sin\ \theta\ \delta\phi}\left[\rho vr\ \delta\theta\ \delta r\left(1-\cos\ \theta\right)\right]r \sin\ \theta\ \delta\phi$$

$$\Rightarrow \frac{\partial \rho}{\partial t}\left[(1-\cos\theta)\,\delta r\; r\delta\theta\; r\sin\theta\;\delta\phi\right]$$

$$+\frac{\partial}{r\partial\theta}\left[\rho u\sin\theta\,(1-\cos\theta)\right]r^2\delta r\;\delta\theta\;\delta\phi$$

$$+\frac{\sin\theta\partial}{r\sin\theta\partial\phi}\left[(\rho v)\,r^2(1-\cos\theta)\,\delta r\;\delta\theta\;\delta\phi\right]=0$$

$$\Rightarrow \frac{\partial \rho}{\partial t}+\frac{1}{r\sin\theta\,(1-\cos\theta)}\frac{\partial}{\partial\theta}\left[\rho u\sin\theta\,(1-\cos\theta)\right]$$

$$+\frac{1}{r\sin\theta}\frac{\partial}{\partial\phi}(\rho v)=0$$

$$\Rightarrow \frac{\partial \rho}{\partial t}+\frac{1}{r\sin\theta\,(1-\cos\theta)}\left[\sin\theta\,(1-\cos\theta)\frac{\partial}{\partial\theta}(\rho u)\right.$$

$$\left.+\rho u\left\{\cos\theta\,(1+\cos\theta)+\sin^2\theta\right\}\right]+\frac{1}{r\sin\theta}\frac{\partial}{\partial\phi}(\rho v)=0$$

$$\Rightarrow \frac{\partial \rho}{\partial t}+\frac{1}{r}\frac{\partial}{\partial\theta}(\rho u)+\frac{\rho u\left(1+\cos\theta-2\cos^2\theta\right)}{r\sin\theta\,(1-\cos\theta)}$$

$$+\frac{1}{r\sin\theta}\frac{\partial}{\partial\phi}(\rho v)=0$$

$$\Rightarrow \frac{\partial \rho}{\partial t}+\frac{1}{r}\frac{\partial}{\partial\theta}(\rho u)+\frac{\rho u\,(1+2\cos\theta)}{r\sin\theta}+\frac{1}{r\sin\theta}\frac{\partial}{\partial\phi}(\rho v)=0$$

$$\Rightarrow r\sin\theta\frac{\partial \rho}{\partial t}+\frac{\partial}{\partial\phi}(\rho v)+\sin\theta\frac{\partial}{\partial\theta}(\rho u)$$

$$+\rho u\,(1+2\cos\theta)=0.$$

18. The particles of fluid move symmetrically in space with regard to a fixed centre, prove that the equation of continuity is $\frac{\partial \rho}{\partial t}+u\frac{\partial \rho}{\partial r}+\frac{\rho}{r^2}\frac{\partial}{\partial r}\left(r^2u\right)=0$ where u is the velocity at a distance r.

Solution: Let $P(r, \theta, w)$ be a point in the fluid. We construct a parallelopiped with $PP^1 = \delta r$, $PQ = r\delta\theta$ and $PS = r\sin\theta\delta\phi$ as edges. Let u, v, w be the components of the respective elements.

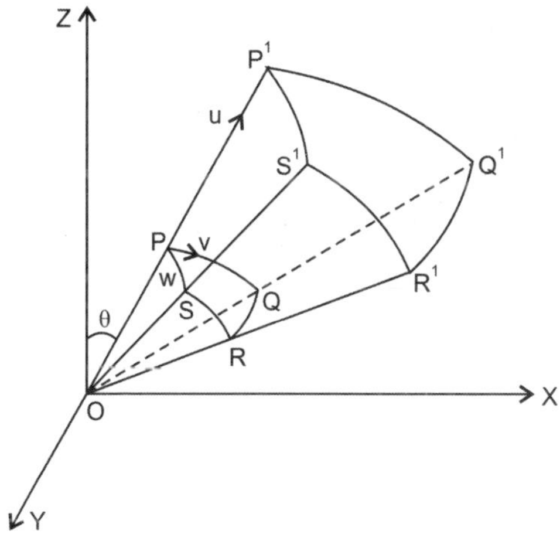

Fig. 1.17

Let the origin O be the fixed centre. Since the fluid particles move symmetrically in space with regard to an origin. It follows that the motion is only along the direction PP^1 and there is no motion along other edges $r\delta\theta$ and $r \sin \theta\, \delta\phi w$.

Excess of mass of flow in over flow out from the faces $PQRS$ and $P^1Q^1R^1S^1$ along PP^1 is

$$= -\,\delta r \frac{\partial}{\partial r}(\rho u r\, \delta\theta r \sin \theta\delta\phi) \text{ per unit time}$$

The excess of mass of flow in over flow out along PQ and PS vanish as there is no motion along these directions.

Mass of the fluid inside the element = $\rho\delta r r\delta\theta r \sin \theta\, \delta\phi$

Hence rate of increase in the mass of the element

$$= \frac{\partial}{\partial t}(\rho\delta r\, r\, \delta\theta\, r \sin \theta\delta\phi) \text{ per unit time}$$

Therefore by the principle of continuity, we have

$$\frac{\partial}{\partial t}(\rho\delta r\, r\delta\theta\, r \sin \theta\delta\phi) = -\,\delta r \frac{\partial}{\partial r}(\rho u r\, \delta\theta\, r \sin \theta\delta\phi)$$

$$\Rightarrow \quad \frac{\partial \rho}{\partial t}(\delta r r\delta\theta\, r \sin \theta\, \delta\phi) + \frac{\partial}{\partial r}\left(\rho u r^2\right)\delta r\delta\theta \sin \theta\delta\phi = 0$$

$$\Rightarrow \quad \frac{\partial \rho}{\partial t} + \frac{1}{r^2}\frac{\partial}{\partial r}\left(\rho u r^2\right) = 0$$

$$\frac{\partial \rho}{\partial t} + u\frac{\partial \rho}{\partial r} + \frac{\rho}{r^2}\frac{\partial}{\partial r}\left(u r^2\right) = 0$$

19. A mass of fluid is in motion so that the lines of motion lie on the surface of coaxial cylinders. Show that the equation of continuity is $\frac{\partial \rho}{\partial t} + \frac{1}{r}\frac{\partial}{\partial \theta}(\rho q_r) + \frac{\partial}{\partial z}(\rho q_z) = 0$, where q_r, q_z are the velocities perpendicular and parallel to z.

Solution: Let $P(r, \theta, z)$ be a point in the fluid. We construct a parallelopiped at P with edges $PS = r\delta\theta$, $PR = \delta r$ and $PP^1 = \delta z$.

Let q_r, q_θ and q_z be the velocity components along PR, PS, PP^1.

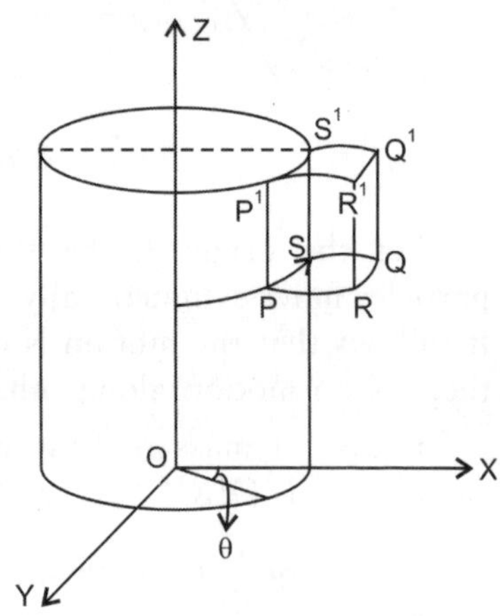

Fig. 1.18

The fluid moves on the surface of coaxial cylinders; hence there is no motion along PR.

Now excess of mass of flow in over flow out along the direction PS

$$= -r\delta\theta \frac{\partial}{r\partial\theta}(\rho q_\theta \delta r \delta z)$$

and excess of mass of flow in over flow out along the direction PP^1

$$= -\delta z \frac{\partial}{\partial z}(\rho q_z \delta r r \delta\theta)$$

Also mass of the fluid in the element = $\rho r\delta\theta\, \delta r \delta z$

Hence rate of increase in the mass of the element

$$= \frac{\partial}{\partial t}(\rho r \delta\theta\, \delta r \delta z)$$

Therefore, by the principle of continuity, we have

$$\frac{\partial}{\partial t}(\rho r\delta\theta\,\delta r\delta z) = -r\delta\theta\,\frac{\partial}{r\partial\theta}(\rho q_\theta \delta r\delta z) - \delta z\,\frac{\partial}{\partial z}(\rho q_z \delta r r\delta\theta)$$

$$\Rightarrow \left[\frac{\partial\rho}{\partial t} + \frac{\partial}{r\partial\theta}(\rho q_\theta) + \frac{\partial}{\partial z}(\rho q_z)\right](r\delta\theta\,\delta r\delta z) = 0$$

$$\Rightarrow \frac{\partial\rho}{\partial t} + \frac{\partial}{r\partial\theta}(\rho q_\theta) + \frac{\partial}{\partial z}(\rho q_z) = 0$$

20. If w is the area of cross section of a stream filament, prove that the equation of continuity is $\frac{\partial}{\partial t}(\rho w) + \frac{\partial}{\partial s}(\rho w q) = 0$.

Solution: Let P be the point, we consider a volume bounded by the cross sections through the points P and Q, where Q is at a distance ds from P.

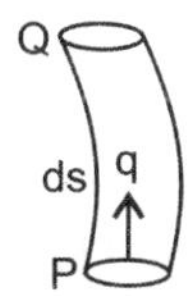

Fig. 1.19

Now excess of flow in over flow out in this volume

$$= -\frac{\partial}{\partial s}(\rho w q)\,ds \text{ per unit time.}$$

Also mass of the fluid in this volume = $\rho w ds$

Hence rate of change of mass

$$= \frac{\partial}{\partial t}(\rho w ds) \text{ per unit time.}$$

Therefore, by the principle of continuity

$$\frac{\partial}{\partial t}(\rho w ds) = -ds\,\frac{\partial}{\partial s}(\rho w q)$$

$$\Rightarrow \frac{\partial}{\partial t}(\rho w) + \frac{\partial}{\partial s}(\rho w q) = 0.$$

21. A mass of fluid moves in such a way that each particle describes a circle in one plane about a fixed axis, show that the equation of continuity is $\frac{\partial\rho}{\partial t} + \frac{\partial}{\partial\theta}(\rho w) = 0$, where w is the angular velocity of a particle whose azimuthal angle is θ at time t.

Solution:

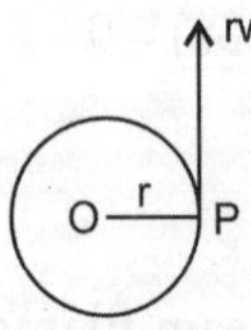

Fig. 1.20

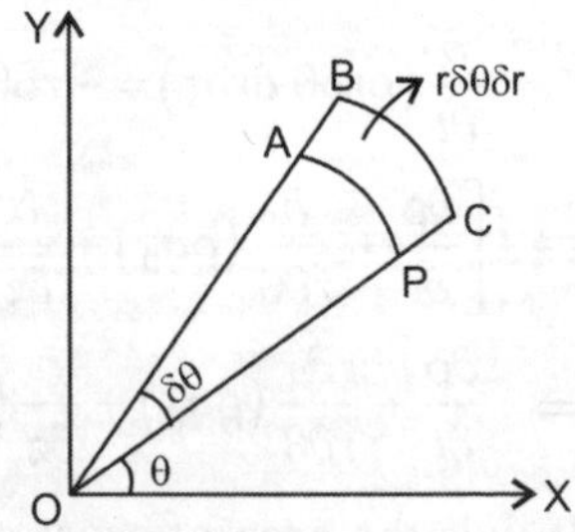

Fig. 1.21

We consider a point P whose polar coordinates are (r, θ). Let there be an elementary area $r\delta\theta\delta r$, when this area is revolved about O, then it describes a circle so that velocity OP vanishes.

The excess of flow in over flow out along PA

$$= -r\delta\theta \frac{\partial}{r\,\partial\theta}(\rho r w \delta r) \text{ per unit time.}$$

Also mass of the fluid inside the element $= \rho\delta r r\delta\theta$

Hence rate of increase in the mass of the element

$$= \frac{\partial}{\partial t}(\rho\delta r r\delta\theta) \text{ per unit time.}$$

Thus by the principle of continuity, we have

$$\frac{\partial}{\partial t}(\rho\delta r r\delta\theta) = -r\delta\theta \frac{\partial}{r\partial\theta}(\rho r w \delta r)$$

$$\Rightarrow \quad \frac{\partial\rho}{\partial t} + \frac{\partial}{\partial\theta}(\rho w) = 0.$$

EXERCISE 1

1. Show that a fluid of constant density can have a velocity $\vec{q}$ given by

$$\vec{q} = \left(\frac{-2xyz}{\left(x^2+y^2\right)^2}, \frac{\left(x^2-y^2\right)z}{\left(x^2+y^2\right)^2}, \frac{y}{x^2+y^2} \right)$$

Find the vorticity vector. [I.A.S. 88]

[**Ans.** $\vec{O}$] [**Hint:** See No. 2]

2. Find the streamlines and paths of the particles for the two dimensional velocity field

$$u = \frac{x}{1+t}, v = y, w = O$$

[**Ans.** Streamlines are $x^{1+t} = ay$, $z = b$

Path lines are $y = be^{\frac{x}{a}-1}, z = C$]

3. Prove that liquid motion is possible when velocity at (x, y, z) is given by

$$u = \frac{3x^2 - r^2}{r^5}, v = \frac{3xy}{r^5}, w = \frac{3xz}{r^5}$$

where $r^2 = x^2 + y^2 + z^2$ and the streamlines are the intersection of the surfaces $(x^2 + y^2 + z^2)^3 = c\,(y^2 + z^2)^2$, by the planes passing through Ox.

4. Given $u = \frac{-c^2 y}{r^2}, v = \frac{c^2 x}{r^2}, w = 0$, where r denotes distance from z-axis. Find the surfaces which are orthogonal to streamlines, the liquid being homogeneous.

[**Ans.** $y = ax$]

5. Show that the ellipsoid

$$\frac{x^2}{a^2 k^2 t^{2n}} + kt^n \left(\frac{y^2}{b^2} + \frac{z^2}{c^2} \right) = 1$$

is a possible form of boundary surface.

6. Prove that a surface of the form $ax^4 + by^4 + cz^4 - x(t) = 0$ is a possible form of boundary surface of a homogeneous liquid at time t, the velocity potential of the liquid motion being

$$\phi = (\beta - \gamma)\,x^2 + (\gamma - \alpha)\,y^2 + (\alpha - \beta)\,z^2$$

where x, α, β, γ are given functions of time and a, b, c are suitable functions of time.

7. Show that $\frac{x^2}{a^2} f(t) + \frac{y^2}{b^2} \phi(t) = 1$, where $f(t)\,\phi(t)$ = const is a possible form of the boundary surface of a liquid.

8. Show that the velocity potential $\phi = \frac{a}{2}\left(x^2 + y^2 + z^2\right)$ satisfy the Laplace equations. Also determine the streamlines.
[**Ans.** $y^2 z = c$]

9. The velocity $\vec{q}$ in a three dimensional flow field for an incompressible fluid is given by $\vec{q} = 2x\vec{i} - y\vec{j} - z\vec{k}$. Is it a possible field? Determine the equations of the streamline passing through (1, 1, 1).
[**Ans.** Yes, streamlines are $xy^2 = 1$, $xz^2 = 1$]

10. Determine the constants l, m and n in order that the velocity

$$\vec{q} = \frac{(x + lr)\vec{i} + (y + mr)\vec{j} + (z + nr)\vec{k}}{r(x + r)}$$

where $r = \sqrt{x^2 + y^2 + z^2}$ may satisfy the equation of continuity for a liquid.
[**Ans.** $l = 1$, $m = 0 = n$]

11. If $u = x^2 + y^2 + z^2$ and $v = -xy - yz - xz$, determine the velocity component w which will satisfy continuity equation for incompressible fluid.

$$\left[\textbf{Ans. } w = -xz + \frac{z^2}{2} + f(x, y, b)\right]$$

12. In a two dimensional flow, the x component of velocity distribution is given by $u = x^2 - 4xy - 3y$. Determine the y component for velocity.
[**Ans.** $v = 2y^2 - 2xy + f(x)$]

2 EQUATIONS OF MOTION

2.1. Euler's Equation of Motion

Let a closed surface S enclosing a volume V of a non-viscous fluid be moving with the fluid so that S contains the same number of fluid particles at any time t. We consider a point P inside S. Let ρ be the fluid density, $\vec{q}$ the fluid velocity and dv the elementary volume enclosing P.

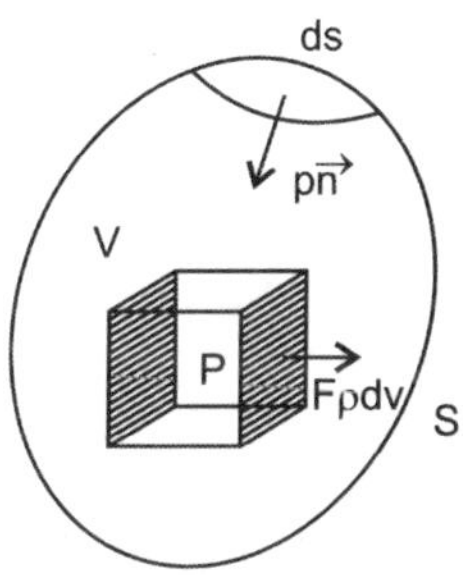

Fig. 2.1

Since the mass ρdv remains unchanged throughout the motion so that

$$\frac{d}{dt}(\rho dv) = 0 \qquad \text{... (1)}$$

The momentum M of the volume V is

$$\vec{M} = \int_V \vec{q}\rho dv$$

$$\therefore \quad \frac{d\vec{M}}{dt} = \int\left[\frac{d\vec{q}}{dt}\rho dv + \frac{d}{dt}(\rho dv)\vec{q}\right] dv$$

$$\Rightarrow \quad \frac{d\vec{M}}{dt} = \int\frac{d\vec{q}}{dt}\rho dv \qquad \text{... (2) [From (1)]}$$

Let $\vec{F}$ be the external force per unit mass acting on the fluid, then the total body force acting on the fluid within the surface S at time t = $\int_V \vec{F}\rho\, dv$

If p is the pressure at a point of the surface elements ds, having outward drawn unit normal $\hat{n}$, then the total surface force S

$$= \int_S p(-\hat{n})\, ds$$

$$= -\int_S p\hat{n} ds$$

Here negative sign is taken as the surface force acts inwards

$$= -\int_V \vec{\nabla} p dv \text{ (By Gauss Theorem)}$$

Therefore, the total force acting on the volume V

$$= \int_V \vec{F}\rho dv - \int_V \vec{\nabla} p dv$$

$$= \int_V \left(\vec{F}\rho - \vec{\nabla} p\right) dv$$

By Newton's second law of motion, the rate of change of linear momentum is equal to the total force acting on the mass of the fluid, we have

$$\int_V \frac{d\vec{q}}{dt} \rho dv = \int_V \left(\vec{F}\rho - \vec{\nabla} p\right) dv$$

$$\Rightarrow \quad \int \left[\frac{d\vec{q}}{dt}\rho - \vec{F}\rho + \vec{\nabla} p\right] dv = 0$$

Since S is arbitrary and so V is arbitrary so that the integrand of the last integral vanishes,

i.e. $$\frac{d\vec{q}}{dt}\rho - \vec{F}\rho + \vec{\nabla} p = 0$$

$$\Rightarrow \quad \frac{d\vec{q}}{dt} = \vec{F} - \frac{1}{\rho}\vec{\nabla} p \qquad \text{... (1)}$$

This equation is known as Euler's equation of motion.

Deductions

(1) If we write $\vec{q} = u\vec{i} + v\vec{j} + w\vec{k}$ and $\vec{F} = X\vec{i} + Y\vec{j} + Z\vec{k}$

Then Euler's equation (1) becomes

$$\frac{d}{dt}\left(u\vec{i}+v\vec{j}+w\vec{k}\right)=\left(X\vec{i}+Y\vec{j}+Z\vec{k}\right)-\frac{1}{\rho}\left(\vec{i}\frac{\partial}{\partial x}+\vec{j}\frac{\partial}{\partial y}+\vec{k}\frac{\partial}{\partial z}\right)p$$

$$\Rightarrow\quad \frac{du}{dt}\vec{i}+\frac{dv}{dt}\vec{j}+\frac{dw}{dt}\vec{k}=\left(X-\frac{1}{\rho}\frac{\partial p}{\partial x}\right)\vec{i}+\left(Y-\frac{1}{\rho}\frac{\partial p}{\partial y}\right)\vec{j}+\left(Z-\frac{1}{\rho}\frac{\partial p}{\partial z}\right)\vec{k}$$

Equating the coefficients of $\vec{i},\vec{j},\vec{k}$, we get

$$\frac{du}{dt}=X-\frac{1}{\rho}\frac{\partial p}{\partial u},\ \frac{dv}{dt}=Y-\frac{1}{\rho}\frac{\partial p}{\partial y},\ \frac{dw}{dt}=Z-\frac{1}{\rho}\frac{\partial p}{\partial z}$$

These equations are the Euler's equation of motion in Cartesian coordinates.

Also $\dfrac{d}{dt}=\dfrac{\partial}{\partial t}+u\dfrac{\partial}{\partial x}+v\dfrac{\partial}{\partial y}+w\dfrac{\partial}{\partial z}$

With these substitutions the above equations become

$$\frac{\partial u}{\partial t}+u\frac{\partial u}{\partial x}+v\frac{\partial u}{\partial y}+w\frac{\partial u}{\partial z}=X-\frac{1}{\rho}\frac{\partial p}{\partial x}$$

$$\frac{\partial v}{\partial t}+u\frac{\partial v}{\partial x}+v\frac{\partial v}{\partial y}+w\frac{\partial v}{\partial z}=Y-\frac{1}{\rho}\frac{\partial p}{\partial y}$$

$$\frac{\partial w}{\partial t}+u\frac{\partial w}{\partial x}+v\frac{\partial w}{\partial y}+w\frac{\partial w}{\partial z}=Z-\frac{1}{\rho}\frac{\partial p}{\partial z}$$

(2) Also $\dfrac{d}{dt}=\dfrac{\partial}{\partial t}+\vec{q}\cdot\vec{\nabla}$

Thus (1) becomes

$$\left[\frac{\partial}{\partial t}+\vec{q}\cdot\vec{\nabla}\right]\vec{q}=\vec{F}-\frac{1}{\rho}\vec{\nabla}p$$

$$\Rightarrow\quad \frac{\partial\vec{q}}{\partial t}+\left(\vec{q}\cdot\vec{\nabla}\right)\vec{q}=\vec{F}-\frac{1}{\rho}\vec{\nabla}p \qquad \dots (2)$$

Also $\vec{\nabla}(\vec{q}\cdot\vec{q}) = 2\left[\vec{q}\times \text{curl}\,\vec{q} + (\vec{q}\cdot\vec{\nabla})\vec{q}\right]$

$$\Rightarrow \quad \vec{\nabla}\left(\frac{1}{2}\vec{q}^2\right) - \vec{q}\times(\text{curl}\,\vec{q}) = (\vec{q}\cdot\vec{\nabla})\vec{q}$$

Hence from (2)

$$\frac{\partial \vec{q}}{\partial t} + \vec{\nabla}\left(\frac{1}{2}\vec{q}^2\right) - \vec{q}\times \text{curl}\,\vec{q} = \vec{F} - \frac{1}{\rho}\vec{\nabla}p$$

Writing $\overrightarrow{W} = \text{curl}\,\vec{q}$

$$\frac{\partial \vec{q}}{\partial t} + \vec{\nabla}\left(\frac{1}{2}\vec{q}^2\right) - \vec{q}\times \overrightarrow{W} = \vec{F} - \frac{1}{\rho}\vec{\nabla}p$$

$$\Rightarrow \quad \frac{\partial \vec{q}}{\partial t} + \vec{\nabla}\left(\frac{1}{2}\vec{q}^2\right) + \overrightarrow{W}\times\vec{q} = \vec{F} - \frac{1}{\rho}\vec{\nabla}p$$

This is known as Lamb's Hydrodynamical equation. The main advantage of it is that it is invariant under a change of coordinate system.

2.2. Euler's Dynamical Equation (Cartesian Coordinates)

Let p be the pressure, ρ be the density at a point $P(x, y, z)$ of the fluid element. Let u, v, w be the components of velocity

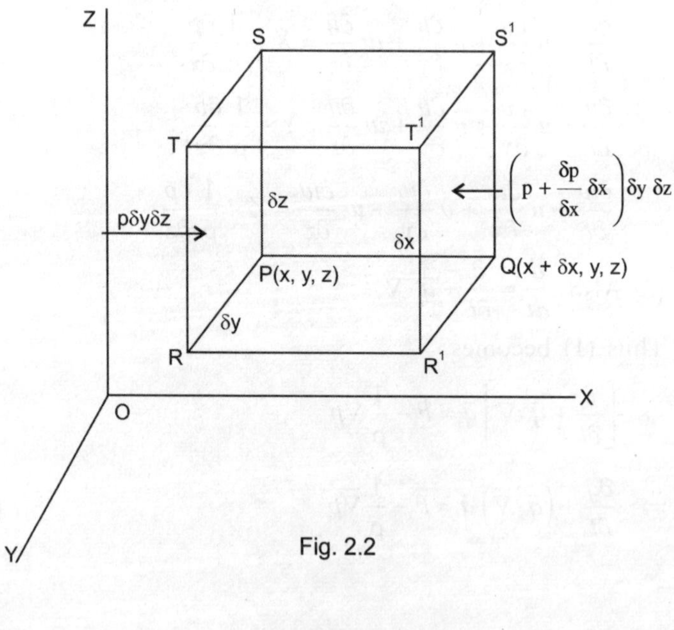

Fig. 2.2

at P and X, Y, Z be the components of external force per unit mass at any time t.

We construct a rectangular parallelopiped having faces parallel to coordinate planes and edges of lengths δx, δy and δz.

Now pressure p on the face $PRTS$

$$= \rho\ \delta y \delta z$$

$$= f(x, y, z)\ \delta y \delta z \quad \text{(say)}$$

Then pressure on the face $QR^1T^1S^1$

$$= f(x + \delta x, y, z)\ \delta y\ \delta z$$

$$= \left[f(x,y,z) + \delta x \frac{\partial f(x,y,z)}{\partial x} \right] \delta y\ \delta z$$

$$= \left(p + \frac{\partial p}{\partial x}\delta x \right) \delta y\ \delta z$$

Also since X is the force per unit mass parallel to x-axis, the force on the solid element parallel to x-axis = $X\rho\delta x\delta y\delta z$.

Again the acceleration along x-axis $= \dfrac{du}{dt}$.

Hence by Newton's second law of motion the product of the mass of the element and its total acceleration in the x-direction must be equal to the sum of the components of force acting on the element in that direction.

Therefore, we have

$$\rho\delta x\delta y\delta z \frac{du}{dt} = X\rho\delta x\delta y\delta z + p\delta y\delta z - \left(p + \frac{\partial p}{\partial x}\delta x \right) \delta y\delta z$$

$$\Rightarrow \quad \frac{du}{dt} = X - \frac{1}{\rho}\frac{\partial p}{\partial x}$$

$$\Rightarrow \quad \frac{\partial u}{\partial t} + u\frac{\partial u}{\partial x} + v\frac{\partial u}{\partial y} + w\frac{\partial u}{\partial z} = X - \frac{1}{\rho}\frac{\partial p}{\partial x}$$

Similarly by the equations in y and z directions are

$$\frac{\partial v}{\partial t} + u\frac{\partial v}{\partial x} + v\frac{\partial v}{\partial y} + w\frac{\partial v}{\partial z} = Y - \frac{1}{\rho}\frac{\partial p}{\partial y}$$

$$\frac{\partial v}{\partial t} + u\frac{\partial w}{\partial x} + v\frac{\partial w}{\partial y} + w\frac{\partial w}{\partial z} = Z - \frac{1}{\rho}\frac{\partial p}{\partial z}$$

2.3. Conservative Field of Force

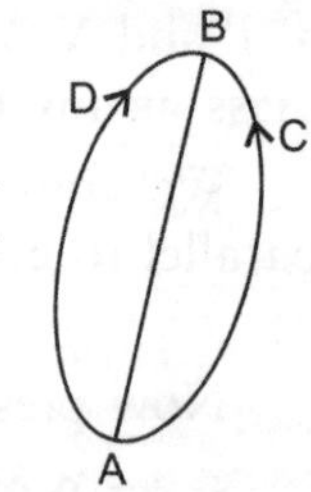

Fig. 2.3

In a conservative field of force the work done by the force $\vec{F}$ of the field in taking a unit mass from A to B is independent of the path of motion.

Thus

$$\int_{ACB} \vec{F} \cdot d\vec{r} = \int_{ADB} \vec{F} \cdot d\vec{r}$$

$$= -\Omega \quad \text{(say)}$$

where Ω is a scalar point function whose value depends on the initial and final position A and B. It can be easily proved that

$$\vec{F} = -\vec{\nabla}\Omega$$

where Ω is known as force potential which measure the potential energy of the field.

2.4. Pressure Equation

When velocity potential exists and forces are conservative and derivable from a potential Ω, the equation of motion can always be integrated and the solution is

$$\int \frac{dp}{\rho} - \frac{\partial \phi}{\partial t} + \frac{1}{2} q^2 + \Omega = F(t)$$

Proof. Since velocity potential exists, so motion is irrotational and $\vec{q} = -\vec{\nabla}\phi$.

Also forces are conservative, so

$$\vec{F} = -\vec{\nabla}\Omega$$

Let $P = \int_{p_0}^{p} \frac{dp}{\rho}$

then $\frac{dP}{dp} = \frac{1}{\rho}$

So that $\vec{\nabla}P = \sum \vec{i}\, \frac{\partial P}{\partial x} = \sum \vec{i}\, \frac{\partial P}{\partial p} \frac{\partial p}{\partial x}$

$$= \sum \frac{\vec{i}}{\rho} \frac{\partial p}{\partial x}$$

$$= \frac{1}{\rho} \sum \vec{i} \frac{\partial p}{\partial x} = \frac{1}{\rho} \vec{\nabla} p$$

By Euler's equation of motion

$$\frac{d\vec{q}}{dt} = \vec{F} - \frac{1}{\rho} \vec{\nabla} p$$

$$\Rightarrow \quad \frac{\partial \vec{q}}{\partial t} + \left(\vec{q} \cdot \vec{\nabla}\right) \vec{q} = -\vec{\nabla}\Omega - \vec{\nabla}P$$

$$\Rightarrow \quad \frac{\partial}{\partial t}\left(-\vec{\nabla}\phi\right) + \vec{\nabla}(\Omega + P) + \left(\vec{q} \cdot \vec{\nabla}\right) \vec{q} = 0$$

$$\Rightarrow \quad \vec{\nabla}\left(\frac{-\partial \phi}{\partial t} + \Omega + P\right) + \left(\vec{q} \cdot \vec{\nabla}\right) \vec{q} = 0 \qquad \dots (1)$$

Also $\vec{\nabla}(\vec{q} \cdot \vec{q}) = 2\left[\vec{q} \times \text{curl}\, \vec{q} + \left(\vec{q} \cdot \vec{\nabla}\right) \vec{q}\right]$

Hence (1) becomes

$$\vec{\nabla}\left(-\frac{\partial \phi}{\partial t} + \Omega + P\right) + \frac{1}{2} \vec{\nabla} q^2 - \vec{q} \times \text{curl}\, \vec{q} = 0$$

$$\Rightarrow \quad \vec{\nabla}\left(\Omega + P + \frac{1}{2} q^2 - \frac{\partial \phi}{\partial t}\right) = \vec{q} \times \text{curl}\, \vec{q} = 0 \qquad \dots (2)$$

Since

$$\text{curl}\, \vec{q} = \text{curl}\left(-\vec{\nabla}\phi\right)$$

$$= -\text{ curl grad } \phi = 0$$

Multiplying (2) scalarly by $d\vec{r}$ and noting that $d\vec{r} \cdot \vec{\nabla} = d$, we get

$$d\left(\Omega + P + \frac{1}{2} q^2 - \frac{\partial \phi}{\partial t}\right) = 0$$

Integrating, we get

$$\Omega + P + \frac{1}{2} q^2 - \frac{\partial \phi}{\partial t} = F(t)$$

Where F(t) is the constant of integration

$$\Rightarrow \quad \Omega + \int \frac{dp}{\rho} + \frac{1}{2} q^2 - \frac{\partial \phi}{\partial t} = F(t) \qquad \dots (3)$$

This equation is known as Bernoulli's equation for unsteady irrotational motion. This is also known as pressure equation. If fluid is incompressible then (3) becomes

$$\Omega+\frac{p}{p}+\frac{1}{2}q^2-\frac{\partial\phi}{\partial t}=F(t)$$

Since $\int\frac{dp}{\rho}=\frac{1}{\rho}\int dp=\frac{p}{\rho}$

Deductions. If the motion is steady then $\frac{\partial\phi}{\partial t}=0$. Hence (3) becomes

$$\Omega+\int\frac{dp}{\rho}+\frac{1}{2}q^2=F(t)$$

$$= C = \text{absolute constant}$$

or $\Omega+\int\frac{dp}{\rho}+\frac{1}{2}q^2=C$

This is known as Bernoulli's equation for steady motion. If ρ = constant, then

$$\Omega+\frac{p}{\rho}+\frac{1}{2}q^2=\text{const.}$$

2.5. Bernoulli's Equation for Steady Motion

Vector Method. Since forces are conservative, so

$$\vec{F}=-\vec{\nabla}\Omega$$

Also motion is steady, so $\frac{\partial\vec{q}}{\partial t}=0$

Also density is a function of pressure p only, hence there exists a relation of the type

$$P=\int_{p_0}^{p}\frac{dp}{\rho}$$

Now $\vec{\nabla}P=\sum\vec{i}\frac{\partial P}{\partial x}=\sum\vec{i}\frac{\partial P}{\partial p}\frac{\partial p}{\partial x}$

$$=\frac{1}{\rho}\sum\vec{i}\frac{\partial p}{\partial x}=\frac{1}{\rho}\vec{\nabla}p$$

Euler's equation of motion is

$$\frac{d\vec{q}}{dt} = -\vec{\nabla}\Omega - \frac{\vec{\nabla}p}{\rho}$$

$$\Rightarrow \quad \frac{\partial \vec{q}}{\partial t} + \left(\vec{q}\cdot\vec{\nabla}\right)\vec{q} = -\vec{\nabla}(\Omega + P) \qquad \ldots (1)$$

Also $\vec{\nabla}(\vec{q}\cdot\vec{q}) = 2\left[\vec{q}\times \text{curl}\,\vec{q} + \left(\vec{q}\cdot\vec{\nabla}\right)\vec{q}\right]$

$$\Rightarrow \quad \left(\vec{q}\cdot\vec{\nabla}\right)\vec{q} = \frac{1}{2}\vec{\nabla}q^2 - \vec{q}\times\left(\vec{\nabla}\times\vec{q}\right) \qquad \ldots (2)$$

From (1) and (2), we get

$$\frac{\partial \vec{q}}{\partial t} + \frac{1}{2}\vec{\nabla}q^2 - \vec{q}\times\left(\vec{\nabla}\times\vec{q}\right) = -\vec{\nabla}(\Omega + P)$$

$$\Rightarrow \quad \frac{1}{2}\vec{\nabla}q^2 - \vec{q}\times\left(\vec{\nabla}\times\vec{q}\right) = -\vec{\nabla}(\Omega + P)\left(\because \frac{\partial \vec{q}}{\partial t} = 0\right)$$

$$\Rightarrow \quad -\vec{q}\times \text{curl}\,\vec{q} = -\vec{\nabla}\left(\Omega + P + \frac{1}{2}q^2\right) \qquad \ldots (3)$$

Multiplying both the sides of (3) scalarly by $\vec{q}$, we get

$$\vec{q}\,(\vec{q}\times \text{curl}\,\vec{q}) = \vec{q}\cdot\vec{\nabla}\left(\Omega + P + \frac{1}{2}q^2\right)$$

$$\Rightarrow \quad 0 = \vec{q}\cdot\vec{\nabla}\left(\Omega + P + \frac{1}{2}q^2\right)$$

Hence $\Omega + P + \frac{1}{2}q^2 = \text{constant} = C$

Alternative Method

Let AB be a streamline in the fluid, we consider an element δs of this streamline. With δs as axis we draw small cylinder of cross section α.

If q be the velocity and S the component of external force per unit mass along the streamline, then the equation of motion gives

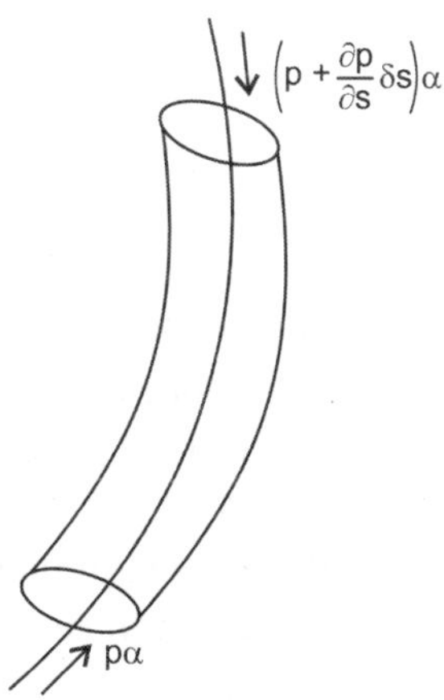

Fig. 2.4

$$\rho\alpha\delta s \frac{dq}{dt} = P\alpha\delta s\, S + p\alpha - \left(p + \frac{\partial p\, \delta s}{\partial s}\right)\alpha$$

$$\Rightarrow \quad \rho \frac{dq}{dt} = PS - \frac{\partial p}{\partial s}\alpha$$

$$\Rightarrow \quad \frac{dq}{dt} = S - \frac{1}{\rho}\frac{\partial p}{\partial s}$$

$$\Rightarrow \quad \frac{\partial q}{\partial t} + q\frac{\partial q}{\partial s} = S - \frac{1}{\rho}\frac{\partial p}{\partial s} \qquad \text{... (1)}$$

Since $\dfrac{dq}{dt} = \dfrac{\partial q}{\partial t} + q\dfrac{\partial q}{\partial s}$

Since the motion is steady $\therefore \dfrac{\partial q}{\partial t} = 0$

Also the external force S is derivable from the potential function V

$$\therefore \quad S = \frac{-\partial v}{\partial s}$$

Hence (1) becomes

$$q\frac{\partial q}{\partial s} = \frac{-\partial v}{\partial s} - \frac{1}{\rho}\frac{\partial p}{\partial s}$$

Integrating it along the streamline, we get

$$\frac{1}{2}q^2 = -V - \int\frac{dp}{\rho} + C$$

$$\Rightarrow \quad \int\frac{dp}{\rho} + \frac{1}{2}q^2 + V = C$$

The constant C depends upon the streamline chosen.

This is Bernoulli's theorem. This holds good when the motion is steady, V exists and ρ is a function of p.

2.6. If the motion of an ideal fluid, for which density is a function of pleasure *p* only, is steady and the external forces are conservative, then there exists a family of surfaces which contain the streamlines and vortex lines

The equation of motion for an ideal fluid are

$$\frac{\partial \vec{q}}{\partial t} + \left(\vec{q}\cdot\vec{\nabla}\right)\vec{q} = -\frac{1}{\rho}\vec{\nabla}p + \vec{F} \qquad \text{... (1)}$$

Since density is a function of pressure only, let

$$P = \int_{p_0}^{p} \frac{dp}{\rho}$$

Now $\vec{\nabla}P = \sum \vec{i}\,\frac{\partial P}{\partial x} = \sum \vec{i}\,\frac{\partial P}{\partial p}\frac{\partial p}{\partial x}$

$$= \frac{1}{\rho}\sum \vec{i}\,\frac{\partial p}{\partial x} = \frac{1}{\rho}\vec{\nabla}p \qquad \text{... (2)}$$

Also $\vec{F}$ is conservative, hence

$$\vec{F} = -\vec{\nabla}\Omega \qquad \text{... (3)}$$

Again

$$\vec{\nabla}(\vec{q}\cdot\vec{q}) = 2\left[\vec{q}\times\text{curl}\,\vec{q} + (\vec{q}\cdot\vec{\nabla})\vec{q}\right]$$

$$\Rightarrow \quad (\vec{q}\cdot\vec{\nabla})\vec{q} = \vec{\nabla}\left(\frac{1}{2}q^2\right) - \vec{q}\times\text{curl}\,\vec{q} \qquad \text{... (4)}$$

Since motion is steady, therefore

$$\frac{\partial\phi}{\partial t} = 0 \qquad \text{... (5)}$$

Using (2), (3), (4) and (5), (1) becomes

$$\frac{1}{2}\vec{\nabla}q^2 - \vec{q}\times(\vec{\nabla}\times\vec{q}) = -\vec{\nabla}(\Omega + P) \qquad \text{... (6)}$$

Taking $\overrightarrow{W} = \text{curl}\,\vec{q}$ = vorticity vector, (6) becomes

$$\frac{1}{2}\vec{\nabla}q^2 - \vec{q}\times\overrightarrow{W} = -\vec{\nabla}(\Omega + P)$$

$$\Rightarrow \quad \vec{\nabla}\left(\Omega + P + \frac{1}{2}q^2\right) = \vec{q}\times\overrightarrow{W}$$

$$\Rightarrow \quad \vec{\nabla}\left(\Omega + P + \frac{1}{2}q^2\right) = \overrightarrow{N}$$

where $\overrightarrow{N} = \vec{q}\times\overrightarrow{W}$

Thus $\overrightarrow{N}\cdot\vec{q} = \vec{q}\cdot(\vec{q}\times\overrightarrow{W}) = 0 = \overrightarrow{N}\cdot\overrightarrow{W}$

Hence $\overrightarrow{N}$ is perpendicular to both $\vec{q}$ and $\overrightarrow{W}$.

Also $\overrightarrow{N}$ is perpendicular to the family of surfaces

$$\Omega + P + \frac{1}{2}q^2 = \text{const.} = C \quad \text{... (7)}$$

(For $\vec{\nabla} f$ is perpendicular everywhere to f = const.)

Therefore (7) is everywhere tangent to the direction of velocity vector $\vec{q}$ and vorticity vector $\overrightarrow{W}$. Hence it contains the streamlines and the vortex lines.

2.7. Integration of the Equation of Motion, Cartesian Form

If a velocity potential ϕ exists, then

$$u = \frac{-\partial \phi}{\partial x}, v = \frac{-\partial \phi}{\partial y}, w = \frac{-\partial \phi}{\partial z}$$

Also if the external forces are derivable from a potential function v, then

$$X = \frac{-\partial v}{\partial x}, Y = \frac{-\partial v}{\partial y}, Z = \frac{-\partial v}{\partial z}$$

Since ϕ exists, liquid motion must be irrotational so that

$$\frac{\partial w}{\partial y} = \frac{\partial v}{\partial z}, \frac{\partial u}{\partial z} = \frac{\partial w}{\partial x} \text{ and } \frac{\partial v}{\partial x} = \frac{\partial u}{\partial y}$$

Putting these values of u, v, w and X, Y, Z in the Euler's equation of motion, we get

$$-\frac{\partial}{\partial t}\left(\frac{\partial \phi}{\partial x}\right) + u\frac{\partial u}{\partial x} + v\frac{\partial v}{\partial x} + w\frac{\partial w}{\partial x} = -\frac{\partial V}{\partial x} - \frac{1}{\rho}\frac{\partial p}{\partial x}$$

$$-\frac{\partial}{\partial t}\left(\frac{\partial \phi}{\partial y}\right) + u\frac{\partial u}{\partial y} + v\frac{\partial v}{\partial y} + w\frac{\partial w}{\partial y} = -\frac{\partial V}{\partial y} - \frac{1}{\rho}\frac{\partial p}{\partial y}$$

$$-\frac{\partial}{\partial t}\left(\frac{\partial \phi}{\partial z}\right) + u\frac{\partial u}{\partial z} + v\frac{\partial v}{\partial z} + w\frac{\partial w}{\partial z} = -\frac{\partial V}{\partial z} - \frac{1}{\rho}\frac{\partial p}{\partial z}$$

or
$$\frac{\partial}{\partial x}\left(\frac{-\partial \phi}{\partial t}\right) + \frac{\partial}{\partial x}\left(\frac{u^2 + v^2 + w^2}{2}\right) = -\frac{\partial V}{\partial x} - \frac{1}{\rho}\frac{\partial p}{\partial x}$$

$$\frac{\partial}{\partial y}\left(\frac{-\partial \phi}{\partial t}\right) + \frac{\partial}{\partial y}\left(\frac{u^2 + v^2 + w^2}{2}\right) = -\frac{\partial V}{\partial y} - \frac{1}{\rho}\frac{\partial p}{\partial y}$$

$$\frac{\partial}{\partial z}\left(\frac{-\partial \phi}{\partial t}\right)+\frac{\partial}{\partial z}\left(\frac{u^2+v^2+w^2}{2}\right)=\frac{-\partial V}{\partial z}-\frac{1}{\rho}\frac{\partial p}{\partial z}$$

Multiplying these by dx, dy, dz and adding, we get

$$\left\{\frac{\partial}{\partial x}\left(\frac{-\partial \phi}{\partial t}\right)dx+\frac{\partial}{\partial y}\left(\frac{-\partial \phi}{\partial t}\right)dy+\frac{\partial}{\partial z}\left(\frac{-\partial \phi}{\partial z}\right)dz\right\}$$

$$+\frac{1}{2}\left\{\frac{\partial}{\partial x}\left(u^2+v^2+w^2\right)dx+\frac{\partial}{\partial y}\left(u^2+v^2+w^2\right)dy\right.$$

$$\left.+\frac{\partial}{\partial z}\left(u^2+v^2+w^2\right)dz\right\}$$

$$=-\left(\frac{\partial V}{\partial x}dx+\frac{\partial V}{\partial y}dy+\frac{\partial V}{\partial z}dz\right)$$

$$-\frac{1}{\rho}\left(\frac{\partial p}{\partial x}dx+\frac{\partial p}{\partial y}dy+\frac{\partial p}{\partial z}dz\right)$$

$$\Rightarrow \quad d\left(-\frac{\partial \phi}{\partial t}\right)+\frac{1}{2}d\left(u^2+v^2+w^2\right)=-dV-\frac{1}{\rho}dp$$

If q be the velocity so that $q^2 = u^2 + v^2 + w^2$, we have

$$-d\left(\frac{\partial \phi}{\partial t}\right)+\frac{1}{2}dq^2=-dV-\frac{1}{\rho}dp$$

$$\Rightarrow \quad -d\left(\frac{\partial \phi}{\partial t}\right)+\frac{1}{2}dq^2+dV+\frac{1}{\rho}dp=0$$

Integrating, we get

$$\int\frac{dp}{\rho}-\frac{\partial \phi}{\partial t}+\frac{1}{2}q^2+V=C$$

Where C is an arbitrary function of time.

If ρ be constant, we get

$$\frac{p}{\rho}-\frac{\partial \phi}{\partial t}+\frac{1}{2}q^2+V=C$$

This is Bernoulli equation for the unsteady, irrotational motion of an incompressible fluid.

For steady motion, $\frac{\partial \phi}{\partial t} = 0$ and therefore

$$\frac{p}{\rho} + \frac{1}{2} q^2 + V = C$$

where C is an absolute constant.

2.8. Lagrange's Equation of Motion

Let initially a fluid particle be at (a, b, c) at time $t = t_0$, when its volume is dv_0 and density is ρ_0. After a lapse of time t, let the same fluid particle be at (x, y, z) when its volume is dV and density is ρ.

Then equation of continuity is $\rho J = \rho_0$ ... (1)

where $J = \frac{\partial(a, b, c)}{\partial(x, y, z)}$

The components of acceleration are $\frac{d^2x}{dt^2}, \frac{d^2y}{dt^2}$ and $\frac{d^2z}{dt^2}$.

Let the external forces be conservative so that

$$\vec{F} = -\vec{\nabla}\Omega$$

Thus Euler's equations of motion are

$$\frac{\partial^2 x}{\partial t^2} = -\frac{\partial \Omega}{\partial x} - \frac{1}{\rho}\frac{\partial p}{\partial x}$$

$$\frac{\partial^2 y}{\partial t^2} = -\frac{\partial \Omega}{\partial y} - \frac{1}{\rho}\frac{\partial p}{\partial y}$$

$$\frac{\partial^2 z}{\partial t^2} = -\frac{\partial \Omega}{\partial z} - \frac{1}{\rho}\frac{\partial p}{\partial z}$$

Multiplying these equations by $\frac{\partial x}{\partial a}, \frac{\partial y}{\partial a}, \frac{\partial z}{\partial a}$ respectively and adding, we get

$$\frac{\partial^2 x}{\partial t^2}\frac{\partial x}{\partial a} + \frac{\partial^2 y}{\partial t^2}\frac{\partial y}{\partial a} + \frac{\partial^2 z}{\partial t^2}\frac{\partial z}{\partial a} = -\frac{\partial \Omega}{\partial a} - \frac{1}{\rho}\frac{\partial p}{\partial a} \quad \text{... (2)}$$

Similarly

$$\frac{\partial^2 x}{\partial t^2}\frac{\partial x}{\partial b} + \frac{\partial^2 y}{\partial t^2}\frac{\partial y}{\partial b} + \frac{\partial^2 z}{\partial t^2}\frac{\partial z}{\partial b} = -\frac{\partial \Omega}{\partial b} - \frac{1}{\rho}\frac{\partial p}{\partial b} \quad \text{... (3)}$$

$$\frac{\partial^2 x}{\partial t^2}\frac{\partial x}{\partial c}+\frac{\partial^2 y}{\partial t^2}\frac{\partial y}{\partial c}+\frac{\partial^2 z}{\partial t^2}\frac{\partial z}{\partial c}=-\frac{\partial \Omega}{\partial c}-\frac{1}{\rho}\frac{\partial p}{\partial c} \quad \ldots (4)$$

These equations (2), (3), (4) are having derivatives with respect to a, b, c and t only.

The equations (1), (2), (3) and (4) together represent Lagrange's Hydrodynamical equations.

2.9. Helmboltz's Vorticity Equation

Statement. If the external forces are conservative and density ρ is a function of p only, then

$$\frac{d}{dt}\left(\frac{\vec{W}}{\rho}\right)=\left(\frac{\vec{W}}{\rho}\cdot\vec{\nabla}\right)\vec{q}$$

Proof. Since $\vec{F}$ is conservative, so

$$\vec{F}=-\vec{\nabla}\Omega$$

Also ρ is a function of p, hence there exists a relation of the type

$$P=\int_{p_0}^{p}\frac{dp}{\rho}$$

Hence $$\vec{\nabla}p=\sum\vec{i}\,\frac{\partial P}{\partial x}=\sum\vec{i}\,\frac{dP}{dp}\frac{\partial p}{\partial x}$$

$$=\sum\vec{i}\,\frac{1}{\rho}\frac{\partial p}{\partial x}$$

$$=\frac{1}{\rho}\sum\vec{i}\,\frac{\partial p}{\partial x}=\frac{\vec{\nabla}p}{\rho}$$

Also Euler's equation of motion is

$$\frac{d\vec{q}}{dt}=\vec{F}-\frac{1}{\rho}\vec{\nabla}p$$

$$\Rightarrow\quad \frac{\partial\vec{q}}{\partial t}+\left(\vec{q}\cdot\vec{\nabla}\right)\vec{q}=-\vec{\nabla}\Omega-\frac{\vec{\nabla}p}{\rho} \quad \ldots (1)$$

But $$\vec{\nabla}(\vec{q}\cdot\vec{q})=2\left[\vec{q}\times\text{curl}\,\vec{q}+\left(\vec{q}\cdot\vec{\nabla}\right)\vec{q}\right]$$

$$\Rightarrow\quad \frac{1}{2}\vec{\nabla}q^2-\vec{q}\times\left(\vec{\nabla}\times\vec{q}\right)=\left(\vec{q}\cdot\vec{\nabla}\right)\vec{q}$$

Hence (1) becomes

$$\frac{\partial \vec{q}}{\partial t}+\frac{1}{2}\vec{\nabla}q^2-\vec{q}\times\left(\vec{\nabla}\times\vec{q}\right)=-\vec{\nabla}\Omega-\vec{\nabla}P$$

$$\Rightarrow \quad \frac{\partial \vec{q}}{\partial t}+\vec{\nabla}\left(\Omega+P+\frac{1}{2}q^2\right)=\vec{q}\times\overrightarrow{W}$$

Where $\overrightarrow{W}=\text{curl}\,\vec{q}=\vec{\nabla}\times\vec{q}$

Taking curl of both the sides, we get

$$\vec{\nabla}\times\left[\frac{\partial \vec{q}}{\partial t}\right]=\text{curl}\left(\vec{q}\times\overrightarrow{W}\right)\text{ for }\overrightarrow{W}\times\vec{\nabla}\phi=0$$

$$\Rightarrow \quad \text{curl}\,\frac{\partial \vec{q}}{\partial t}=\vec{\nabla}\times\left(\vec{q}\times\overrightarrow{W}\right)$$

$$\Rightarrow \quad \frac{\partial}{\partial t}(\text{curl}\,\vec{q})=\left(\vec{\nabla}\cdot\overrightarrow{W}\right)\vec{q}-\left(\vec{\nabla}\cdot\vec{q}\right)\overrightarrow{W}+\vec{q}\left(\overrightarrow{W}\cdot\vec{\nabla}\right)-\left(\vec{q}\cdot\vec{\nabla}\right)\overrightarrow{W}$$

$$\Rightarrow \quad \frac{\partial \overrightarrow{W}}{\partial t}=\left(\vec{\nabla}\cdot\overrightarrow{W}\right)\vec{q}-\left(\vec{\nabla}\cdot\vec{q}\right)\overrightarrow{W}+\vec{q}\left(\overrightarrow{W}\cdot\vec{\nabla}\right)-\left(\vec{q}\cdot\vec{\nabla}\right)\overrightarrow{W}$$

But $\vec{\nabla}\cdot\overrightarrow{W}=\text{div curl}\,\vec{q}=0$

Also $\dfrac{\partial \overrightarrow{W}}{\partial t}=\dfrac{d\overrightarrow{W}}{dt}-\left(\vec{q}\cdot\vec{\nabla}\right)\overrightarrow{W}$

Hence $\dfrac{d\overrightarrow{W}}{dt}-\left(\vec{q}\cdot\vec{\nabla}\right)\overrightarrow{W}=-\left(\vec{\nabla}\cdot\vec{q}\right)\overrightarrow{W}+\vec{q}\left(\overrightarrow{W}\cdot\vec{\nabla}\right)-\left(\vec{q}\cdot\vec{\nabla}\right)\overrightarrow{W}$

$$\frac{d\overrightarrow{W}}{dt}=\vec{q}\left(\overrightarrow{W}\cdot\vec{\nabla}\right)-\left(\vec{\nabla}\cdot\vec{q}\right)\overrightarrow{W} \qquad \dots (2)$$

Also equation of continuity is

$$\frac{d\rho}{dt}+\rho\left(\vec{\nabla}\cdot\vec{q}\right)=0$$

Hence $\vec{\nabla}\cdot\vec{q}=-\dfrac{1}{\rho}\dfrac{d\rho}{dt}$

Thus (2) becomes

$$\frac{d\overrightarrow{W}}{dt}=\vec{q}\left(\overrightarrow{W}\cdot\vec{\nabla}\right)-\frac{\overrightarrow{W}}{\rho}\frac{d\rho}{dt}$$

$$\Rightarrow \quad \frac{1}{\rho}\frac{d\vec{W}}{dt} - \frac{\vec{W}}{\rho^2}\frac{d\rho}{dt} = \frac{\vec{q}}{\rho}\left(\vec{W}\cdot\vec{\nabla}\right)$$

$$\Rightarrow \quad \frac{d}{dt}\left(\frac{\vec{W}}{\rho}\right) = \left(\frac{\vec{W}}{\rho}\cdot\vec{\nabla}\right)\vec{q} \qquad \ldots (3)$$

This is called Helmholtz vorticity equation.

If $\vec{q} = u\vec{i} + v\vec{j} + w\vec{k}$ and $\vec{w} = \xi\vec{i} + \eta\vec{j} + \zeta\vec{k}$ then (3) becomes

$$\left(u\frac{\partial}{\partial x} + v\frac{\partial}{\partial y} + w\frac{\partial}{\partial z} + \frac{\partial}{\partial t}\right)\left(\frac{\xi}{\rho}\right) = \frac{1}{\rho}\left(\xi\frac{\partial}{\partial x} + \eta\frac{\partial}{\partial y} + \zeta\frac{\partial}{\partial z}\right)u$$

$$\left(u\frac{\partial}{\partial x} + v\frac{\partial}{\partial y} + w\frac{\partial}{\partial z} + \frac{\partial}{\partial t}\right)\left(\frac{\eta}{\rho}\right) = \frac{1}{\rho}\left(\xi\frac{\partial}{\partial x} + \eta\frac{\partial}{\partial y} + \zeta\frac{\partial}{\partial z}\right)v$$

$$\left(u\frac{\partial}{\partial x} + v\frac{\partial}{\partial y} + w\frac{\partial}{\partial z} + \frac{\partial}{\partial t}\right)\left(\frac{\zeta}{\rho}\right) = \frac{1}{\rho}\left(\xi\frac{\partial}{\partial x} + \eta\frac{\partial}{\partial y} + \zeta\frac{\partial}{\partial z}\right)w$$

2.10. Cauchy's Integrals

Lagrange's hydrodynamical equations are

$$\frac{\partial^2 x}{\partial t^2}\frac{\partial x}{\partial a} + \frac{\partial^2 y}{\partial t^2}\frac{\partial y}{\partial a} + \frac{\partial^2 z}{\partial t^2}\frac{\partial z}{\partial a} = \frac{-\partial\Omega}{\partial a} - \frac{1}{\rho}\frac{\partial p}{\partial x} \qquad \ldots (1)$$

and two similar equations.

If we suppose

$$Q = \Omega + \int\frac{dp}{\rho}$$

then (1) can be written as

$$\frac{\partial^2 x}{\partial t^2}\frac{\partial x}{\partial a} + \frac{\partial^2 y}{\partial t^2}\frac{\partial y}{\partial a} + \frac{\partial^2 z}{\partial t^2}\frac{\partial z}{\partial a} = -\frac{\partial Q}{\partial a} \qquad \ldots (2)$$

Similarly we have

$$\frac{\partial^2 x}{\partial t^2}\frac{\partial x}{\partial b}+\frac{\partial^2 y}{\partial t^2}\frac{\partial y}{\partial b}+\frac{\partial^2 z}{\partial t^2}\frac{\partial z}{\partial b}=-\frac{\partial Q}{\partial b} \quad \ldots (3)$$

and $$\frac{\partial^2 x}{\partial t^2}\frac{\partial x}{\partial c}+\frac{\partial^2 y}{\partial t^2}\frac{\partial y}{\partial c}+\frac{\partial^2 z}{\partial t^2}\frac{\partial z}{\partial c}=-\frac{\partial Q}{\partial c} \quad \ldots (4)$$

Writing $u=\dfrac{\partial x}{\partial t}, v=\dfrac{\partial y}{\partial t}, w=\dfrac{\partial z}{\partial t}$

Equations (3) and (4) become

$$\frac{\partial u}{\partial t}\frac{\partial x}{\partial b}+\frac{\partial v}{\partial t}\frac{\partial y}{\partial b}+\frac{\partial w}{\partial t}\frac{\partial z}{\partial b}=-\frac{\partial Q}{\partial b} \quad \ldots (5)$$

$$\frac{\partial u}{\partial t}\frac{\partial x}{\partial c}+\frac{\partial v}{\partial t}\frac{\partial y}{\partial c}+\frac{\partial w}{\partial t}\frac{\partial z}{\partial c}=-\frac{\partial Q}{\partial c} \quad \ldots (6)$$

Eliminating Q from (5) and (6), we get

$$\frac{\partial^2 u}{\partial c\partial t}\frac{\partial x}{\partial b}+\frac{\partial u}{\partial t}\frac{\partial^2 x}{\partial b\partial c}+\frac{\partial^2 v}{\partial c\partial t}\frac{\partial y}{\partial b}+\frac{\partial v}{\partial t}\frac{\partial^2 y}{\partial b\partial c}$$

$$+\frac{\partial^2 w}{\partial c\partial t}\frac{\partial z}{\partial b}+\frac{\partial w}{\partial t}\frac{\partial^2 z}{\partial b\partial c}=\frac{\partial^2 u}{\partial b\partial t}\frac{\partial x}{\partial c}+\frac{\partial u}{\partial t}\frac{\partial^2 x}{\partial b\partial c}+\frac{\partial^2 v}{\partial b\partial t}\frac{\partial y}{\partial c}$$

$$+\frac{\partial v}{\partial t}\frac{\partial^2 y}{\partial b\partial c}+\frac{\partial^2 w}{\partial b\partial t}\frac{\partial z}{\partial c}+\frac{\partial w}{\partial t}\frac{\partial^2 z}{\partial b\partial c}$$

$$\Rightarrow \left(\frac{\partial^2 u}{\partial b\partial t}\frac{\partial x}{\partial c}-\frac{\partial^2 u}{\partial c\partial t}\frac{\partial x}{\partial b}\right)+\left(\frac{\partial^2 v}{\partial b\partial t}\frac{\partial y}{\partial c}-\frac{\partial^2 v}{\partial c\partial t}\frac{\partial y}{\partial b}\right)$$

$$+\left(\frac{\partial^2 w}{\partial b\partial t}\frac{\partial z}{\partial c}-\frac{\partial^2 w}{\partial c\partial t}\frac{\partial z}{\partial b}\right)=0$$

$$\Rightarrow \frac{\partial}{\partial b}\left(\frac{\partial u}{\partial b}\frac{\partial x}{\partial c}-\frac{\partial u}{\partial c}\frac{\partial x}{\partial b}\right)-\frac{\partial u}{\partial b}\frac{\partial^2 x}{\partial t\partial c}+\frac{\partial u}{\partial c}\frac{\partial^2 x}{\partial t\partial b}$$

$$+\frac{\partial}{\partial t}\left(\frac{\partial v}{\partial b}\frac{\partial y}{\partial c}-\frac{\partial v}{\partial c}\frac{\partial y}{\partial b}\right)-\frac{\partial v}{\partial b}\frac{\partial^2 y}{\partial t\partial c}+\frac{\partial v}{\partial c}\frac{\partial^2 y}{\partial t\partial b}$$

$$+\frac{\partial}{\partial t}\left(\frac{\partial w}{\partial b}\frac{\partial z}{\partial c}-\frac{\partial w}{\partial c}\frac{\partial z}{\partial b}\right)-\frac{\partial w}{\partial b}\frac{\partial^2 z}{\partial t\partial c}+\frac{\partial w}{\partial c}\frac{\partial^2 z}{\partial t\partial b}=0 \quad \ldots (7)$$

Now since $\frac{\partial u}{\partial c} = \frac{\partial^2 x}{\partial c \partial t}, \frac{\partial u}{\partial b} = \frac{\partial^2 x}{\partial b \partial t}$

Hence terms which are not in bracket in (7) cancel so that

$$\frac{\partial}{\partial t}\left(\frac{\partial u}{\partial b}\frac{\partial x}{\partial c} - \frac{\partial u}{\partial c}\frac{\partial x}{\partial b}\right) + \frac{\partial}{\partial t}\left(\frac{\partial v}{\partial b}\frac{\partial y}{\partial c} - \frac{\partial v}{\partial c}\frac{\partial y}{\partial b}\right) + \frac{\partial}{\partial t}\left(\frac{\partial w}{\partial b}\frac{\partial z}{\partial c} - \frac{\partial w}{\partial c}\frac{\partial z}{\partial b}\right) = 0$$

Integrating with respect to t, we get

$$\frac{\partial u}{\partial b}\frac{\partial x}{\partial c} - \frac{\partial u}{\partial c}\frac{\partial x}{\partial b} + \frac{\partial v}{\partial b}\frac{\partial y}{\partial c} - \frac{\partial v}{\partial c}\frac{\partial y}{\partial b} + \frac{\partial w}{\partial b}\frac{\partial z}{\partial c} - \frac{\partial w}{\partial c}\frac{\partial z}{\partial b} = C \quad \ldots (8)$$

Where the constant C is independent of t.

Now initially $x = a$, $y = b$, $z = c$ so that

$$\frac{\partial x}{\partial a} = 1, \frac{\partial x}{\partial b} = 0, \frac{\partial x}{\partial c} = 0$$

$$\frac{\partial y}{\partial a} = 0, \frac{\partial y}{\partial b} = 1, \frac{\partial y}{\partial c} = 0$$

$$\frac{\partial z}{\partial a} = 0, \frac{\partial z}{\partial b} = 0, \frac{\partial z}{\partial c} = 1$$

With these above values (8) becomes

$$\frac{\partial u}{\partial b}\frac{\partial x}{\partial c} - \frac{\partial u}{\partial c}\frac{\partial x}{\partial b} + \frac{\partial v}{\partial b}\frac{\partial y}{\partial c} - \frac{\partial v}{\partial c}\frac{\partial y}{\partial b} + \frac{\partial w}{\partial b}\frac{\partial z}{\partial c} - \frac{\partial w}{\partial c}\frac{\partial z}{\partial b} = \left(-\frac{\partial v}{\partial c}\right)_0 + \left(\frac{\partial w}{\partial b}\right)_0 = 2\xi_0 \quad \ldots (9)$$

where (ξ_0, η_0, ζ_0) are the initial values of (ξ, η, ζ) and $\vec{W} = \text{curl}\,\vec{q} = \xi\vec{i} + \eta\vec{j} + \zeta\vec{k}$.

Again $\frac{\partial u}{\partial b} = \frac{\partial u}{\partial x}\frac{\partial x}{\partial b} + \frac{\partial u}{\partial y}\frac{\partial y}{\partial b} + \frac{\partial u}{\partial z}\frac{\partial z}{\partial b}$ and so on. With these substitutions (9) become,

$$\frac{\partial x}{\partial c}\left(\frac{\partial u}{\partial x}\frac{\partial x}{\partial b}+\frac{\partial u}{\partial y}\frac{\partial y}{\partial b}+\frac{\partial u}{\partial z}\frac{\partial z}{\partial b}\right)$$

$$-\frac{\partial x}{\partial b}\left(\frac{\partial u}{\partial x}\frac{\partial x}{\partial c}+\frac{\partial u}{\partial y}\frac{\partial y}{\partial c}+\frac{\partial u}{\partial z}\frac{\partial z}{\partial c}\right)$$

$$+\frac{\partial y}{\partial c}\left(\frac{\partial v}{\partial x}\frac{\partial x}{\partial b}+\frac{\partial v}{\partial y}\frac{\partial y}{\partial b}+\frac{\partial v}{\partial z}\frac{\partial z}{\partial b}\right)$$

$$-\frac{\partial y}{\partial b}\left(\frac{\partial v}{\partial x}\frac{\partial x}{\partial c}+\frac{\partial v}{\partial y}\frac{\partial y}{\partial c}+\frac{\partial v}{\partial z}\frac{\partial z}{\partial c}\right)$$

$$+\frac{\partial z}{\partial c}\left(\frac{\partial w}{\partial x}\frac{\partial x}{\partial b}+\frac{\partial w}{\partial y}\frac{\partial y}{\partial b}+\frac{\partial w}{\partial z}\frac{\partial z}{\partial b}\right)$$

$$-\frac{\partial z}{\partial b}\left(\frac{\partial w}{\partial x}\frac{\partial x}{\partial c}+\frac{\partial w}{\partial y}\frac{\partial y}{\partial c}+\frac{\partial w}{\partial z}\frac{\partial z}{\partial c}\right)=2\xi_0$$

$$\Rightarrow \left(\frac{\partial y}{\partial b}\frac{\partial z}{\partial c}-\frac{\partial y}{\partial c}\frac{\partial z}{\partial b}\right)\left(\frac{\partial w}{\partial y}-\frac{\partial v}{\partial z}\right)+\ldots\ldots=2\xi_0$$

$$\Rightarrow \frac{\partial(y, z)}{\partial(b, c)}2\xi+\ldots\ldots+\ldots\ldots=2\xi_0$$

$$\Rightarrow \xi\frac{\partial(y, z)}{\partial(b, c)}+\eta\frac{\partial(z, x)}{\partial(b, c)}+\zeta\frac{\partial(x, y)}{\partial(b, c)}=\xi_0 \qquad \ldots (10)$$

Similarly

$$\xi\frac{\partial(y, z)}{\partial(c, a)}+\eta\frac{\partial(z, x)}{\partial(c, a)}+\zeta\frac{\partial(x, y)}{\partial(c, a)}=\eta_0 \qquad \ldots (11)$$

$$\xi\frac{\partial(y, z)}{\partial(a, b)}+\eta\frac{\partial(z, x)}{\partial(a, b)}+\zeta\frac{\partial(x, y)}{\partial(a, b)}=\zeta_0 \qquad \ldots (12)$$

Multiplying (10), (11), (12) by $\frac{\partial x}{\partial a}, \frac{\partial x}{\partial b}, \frac{\partial x}{\partial c}$ respectively and adding, we get

$$\xi\frac{\partial(x,y,z)}{\partial(a,b,c)}+\eta\frac{\partial(z,x,x)}{\partial(a,b,c)}+\zeta\frac{\partial(x,y,y)}{\partial(a,b,c)}$$

$$=\xi_0\frac{\partial x}{\partial a}+\eta_0\frac{\partial x}{\partial b}+\zeta_0\frac{\partial x}{\partial c} \qquad \ldots (13)$$

Equation of continuity is

$$\rho \frac{\partial(x,y,z)}{\partial(a,b,c)} = \rho_0$$

Hence (13) becomes

$$\xi \frac{\rho_0}{\rho} = \xi_0 \frac{\partial x}{\partial a} + \eta_0 \frac{\partial x}{\partial b} + \zeta_0 \frac{\partial x}{\partial c}$$

$$\Rightarrow \quad \frac{\xi}{\rho} = \frac{\xi_0}{\rho_0} \frac{\partial x}{\partial a} + \frac{\eta_0}{\rho_0} \frac{\partial x}{\partial b} + \frac{\zeta_0}{\rho_0} \frac{\partial x}{\partial c} \qquad \text{... (14)}$$

Similarly

$$\frac{\eta}{\rho} = \frac{\xi_0}{\rho_0} \frac{\partial y}{\partial a} + \frac{\eta_0}{\rho_0} \frac{\partial y}{\partial b} + \frac{\zeta_0}{\rho_0} \frac{\partial y}{\partial c} \qquad \text{... (15)}$$

$$\frac{\zeta}{\rho} = \frac{\xi_0}{\rho_0} \frac{\partial z}{\partial a} + \frac{\eta_0}{\rho_0} \frac{\partial z}{\partial b} + \frac{\zeta_0}{\rho_0} \frac{\partial z}{\partial c} \qquad \text{... (16)}$$

The equations (14), (15), (16) are called Cauchy's integrals. The vector form of these equations is

$$\frac{\vec{W}}{\rho} = \left(\frac{\xi_0}{\rho_0} \frac{\partial}{\partial a} + \frac{\eta_0}{\rho_0} \frac{\partial}{\partial b} + \frac{\zeta_0}{\rho_0} \frac{\partial}{\partial c} \right) \vec{r}$$

$$\Rightarrow \quad \frac{\vec{W}}{\rho} = \left(\frac{\vec{W}}{\rho} \cdot \vec{\nabla} \right)_0 \vec{r}$$

Deductions

(1) To prove the principle of permanence of irrotational motion.

Proof. If $\vec{W_0} = 0$ i.e. if $\xi_0 = \eta_0 = \zeta_0 = 0$, then (14), (15), (16) gives

$$\xi = \eta = \zeta = 0$$

i.e. if $W_0 = 0$ then $W = 0$

So that if at any time the motion is irrotational, then it is always rotational.

(2) Cauchy's integrals are the integrals of Helmholtz's vorticity equation or, to deduce Helmholtz equations with the help of Cauchy's integrals.

Proof.

$(14)\times\dfrac{\partial u}{\partial x}+(15)\times\dfrac{\partial u}{\partial y}+(16)\times\dfrac{\partial u}{\partial z}$ gives

$$\frac{1}{\rho}\left(\xi\frac{\partial u}{\partial x}+\eta\frac{\partial u}{\partial y}+\zeta\frac{\partial u}{\partial z}\right)$$

$$=\frac{\xi_0}{\rho_0}\left(\frac{\partial u}{\partial x}\frac{\partial x}{\partial a}+\frac{\partial u}{\partial y}\frac{\partial y}{\partial a}+\frac{\partial u}{\partial z}\frac{\partial z}{\partial a}\right)$$

$$+\frac{\eta_0}{\rho_0}\left(\frac{\partial u}{\partial x}\frac{\partial x}{\partial b}+\frac{\partial u}{\partial y}\frac{\partial y}{\partial b}+\frac{\partial u}{\partial z}\frac{\partial z}{\partial b}\right)$$

$$+\frac{\zeta_0}{\rho_0}\left(\frac{\partial u}{\partial x}\frac{\partial x}{\partial c}+\frac{\partial u}{\partial y}\frac{\partial y}{\partial c}+\frac{\partial u}{\partial z}\frac{\partial z}{\partial c}\right)$$

$$=\frac{\xi_0}{\rho_0}\frac{\partial u}{\partial a}+\frac{\eta_0}{\rho_0}\frac{\partial u}{\partial b}+\frac{\zeta_0}{\rho_0}\frac{\partial u}{\partial c}$$

$$=\frac{d}{dt}\left(\frac{\xi}{\rho}\right) \qquad \text{From (14)}$$

Hence $\dfrac{1}{\rho}\left(\xi\dfrac{\partial u}{\partial x}+\eta\dfrac{\partial u}{\partial y}+\zeta\dfrac{\partial u}{\partial z}\right)=\dfrac{d}{dt}\left(\dfrac{\xi}{\rho}\right)$

Similarly $\dfrac{1}{\rho}\left(\xi\dfrac{\partial v}{\partial x}+\eta\dfrac{\partial v}{\partial y}+\zeta\dfrac{\partial v}{\partial z}\right)=\dfrac{d}{dt}\left(\dfrac{\eta}{\rho}\right)$

$$\frac{1}{\rho}\left(\xi\frac{\partial v}{\partial x}+\eta\frac{\partial w}{\partial y}+\zeta\frac{\partial w}{\partial z}\right)=\frac{d}{dt}\left(\frac{\zeta}{\rho}\right)$$

In vector form these equations can be written as

$$\frac{1}{\rho}\left(\xi\frac{\partial\vec{q}}{\partial x}+\eta\frac{\partial\vec{q}}{\partial y}+\zeta\frac{\partial\vec{q}}{\partial z}\right)=\frac{d}{dt}\left(\frac{\vec{w}}{\rho}\right)$$

$$\Rightarrow\quad \frac{1}{\rho}\left(\vec{w}\cdot\vec{\nabla}\right)\vec{q}=\frac{d}{dt}\left(\frac{\vec{w}}{\rho}\right)$$

$$\Rightarrow\quad \frac{d}{dt}\left(\frac{\vec{w}}{\rho}\right)=\left(\frac{\vec{w}}{\rho}\cdot\vec{\nabla}\right)\vec{q}$$

which is Helmholtz vorticity equation.

2.11. General Equation of Impulsive Motion

We consider an arbitrary closed surface S moving with a non-viscous fluid such that it encloses a volume V. Let $\vec{q}_1$ and $\vec{q}_2$ be the fluid velocities at P within S just before the impulse and just after the impulse. Let I be the external impulse per unit mass and $\bar{w}$ the impulsive pressure on the surface element ds. Let $\hat{n}$ be the unit outward normal vector.

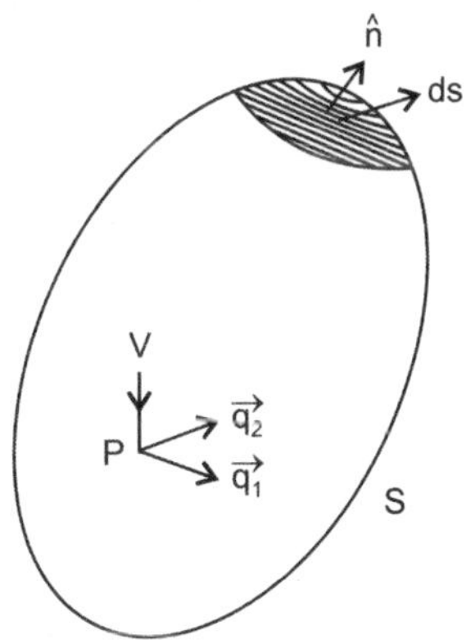

Fig. 2.5

Now by Newton's second law of motion applied to the fluid within closed surface S.

Change in momentum = Total body force

i.e. $$\int_V \rho(\vec{q}_2 - \vec{q}_1)\, dv = \int_V \vec{I}\,\rho dv + \int_S (-\hat{n}\bar{w})\, ds$$

($\because \hat{n}$ is the unit outward drawn normal)

$$= \int_V \vec{I}\rho dv = \int_V \vec{\nabla}\bar{w}\, dv$$

(By Gauss Theorem)

$$= \int_V \left(\vec{I}\rho - \vec{\nabla}\bar{w}\right) dv$$

Since V is an arbitrary volume

$$\rho\,(\vec{q}_2 - \vec{q}_1) = \vec{I}\rho - \vec{\nabla}\bar{w}$$

or $$\vec{q}_2 - \vec{q}_1 = \vec{I} - \frac{1}{\rho}\vec{\nabla}\bar{w} \qquad \text{... (1)}$$

This is the required equation for impulsive action.

If $\vec{I} = X\vec{i} + Y\vec{j} + Z\vec{k}$, $\vec{q}_2 = u\vec{i} + v\vec{j} + w\vec{k}$, $\vec{q}_1 = u_0\vec{i} + v_0\vec{j} + w_0\vec{k}$, then equating the coefficients of $\vec{i}, \vec{j}, \vec{k}$ in (1) we get Cartesian equation of impulsive motion as

$$u - u_0 = X - \frac{1}{\rho}\frac{\partial \bar{w}}{\partial x}$$

$$v - v_0 = Y - \frac{1}{\rho}\frac{\partial \overline{w}}{\partial y}$$

$$w - w_0 = Z - \frac{1}{\rho}\frac{\partial \overline{w}}{\partial z}$$

Deductions

(1) Vorticity in a non-viscous incompressible fluid is never generated by impulses if the external forces are conservative.

Proof. Since external impulses are conservative so $I = -\vec{\nabla}\Omega$

Also fluid is incompressible, so ρ is constant.

Hence from (1), we have

$$\vec{q}_2 - \vec{q}_1 = -\vec{\nabla}\left(\Omega + \frac{\overline{w}}{\rho}\right)$$

$$\Rightarrow \quad \vec{\nabla} \times (\vec{q}_2 - \vec{q}_1) = 0$$

$$\Rightarrow \quad \text{curl } \vec{q}_2 = \text{curl } \vec{q}_1$$

or $\quad \overrightarrow{W_2} = \overrightarrow{W_1}$

Hence the required result follows.

(2) $\vec{\nabla}^2 \overline{w} = 0$ under suitable conditions.

Proof. Let the external impulse be absent so that $\vec{I} = 0$. Also let ρ be constant.

Then (1) gives

$$\vec{q}_2 - \vec{q}_1 = -\vec{\nabla}\left(\frac{\overline{w}}{\rho}\right) \quad \dots (2)$$

$$\Rightarrow \quad \vec{\nabla}\cdot(\vec{q}_2 - \vec{q}_1) = -\vec{\nabla}\cdot\vec{\nabla}\left(\frac{\overline{w}}{\rho}\right)$$

$$= -\vec{\nabla}^2\left(\frac{\overline{w}}{\rho}\right)$$

$$\Rightarrow \quad \vec{\nabla}^2 \overline{w} = \rho\left[-\vec{\nabla}\cdot\vec{q}_2 + \vec{\nabla}\cdot\vec{q}_1\right]$$

$$= 0$$

Since $\vec{\nabla}\cdot\vec{q}_1 = 0 = \vec{\nabla}\cdot\vec{q}_2$ is the equation of continuity.

If the motion is irrotational then

$$\vec{q}_1 = -\vec{\nabla}\phi_1, \vec{q}_2 = -\vec{\nabla}\phi_2$$

Hence from (2)

$$-\vec{\nabla}\phi_2 + \vec{\nabla}\phi_1 = -\vec{\nabla}\left(\frac{\bar{w}}{\rho}\right)$$

$$\Rightarrow \quad \vec{\nabla}\left[\rho\left(\phi_2 - \phi_1\right) - \bar{w}\right] = 0$$

Integrating, we get

$\rho\left(\phi_2 - \phi_1\right) - \bar{w} = 0$, neglecting constant of integration

$$\Rightarrow \quad \bar{w} = \rho\left(\phi_2 - \phi_1\right)$$

(3) $\bar{w} = \rho\psi$ under suitable conditions.

Let the external impulse be absent so that $\vec{I} = 0$. Also let ρ be constant and motion start from rest. Then (1) becomes

$$q_2 - 0 = 0 - \frac{1}{\rho}\vec{\nabla}\bar{w}$$

Since the motion start from rest by the application of impulsive pressure and hence it must be irrotational, then

$$\vec{q}_2 = -\vec{\nabla}\phi$$

Hence

$$-\vec{\nabla}\phi = -\frac{1}{\rho}\vec{\nabla}\bar{w}$$

or $\quad \vec{\nabla}\left(\rho\phi - \bar{w}\right) = 0$

Integrating and neglecting constant of integration, we get

$$\bar{w} = \rho\phi$$

If $\rho = 1$ then $\bar{w} = \phi$

(4) If $\vec{I} = 0, \vec{q}_2 = 0$ then

$$-\vec{q}_1 = -\frac{1}{\rho}\vec{\nabla}\bar{w}$$

$$\Rightarrow \quad \vec{q}_1 = \frac{1}{\rho}\vec{\nabla}\bar{w}$$

Further if velocity has one component, then this gives

$$u = \frac{1}{\rho}\frac{\partial\bar{w}}{\partial x} = \frac{1}{\rho}\frac{d\bar{w}}{dx}$$

$$d\bar{w} = \rho u dx$$

2.12. Equation of Motion under Impulsive Forces (Cartesian Form)

Let at a point P in the fluid, X^1, Y^1, Z^1 be the components of external impulses per unit mass parallel to axes.

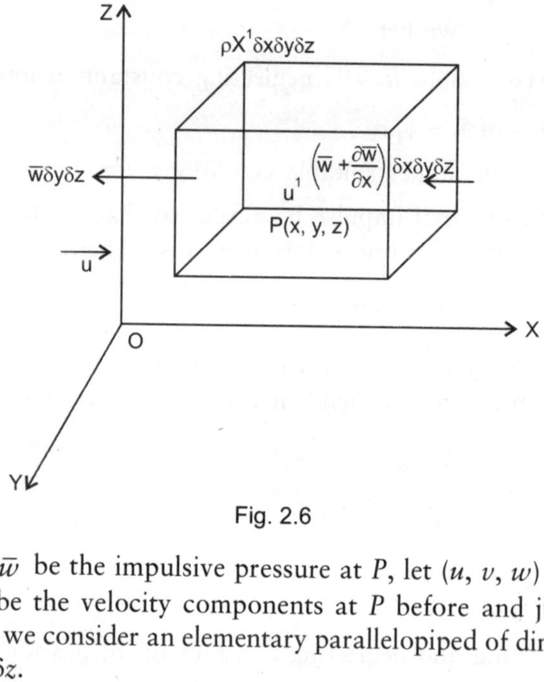

Fig. 2.6

Let $\bar{w}$ be the impulsive pressure at P, let (u, v, w) and (u^1, v^1, w^1) be the velocity components at P before and just after impulse, we consider an elementary parallelopiped of dimensions δx, δy, δz.

Then change in momentum in x direction

$$= \rho\delta x\delta y\delta z\ (u^1 - u)$$

By Newton's second law of motion

Change in momentum along x direction

$$= \text{resolved impulses in } x \text{ direction}$$

$$\rho\delta x\delta y\delta z\ (u^1 - u)$$

$$= \rho X^1\delta x\delta y\delta z + \bar{w}\delta y\delta z - \left(\bar{w} + \frac{\partial \bar{w}}{\partial x}\delta x\right)\delta y\delta z$$

$$\Rightarrow \rho\delta x\delta y\delta z\left(u^1 - u\right) = \left(\rho X^1 - \frac{\partial \bar{w}}{\partial x}\right)\delta x\delta y\delta z$$

$$\Rightarrow \quad \rho\left(u^1 - u\right) = \rho X^1 - \frac{\partial \bar{w}}{\partial x}$$

Similarly

$$\rho\left(v^1 - v\right) = \rho Y^1 - \frac{\partial \bar{w}}{\partial y}$$

$$\rho\left(w^1 - w\right) = \rho Z^1 - \frac{\partial \bar{w}}{\partial z}$$

These are the equations of motion for impulsive forces.

Multiplying these equations by dx, dy, dz respectively and adding, we have

$$\rho\left[\left(u^1 - u\right)dx + \left(v^1 - v\right)dy + \left(w^1 - w\right)dz\right]$$
$$= \rho\left(X^1 dx + Y^1 dy + Z^1 dz\right)$$
$$+\left(\frac{\partial \bar{w}}{\partial x}dx + \frac{\partial \bar{w}}{\partial y}dy + \frac{\partial \bar{w}}{\partial z}dz\right)$$

$$\Rightarrow \quad d\bar{w} = \rho\left[\left(u^1 - u\right)dx + \left(v^1 - v\right)dy + \left(w^1 - w\right)dz\right]$$
$$+ \rho\left(X^1 dx + Y^1 dy + Z^2 dz\right)$$

If ϕ, ϕ^1 denote the velocity potential just before and just after the impulsive action, then $udx + vdy + wdz = -\, d\phi$, $u^1dx + v^1dy + w^1dz = -\, d\phi$ and if the external impulses are derived from a potential V^1, so that

$$X^1dx + Y^1dy + Z^1dz = -\, dV^1$$

then $d\bar{w} = \rho\left(d\phi^1 - d\phi - dV^1\right)$

If ρ be constant then integrating it, we get

$$\bar{w} = \rho\left(\phi^1 - \phi - V^1\right) + C$$

Neglecting C as it may be included either in ϕ or in ϕ^1 then

$$\bar{w} = \rho\left(\phi^1 - \phi - V^1\right)$$

If $\quad V^1 = 0, \bar{w} = \rho\left(\phi^1 - \phi\right)$

If the liquid be at rest i.e. $\phi = 0$, then $\bar{w} = \rho\phi^1$ so that by applying suitable impulse we can produce a liquid motion whose velocity potential is ϕ^1 and if the liquid be of unit density, velocity potential becomes equal to the impulsive pressure.

Further from the equation of continuity

$$\frac{\partial u}{\partial x}+\frac{\partial v}{\partial y}+\frac{\partial w}{\partial z}=0=\frac{\partial u^1}{\partial x}+\frac{\partial v^1}{\partial y}+\frac{\partial w^1}{\partial z}$$

we have $\frac{\partial^2 \phi}{\partial x^2}+\frac{\partial^2 \phi}{\partial y^2}+\frac{\partial^2 \phi}{\partial z^2}=0=\frac{\partial^2 \phi^1}{\partial x^2}+\frac{\partial^2 \phi^1}{\partial y^2}+\frac{\partial^2 \phi^1}{\partial z^2}$

Hence $\frac{\partial^2 \bar{w}}{\partial x^2}+\frac{\partial^2 \bar{w}}{\partial y^2}+\frac{\partial^2 \bar{w}}{\partial z^2}=0$

2.13. Energy Equation

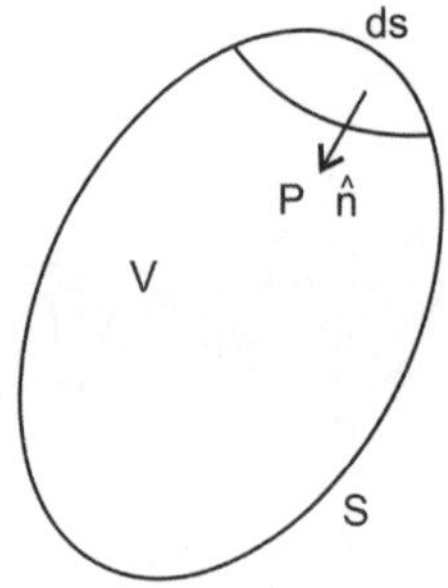

Fig. 2.7

The principle of energy enunciates that the change in energy is equal to work done by the extraneous forces. The potential due to external forces is supposed to the independent of time.

We consider an arbitrary closed surface S moving with a non-viscous fluid such that it encloses a volume V. Let $\hat{n}$ be the unit inward drawn normal vector on an element ds. Let the force be conservative so that $\vec{F}=-\vec{\nabla}\Omega$. Since force potential Ω is supposed to be independent of time t so that $\frac{\partial \Omega}{\partial t}=0$.

Also $\frac{d}{dt}=\frac{\partial}{\partial t}+\left(\vec{q}\cdot\vec{\nabla}\right)$

Hence $\frac{d\Omega}{dt}=\left(\vec{q}\cdot\vec{\nabla}\right)\Omega$ ... (1)

Let T, W, I denote kinetic energy, potential energy, intrinsic energy respectively. Since Ω is a force potential per unit mass and hence

$$W=\int \Omega dm=\int \Omega \rho dv$$

$$T=\int \frac{1}{2}\rho dv q^2=\frac{1}{2}\int q^2 \rho dv$$

Since elementary mass remains invariant throughout the motion and so

$$\frac{d}{dt}(\rho dv) = 0$$

Now

$$\frac{dT}{dt} = \frac{1}{2}\int \frac{dq^2}{dt}\rho dv + \frac{1}{2}\int q^2 \cdot 0$$

$$= \int \vec{q} \cdot \frac{d\vec{q}}{dt}\rho dv$$

Also

$$\frac{dW}{dt} = \int \frac{d\Omega}{dt}\rho dv + \int \Omega \cdot 0$$

$$= \int \frac{d\Omega}{dt}\rho dv$$

Intrinsic energy E per unit mass of the fluid is defined as the work done by the unit mass of the fluid against external pressure p under the supposed relation between pressure and density from its actual state to some standard state in which pressure and density are p_0 and ρ_0 respectively. Then

$$I = \int E\rho dv$$

and $$E = \int_V^{V_0} p dv \qquad \text{where } \rho V = 1$$

$$= \int_\rho^{\rho_0} p d\left(\frac{1}{\rho}\right)$$

$$= -\int_\rho^{\rho_0} \frac{p}{\rho^2} d\rho$$

or $$E = \int_{\rho_0}^{\rho} \frac{p}{\rho^2} d\rho$$

$$\therefore \quad dE = \frac{p}{\rho^2} d\rho$$

$$\begin{aligned}
\frac{dI}{df} &= \int\left[\frac{dE}{dt}\rho dv + E\frac{d}{dt}(\rho dv)\right] \\
&= \int\frac{dE}{dt}\rho dv = \int\frac{dE}{d\rho}\frac{d\rho}{dt}\rho dv \\
&= \int\frac{p}{p^2}\frac{d\rho}{dt}\rho dv \\
&= \int\frac{p}{\rho}\frac{d\rho}{dt}dv \\
&= \int\frac{p}{\rho}\left(-\rho\,\vec{\nabla}\cdot\vec{q}\right)dv
\end{aligned}$$

as $\frac{d\rho}{dt} + \rho\,\vec{\nabla}\cdot\vec{q} = 0$ is the equation of continuity

or $\frac{dI}{dt} = -\int p\left(\vec{\nabla}\cdot\vec{q}\right)dv$

Hence $\frac{dT}{dt} = \int\vec{q}\,\frac{d\vec{q}}{dt}\rho dv, \quad \frac{dW}{dt} = \int\frac{d\Omega}{dt}\rho dv$ ... (2)

$$\frac{dI}{dt} = -\int p\left(\vec{\nabla}\cdot\vec{q}\right)dv \qquad \text{... (3)}$$

By Euler's equation of motion

$$\frac{d\vec{q}}{dt} = -\vec{\nabla}\Omega - \frac{1}{\rho}\vec{\nabla}p$$

$$\Rightarrow \quad \vec{q}\,\frac{d\vec{q}}{dt}\rho dv = -\left[\left(\vec{q}\cdot\vec{\nabla}\Omega\right)\right]\rho dv - \left(\vec{q}\cdot\vec{\nabla}p\right)dv$$

$$\Rightarrow \quad \int_V \vec{q}\cdot\frac{d\vec{q}}{dt}\rho dv = -\int_V\left(\vec{q}\cdot\vec{\nabla}\Omega\right)\rho dv - \int_V\left(\vec{q}\cdot\vec{\nabla}p\right)dv$$

$$\Rightarrow \quad \frac{dT}{dt} = -\int_V\frac{d\Omega}{dt}\rho dv - \int_V\left(\vec{q}\cdot\vec{\nabla}p\right)dv \qquad \text{[From (1)]}$$

$$\Rightarrow \quad \frac{dT}{dt} + \frac{dW}{dt} + \int_V\left(\vec{q}\cdot\vec{\nabla}p\right)dv \qquad \text{... (4) [From (2)]}$$

Also $\vec{\nabla}\cdot(p\vec{q}) = p\vec{\nabla}\cdot\vec{q} + \vec{q}\cdot\vec{\nabla}p$

$$\Rightarrow \quad \int \vec{\nabla}\cdot(p\vec{q})dv - \int p\vec{\nabla}\cdot\vec{q}dv = \int\left(\vec{q}\cdot\vec{\nabla}p\right)dv$$

$$\Rightarrow \quad \int -\hat{n}\cdot(p\vec{q})\,ds + \frac{dI}{dt} = \int_V\left(\vec{q}\cdot\vec{\nabla}p\right)dv \qquad \text{[From (3)]}$$

[By Gauss Theorem and $\hat{n}$ is the unit inward normal]

Hence (4) becomes

$$\frac{dT}{dt} + \frac{dW}{dt} + \frac{dT}{dt} = \int \hat{n}(p\vec{q})\,ds$$

$$\Rightarrow \quad \frac{d}{dt}(T + W + I) = \int_S \hat{n}\cdot(p\vec{q})\,ds$$

This proves that the rate of change of total energy of any portion of the fluid is equal to the work done by external pressure on the boundary provided by external forces are conservative.

Corollary. *Principle of energy for incompressible fluids*

In this case I = 0. Hence rate of change of total energy is equal to the work done by the pressure on the boundary.

Working Rules for Solving Problems

In order to solve the equations of motion, we adopt the following methods:

(1) Equation of motion is $\dfrac{\partial v}{\partial t} + v\dfrac{\partial v}{\partial x} = F - \dfrac{1}{\rho}\dfrac{\partial p}{\partial x}$

where $F = -\dfrac{\partial \Omega}{\partial x}$.

(2) Equation of continuity is

(i) $x^2 v = F(t)$ for spherical symmetry

(ii) $xv = F(t)$ for cylindrical symmetry

(iii) $\dfrac{\partial \rho}{\partial t} + \dfrac{\partial}{\partial x}(\rho v) = 0$ in general.

(3) Generally the fluid is assumed to be rest at infinity i.e. $x = \infty$, $v = 0$, $p = \Pi$ (say)

(4) If r be the radius of cavity (or hollow sphere) then $x = r$, $v = r$, $p = 0$.

(5) When $r = a$, $v = 0$ so that $F(t) = 0$

(6) Boyle's law: $p_1v_1 = p_2v_2 =$ const. Its alternate form is $p = k\rho$.

(7) Flux = cross-sectional area · normal velocity · density.

(8) Equation of impulsive action is $d\bar{w} = \rho v dx = \rho v^1 dr^1$

$$\int_0^{\Pi/2} \sin^p\theta\cos^q\theta\, d\theta = \frac{\Gamma\left(\frac{p+1}{2}\right)\cdot\Gamma\left(\frac{q+1}{2}\right)}{2\,\Gamma\left(\frac{p+q+2}{2}\right)}$$

$$\Gamma(n)\,\Gamma(1-n) = \frac{\Pi}{\sin n\Pi} \text{ and } \Gamma(n)\,\Gamma\left(n+\frac{1}{2}\right) = \frac{\sqrt{\Pi}\,\Gamma(2n)}{2^{2n-1}}$$

(9) K.E. of the liquid = work done = $\int -p\,dv$

(10) If a sphere of radius a is annihilated then when $x = a$, $p = 0$ so that $v = \dot{x} = 0$.

(11) If a problem contains external and internal radii i.e. R and r, then subject the result (which is obtained from the integration of the equation of motion) to the two boundary conditions for R and r. In this way we obtain an equation free from constant of integration and pressure p. Again we integrate this equation to obtain the required result.

WORKED OUT EXAMPLES

1. Show that the velocity field

$$u(x,y) = \frac{A\left(x^2 - y^2\right)}{\left(x^2 + y^2\right)^2},\ v(x,y) = \frac{2Axy}{\left(x^2 + y^2\right)^2},\ w = 0$$

satisfies the equation of motion for inviscid, incompressible flow. Determine the pressure associated with this velocity field.

Solution: Here

$$u(x,y) = \frac{A\left(x^2 - y^2\right)}{\left(x^2 + y^2\right)^2},\ v(x,y) = \frac{2Axy}{\left(x^2 + y^2\right)^2},\ w = 0$$

$$\frac{\partial u}{\partial x}=\frac{2Ax\left(3y^2-x^2\right)}{\left(x^2+y^2\right)^3}$$

$$\frac{\partial u}{\partial y}=\frac{-2Ay\left(3x^2-y^2\right)}{\left(x^2+y^2\right)^3}$$

$$\frac{\partial u}{\partial z}=0$$

$$\frac{\partial v}{\partial x}=\frac{2Ay\left(y^2-3x^2\right)}{\left(x^2+y^2\right)^3}$$

$$\frac{\partial v}{\partial y}=\frac{2Ax\left(x^2-3y^2\right)}{\left(x^2+y^2\right)^3}$$

$$\frac{\partial v}{\partial z}=0$$

The equations of motion for steady flow are given by

$$u\frac{\partial u}{\partial x}+v\frac{\partial u}{\partial y}+w\frac{\partial u}{\partial z}=-\frac{1}{\rho}\frac{\partial p}{\partial x} \quad \text{... (1)}$$

$$u\frac{\partial v}{\partial x}+v\frac{\partial v}{\partial y}+w\frac{\partial v}{\partial z}=-\frac{1}{\rho}\frac{\partial p}{\partial y} \quad \text{... (2)}$$

$$u\frac{\partial w}{\partial x}+v\frac{\partial w}{\partial y}+w\frac{\partial w}{\partial z}=-\frac{1}{\rho}\frac{\partial p}{\partial z} \quad \text{... (3)}$$

Putting the values of $\frac{\partial u}{\partial x}, \frac{\partial u}{\partial y}, \frac{\partial u}{\partial z}$ etc. in (1), (2) and (3) the equations become

$$\frac{2A^2x}{\left(x^2+y^2\right)^3}=\frac{1}{\rho}\frac{\partial p}{\partial x} \quad \text{... (4)}$$

$$\frac{2A^2y}{\left(x^2+y^2\right)^3}=\frac{1}{\rho}\frac{\partial p}{\partial y} \quad \text{... (5)}$$

$$0=-\frac{1}{\rho}\frac{\partial p}{\partial z} \quad \text{... (6)}$$

Equation (6) shows that the pressure p is inependent of z i.e. $p = p\ (x, y)$

Therefore,

$$dp = \frac{\partial p}{\partial x} dx + \frac{\partial p}{\partial y} dy$$

$$dp = \frac{2A^2\rho\left(xdx + ydy\right)}{\left(x^2 + y^2\right)^3}$$

Integrating, we get

$$p = -\frac{A^2\rho}{\left(x^2 + y^2\right)^2} + C$$

2. The particle velocity for a fluid motion referred to rectangular axes is given by the components

$$u = A\cos\frac{\Pi x}{2a}\cos\frac{\Pi z}{2a},\ v = 0,\ w = A\sin\frac{\Pi x}{2a}\sin\frac{\Pi z}{2a}$$

where A is a constant. Show that this is a possible motion of an incompressible fluid under no body forces in an infinite fixed rigid tube, $-a \le x \le a, 0 \le z \le 2a$. Also, find the pressure associated with this velocity field. [I.A.S. 94]

Solution: The equations of motion for a two dimensional steady inviscid, incompressible flow under no body force, in Cartesian coordinates are given by

$$u\frac{\partial u}{\partial x} + v\frac{\partial u}{\partial y} + w\frac{\partial u}{\partial z} = -\frac{1}{\rho}\frac{\partial p}{\partial x} \qquad \text{... (1)}$$

$$0 = -\frac{1}{\rho}\frac{\partial p}{\partial y} \qquad \text{... (2)}$$

$$u\frac{\partial w}{\partial x} + v\frac{\partial w}{\partial y} + w\frac{\partial w}{\partial z} = -\frac{1}{\rho}\frac{\partial p}{\partial z} \qquad \text{... (3)}$$

Here

$$u = A\cos\frac{\Pi x}{2a}\cos\frac{\Pi z}{2a},\ v = 0,\ w = A\sin\frac{\Pi x}{2a}\sin\frac{\Pi z}{2a} \qquad \text{... (4)}$$

From (2), it follows that the pressure p is independent of y i.e. $p = p\ (x, z)$.

Using (4) into (1) and (3), we get

$$\left(A\cos\frac{\Pi x}{2a}\cos\frac{\Pi z}{2a}\right)\left(-\frac{\Pi A}{2a}\sin\frac{\Pi x}{2a}\cos\frac{\Pi z}{2a}\right)$$

$$+\left(A\sin\frac{\Pi x}{2a}\sin\frac{\Pi z}{2a}\right)\left(-\frac{\Pi A}{2a}\cos\frac{\Pi x}{2a}\sin\frac{\Pi z}{2a}\right)=-\frac{1}{\rho}\frac{\partial p}{\partial x}$$

$$\Rightarrow \quad \frac{\Pi A^2}{2a}\left[\cos\frac{\Pi x}{2a}\sin\frac{\Pi x}{2a}\cos^2\frac{\Pi z}{2a}\right.$$

$$\left.+\sin^2\frac{\Pi z}{2a}\sin\frac{\Pi x}{2a}\cos\frac{\Pi x}{2a}\right]=\frac{1}{\rho}\frac{\partial p}{\partial x}$$

$$\Rightarrow \quad \frac{\Pi A^2}{2a}\cos\frac{\Pi x}{2a}\sin\frac{\Pi x}{2a}=\frac{1}{\rho}\frac{\partial p}{\partial x} \qquad \ldots (5)$$

and $\left(A\cos\frac{\Pi x}{2a}\cos\frac{\Pi z}{2a}\right)\left(\frac{\Pi A}{2a}\cos\frac{\Pi x}{2a}\sin\frac{\Pi z}{2a}\right)$

$$+\left(A\sin\frac{\Pi x}{2a}\sin\frac{\Pi z}{2a}\right)\left(\frac{\Pi A}{2a}\sin\frac{\Pi x}{2a}\cos\frac{\Pi z}{2a}\right)=-\frac{1}{\rho}\frac{\partial p}{\partial z}$$

$$\Rightarrow \quad \frac{\Pi A^2}{2a}\left[\cos^2\frac{\Pi x}{2a}\cos\frac{\Pi z}{2a}\sin\frac{\Pi z}{2a}\right.$$

$$\left.+\sin^2\frac{\Pi x}{2a}\sin\frac{\Pi z}{2a}\cos\frac{\Pi z}{2a}\right]=-\frac{1}{\rho}\frac{\partial p}{\partial z}$$

$$\Rightarrow \quad \frac{\Pi A^2}{2a}\sin\frac{\Pi z}{2a}\cos\frac{\Pi z}{2a}=-\frac{1}{\rho}\frac{\partial p}{\partial z} \qquad \ldots (6)$$

The relations (5) and (6) show that the velocity components satisfy the equations of motion.

Again $dp=\frac{\partial p}{\partial x}dx+\frac{\partial p}{\partial z}dz$

$$\Rightarrow \quad dp=\frac{\Pi\rho A^2}{2a}\left[\cos\frac{\Pi x}{2a}\sin\frac{\Pi x}{2a}dx-\cos\frac{\Pi z}{2a}\sin\frac{\Pi z}{2a}dz\right]$$

Integrating, we get

$$p=\frac{1}{2}\rho A^2\left[\cos^2\frac{\Pi z}{2a}-\cos^2\frac{\Pi x}{2a}\right]+C$$

where C is an integration constant. This gives the required pressure distribution.

3. Prove that the equation of motion is satisfied for an inviscid, incompressible, steady flow with negligible body force whose velocity components are given by

$$q_r = u\left(1 - \frac{A^2}{r^3}\right)\cos\theta,\ q_0 = -u\left(1 + \frac{A^3}{2r^3}\right)\sin\theta,\ q_\phi = 0$$

where A is a constant. Find the resultant velocity when $r \to \infty$.

Solution: The equations of motion for an inviscid, incompressible and steady flow with negligible external force, in spherical polar coordinates, are given by

$$q_r\frac{\partial q_r}{\partial r} + \frac{q_\theta}{r}\frac{\partial q_r}{\partial \theta} + \frac{q_\phi}{r\sin\theta}\frac{\partial q_r}{r\phi} - \frac{q_\theta^2 + q_\phi^2}{r} = -\frac{1}{\rho}\frac{\partial p}{\partial r} \qquad \dots (1)$$

$$q_r\frac{\partial q_\theta}{\partial r} + \frac{q_\theta}{r}\frac{\partial q_\theta}{\partial \theta} + \frac{q_\phi}{\partial\sin\theta}\frac{\partial q_\theta}{\partial \phi} + \frac{q_r q_\theta}{r} - \frac{q_\phi^2\cot\theta}{r}$$

$$= -\frac{1}{\rho}\frac{\partial p}{r\partial\theta} \qquad \dots (2)$$

$$q_r\frac{\partial q_\phi}{\partial r} + \frac{q_\theta}{r}\frac{\partial q_\phi}{\partial \theta} + \frac{q_\phi}{r\sin\theta}\frac{\partial q_\phi}{\partial \phi} + \frac{q_\phi q_r}{r} + \frac{q_\theta q_\phi\cot\theta}{r}$$

$$= -\frac{1}{\rho}\frac{1}{r\sin\phi}\frac{\partial p}{\partial\phi} \qquad \dots (3)$$

Here

$$q_r = u\left(1 - \frac{A^3}{r^3}\right)\cos\theta,\ q_\theta = -u\left(1 + \frac{A^3}{2r^3}\right)\sin\theta,\ q_\phi = 0 \qquad \dots (4)$$

Using (4) the equations (1), (2), (3) reduces to

$$u\left(1 - \frac{A^3}{r^3}\right)\cos\theta\left(\frac{3uA^2}{r^4}\right)\cos\theta + \frac{u}{r}\left(1 + \frac{A^3}{2r^3}\right)\sin\theta$$

$$\times u\left(1 - \frac{A^3}{r^3}\right)\sin\theta - \frac{u^2}{r}\left(1 + \frac{A^3}{2r^3}\right)^2\sin^2\theta = -\frac{1}{\rho}\frac{\partial p}{\partial r} \qquad \dots (5)$$

$$\left(1-\frac{A^3}{r^3}\right)\cos\theta\,\frac{3uA^3}{2r^4}\sin\theta+\frac{u}{r}\left(1+\frac{A^3}{2r^3}\right)\sin\theta u$$

$$\left(1+\frac{A^3}{2r^3}\right)\cos\theta-\frac{u^2}{r}\left(1-\frac{A^3}{r^3}\right)\left(1+\frac{A^3}{2r^3}\right)\sin\theta\cos\theta$$

$$=-\frac{1}{\rho}\frac{\partial p}{r\partial\theta} \qquad \ldots (6)$$

$$0=-\frac{1}{\rho}\frac{1}{\partial\sin\phi}\frac{\partial p}{\partial\phi} \qquad \ldots (7)$$

Equation (7) shows that the pressure p is independent of ϕ, hence $p = p(r, \theta)$.

Simplifying the equations (5) and (6), we get

$$\frac{3u^2A^3}{r^4}\left(1-\frac{A^3}{r^3}\right)\cos^2\theta+\frac{u^2}{r}\left(1-\frac{A^3}{2r^3}-\frac{A^6}{2r^6}\right)\sin^2\theta$$

$$-\frac{u^2}{r}\left(1+\frac{A^3}{r^3}+\frac{A^6}{4r^6}\right)\sin^2\theta=-\frac{1}{\rho}\frac{\partial p}{\partial r}$$

$$\Rightarrow \quad \frac{3u^2A^3}{r^4}\left(1-\frac{A^3}{r^3}\right)\cos^2\theta+\frac{u^2}{r}\left(-\frac{3A^3}{2r^3}-\frac{3A^6}{4r^6}\right)\sin^2\theta$$

$$=-\frac{1}{\rho}\frac{\partial p}{\partial r}$$

$$\Rightarrow \quad \frac{3u^2A^3}{r^4}\left(1-\frac{A^3}{r^3}\right)\cos^2\theta-\frac{3u^2A^2}{2r^4}\left(1+\frac{A^3}{2r^3}\right)\sin^2\theta$$

$$=-\frac{1}{\rho}\frac{\partial p}{\partial r} \qquad \ldots (8)$$

and

$$\frac{3u^2A^3}{2r^4}\left(1-\frac{A^3}{r^3}\right)\sin\theta\cos\theta+\frac{u^2}{r}\left(1+\frac{A^3}{r^3}+\frac{A^6}{4r^6}\right)\sin\theta\cos\theta$$

$$-\frac{u^2}{r}\left(1-\frac{A^3}{2r^3}-\frac{A^6}{2r^6}\right)\sin\theta\cos\theta=-\frac{1}{\rho}\frac{\partial p}{r\partial\theta}$$

$$\Rightarrow \quad \frac{3u^2A^3}{2r^3}\left(1-\frac{A^3}{r^3}\right)\sin\theta\cos\theta$$

$$+\frac{3u^2A^3}{2r^4}\left(1+\frac{A^3}{2r^3}\right)\sin\theta\cos\theta = -\frac{1}{\rho}\frac{\partial p}{r\partial\theta} \qquad \ldots (9)$$

Differentiating (8) with respect to θ, we have

$$\frac{-6u^2A^3}{r^4}\left(1-\frac{A^3}{r^3}\right)\sin\theta\cos\theta$$

$$-\frac{3u^2A^2}{2r^4}\left(1+\frac{A^3}{2r^3}\right)\sin\theta\cos\theta = -\frac{1}{\rho}\frac{\partial^2 p}{\partial r\,\partial\theta}$$

$$\Rightarrow \quad \left(\frac{-9u^2A^3}{r^4}+\frac{9u^2A^6}{2r^7}\right)\sin\theta\cos\theta = -\frac{1}{\rho}\frac{\partial^2 p}{\partial r\partial\theta} \qquad \ldots (10)$$

Differentiating (9) with respect to r, we have

$$\frac{3u^2A^3}{2}\left(-\frac{3}{r^4}+\frac{6A^3}{r^7}\right)\cos\theta\sin\theta$$

$$+\frac{3u^2A^3}{2}\left(-\frac{3}{r^4}-\frac{3A^3}{r^7}\right)\sin\theta\cos\theta = -\frac{1}{\rho}\frac{\partial^2 p}{\partial r\partial\theta}$$

$$\Rightarrow \quad \frac{-9}{2}\frac{u^2A^3}{r^4}\left(1-\frac{2A^3}{r^3}\right)\sin\theta\cos\theta$$

$$-\frac{9}{2}\frac{u^2A^3}{r^4}\left(1+\frac{A^3}{r^3}\right)\sin\theta\cos\theta = -\frac{1}{\rho}\frac{\partial^2 p}{\partial r\partial\theta}$$

$$\Rightarrow \quad \left(\frac{-9u^2A^3}{r^4}+\frac{9u^2A^6}{2r^7}\right)\sin\theta\cos\theta = -\frac{1}{\rho}\frac{\partial^2 p}{\partial r\partial\theta} \qquad \ldots (11)$$

Equations (10) and (11) are identical. Hence the equation of motion is satisfied. When $r \to \infty$, the resultant velocity is equal to u.

4. Stream is rushing from a boiler through a conical pipe, the diameters of the end of which are D and d; if V and υ be the corresponding velocities of the stream, and if the motion is supposed to be that of divergence from the vertex of the cone, prove that

$$\frac{\upsilon}{V} = \frac{D^2}{d^2} e^{(\upsilon^2 - V^2)/2K}$$

where K is the pressure divided by the density and supposed constant. [I.A.S. 93]

Solution: Let AB and CD be the ends of the conical pipe. Let ρ_1 and ρ_2 be the densities at these ends.

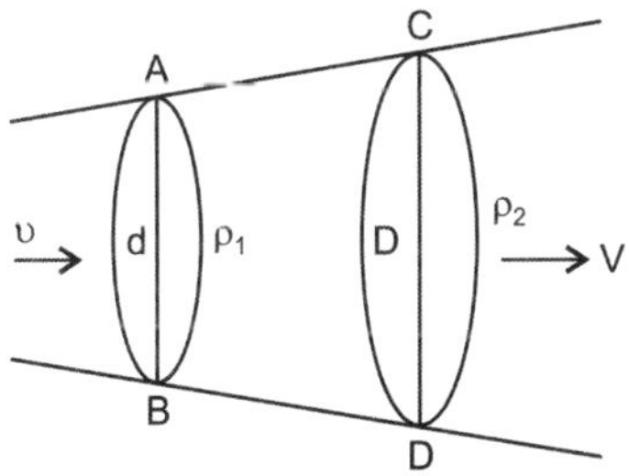

Fig. 2.8

By the principle of conservation of mass the masses of the stream that enter and leave at the ends AB and CD are the same. Thus we have

$$\Pi\left(\frac{1}{2}d\right)^2 \upsilon\rho_1 = \Pi\left(\frac{1}{2}D\right)^2 V\rho_2$$

$$\Rightarrow \quad \frac{\upsilon}{V} = \frac{D^2}{d^2}\frac{\rho_2}{\rho_1} \qquad \text{... (1)}$$

Now let p be the pressure, ρ be the density and u be the velocity at a distance r from AB, then the equation of motion is

$$u\frac{\partial u}{\partial r} = -\frac{1}{\rho}\frac{\partial p}{\partial r}$$

$$\Rightarrow \quad u\frac{\partial u}{\partial r} = -\frac{K}{\rho}\frac{\partial \rho}{\partial r} \quad \text{Since } p = K\rho$$

Integrating, we have

$$\frac{u^2}{2} = -K[\log\rho - \log C]$$

$$\Rightarrow \quad \frac{-u^2}{2K} = \log\frac{\rho}{C}$$

$$\rho = Ce^{-\frac{u^2}{2K}}$$

Boundary conditions are

(i) $\rho = \rho_1$ when $u = \upsilon$

(ii) $\rho = \rho_2$ when $u = V$

So that

$$\frac{\rho_1}{\rho_2} = \frac{e^{-\frac{\upsilon^2}{2K}}}{e^{-\frac{V^2}{2K}}} = e^{\frac{V^2-\upsilon^2}{2K}}$$

Hence (1) becomes

$$\frac{\upsilon}{V} = \frac{D^2}{d^2} e^{\frac{\upsilon^2 - V^2}{2K}}$$

5. A sphere of radius a is surrounded by an infinite liquid of density ρ, the pressure at infinity being Π. The sphere is suddenly annihilated. Show that the pressure at a distance r from the centre immediately falls to $\Pi\left(1-\frac{a}{r}\right)$. [I.A.S. 86]

Show further that if the liquid is brought to rest by impinging on a concentric sphere of radius $\frac{a}{2}$, the impulsive pressure sustained by the surface of this sphere is $\left[\frac{7}{6}\Pi\rho a^2\right]^{\frac{1}{2}}$.

Solution: Let υ be the velocity and p the pressure at a distance x from the centre of the sphere. The equation of continuity is

$$\chi^2\upsilon = F(t)$$

So that $\dfrac{\partial \upsilon}{\partial t} = \dfrac{F^1(t)}{x^2}$

Equation of motion is

$$\frac{\partial \upsilon}{\partial t} + \upsilon\frac{\partial \upsilon}{\partial x} = -\frac{1}{\rho}\frac{\partial p}{\partial x}$$

$$\Rightarrow \quad \frac{F^1(t)}{x^2} + \frac{\partial}{\partial x}\left(\frac{1}{2}\upsilon^2\right) = -\frac{\partial}{\partial x}\left(\frac{p}{\rho}\right)$$

Integrating w.r.t. x, we get

$$-\frac{F^1(t)}{x}+\frac{1}{2}\upsilon^2=-\frac{p}{\rho}+C \quad \text{... (1)}$$

Initially when $x^1 = \infty$, $p = \Pi$, $\upsilon = 0$

Hence, from (1)

$$C=\frac{\Pi}{\rho}$$

Thus

$$\frac{-F^1(t)}{x}+\frac{1}{2}\upsilon^2=\frac{1}{\rho}(\Pi-p) \quad \text{... (2)}$$

At the time of annihilation

$$x = a,\ p = 0,\ \upsilon = x = 0,\ t = 0$$

Therefore, from (2)

$$\frac{-F^1(0)}{a}=\frac{\Pi}{\rho}=C \quad \text{... (3)}$$

Just after annihilation, when $t = 0$, $x = r$, $\upsilon = 0$, $p = p_0$, $r > a$.

Hence, from (1)

$$\frac{-F^1(0)}{r}=-\frac{p_0}{\rho}+C$$

$$\Rightarrow \quad \frac{a\Pi}{\rho}\frac{1}{r}=\frac{-p_0}{\rho}+\frac{\Pi}{\rho} \qquad \text{From (3)}$$

$$\Rightarrow \quad p_0=\Pi\left(1-\frac{a}{r}\right)$$

Second Part. Firstly we shall determine velocity on the inner surface. Let r be the radius of inner surface. Then

when $x = r$, $\upsilon = \dot{r} = u$ (say), $p = 0$, when $r < a$, as pressure vanishes on the inner surface.

Hence, with these conditions (1) become

$$\frac{-F^1(t)}{r}+\frac{1}{2}u^2=C=\frac{\Pi}{\rho}$$

$$\Rightarrow \quad \frac{-F^1(t)}{r} + \frac{1}{2}\frac{F^2}{r^4} = \frac{\Pi}{\rho} \qquad \text{as } r^2 u = F(t)$$

Multiplying it by $2F(t)\ dt = 2r^2udt$

$= 2r^2dr$, we get

$$-\frac{2FF^1dt}{r} + \frac{F^2}{r^2}dr = \frac{\Pi}{\rho}2r^2dr$$

$$\Rightarrow \quad d\left(\frac{-F^2}{r}\right) = \frac{\Pi}{\rho}2r^2dr$$

Integrating, we get

$$\frac{-F^2}{r} = \frac{2}{3}\frac{\Pi}{\rho}r^3 + C_1$$

$$\Rightarrow \quad -r^3u^2 = \frac{2}{3}\frac{\Pi}{\rho}r^3 + C_1 \qquad \text{... (4)}$$

But when $x = a$, $\upsilon = 0 = u = r$

Hence $0 = \frac{2}{3}\frac{\Pi}{\rho}a^3 + C_1$

Hence (4) becomes

$$-r^3u^2 = \frac{2}{3}\frac{\Pi}{\rho}r^3 - \frac{2}{3}\frac{\Pi}{\rho}a^3$$

$$\Rightarrow \quad u^2 = \frac{2\Pi}{3\rho}\left(\frac{a^3}{r^3} - 1\right)$$

when $r = \frac{a}{2}$,

$$u^2 = \frac{2\Pi}{3\rho}\left(\frac{a^3 8}{a^3} - 1\right)$$

$$= \frac{14\Pi}{3\rho}$$

$$\Rightarrow \quad u = \left[\frac{14\Pi}{3\rho}\right]^{\frac{1}{2}}$$

The liquid is brought to rest by striking a sphere, let v^1 be the velocity at a point r just before striking and $\bar{w}$ be the impulsive pressure, then

$$\left(0 - v^1\right) = -\frac{1}{\rho}\frac{d\bar{w}}{dr^1}$$

$$\Rightarrow \quad d\bar{w} = \rho v^1 dr^1 = \rho u dr$$

$$\Rightarrow \quad \int_0^{\bar{w}} d\bar{w} = \rho u \int_0^{\frac{a}{2}} dr$$

$$\Rightarrow \quad \bar{w} = \rho\left[\frac{14\Pi}{3\rho}\right]^{\frac{1}{2}}\frac{a}{2} = \left[\frac{7}{6}\rho\Pi a^2\right]^{\frac{1}{2}}$$

6. An infinite mass of homogeneous incompressible fluid is at rest subject to a uniform pressure Π and contains a spherical cavity of radius a, filled with gas at a pressure $m\Pi$, pove that if the inertia of the gas be neglected, and Boyle's law be supposed to hold throughout the ensuring motion, the radius of the sphere will oscillate between the values a and na, where n is determined by the equation $1 + 3m \log n - n^3 = 0$.

If m be nearly equal to 1, the time of an oscillation will be

$2\Pi\left(\frac{a^2\rho}{3\Pi}\right)^{\frac{1}{3}}$, ρ being the density of the fluid. [I.A.S. 84, 71]

Solution: At time t, let p be the pressure and v^1 the velocity at a distance r^1, then the equation of motion and continuity are

$$\frac{\partial \upsilon^1}{\partial t} + \upsilon^1\frac{\partial \upsilon^1}{\partial r^1} = -\frac{1}{\rho}\frac{\partial p}{\partial r^1} \quad \text{... (1)}$$

and $\quad r^{1^2}\upsilon^1 = f(t) = r^2\upsilon \quad \text{... (2)}$

From (2)

$$\frac{\partial \upsilon^1}{\partial t} = \frac{f^1(t)}{r^{1^2}}$$

Hence (1) becomes

$$\frac{f^1(t)}{r^{1^2}} + \upsilon^1 \frac{\partial \upsilon^1}{\partial r^1} = -\frac{1}{\rho}\frac{\partial p}{\partial r^1}$$

$$\Rightarrow \quad \frac{f^1(t)}{r^{1^2}} + \frac{\partial}{\partial r^1}\left(\frac{1}{2}\upsilon^{1^2}\right) = -\frac{1}{\rho}\frac{\partial p}{\partial r^1}$$

Integrating, we get

$$-\frac{f^1(t)}{r^1} + \frac{1}{2}\upsilon^{1^2} = -\frac{p}{\rho} + C \qquad \text{... (3)}$$

Initially when $r^1 = \infty$, $v^1 = 0$, $C = \dfrac{\Pi}{\rho}$, $p = \Pi$

So from (3) we have

$$\frac{-f^1(t)}{r^1} + \frac{1}{2}v^{1^2} = \frac{\Pi - p}{\rho} \qquad \text{... (4)}$$

From Boyle's law, we have

$pv = \text{const.}$

$$\Rightarrow \quad \frac{4}{3}\Pi r^3 p = \frac{4}{3}\Pi a^3 m\Pi$$

$$\Rightarrow \quad p = \frac{ma^3\Pi}{r^3}$$

When $r^1 = r$, then $\upsilon^1 = \upsilon$, $p^1 = p = \dfrac{ma^3\Pi}{r^3}$. So (4) becomes

$$\frac{-f^1(t)}{r} + \frac{1}{2}\upsilon^2 = \frac{\Pi}{\rho} - \frac{a^3\Pi m}{\rho}\frac{1}{r^3} \qquad \text{... (5)}$$

From (2)

$$\begin{aligned} f^1(t) &= \frac{2rdr}{dt}\upsilon + r^2\frac{d\upsilon}{dt} \\ &= 2r\upsilon^2 + r^2\frac{d\upsilon}{dr}\frac{dr}{dt} \\ &= 2r\upsilon^2 + r^2\upsilon\frac{d\upsilon}{dr} \end{aligned}$$

Hence (5) becomes

$$-\frac{1}{r}\left[2r\upsilon^2 + r^2\upsilon\frac{d\upsilon}{dr}\right] + \frac{\upsilon^2}{2} = \frac{\Pi}{\rho} - \frac{a^3 m\Pi}{\rho}\frac{1}{r^3}$$

$$\Rightarrow \quad r\upsilon\frac{d\upsilon}{dr} + \frac{3}{2}\upsilon^2 = -\frac{\Pi}{\rho} + \frac{a^3 m\Pi}{\rho r^3} \qquad \dots (6)$$

Multiplying both sides of (6) by $2r^2 dr$, we get

$$2r^3\upsilon d\upsilon + 3r^2\upsilon^2 dr = \left(-\frac{2\Pi r^2}{\rho} + \frac{2a^3 m\Pi}{\rho r}\right)dr$$

$$\Rightarrow \quad d\left(r^3\upsilon^2\right) = \left(\frac{-2\Pi r^2}{\rho} + \frac{2a^3 m\Pi}{\rho r}\right)dr$$

Integrating, we get

$$r^3\upsilon^2 = \frac{-2\Pi}{3\rho}r^3 + \frac{2a^3 m\Pi}{\rho}\log r + C_1 \qquad \dots (7)$$

Initially when $r = a$, $\upsilon = 0$

$$0 = -\frac{2\Pi}{3\rho}a^3 + \frac{2a^3 m\Pi}{\rho}\log a + C_1$$

$$\Rightarrow \quad C_1 = \frac{2\Pi a^3}{3\rho} = \frac{2a^3 m\Pi}{\rho}\log a$$

Hence from (7), we have

$$r^3\,\upsilon^2 = -\frac{2\Pi}{3\rho}r^3 + \frac{2a^3 m\Pi}{\rho}\log r + \frac{2\Pi a^3}{3\rho} - \frac{2a^3 m\Pi}{\rho}\log a$$

$$= \frac{2\Pi}{3\rho}\left(a^3 - r^3\right) + \frac{2a^3 m\Pi}{\rho}\log\frac{r}{a} \qquad \dots (8)$$

Since the radius of the sphere oscillates between a and na, it follows that at $r = a$, $\upsilon = 0$ and $r = na$, $\upsilon = 0$.

Putting $\upsilon = 0$ and $r = na$ in (8), we get

$$0 = \frac{2\Pi}{3\rho}\left[a^3 - n^3a^3 + 3a^3 m\log n\right]$$

$$\Rightarrow \quad 1 + 3m\log n - n^3 = 0 \qquad \text{Since } a \neq 0$$

Second Part. When m is nearly equal to (1).

Let $r = a + y$ where y is small.

Again $\upsilon = \dfrac{dr}{dt} = \dfrac{dy}{dt} = \dot{y}$

Hence taking $m = 1$, (8) reduces to

$$(a+y)^3 \dot{y}^2 = \frac{2\Pi}{3\rho}\left[a^3 - (a+y)^3\right] + \frac{2a^3\Pi}{\rho}\log\frac{a+y}{a}$$

$$\Rightarrow a^3\left(1+\frac{y}{a}\right)^3 y^2 = \frac{2\Pi a^3}{3\rho}\left[1-\left(1+\frac{y}{a}\right)^3\right] + \frac{2a^3\Pi}{\rho}\log\left(1+\frac{y}{a}\right)$$

$$\Rightarrow \dot{y}^2 = \frac{2\Pi}{3\rho}\left[1-\left(1+\frac{y}{a}\right)^3 + 3\log\left(1+\frac{y}{a}\right)\right]\left(1+\frac{y}{a}\right)^3$$

$$= \frac{2\Pi}{3\rho}\left[1-\left(1+\frac{3y}{a}+\frac{3y^2}{a^2}\right)+3\left(\frac{y}{a}-\frac{y^2}{2a^2}\right)\right]$$

$$\times\left(1-\frac{3y}{a}+.....\right)$$

$$\left(\text{Neglecting higher powers of } \frac{y}{a}\right)$$

$$= \frac{2\Pi}{3\rho}\left(-\frac{9y^2}{2a^2}\right)$$

$$\dot{y}^2 = -\frac{3\Pi y^2}{\rho a^2}$$

Differentiating with respect to y, we get

$$2\dot{y}\ddot{y} = \frac{-3\Pi}{\rho a^2}2y\dot{y}$$

$$\Rightarrow \ddot{y} = -\frac{3\Pi}{\rho a^2}y$$

Which is a S.H.M.

Hence time period $= T = 2\Pi\sqrt{\dfrac{a^2\rho}{3\Pi}}$

7. An infinite fluid in which a spherical hollow shell of radius a is initially at rest under the action of no forces. If a constant pressure Π is applied at infinity, show that the time of filling up the cavity is $\Pi^2 a\left(\frac{\rho}{\Pi}\right)^{\frac{1}{2}} 2^{\frac{5}{6}}\left[\Gamma\left(\frac{1}{3}\right)\right]^{-3}$. [I.A.S. 83]

Solution: Let v^1 be the velocity at a distance x at any time t, and let p be the pressure there.

The equations of motion and continuity are

$$\frac{\partial \upsilon}{\partial t}+\upsilon \frac{\partial \upsilon}{\partial x}=-\frac{1}{\rho} \frac{\partial p}{\partial x} \quad \ldots (1)$$

and

$$x^2 \upsilon = F(t) \quad \ldots (2)$$

$$\frac{\partial \upsilon}{\partial t}=\frac{F^1(t)}{x^2}$$

Hence from (1)

$$\frac{F^1(t)}{x^2}+\upsilon \frac{\partial \upsilon}{\partial x}=-\frac{1}{\rho} \frac{\partial p}{\partial x}$$

Integrating with respect to x, we get

$$-\frac{F^1(t)}{x}+\frac{\upsilon^2}{2}=-\frac{p}{\rho}+C \quad \ldots (3)$$

Initially when $x=\infty$, $\upsilon=0$, $p=\Pi$

Hence from (3)

$$C=\frac{\Pi}{\rho}$$

Thus (3) becomes

$$-\frac{F^1(t)}{x}+\frac{\upsilon^2}{2}=\frac{\Pi-p}{\rho} \quad \ldots (4)$$

Let u be the velocity and r be the radius of spherical cavity at any time t then $\dot{\upsilon}=u$, $x=r$, $p=0$, as pressure vanishes on the surface of cavity.

Hence from (4)

$$-\frac{F^1(t)}{r}+\frac{u^2}{2}=C=\frac{\Pi}{\rho} \quad \text{... (5)}$$

Also from (2) $r^2u = F(t)$

Hence (5) becomes

$$-\frac{F^1(t)}{r}+\frac{1}{2}\frac{F^2}{r^4}=\frac{\Pi}{\rho}$$

Multiplying both the sides by $2F\ dt = 2r^2dr$

$$-\frac{2FF^1}{r}+\frac{F^2}{r^2}dr=\frac{\Pi}{\rho}2r^2dr$$

$$\Rightarrow \frac{2FF^1}{r}-\frac{F^2}{r^2}dr=-\frac{\Pi}{\rho}2r^2dr$$

$$\Rightarrow d\left(\frac{F^2}{r}\right)=-\frac{\Pi}{\rho}2r^2dr$$

Integrating, we get

$$\frac{F^2}{r}=-\frac{2\Pi}{3\rho}r^3+C_1 \quad \text{... (6)}$$

Initially when $r = a$, so that $\dot{r}=u=0, F(t)=0$

Thus from (6)

$$0=-\frac{2\Pi a^3}{3\rho}+C_1$$

$$\Rightarrow C_1=\frac{2\Pi a^3}{3\rho}$$

Therefore, (6) becomes

$$\frac{F^2}{r}=-\frac{2\Pi}{3\rho}r^3+\frac{2\Pi a^3}{3\rho}$$

$$\Rightarrow \frac{F^2}{r}=\frac{2\Pi}{3\rho}\left(a^3-r^3\right)$$

$$\Rightarrow \quad \frac{r^4}{r}\left(\frac{dr}{dt}\right)^2 = \frac{2\Pi}{3\rho}\left(a^3 - r^3\right) \qquad \text{as } Fdt = r^2dr$$

$$\Rightarrow \quad r^3\left(\frac{dr}{dt}\right)^2 = \frac{2\Pi}{3\rho}\left(a^3 - r^3\right)$$

$$\Rightarrow \quad \frac{dr}{dt} = -\left[\frac{2\Pi}{3\rho}\left(a^3 - r^3\right)\right]^{\frac{1}{2}}$$

Here negative sign is taken as velocity increases when r decreases.

Let t be the time of filling up the cavity

$$\int_0^t dt = -\int_a^0 \sqrt{\frac{3\rho}{2\Pi}}\sqrt{\frac{r^3}{a^3 - r^3}}\,dr$$

$$\Rightarrow \quad t = \sqrt{\frac{3\rho}{2\Pi}}\int_0^a \sqrt{\frac{r^3}{a^3 - r^3}}\,dr$$

Let $r^3 = a^3 \sin^2 \theta$

$$3r^2dr = 2a^3 \sin\theta \cos\theta \, d\theta$$

When $r = a$, $\theta = \frac{\Pi}{2}$, $r = 0$, $\theta = 0$

Hence

$$t = \sqrt{\frac{3\rho}{2\Pi}}\int_0^{\frac{\Pi}{2}} \sqrt{\frac{a^3 \sin^2\theta}{a^3 - a^3\sin^2\theta}} \times \frac{2a^3}{3r^2}\sin\theta\cos\theta\, d\theta$$

$$t = \sqrt{\frac{3\rho}{2\Pi}}\,\frac{2a^3}{3}\int_0^{\frac{\Pi}{2}} \frac{\sin^2\theta}{a^2\sin^{\frac{4}{3}}\theta}\,d\theta$$

$$= \sqrt{\frac{3\rho}{2\Pi}}\,\frac{2a}{3}\int_0^{\frac{\Pi}{2}} (\sin\theta)^{\frac{2}{3}}$$

By applying Gamma function, we get

$$t = \sqrt{\frac{3\rho}{2\Pi}}\,\frac{2a}{3}\,\frac{\Gamma\left(\frac{5}{6}\right)}{2}\,\frac{\Gamma\left(\frac{1}{2}\right)}{\Gamma\left(\frac{5}{6}+\frac{1}{2}\right)}$$

$$\Rightarrow\quad t = \sqrt{\frac{3\rho}{2\Pi}}\,\frac{2a}{3}\,\frac{\Gamma\left(\frac{5}{6}\right)\Gamma\left(\frac{1}{2}\right)}{2\,\Gamma\left(\frac{4}{3}\right)}$$

$$\Rightarrow\quad t = \frac{a}{3}\sqrt{\frac{3\rho}{2\Pi}}\,\frac{\sqrt{\Pi}\,\Gamma\left(\frac{1}{2}+\frac{1}{3}\right)}{\Gamma\left(1+\frac{1}{3}\right)}$$

$$\Rightarrow\quad t = \frac{a}{3}\sqrt{\frac{3\rho}{2\Pi}}\,\frac{\sqrt{\Pi}\,\Gamma\left(\frac{1}{2}+\frac{1}{3}\right)}{\Gamma\left(1+\frac{1}{3}\right)} \qquad \ldots (7)$$

Also

$$\Gamma(n)\,\Gamma(1-n) = \frac{\Pi}{\sin n\Pi}\qquad \Gamma(n)\,\Gamma\left(n+\frac{1}{2}\right) = \frac{\sqrt{\Pi}\,\Gamma(2n)}{2^{2n-1}}$$

$$\text{For } n = \frac{1}{3},\ \Gamma\left(\frac{1}{3}\right)\Gamma\left(\frac{2}{3}\right) = \frac{\Pi}{\sin\frac{\Pi}{3}} = \frac{2\Pi}{\sqrt{3}}$$

and

$$\Gamma\left(\frac{1}{3}\right)\Gamma\left(\frac{5}{6}\right) = \frac{\sqrt{\Pi}\,\Gamma\left(\frac{2}{3}\right)}{2^{-\frac{1}{3}}}$$

$$\Rightarrow\quad \Gamma\left(\frac{5}{6}\right) = \frac{\sqrt{\Pi}\,\Gamma\left(\frac{2}{3}\right)2^{\frac{1}{3}}}{\Gamma\left(\frac{1}{3}\right)}$$

$$= \frac{\sqrt{\Pi}\, 2^{\frac{1}{3}}\, 2\Pi}{\sqrt{3}\left(\left|\frac{1}{3}\right.\right)^2} = \frac{2^{\frac{4}{3}}\Pi^{\frac{3}{2}}}{\sqrt{3}\left(\left|\frac{1}{3}\right.\right)^2}$$

Using this in (7), we get

$$t = \frac{a}{3}\sqrt{\frac{3\rho}{2\Pi}}^{\sqrt{\Pi}} \frac{2^{\frac{4}{3}}\Pi^{\frac{3}{2}}}{\sqrt{3}\left(\left|\frac{1}{3}\right.\right)^2}$$

$$= \Pi^2 a\left(\frac{\rho}{\Pi}\right)^{\frac{1}{2}} 2^{\frac{5}{6}}\left(\left|\frac{1}{3}\right.\right)^{-2}$$

8. A sphere is at rest in an infinite mass of homogeneous liquid of density ρ, the pressure at infinity being Π, show that if the radius R of the sphere varies in any manner, the pressure at the surface of the sphere at any time is

$$\Pi + \frac{1}{2}\rho\left[\frac{d^2R^2}{dt^2} + \left(\frac{dR}{dt}\right)^2\right]$$

If $R = a\,(2 + \cos nt)$, show that, to prevent cavitation in the fluid, Π must not be less than $3\rho a^2 n^2$.

Solution: Let υ be the velocity of a fluid particle at a distance r from the centre at any time t.

Equations of motion and continuity are

$$\frac{\partial v}{\partial l} + \upsilon\frac{\partial \upsilon}{\partial r} = -\frac{1}{\rho}\frac{\partial p}{\partial r} \qquad \text{... (1)}$$

and $r^2\upsilon$ = const = $F(t)$ (Let)

Differentiating it with respect to t, we get

$$r^2\frac{\partial \upsilon}{\partial t} = F^1(t)$$

$$\Rightarrow \quad \frac{\partial \upsilon}{\partial t} = \frac{F^1(t)}{r^2}$$

So from (1), we have

$$\frac{F^1(t)}{r^2} + \upsilon \frac{\partial \upsilon}{\partial r} = -\frac{1}{\rho}\frac{\partial p}{\partial r}$$

$$\Rightarrow \quad \frac{F^1(t)}{r^2} + \frac{\partial}{\partial r}\left(\frac{1}{2}\upsilon^2\right) = -\frac{\partial}{\partial r}\left(\frac{p}{\rho}\right) \quad \text{as } \rho \text{ is constant}$$

Integrating with respect to r, we get

$$\frac{-F^1(t)}{r} + \frac{1}{2}\upsilon^2 = -\frac{p}{\rho} + C \qquad \dots (2)$$

Initially when $r = \infty$, $\upsilon = 0$, $p = \Pi$

Thus from (2)

$$C = \frac{\Pi}{\rho}$$

Hence (2) becomes

$$\frac{-F^1(t)}{r} + \frac{1}{2}\upsilon^2 = \frac{\Pi - p}{\rho}$$

$$\Rightarrow \quad \rho\left[\frac{-F^1(t)}{r} + \frac{1}{2}\upsilon^2\right] = \Pi - p$$

$$p = \Pi - \rho\left[\frac{-F^1(t)}{r} + \frac{1}{2}\upsilon^2\right]$$

$$\Rightarrow \quad p = \Pi + \frac{\rho}{2}\left[\frac{2F^1(t)}{r} - \upsilon^2\right] \qquad \dots (3)$$

Let P be the pressure on the surface of the sphere of radius R and v be the velocity then from (3), we get

$$p = \Pi + \frac{\rho}{2}\left[\frac{2F^1(t)}{R} - V^2\right] \qquad \dots (4)$$

Also equation of continuity becomes

$$R^2 V = F(t)$$

$$\Rightarrow \quad F(t) = R^2 V = R^2 \frac{dR}{dt}$$

Differentiating with respect to t, we get

$$F^1(t) = R^2 \frac{d^2R}{dt^2} + 2R\left(\frac{dR}{dt}\right)^2$$

Hence from (4), we get

$$P = \Pi + \frac{\rho}{2}\left[2R\frac{d^2R}{dt^2} + 4\left(\frac{dR}{dt}\right)^2 - V^2\right]$$

$$= \Pi + \frac{\rho}{2}\left[2R\frac{d^2R}{dt^2} + 4\left(\frac{dR}{dt}\right)^2 - \left(\frac{dR}{dt}\right)^2\right]$$

$$= \Pi + \frac{\rho}{2}\left[2R\frac{d^2R}{dt^2} + 3\left(\frac{dR}{dt}\right)^2\right] \qquad \dots (5)$$

$$= \Pi + \frac{\rho}{2}\left[2\left\{R\frac{d^2R}{dt^2} + \left(\frac{dR}{dt}\right)^2\right\} + \left(\frac{dR}{dt}\right)^2\right]$$

$$= \Pi + \frac{\rho}{2}\left[\frac{d^2}{dt^2}\left(R^2\right) + \left(\frac{dR}{dt}\right)^2\right]$$

Second Part. Here $R = a\ (2 + \text{const.})$

Let there be no cavitation in the fluid everywhere on the surface so that $p > 0$.

Then we are to prove that $\Pi > 3\rho a^2 n^2$

Since

$$R = a\ (2 + \cos nt)$$

$$\dot{R} = -\ na \sin nt$$

$$\ddot{R} = -\ an^2 \cos nt$$

Now

$$2R\ddot{R} + 3\dot{R}^2$$

$$= 2a(2 + \cos nt)\left(-an^2 \cos nt\right) + 3n^2a^2 \sin^2 nt$$

$$= a^2n^2\left[-4 \cos nt - 2\cos^2 nt + 3 \sin^2 nt\right]$$

$$= a^2n^2\left[-4\cos nt - 2\left(1-\sin^2 nt\right) + 3\sin^2 nt\right]$$

$$= a^2n^2\left[-4\cos nt - 2 + 5\sin^2 nt\right]$$

Using it in (5), it becomes

$$p = \Pi + \frac{1}{2}\rho a^2n^2\left(-4\cos nt - 2 + 4\sin^2 nt\right) \qquad \text{... (6)}$$

As cos nt varies from -1 to 1 so R varies from a to $3a$. The sphere shrinks from $R = 3a$ to $R = a$ and so there is a possibility of cavitation. Also p is minimum when $nt = 0$ or $2m\Pi$.

$$\text{Hence min } p = \Pi + \frac{1}{2}\rho a^2n^2(-4-2)$$

$$= \Pi - 3\rho a^2n^2$$

Also $p > 0$

$\Rightarrow \Pi - 3\rho\, a^2n^2 > 0$

$\Rightarrow \Pi > 3\rho\, a^2n^2$

9. Air obeying Boyle's law, is in motion in a uniform tube of small section, prove that if ρ be the density and υ be the velocity at a distance x from a fixed point at time t

$$\frac{\partial^2\rho}{\partial t^2} = \frac{\partial^2}{\partial x^2}\left\{\rho\left(\upsilon^2 + K\right)\right\}$$

Solution: Let p be the pressure and υ be the velocity at a distance x from the end of the tube at any time t. The equations of motion and continuity are

$$\frac{\partial\upsilon}{\partial t} + \upsilon\frac{\partial\upsilon}{\partial x} = -\frac{1}{\rho}\frac{\partial p}{\partial x} \qquad \text{... (1)}$$

and

$$\frac{\partial\rho}{\partial t} + \frac{\partial}{\partial x}(\rho\upsilon) = 0 \qquad \text{... (2)}$$

Since the air obeys Boyle's law, hence $p = K\rho$ ... (3)

From (1) and (3), we get

$$\frac{\partial \upsilon}{\partial t} + \upsilon \frac{\partial \upsilon}{\partial x} = \frac{-K}{\rho} \frac{\partial \rho}{\partial x} \qquad \dots (4)$$

Differentiating (2) with respect to t, we get

$$\frac{\partial^2 \rho}{\partial t^2} + \frac{\partial}{\partial t}\left\{\frac{\partial}{\partial x}(\rho\upsilon)\right\} = 0$$

$$\Rightarrow \quad \frac{\partial^2 \rho}{\partial t^2} + \frac{\partial}{\partial x}\left\{\frac{\partial}{\partial t}(\rho\upsilon)\right\} = 0$$

$$\Rightarrow \quad \frac{\partial^2 \rho}{\partial t^2} + \frac{\partial}{\partial x}\left(\rho \frac{\partial \upsilon}{\partial t} + \upsilon \frac{\partial \rho}{\partial t}\right) = 0 \qquad \dots (5)$$

From (2) and (4), (5) becomes

$$\frac{\partial^2 \rho}{\partial t^2} + \frac{\partial}{\partial x}\left[\rho\left(-\frac{K}{\rho}\frac{\partial \rho}{\partial x} - \upsilon \frac{\partial \upsilon}{\partial x}\right) + \upsilon\left\{\frac{-\partial}{\partial x}(\rho\upsilon)\right\}\right] = 0$$

$$\Rightarrow \quad \frac{\partial^2 \rho}{\partial t^2} = \frac{\partial}{\partial x}\left[K \frac{\partial \rho}{\partial x} + \rho\upsilon \frac{\partial \upsilon}{\partial x} + \upsilon \frac{\partial}{\partial x}(\rho\upsilon)\right]$$

$$\Rightarrow \quad \frac{\partial^2 \rho}{\partial t^2} = \frac{\partial}{\partial x}\left[\frac{\partial}{\partial x}(\rho\upsilon \cdot \upsilon) + K \frac{\partial \rho}{\partial x}\right]$$

$$= \frac{\partial}{\partial x}\left\{\frac{\partial}{\partial x}\left[K\rho + \rho\upsilon^2\right]\right\}$$

$$= \frac{\partial^2}{\partial x^2}\left[\rho\left(K + \upsilon^2\right)\right]$$

10. A pulse travelling along a fine straight uniform tube filled with gas causes the density at time t and distance x from the origin where the velocity is u_0 to become $\rho_0\, \phi\, (\upsilon t - x)$. Prove that the velocity u (at time t and distance x from the origin) is given by $\upsilon + \dfrac{(u_0 - \upsilon)\,\phi\,(\upsilon t)}{\phi\,(\upsilon t - x)}$.

Solution: Let u be the velocity and ρ be the density of the gas at a distance x then

$$\rho = \rho_0 \phi(\upsilon t - x) \qquad \dots (1)$$

The equation of continuity is

$$\frac{\partial \rho}{\partial t} + \frac{\partial}{\partial x}(\rho u) = 0$$

$$\Rightarrow \quad \frac{\partial \rho}{\partial t} + \rho \frac{\partial u}{\partial x} + u \frac{\partial \rho}{\partial x} = 0 \qquad \text{... (2)}$$

Differentiating (1) partially with respect to t and x respectively, we get

$$\frac{\partial \rho}{\partial t} = \rho_0 \upsilon \phi'(\upsilon t - x)$$

$$\frac{\partial \rho}{\partial x} = -\rho_0 \phi'(\upsilon t - x)$$

Putting these values in (2), we get

$$\rho_0 \upsilon \phi'(\upsilon t - x) + \rho_0 \phi(\upsilon t - x) \frac{\partial u}{\partial x} - u \rho_0 \phi'(\upsilon t - x) = 0$$

$$\Rightarrow \quad (\upsilon - u) \rho_0 \phi'(\upsilon t - x) + \rho_0 \phi(\upsilon t - x) \frac{\partial u}{\partial x} = 0$$

$$\Rightarrow \quad (\upsilon - u) \phi'(\upsilon t - x) + \phi(\upsilon t - x) \frac{\partial u}{\partial x} = 0$$

$$\Rightarrow \quad \frac{du}{\upsilon - u} + \frac{\phi'(\upsilon t - x)}{\upsilon t - x} dx = 0$$

Integrating, we get

$$-\log(\upsilon - u) - \log\{\phi(\upsilon t - x)\} = -\log A$$

$$\Rightarrow \quad (\upsilon - u) \phi(\upsilon t - x) = A$$

When $x = 0$, $u = u_0$

$$(\upsilon - u_0)\, \phi(\upsilon t) = A$$

Hence

$$(\upsilon - u) \phi(\upsilon t - x) = (\upsilon - u_0) \phi(\upsilon t)$$

$$\Rightarrow \quad \upsilon - u = \frac{(\upsilon - u_0) \phi(\upsilon t)}{\phi(\upsilon t - x)}$$

$$\Rightarrow \quad u = \upsilon + \frac{(u_0 - \upsilon) \phi(\upsilon t)}{\phi(\upsilon t - x)}$$

11. A stream in a horizontal pipe after passing a contraction in the pipe at which its sectional area is A, is delivered at atmospheric pressure at a place where the sectional area is B. Show that if a side tube is connected with the pipe at the former place, water will be sucked up through it into the pipe from a reservoir at a depth $\frac{S^2}{2g}\left(\frac{1}{A^2}-\frac{1}{B^2}\right)$ below the pipe, S being the delivery per second.

Solution: Let υ and V be the velocities of the stream at two cross sections. The equation of continuity is given by

Flux at the first cross section = flux at the second cross section

$$\Rightarrow \quad A\upsilon\rho = BV\rho = S \text{ (given)}$$

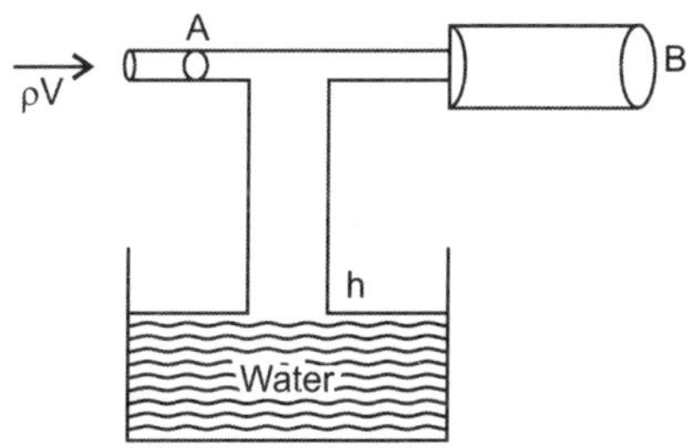

Fig. 2.9

Also $\rho = 1$, for stream

Hence $\upsilon = \frac{S}{A}, V = \frac{S}{B}$

The equation of motion is

$$\upsilon\frac{\partial \upsilon}{\partial r} = -\frac{1}{\rho}\frac{\partial p}{\partial r}$$

Integrating, we get

$$\frac{\upsilon^2}{2} = -\frac{p}{\rho} + C \quad \text{... (1)}$$

When $\upsilon = V$, $p = \Pi$

Thus from (1)

$$\frac{V^2}{2} = -\frac{\Pi}{\rho} + C \qquad \text{... (2)}$$

Subtracting (2) from (1), we get

$$\frac{1}{2}\left(\upsilon^2 - V^2\right) = \frac{1}{\rho}(\Pi - p) = \Pi - \rho \qquad \text{... (3)}$$

Since $\rho = 1$

Let h be the height of water column in the side table which is sucked from a reservoir, then $\Pi - p$ = difference of pressure $= \rho g h = gh$ as $\rho = 1$

Hence from (3), we have

$$\frac{1}{2}\left(\upsilon^2 - V^2\right) = gh$$

$$\Rightarrow \quad h = \frac{1}{2g}\left(\upsilon^2 - V^2\right)$$

$$= \frac{S^2}{2g}\left(\frac{1}{A^2} - \frac{1}{B^2}\right) = \frac{1}{2g}\left(\frac{S^2}{A^2} - \frac{S^2}{B^2}\right)$$

12. An infinite mass of fluid is acted on by a force $\mu r^{-3/2}$ per unit mass directed to the origin. If initially the fluid is at rest and there is a cavity in the form of the sphere $r = c$ in it, show that the cavity will be filled up after an interval of time $\left(\frac{2}{5\mu}\right)^{\frac{1}{2}} c^{5/4}$

Solution: Let υ be the velocity, p the pressure at a distance x from the origin, then the equations of motion and continuity are

$$\frac{\partial \upsilon}{\partial t} + \upsilon\frac{\partial \upsilon}{\partial x} = -\mu x^{-3/2} - \frac{1}{\rho}\frac{\partial p}{\partial x} \qquad \text{... (1)}$$

and

$$x^2\upsilon = F(t) \qquad \text{... (2)}$$

$$\Rightarrow \quad \upsilon = \frac{F(t)}{x^2}$$

$$\therefore \quad \frac{\partial \upsilon}{\partial t} = \frac{F^1(t)}{x^2}$$

Thus (1) becomes

$$\frac{F^1(t)}{x^2} + \frac{\partial}{\partial x}\left(\frac{1}{2}\upsilon^2\right) = -\mu x^{-3/2} - \frac{1}{\rho}\frac{\partial p}{\partial x}$$

Integrating, we get

$$-\frac{F^1(t)}{x} + \frac{1}{2}\upsilon^2 = 2\mu x^{-1/2} - \frac{p}{\rho} + C \qquad \text{... (3)}$$

Initially when $r^1 = \infty$, $\upsilon^1 = 0$, $p = 0$

Thus from (3) $C = 0$

Hence (3) becomes

$$-\frac{F^1(t)}{x} + \frac{1}{2}\upsilon^2 = 2\mu x^{-1/2} - \frac{p}{\rho}$$

Again when $x = r$ (radius of cavity), $p = 0$, $\upsilon = \frac{dr}{dt} = \dot{r}$

Therefore, from (3)

$$\frac{-F^1(t)}{r} + \frac{1}{2}r^2 = \frac{2\mu}{\sqrt{r}} - \frac{p}{\rho} \qquad \text{... (4)}$$

Also from (2) $r^2\dot{r} = F(t)$

$\Rightarrow$ $r^2 dr = F(t)dt$

Multiplying both sides of (4) by $2F(t)dt = 2r^2 dr$

We get

$$\frac{-2F^1(t)\,F(t)dt}{r} + \frac{1}{2}r^2 2r^2 dr = \frac{2\mu}{\sqrt{r}}2r^2 dr$$

$$\Rightarrow \quad \frac{-2F^1(t)F(t)dt}{r} + \frac{1}{2}\frac{F^2(t)dr}{r^2} = \frac{4\mu}{\sqrt{r}}r^2 dr$$

$$\Rightarrow \quad d\left[\frac{-F^2(t)}{r}\right] = 4\mu r^{3/2} dr$$

Integrating, we get

$$\frac{-F^2(t)}{r} = 4\mu\frac{2}{5}r^{5/2} + C_1 \qquad \text{... (5)}$$

When $r = C$, $\upsilon = 0$, so that $F(t) = 0$

$$C_1 = -\frac{8\mu}{5}C^{\frac{5}{2}}$$

Hence from (5)

$$-\frac{F^2(t)}{r} = \frac{8\mu}{5}\left(r^{\frac{5}{2}} - C^{\frac{5}{2}}\right)$$

$$\Rightarrow \quad \frac{-r^4 r^2}{r} = \frac{8\mu}{5}\left(r^{\frac{5}{2}} - C^{\frac{5}{2}}\right)$$

$$\Rightarrow \quad \left(\frac{dr}{dt}\right)^2 = \frac{8\mu}{5r^3}\left(C^{\frac{5}{2}} - r^{\frac{5}{2}}\right)$$

$$\Rightarrow \quad \frac{dr}{dt} = -\left[\frac{8\mu}{5r^3}\left(C^{\frac{5}{2}} - r^{\frac{5}{2}}\right)\right]^{\frac{1}{2}}$$

Here negative sign is taken as υ increases when r decreases. If t be the time of filling the cavity then

$$\int_0^t dt = -\sqrt{\frac{5}{8\mu}}\int_C^0 \frac{r^{3/2}dr}{\sqrt{C^{5/2} - r^{5/2}}}$$

$$\text{or} \quad t = \sqrt{\frac{5}{8\mu}}\int_0^C \frac{r^{3/2}dr}{\sqrt{C^{5/2} - r^{5/2}}}$$

Let $r^{5/2} = C^{5/2}\sin^2\theta$

When $r = C$, $\theta = \dfrac{\Pi}{2}$

$r = 0$, $\theta = 0$

$$\frac{5}{2}r^{\frac{3}{2}}dr = C^{\frac{5}{2}}2\sin\theta\cos\theta\, d\theta$$

$$t=\sqrt{\frac{5}{8\mu}}\int_0^{\frac{\Pi}{2}}\frac{\frac{4}{5}C^{5/2}\sin\theta\cos\theta\,d\theta}{C^{5/4}\cos\theta}$$

$$=\frac{4}{5}\sqrt{\frac{5}{8\mu}}\,C^{\frac{5}{4}}\int_0^{\frac{\Pi}{2}}\sin\theta\,d\theta$$

$$\Rightarrow\quad t-\frac{4}{5}C^{5/4}\sqrt{\frac{5}{8\mu}}$$

$$\Rightarrow\quad t=\sqrt{\frac{2}{5\mu}}\,C^{\frac{5}{4}}$$

13. A mass of homogeneous liquid is moving so that the velocity at any point is proportional to the time, and that the pressure is given by

$$\frac{p}{\rho}=\mu xyz-\frac{1}{2}t^2\left(y^2z^2+z^2x^2+x^2y^2\right)$$

Prove that this motion may have been generated from rest by finite natural forces independent of the time, and show that if the direction of motion at every point coincides with the direction of acting forces, each particle of the liquid describes a curve which is the intersection of two hyperbolic cylinders.

Solution: Given, velocity is proportional to time i.e.

$$\vec{q}=\lambda t \qquad \dots(1)$$

Also given

$$\frac{p}{\rho}=\mu xyz-\frac{1}{2}t^2\left(y^2z^2+z^2x^2+x^2y^2\right) \qquad \dots(2)$$

Let the motion be generated from rest by finite natural forces $\vec{F}$ (conservative), then there exists velocity potential ϕ such that $\vec{q}=-\vec{\nabla}\phi$. To prove that $\vec{F}$ is independent of time t.

By pressure equation

$$\frac{p}{\rho}+\frac{1}{2}q^2-\frac{\partial\phi}{\partial t}+V=F(t)$$

$$\Rightarrow \quad \frac{p}{\rho} = \frac{\partial \phi}{\partial t} - \frac{1}{2}\lambda^2 t^2 - V + F(t) \qquad \text{... (3)}$$

Comparing (2) and (3), we have on equating the coefficients of t^2,

$$\lambda^2 = y^2 z^2 + z^2 x^2 + x^2 y^2$$

and then

$$q^2 = t^2 \lambda^2 = t^2 \left(y^2 z^2 + z^2 x^2 + x^2 y^2\right) \qquad \text{... (4)}$$

Also

$$q^2 = \left(\frac{\partial \phi}{\partial x}\right)^2 + \left(\frac{\partial \phi}{\partial y}\right)^2 + \left(\frac{\partial \phi}{\partial z}\right)^2 \qquad \text{... (5)}$$

Comparing (4) and (5), we get

$$\phi = t\, xyz$$

Thus (3) becomes

$$\frac{p}{\rho} = xyz - \frac{1}{2} t^2 \left(y^2 z^2 + z^2 x^2 + x^2 y^2\right) - V + F(t) \qquad \text{... (6)}$$

Comparing (2) and (6), we get

$$F(t) = 0, \; xyz - V = \mu xyz$$

or $V = xyz\,(1 - \mu)$

If u, v, w are velocity components and X, Y, Z be the components of the forces, then

$$u = \frac{-\partial \phi}{\partial x} = -\, tyz, \; v = \frac{-\partial \phi}{\partial y} = -\, txz, \; w = \frac{-\partial \phi}{\partial z} = -\, txy$$

Also

$$X = -\frac{\partial V}{\partial x} = (\mu - 1)\, yz$$

$$Y = -\frac{\partial V}{\partial y} - (\mu - 1)\, xz$$

$$Z = -\frac{\partial V}{\partial z} = (\mu - 1)\, xy$$

It is given that the direction of motion coincides with the direction of acting force.

$$\therefore \quad \frac{u}{X} = \frac{v}{Y} = \frac{w}{Z} \qquad \text{... (7)}$$

Equations of streamlines are

$$\frac{dx}{u} = \frac{dy}{v} = \frac{dz}{w}$$

Using (7), we get

$$\frac{dx}{X} = \frac{dy}{Y} = \frac{dz}{Z}$$

$$\Rightarrow \quad \frac{dx}{(-1+\mu)\,yz} = \frac{dy}{(\mu-1)\,zx} = \frac{dz}{(\mu-1)\,xy}$$

$$\Rightarrow \quad \frac{dx}{yz} = \frac{dy}{zx} = \frac{dz}{xy}$$

$$\Rightarrow \quad xdx = ydy = zdz \qquad \text{... (8)}$$

Taking first two members of (8), we get

$$xdx = ydy$$

Integrating $x^2 - y^2 = a^2$

Taking first and last members of (8), we have

$$xdx = zdx$$

Integrating, we get

$$x^2 - z^2 = b^2$$

The equations $x^2 - y^2 = a^2$, $x^2 - z^2 = b^2$ represent two distinct hyperbolic cylinders. Hence the result.

14. A mass of liquid surrounds a solid sphere of radius a, and its outer surface, which is a concentric sphere of radius b, is subject to a given constant pressure Π, no other forces being in action on the liquid. The solid sphere suddenly shrinks into a concentric sphere, it is required to determine the subsequent motion and the impulsive action on the sphere.

Solution: At time t, let υ be the velocity at a distance x from the centre, p be the pressure and ρ be density of the fluid.

The equations of motion and continuity are

$$\frac{\partial \upsilon}{\partial t} + \upsilon \frac{\partial \upsilon}{\partial x} = -\frac{1}{\rho}\frac{\partial p}{\partial x} \qquad \text{... (1)}$$

and

$$x^2\upsilon = F(t) \qquad \text{... (2)}$$

$$\frac{\partial \upsilon}{\partial t} = \upsilon^1 = \frac{F^1(t)}{x^2}$$

Hence (1) becomes

$$\frac{F^1(t)}{x^2} + \frac{\partial}{\partial x}\left(\frac{1}{2}\upsilon^2\right) = -\frac{\partial}{\partial x}\left(\frac{p}{\rho}\right)$$

Integrating with respect to x, we get

$$-\frac{F^1(t)}{x} + \frac{1}{2}\upsilon^2 = \frac{-p}{\rho} + C \qquad \text{... (3)}$$

Since the liquid is contained between two spheres $r = a$, $r = b$, so we assume that r and R are internal and external radii at any time t and the corresponding velocities are u and U respectively.

When $x = r,\ \upsilon = u = r,\ p = 0$

as pressure vanishes on the internal boundary.

Hence from (3), we get

$$\frac{-F^1(t)}{x} + \frac{1}{2}u^2 = 0 + C \qquad \text{... (4)}$$

Also when $x = R,\ v = R = U,\ p = \Pi$

Hence from (3), we have

$$\frac{-F^1(t)}{R} + \frac{1}{2}U^2 = -\frac{\Pi}{\rho} + C \qquad \text{... (5)}$$

Also from (2)

$r^2U = F(t) = R^2U$

Substracting (5) from (4), we get

$$F^1(t)\left(\frac{1}{R} - \frac{1}{r}\right) + \frac{1}{2}\left(u^2 - U^2\right) = \frac{\Pi}{\rho}$$

$$\Rightarrow\ F^1(t)\left(\frac{1}{R} - \frac{1}{r}\right) + \frac{1}{2}\left[\frac{F^2(t)}{r^4} - \frac{F^2(t)}{R^4}\right] = \frac{\Pi}{\rho}$$

$$\Rightarrow \quad F^1(t)\left(\frac{1}{R}-\frac{1}{r}\right)+\frac{F^2}{2}\left(\frac{1}{r^4}-\frac{1}{R^4}\right)=\frac{\Pi}{\rho}$$

Multiplying both the sides by

$$2r^2u = 2F(t) = 2R^2U$$

$$\Rightarrow \quad 2r^2dr = 2F(t)dt = 2R^2dR \qquad \text{as } u = \frac{dr}{dt}$$

$$\Rightarrow \quad 2F(t)F^1(t)\left(\frac{1}{R}-\frac{1}{r}\right)+F^2\left(\frac{dr}{r^2}-\frac{dR}{R^2}\right)=\frac{\Pi}{\rho}2r^2dr$$

$$\Rightarrow \quad d\left[\left(\frac{1}{R}-\frac{1}{r}\right)F^2(t)\right]=\frac{2\Pi}{\rho}r^2dr$$

Integrating, we get

$$\left(\frac{1}{R}-\frac{1}{r}\right)F^2(t)=\frac{2}{3}\frac{\Pi}{\rho}r^3+C \qquad \dots (6)$$

When $r = a$, $\upsilon = 0$, so that $F(t) = 0$

Hence from (6)

$$C=\frac{-2\Pi}{3\rho}a^3$$

Therefore, from (6), we get

$$\left(\frac{1}{R}-\frac{1}{r}\right)F^2(t)=\frac{2\Pi}{3\rho}\left(r^3-a^3\right)$$

$$\Rightarrow \quad \frac{R-r}{rR}\left(r^2u\right)^2=\frac{2\Pi}{3\rho}\left(a^3-r^3\right)$$

$$\Rightarrow \quad \left(r^3u^2\right)\left(\frac{R-r}{r}\right)=\frac{2\Pi}{3\rho}\left(a^3-r^3\right)$$

With $R^3 - r^3 = b^3 - a^3$

Since total mass of liquid is constant

So volume of the liquid of any time t = volume of the liquid initially

$$\Rightarrow \quad \frac{4}{3}\Pi R^3-\frac{4}{3}\Pi r^3=\frac{4}{3}\Pi b^3-\frac{4}{3}\Pi a^3$$

To find the equation of impulsive action

Let x be the radius of the solid sphere and $\overline{\omega}$ be the impulsive pressure at a distance r, then equation of impulsive action is

$$d\overline{\omega} = \rho v dx = \frac{\rho F}{x^2} dx$$

$$\Rightarrow \quad \int_0^{\overline{\omega}} d\omega = \int_r^R \frac{F dx}{x^2}$$

$$\Rightarrow \quad \overline{\omega} = -\rho F\left(\frac{1}{R} - \frac{1}{r}\right)$$

$$\Rightarrow \quad \overline{\omega} = \rho r^2 u\left(\frac{1}{r} - \frac{1}{R}\right) \qquad \dots (7)$$

The whole impulse on the surface of the sphere

$$= 4\Pi r^2 \overline{\omega}$$

$$= 4\Pi r^2 \rho r^2 u\left(\frac{1}{r} - \frac{1}{R}\right) \quad \text{From (7)}$$

$$= 4\Pi \rho u r^3\left(\frac{R - r}{R}\right)$$

15. A portion of homogeneous fluid is confined between two concentric spheres of radii A and a and is attracted towards their centre by a force varying inversely as the square of the distance. The inner spherical surface is suddenly annihilated and when the radii of the inner and outer surfaces of the fluid are r and R, the fluid impinges on a solid ball concentric with their surfaces, prove that the impulsive pressure at any point of the ball for different values of R and r varies as

$$\left[\left(a^2 - r^2 - A^2 + R^2\right)\left(\frac{1}{r} - \frac{1}{R}\right)\right]^{\frac{1}{2}}$$

Solution: Let at any time t, p be the pressure and υ be the velocity at a distance x from the centre. The equations of motion and continuity are

$$\frac{\partial \upsilon}{\partial t} + \upsilon \frac{\partial \upsilon}{\partial x} = F - \frac{1}{\rho}\frac{\partial p}{\partial x} = \frac{-\mu}{x^2} - \frac{1}{\rho}\frac{\partial p}{\partial x} \qquad \text{... (1)}$$

and

$$x^2\upsilon = F(t)$$

[as μ/x^2 is a force towards the centre]

$$\therefore \quad \frac{\partial \upsilon}{\partial t} = \frac{F^1(t)}{x^2}$$

Hence (1) becomes

$$\frac{F^1(t)}{x^2} + \upsilon\frac{\partial \upsilon}{\partial x} = \frac{-\mu}{x^2} - \frac{1}{\rho}\frac{\partial p}{\partial x}$$

Integating with respect to x, we get

$$\frac{-F^1(t)}{x} + \frac{\upsilon^2}{2} = \frac{\mu}{x} - \frac{p}{\rho} + C \qquad \text{... (2)}$$

Let R and r be the external and internal radii at any time t. Let u and U be the velocities thereat.

When $x = r$, $\upsilon = r = u$, $p = 0$

Hence (2) becomes

$$\frac{-F^1(t)}{r} + \frac{u^2}{2} = \frac{\mu}{r} + C \qquad \text{... (3)}$$

Again when $x = R$, $\upsilon = R = U$, $p = 0$

Thus (2) becomes

$$\frac{-F^1(t)}{R} + \frac{U^2}{2} = \frac{\mu}{R} + C \qquad \text{... (4)}$$

Subtracting (4) from (3), we get

$$F^1(t)\left(\frac{1}{R} - \frac{1}{r}\right) + \frac{1}{2}\left(u^2 - U^2\right) = \mu\left(\frac{1}{r} - \frac{1}{R}\right) \qquad \text{... (5)}$$

Also

$x^2\upsilon = F(t)$ gives

$r^2u = F(t) = R^2U$

Therefore, from (5), we have

$$F^1(t)\left(\frac{1}{R}-\frac{1}{r}\right)+\frac{F}{2}\left(\frac{1}{r^4}-\frac{1}{R^4}\right)=\mu\left(\frac{1}{r}-\frac{1}{R}\right) \qquad \dots (6)$$

Multiplying both sides by $2r^2u = 2F(t) = 2R^2U$

$$2r^2dr = 2F(t)dt = 2R^2dR \qquad \text{as } u = \frac{dr}{dt}$$

So from (6), we get

$$2FF^1(t)\left(\frac{1}{R}-\frac{1}{r}\right)dt + F^2\left(\frac{dr}{r^2}-\frac{dR}{R^2}\right)=2\mu\,(rdr - RdR)$$

$$\Rightarrow \quad d\left[\left(\frac{1}{R}-\frac{1}{r}\right)F^2\right]=2\mu\,(rdr - RdR)$$

Integrating, we get

$$\left(\frac{1}{R}-\frac{1}{r}\right)F^2=\mu\left(r^2-R^2\right)+C_1 \qquad \dots (7)$$

When $r = a$, $R = A$, $F(t) = 0$

Hence from (7), we get

$$C_1 = -\,\mu\,(a^2 - A^2)$$

Thus (7) becomes

$$\left(\frac{1}{R}-\frac{1}{r}\right)F^2=\mu\left(r^2-R^2-a^2+A^2\right) \qquad \dots (8)$$

Let $\bar{\omega}$ be the impulsive pressure at a distance x, then

$$d\bar{\omega}=\rho\upsilon dx=\frac{\rho F(t)}{x^2}dx$$

$$\Rightarrow \quad \int_0^{\bar{\omega}} d\bar{\omega}=\int_r^R \rho F(t)\frac{dx}{x^2}$$

$$\Rightarrow \quad \bar{\omega}=\rho F(t)\left(\frac{1}{r}-\frac{1}{R}\right) \qquad \dots (9)$$

Putting the values of $F(t)$ from (8), (9) becomes

$$\bar{\omega}=\rho\left(\frac{1}{r}-\frac{1}{R}\right)\left[\frac{\mu\left(r^2-R^2-a^2+A^2\right)}{\frac{1}{R}-\frac{1}{r}}\right]^{\frac{1}{2}}$$

$$\Rightarrow \quad \bar{\omega} = \rho\left[\mu\left(a^2 - r^2 - A^2 + R^2\right)\left(\frac{1}{r} - \frac{1}{R}\right)\right]^{\frac{1}{2}}$$

$$\Rightarrow \quad \bar{\omega} \text{ varies as } \left[\left(a^2 - r^2 - A^2 + R^2\right)\left(\frac{1}{r} - \frac{1}{R}\right)\right]^{\frac{1}{2}}$$

16. A mass of liquid of density ρ whose external surface is a long circular cylinder of radius *a*, which is subject to a constant pressure Π, surrounds a coaxial long circular cylinder of radius *b*. The internal cylinder is suddenly destroyed show that if υ is the velocity at the internal surface when its radius is *r*, then

$$\upsilon^2 = \frac{2\Pi\left(b^2 - r^2\right)}{\rho r^2 \log \dfrac{r^2 + a^2 - b^2}{r^2}}$$

Solution: When the internal cylinder is destroyed suddenly the motion of the liquid will be along the radii of the normal sections of the cylinder. The velocity will be a function of *r* and time *t* only. Let *u* be the velocity and *p* be the pressure at any point distance *x* from the axis of the cylinder.

Hence the equations of continuity and motion are

$$xu = F(t) \qquad \dots (1)$$

and

$$\frac{\partial u}{\partial t} + u\frac{\partial u}{\partial x} = -\frac{1}{\rho}\frac{\partial p}{\partial x} \qquad \dots (2)$$

From (1)

$$\frac{\partial u}{\partial t}x = F^1(t)$$

$$\Rightarrow \quad \frac{\partial u}{\partial t} = \frac{F^1(t)}{x}$$

Hence (2) becomes

$$\frac{F^1(t)}{x} + u\frac{\partial u}{\partial x} = -\frac{1}{\rho}\frac{\partial p}{\partial x}$$

$$\Rightarrow \quad \frac{F^1(t)}{x}+\frac{\partial}{\partial x}\left(\frac{1}{2}u^2\right)=-\frac{\partial}{\partial x}\left(\frac{p}{\rho}\right)$$

Integrating, we get

$$F^1(t)\log x+\frac{u^2}{2}=C-\frac{1}{\rho}$$

$$\Rightarrow \quad F^1(t)\log x+\frac{1}{2}\frac{F^2}{x^2}=C-\frac{p}{\rho} \qquad \text{... (3)}$$

From (1) $xu = F(t)$

$\Rightarrow \; xdx = F(t)\; dt$

Let R and r be the external and internal radii at any time t.

When $x = R$, $u = R$, $p = \Pi$

Hence from (3)

$$F^1(t)\log R+\frac{1}{2}\frac{F^2}{R^2}=-\frac{\Pi}{\rho}+C \qquad \text{... (4)}$$

Again when $x = r$, $u = r = \upsilon$, $p = 0$

$$F^1(t)\log r+\frac{1}{2}\frac{F^2}{r^2}=C \qquad \text{... (5)}$$

Subtracting (5) from (4), we get

$$F^1(t)(\log R-\log r)+\frac{F^2}{2}\left(\frac{1}{R^2}-\frac{1}{r^2}\right)=-\frac{\Pi}{\rho}$$

Multiplying by $2ru = 2F(t) = 2Ru$

$$\Rightarrow \quad 2rdr = 2F(t)dt = 2RdR \qquad \text{as } u=\frac{dr}{dt}$$

$$2F^1(t)\,F(t)(\log R-\log r)dt$$

$$+F^2\left(\frac{dR}{R}-\frac{dr}{r}\right)=-\frac{\Pi}{\rho}2rdr$$

$$\Rightarrow \quad d\left[(\log R-\log r)F^2\right]=\frac{-2r\Pi}{\rho}dr$$

Integrating, we get

$$(\log R-\log r)F^2=-r^2\frac{\Pi}{\rho}+C_1 \qquad \text{... (6)}$$

Again from the equation of continuity, we have

$$2ru = 2F(t) = 2RU$$

or $2rdr = 2F(t)dt = 2RdR$

Integrating, we get

$$r^2 = R^2 + C_2$$

Initially $r = b$, $R = C_1$

Hence

$$b^2 = a^2 + C_2$$

$$b^2 - a^2 = C_2$$

Hence

$$R^2 = r^2 - b^2 + a^2$$

Therefore, from (6), we get

$$(\log R - \log r) F^2 = -\frac{r^2\Pi}{\rho} + C_1$$

$$\left[\log \frac{(r^2 + a^2 - b^2)^{\frac{1}{2}}}{r}\right] F^2 = \frac{-r^2\Pi}{\rho} + C_1 \qquad \dots (7)$$

At $r = b$, $u = 0$, $F(t) = 0$

$$0 = \frac{-b^2\Pi}{\rho} + C_1$$

$$C_1 = \frac{b^2\Pi}{\rho}$$

Thus from (7), we have

$$\left[\log \frac{(r^2 + a^2 - b^2)^{\frac{1}{2}}}{r}\right] F^2 = \frac{\Pi}{\rho}(b^2 - r^2)$$

$$\Rightarrow \log\left(\frac{r^2 + a^2 - b^2}{r^2}\right)^{\frac{1}{2}} (rv)^2 = (b^2 - r^2)\frac{\Pi}{\rho}$$

$$\Rightarrow \frac{1}{2}\left[\log \frac{r^2 + a^2 - b^2}{r}\right](rv)^2 = (b^2 - r^2)\frac{\Pi}{\rho}$$

$$\Rightarrow \ (rv)^2 = \frac{2(b^2 - r^2)\Pi}{\rho \log \dfrac{r^2 + a^2 - b^2}{r^2}}$$

$$\Rightarrow \ \upsilon^2 = \frac{2\Pi(b^2 - r^2)}{\rho r^2} \frac{1}{\log \dfrac{r^2 + a^2 - b^2}{r^2}}$$

Alternative Method

Equation of continuity is

$$xu = F(t) \text{ where } x = u$$

Let R and r be external and internal radii. Since total mass of the liquid is constant and hence

$$\left(\pi R^2 h - \pi r^2 h\right)\rho = \left(\pi a^2 h - \pi b^2 h\right)\rho$$

$$\Rightarrow \ R^2 - r^2 = a^2 - b^2$$

$$\Rightarrow \ R^2 = r^2 + a^2 - b^2 \qquad \dots (1)$$

Also K.E. of the liquid

$$= \frac{1}{2}\int_r^R (2\pi x dx)\, \rho u^2$$

$$= \pi\rho \int_r^R x dx \, \frac{F^2}{x^2}$$

$$= \pi\rho F^2 \int_r^R \frac{dx}{x}$$

$$= \pi\rho F^2 \log \frac{R}{r}$$

Work done by the outer pressure

$$= \int_a^R -p dV = \int_a^R -2\pi x dx \, \Pi$$

$$= \pi\Pi\left(a^2 - R^2\right) = \pi\Pi\left(b^2 - r^2\right) \qquad \text{From (1)}$$

Now work done = K.E.

$$\Rightarrow \quad \pi\Pi\left(b^2 - r^2\right) = \Pi\rho F^2 \log\frac{R}{r}$$

$$\Rightarrow \quad 2\pi\Pi\left(b^2 - r^2\right) = \Pi\rho r^2 \upsilon^2 \log\frac{R^2}{r^2}$$

$$\Rightarrow \quad \upsilon^2 = \frac{2\pi\Pi\left(b^2 - r^2\right)}{\rho r^2}\,\frac{1}{\log\left[\dfrac{r + a^2 - b^2}{r^2}\right]}$$

17. A solid sphere of radius a is surrounded by a mass of liquid whose volume is $\frac{4}{3}\pi c^3$ and its centre is a centre of attractive force varying directly as the square of the distance. If the solid sphere be suddenly annihilated, show that the velocity of the inner surface; when its radius is x, is given by

$$x^2 x^3\left[\left(x^3 + c^3\right)^{\frac{1}{3}} - x\right] = \left(\frac{2\Pi}{3\rho} + \frac{2\mu c^3}{a}\right)\left(a^3 - x^3\right)\left(c^3 + x^3\right)^{\frac{1}{3}}$$

where ρ is the density, Π the external pressure and μ the distance.

Solution: Let υ be the velocity at a distance x at any time t. Also let p be the pressure there.

The force $F = -\mu x^2$ as μx^2 is a force directed towards the origin i.e. in the negative direction.

The equation of continuity and motion are

$$x^2\upsilon = F(t)$$

So that

$$\frac{\partial\upsilon}{\partial t} = \frac{F^1(t)}{x^2}$$

and

$$\frac{\partial\upsilon}{\partial t} + \upsilon\frac{\partial\upsilon}{\partial x} = -\mu x^2 - \frac{1}{\rho}\frac{\partial p}{\partial x}$$

$$\Rightarrow \quad \frac{F^1(t)}{x^2} + \upsilon\frac{\partial\upsilon}{\partial x} = -\mu x^2 - \frac{\partial}{\partial x}\left(\frac{p}{\rho}\right)$$

Integrating w.r.t. x, we get

$$\frac{-F^1(t)}{x} + \frac{\upsilon^2}{2} = -\frac{\mu x^3}{3} - \frac{p}{\rho} + C \qquad \text{... (1)}$$

Let r and R be the internal and external radii respectively at any time t. Since the total mass of the liquid is constant and so

$$\left(\frac{4}{3}\Pi R^3 - \frac{4}{3}\Pi r^3\right)\rho = \frac{4}{3}\Pi c^3 \rho$$

$\Rightarrow$ $R^3 - r^3 = C^3$ ... (2)

When $x = R$, $\upsilon = R = U$, $p = \Pi$ ... (3)

and when $x = r$, $\upsilon = r = u$, $p = 0$ ... (4)

as pressure vanishes on the surface of inner sphere.

Also

$x^2\upsilon = R^2U = r^2u = F(t)$ ... (5)

Applying the conditions (3) and (4) on (1), we get

$$\frac{-F^1(t)}{R} + \frac{U^2}{2} = \frac{-\mu R^3}{3} - \frac{\Pi}{\rho} + C \qquad \text{... (6)}$$

and

$$\frac{-F^1(t)}{r} + \frac{u^2}{2} = \frac{-\mu}{3} r^3 + C \qquad \text{... (7)}$$

Subtracting (7) from (6), we get

$$\left(\frac{1}{r} - \frac{1}{R}\right) F^1(t) + \frac{1}{2}\left(U^2 - u^2\right) = \frac{\mu}{3}\left(r^3 - R^3\right) - \frac{\Pi}{\rho}$$

$$\Rightarrow \quad \left(\frac{1}{r} - \frac{1}{R}\right) F^1(t) + \frac{F^2}{2}\left[\frac{1}{R^4} - \frac{1}{r^4}\right] = \frac{\mu}{3}\left(r^3 - R^3\right) - \frac{\Pi}{\rho}$$

From (5)

$$\Rightarrow \quad \left(\frac{1}{r} - \frac{1}{R}\right) F^1(t) + \frac{F^2}{2}\left(\frac{1}{R^4} - \frac{1}{r^4}\right) = \frac{-\mu}{3} C^3 - \frac{\Pi}{\rho}$$

Multiplying by $2F = 2R^2U = 2r^2u$

$\Rightarrow$ $2Fdt = 2R^2dR = 2r^2dr$

$$\Rightarrow \left(\frac{1}{r}-\frac{1}{R}\right)2FF^1dt + F^2\left(\frac{dR}{R^2}-\frac{dr}{r^2}\right)$$

$$= \frac{-\mu C^3}{3}2r^2dr - \frac{\Pi}{\rho}2r^2dr$$

$$\Rightarrow d\left[\left(\frac{1}{r}-\frac{1}{R}\right)F^2\right] = \left(\frac{-\mu C^3}{3}-\frac{\Pi}{\rho}\right)2r^2dr$$

Integrating, we get

$$\left(\frac{1}{r}-\frac{1}{R}\right)F^2 = \left(-\frac{\mu C^3}{3}-\frac{\Pi}{\rho}\right)\frac{2}{3}r^3 + C_1 \qquad \text{... (8)}$$

Again when $r = a$, $\upsilon = 0$, $F(t) = 0$

Thus from (8), we have

$$0 = \frac{-2\mu C^3a^3}{9} - \frac{2\Pi}{3\rho}a^3 + C_1 \qquad \text{... (9)}$$

(8) – (9) gives

$$\left(\frac{1}{r}-\frac{1}{R}\right)r^4u^2 = \frac{2\mu C^3}{9}\left(a^3-r^3\right) + \frac{2\Pi}{3\rho}\left(a^3-r^3\right)$$

$$\Rightarrow (R-r)r^3u^2 = \left(\frac{2\mu C^3}{9}+\frac{2\Pi}{3\rho}\right)\left(a^3-r^3\right)R$$

$$\Rightarrow \left[\left(C^3+r^3\right)^{\frac{1}{3}} - r\right]r^3\dot{r}^2$$

$$= \left(\frac{2\mu C^3}{9}+\frac{2\Pi}{3\rho}\right)\left(a^3-r^3\right)\left(C^3+r^3\right)^{\frac{1}{3}}$$

Replacing r by x, we get

$$\left[\left(x^3+C^3\right)^{\frac{1}{3}} - x\right]x^3\dot{x}^2$$

$$= \left(\frac{2\pi C^3}{9}+\frac{2\Pi}{3\rho}\right)\left(a^3-x^3\right)\left(C^3+x^3\right)^{\frac{1}{3}}$$

18. A mass of liquid of density ρ and volume $\frac{4}{3}\Pi c^3$ is in the form of a spherical shell, a constant pressure Π is exerted on the external surface of the shell, there is no pressure on the internal surface and no other forces act on the liquid, initially the liquid is at rest and the internal radius of the shell is 2C, prove that the velocity of the internal surface, when its radius is C, is $\left(\dfrac{14\Pi}{3\rho}\dfrac{2^{1/3}}{2^{1/3}-1}\right)^{\frac{1}{2}}$

Solution: Let υ be the velocity at a distance x from the centre and R and r be the radii of the external and internal boundaries at any time t.

Kinetic energy of the mass of the liquid

$$=\frac{1}{2}\int_r^R\left(4\Pi x^2 dx\rho\right)x^2$$

$$=2\Pi\rho\int_r^R x^2 dx\frac{r^4\upsilon^2}{x^4} \qquad \text{as } r^2\upsilon = x^2\dot{x}$$

$$=2\Pi\rho r^4\upsilon\left(\frac{1}{r}-\frac{1}{R}\right)$$

Let us suppose that initially a and b are the radii of the internal and external boundaries of the liquid.

Then work done by the outer pressure

$$=\int_b^R 4\Pi x^2 dx(-\Pi)$$

$$=\frac{4\pi\Pi}{3}\left(b^3-R^3\right)$$

$$=\frac{4\pi\Pi}{3}\left(a^3-r^3\right)$$

Since $\dfrac{4}{3}\Pi R^3-\dfrac{4}{3}\Pi r^3=\dfrac{4}{3}\Pi b^3-\dfrac{4}{3}\Pi a^3$

Since the change in K.E. is equal to the work done by the force, we have

$$2\Pi\rho r^4\upsilon^2\left(\frac{1}{r}-\frac{1}{R}\right)=\frac{4\pi\Pi}{3}\left(a^3-r^3\right)$$

$$\Rightarrow \quad \upsilon^2 = \frac{2\Pi}{3\rho r^4} \frac{a^3 - r^3}{\frac{1}{r} - \frac{1}{R}} \qquad \dots (1)$$

But here $a = 2c$ and velocity is to be found for $r = c$. Also since the volume of the liquid is $\frac{4}{3}\Pi c^3$.

We have

$$\frac{4}{3}\Pi R^3 - \frac{4}{3}\Pi r^3 = \frac{4}{3}\Pi c^3$$

$$\Rightarrow \quad R^3 - r^3 = c^3$$

$$\Rightarrow \quad R^3 - c^3 = c^3$$

$$R^3 = 2c^3$$

$$\Rightarrow \quad R = 2^{\frac{1}{3}} c$$

Hence from (1), we have

$$\upsilon^2 = \frac{2\Pi}{3\rho c^4} \frac{8c^3 - c^3}{\frac{1}{c} - \frac{1}{c \cdot 2^{1/3}}}$$

$$\upsilon = \sqrt{\frac{14\Pi}{3\rho} \frac{2^{1/3}}{2^{1/3} - 1}}$$

19. A mass of gravitating fluid is at rest under its own attraction only, the free surface being a sphere of radius b and the inner surface a rigid concentric shell of radius a. Show that if this shell suddenly disappear, the initial pressure at any point of the fluid at distance r from the centre is

$$\frac{2}{3}\pi\rho^2 (b - r)(r - a)\left(\frac{a + b}{r} + 1\right)$$

Solution: Let r be the radius of inner surface at any time t. The force F of attraction at a distance x from the centre of the liquid is

$$\frac{4}{3}\pi\rho \frac{\gamma\left(x^3 - r^3\right)}{x^2} \qquad \left[\text{Since } F = \frac{\gamma m_1 m_2}{d^2}\right]$$

Equations of continuity and motion are

$$x^2\upsilon = F(t) \quad \text{... (1)}$$

and

$$\frac{\partial \upsilon}{\partial t} + \upsilon \frac{\partial \upsilon}{\partial x} = -\frac{4}{3}\frac{\pi\rho\gamma\left(x^3 - r^3\right)}{x^2} - \frac{1}{\rho}\frac{\partial p}{\partial x} \quad \text{... (2)}$$

Here negative sign is taken as force is directed towards origin

From (1)

$$\frac{x^2 \partial \upsilon}{\partial t} = F^1(t)$$

$$\Rightarrow \quad \frac{\partial \upsilon}{\partial t} = \frac{F^1(t)}{x^2}$$

Hence from (2)

$$\frac{F^1(t)}{x^2} + \upsilon \frac{\partial \upsilon}{\partial x} = -\frac{4}{3}\frac{\pi\gamma\rho\left(x^3 - r^3\right)}{x^2} - \frac{1}{\rho}\frac{\partial p}{\partial x}$$

Integrating, we get

$$\frac{-F^1(t)}{x} + \frac{\upsilon^2}{2} = -\frac{4}{3}\Pi\gamma\rho\left(\frac{x^2}{2} + \frac{r^3}{x}\right) - \frac{p}{\rho} + C \quad \text{... (3)}$$

When $t = 0$, $\upsilon = 0$, $r = a$, $p = p_0$ (say)

For initially the radius of the inner surface is a and also this surface contains gravitating mass and so there will be pressure on it.

Thus from (3), we get

$$\frac{-F^1(0)}{x} = -\frac{4}{3}\pi\gamma\rho\left(\frac{x^2}{2} + \frac{a^3}{x}\right) - \frac{p_0}{\rho} + C \quad \text{... (4)}$$

Again when $t = 0$, $x = a$, $\upsilon = 0$, $p_0 = 0$

$$\frac{-F^1(0)}{a} = -\frac{4}{3}\pi\gamma\rho\left(\frac{a^2}{2} + \frac{a^3}{a}\right) + C \quad \text{... (5)}$$

Again when $t = 0$, $x = b$, $\upsilon = 0$, $p_0 = 0$, we get from (4)

$$\frac{-F^1(0)}{b} = -\frac{4}{3}\pi\gamma\rho\left(\frac{b^2}{2} + \frac{a^3}{b}\right) + C \quad \text{... (6)}$$

(5) – (6) gives

$$\left(\frac{1}{b}-\frac{1}{a}\right)F^1(0)=-\frac{4}{3}\pi\rho\gamma\left[\frac{a^2-b^2}{2}+a^3\left(\frac{1}{a}-\frac{1}{b}\right)\right]$$

$$\Rightarrow\quad F^1(0)=-\frac{4}{3}\pi\rho\gamma\, ab\left[\frac{a+b}{2}-\frac{a^2}{b}\right]$$

$$\Rightarrow\quad F^1(0)=-\frac{2}{3}\pi\gamma\rho a\left[b(a+b)-2a^2\right] \qquad \ldots (7)$$

Again (4) – (5) gives

$$F^1(0)\left\{\frac{1}{a}-\frac{1}{x}\right\}=-\frac{4}{3}\pi\gamma\rho\left[\frac{x^2-a^2}{2}+a^3\left(\frac{1}{x}-\frac{1}{a}\right)\right]-\frac{p_0}{\rho}$$

$$\Rightarrow\quad \frac{p_0}{\rho}=-\frac{4}{3}\pi\rho\gamma(x-a)\left[\frac{x+a}{2}-\frac{a^2}{x}\right]-F^1(0)\frac{(x-a)}{xa}\rho$$

$$\Rightarrow\quad p_0=-\frac{2}{3}\pi\rho^2\gamma(x-a)\left[2\left(\frac{x+a}{2}-\frac{a^2}{x}\right)+\frac{F^1(0)}{xa\frac{2}{3}\pi\rho\gamma}\right]$$

$$=-\frac{2}{3}\pi\rho^2\gamma(x-a)\left[\frac{(x+a)x-2a^2}{x}-\frac{a}{xa}\left\{b(a+b)-2a^2\right\}\right] \quad \text{From (7)}$$

$$=-\frac{2}{3}\pi\rho^2\gamma(x-a)\left[\frac{x^2+ax-2a^2-ab-b^2+2a^2}{x}\right]$$

$$=\frac{2}{3}\pi\rho^2\gamma(x-a)\left[\frac{x^2-b^2+ax-ab}{x}\right]$$

$$=\frac{2}{3}\pi\rho^2\gamma(x-a)(b-x)\left[\frac{x+b+a}{x}\right]$$

$$p_0=\frac{2}{3}\pi\rho^2\gamma(x-a)(b-x)\left[1+\frac{a+b}{x}\right]$$

Replacing x by r, we get

$$p_0=\frac{2}{3}\pi\rho^2\gamma(r-a)(b-r)\left(1+\frac{a+b}{r}\right)$$

20. A volume $\frac{4}{3}\pi c^3$ of gravitating liquid of density ρ is initially in the form of a spherical shell of infinitely great radius. If the liquid shell contracts under the influence of its own attraction, there being no external or internal pressure, show that when the radius of the inner spherical surface is x, its velocity will be given by

$$V^2 = \frac{4\pi\rho\gamma z}{15x^3}\left[2z^4 + 2z^3x + 2z^3x^2 - 3zx^3 - 3x^4\right]$$

where γ is the constant of gravitation and $z^3 = x^3 + c^3$.

Solution: Let r be the radius of inner surface at any time t. The force F of attraction at a distance x from the centre of the liquid is

$$\frac{4}{3}\pi\rho\gamma\frac{(x^3 - r^3)}{x^2}$$

Since $F = \frac{\gamma m_1 m_2}{d^2}$

Also equations of continuity and motion are

$$x^2 \upsilon = F(t) \qquad \dots (1)$$

and

$$\frac{\partial \upsilon}{\partial t} + \upsilon\frac{\partial \upsilon}{\partial x} = -\frac{4}{3}\pi\gamma\rho\frac{(x^3 - r^3)}{x^2} - \frac{1}{\rho}\frac{\partial p}{\partial x} \qquad \dots (2)$$

From (1)

$$\frac{\partial \upsilon}{\partial t} = \frac{F^1(t)}{x^2}$$

So from (2), we get

$$\frac{F^1(t)}{x^2} + \upsilon\frac{\partial \upsilon}{\partial x} = -\frac{4}{3}\frac{\pi\gamma\rho(x^3 - r^3)}{x^2} - \frac{1}{\rho}\frac{\partial p}{\partial x}$$

Integrating, we get

$$\frac{-F^1(t)}{x} + \frac{\upsilon^2}{2} = -\frac{4}{3}\pi\gamma\rho\left(\frac{x^2}{2} + \frac{r^3}{x}\right) - \frac{p}{\rho} + c \qquad \dots (3)$$

Let R and r be the external and internal radii at any time t. Since the total mass of the liquid is constant.

$$\frac{4}{3}\pi R^3\rho - \frac{4}{3}\pi r^3\rho = \frac{4}{3}\pi c^3\rho$$

$$\Rightarrow \quad R^3 - r^3 = c^3 \qquad \ldots (4)$$

When $x = R$, $\upsilon = \dot{R} = U$, $p = 0$, as there is no external pressure.

Hence from (3)

$$\frac{-F^1(t)}{R} + \frac{1}{2}U^2 = -\frac{4}{3}\pi\rho\gamma\left(\frac{R^2}{2} + \frac{r^3}{R}\right) + C \qquad \ldots (5)$$

Again when $x = r$, $\upsilon = \dot{r} = u$, $p = 0$

Thus from (3)

$$\frac{-F^1(t)}{r} + \frac{1}{2}u^2 = -\frac{4}{3}\pi\rho\gamma\left(\frac{r^2}{2} + \frac{r^3}{2}\right) + C \qquad \ldots (6)$$

(5) – (6) gives

$$F^1(t)\left(\frac{1}{r} - \frac{1}{R}\right) + \frac{1}{2}\left(U^2 - u^2\right)$$

$$= -\frac{4}{3}\pi\rho\gamma\left[\frac{R^2 - r^2}{2} + r^3\left(\frac{1}{R} - \frac{1}{r}\right)\right]$$

Also from (1)

$$r^2 u = F = R^2 U$$

Hence

$$F^1(t)\left(\frac{1}{r} - \frac{1}{R}\right) + \frac{F^2}{2}\left(\frac{1}{R^4} - \frac{1}{r^4}\right)$$

$$= -\frac{4}{3}\pi\rho\gamma\left[\frac{R^2 - r^2}{2} + r^3\left(\frac{1}{R} - \frac{1}{r}\right)\right]$$

Multiplying both the sides by

$$2r^2 u = 2F(t) = 2R^2 U$$

or $2r^2 dr = 2F(t)dt = 2R^2 dR$, we get

$$2FF^1\left(\frac{1}{r}-\frac{1}{R}\right)+F^2\left(\frac{dR}{R^2}-\frac{dr}{r^2}\right)$$

$$=-\frac{4}{3}\pi\rho\gamma\left[R^4dR-r^4dr+r^3 2RdR-2r^4dr\right]$$

$$\Rightarrow\ d\left[\left(\frac{1}{r}-\frac{1}{R}\right)F^2\right]=-\frac{4}{3}\pi\rho\gamma\left[\left(R^4dR-r^4dr\right)\right.$$

$$\left.+2r^3\left(RdR-rdr\right)\right]$$

$$=-\frac{4}{3}\pi\rho\gamma\left[\left(R^4dR-r^4dr\right)\right.$$

$$\left.+2\left(R^3-c^3\right)RdR-2r^4dr\right]$$

From (4)

Integrating, we get

$$\left(\frac{1}{r}-\frac{1}{R}\right)F^2=-\frac{4}{3}\pi\rho\gamma\left[\frac{R^5-r^5-2r^5}{5}+\frac{2R^5}{5}-\frac{2c^3R^2}{2}\right] \quad \text{... (7)}$$

Neglecting constant of integration

Also from (1) $r^2u = F(t)$

Hence (7) becomes

$$u^2r^4\frac{(R-r)}{rR}=-\frac{4}{15}\pi\rho\gamma\left[3\left(R^5-r^5\right)-5c^3R^2\right]$$

$$u^2=\frac{4\pi\rho\gamma R}{15r^3}\left[\frac{3\left(r^5-R^5\right)+5R^2\left(R^3-r^3\right)}{R-r}\right] \quad \text{From (4)}$$

$$=\frac{4}{15}\frac{\pi\rho\gamma R}{r^3(R-r)}\left[2R^5-5R^2r^3+3r^5\right]$$

$$=\frac{4}{15}\frac{\pi\rho\gamma R}{r^3(R-r)}\left[2R^5-2R^2r^3-3R^2r^3+3r^5\right]$$

$$=\frac{4}{15}\frac{\pi\rho\gamma R}{r^3(R-r)}\left[2R^2\left(R^3-r^3\right)-3r^3\left(R^2-r^2\right)\right]$$

$$=\frac{4}{15}\frac{\pi\rho\gamma R}{r^3}\left[2R^2\left(R^2+rR+r^2\right)-3r^3\left(R+r\right)\right]$$

$$u^2 = \frac{4}{15}\frac{\pi\rho\gamma R}{r^3}\left[2R^4 + 2R^3 r + 2R^2 r^2 - 3Rr^3 - 3r^4\right]$$

Putting $R = z$, $r = x$ and $u = V$, we get

$$V^2 = \frac{4}{15}\frac{\pi\rho\gamma z}{x^3}\left[2z^4 + 2z^3 x + 2z^2 x^2 - 3zx^3 - 3x^4\right]$$

21. A mass of perfect incompressible fluid of density ρ is bounded by concentric spherical surfaces. The outer surface is contained by a flexible envelope which exerts continuously a uniform pressure Π and contracts from radius R_1 to radius R_2. The hollow is filled with a gas obeying Boyle's law, its radius contracts from c_1 to c_2 and the pressure of the gas is initially p_1. Initially the whole mass is at rest. Prove that, neglecting the mass of the gas, the velocity υ of the inner surface when the configuration (R_2, c_2) is reached is given by

$$\frac{1}{2}\upsilon^2 = \frac{c_1^3}{c_2^3}\left\{\frac{1}{3}\left(1 - \frac{c_2^3}{c_1^3}\right)\frac{\Pi}{\rho} - \frac{p_1}{\rho}\log\frac{c_1}{c_2}\right\}\Big/\left(1 - \frac{c_2}{R_2}\right)$$

Solution: At any time t, let υ^1 be the velocity at a distance r^1 and p be the pressure there.

Then equation of continuity is

$$r^{1^2}\upsilon^1 = F(t) = c_2^2\upsilon \qquad \dots (1)$$

When the internal radius is c_2, let p_2 be the pressure of the inside gas.

By Boyle's law $\frac{4}{3}\pi c_1^3 p_1 = \frac{4}{3}\pi c_2^3 p_2$

$$\Rightarrow \quad p_2 = \frac{c_1^3}{c_2^3}p_1 \qquad \dots (2)$$

Initial kinetic energy = 0

$$\text{Final kinetic energy} = \frac{1}{2}\int_{c_2}^{R_2} 4\pi r^{1^2}\, dr^1 \rho \upsilon^{1^2}$$

$$= 2\pi\rho \int_{c_2}^{R_2} r^{1^2} \upsilon^{1^2} dr^1$$

$$= 2\pi\rho \int_{c_2}^{R_2} \frac{r^{1^2} c_2^4 \upsilon^2}{r^{1^4}} dr^1 \text{ From (1)}$$

$$= 2\pi\rho c_2^4 \upsilon^2 \int_{c_2}^{R_2} \frac{dr^1}{r^{1^2}}$$

$$= 2\pi\rho c_2^4 \upsilon^2 \left(\frac{1}{c_2} - \frac{1}{R_2} \right)$$

$$= 2\pi\rho c_2^3 \upsilon^2 \left(1 - \frac{c_2}{R_2} \right)$$

Also work done by the external pressure Π and the internal pressure p_2

$$= \int_{R_1}^{R_2} 4\pi R_2^2 \Pi (-dR_2) + \int_{c_1}^{c_2} 4\pi c_2^2 p_2 dc_2$$

$$= \frac{4}{3} \pi\Pi \left(R_1^3 - R_2^3 \right) + 4\pi \int_{c_1}^{c_2} c_2^2 \frac{c_1^3}{c_2^3} p_1 dc_2 \qquad \text{[From (2)]}$$

$$= \frac{4}{3} \pi\Pi \left(R_1^3 - R_2^3 \right) + 4\pi p_1 c_1^3 \int_{c_1}^{c_2} \frac{dc_2}{c_2}$$

$$= \frac{4}{3} \pi\Pi \left(R_1^3 - R_2^3 \right) + 4\pi p_1 c_1^3 \log \frac{c_2}{c_1} \qquad \dots (3)$$

Again since mass is constant

$$\frac{4}{3} \pi \left(R_2^3 - c_2^3 \right) = \frac{4}{3} \pi \left(R_1^3 - c_1^3 \right)$$

$$\Rightarrow \ R_1^3 - R_2^3 = c_1^3 - c_2^3$$

Hence from (3)

$$\text{Work done} = \frac{4}{3} \pi\Pi \left(c_1^3 - c_2^3 \right) + 4\pi p_1 c_1^3 \log \frac{c_2}{c_1}$$

Hence from the principle of energy

Kinetic energy = work done

$$\Rightarrow \quad 2\pi\rho c_2^3 \upsilon^2 \left(1 - \frac{c_2}{R_2}\right) = 4\pi c_1^3 \left[\frac{\Pi}{3}\left(1 - \frac{c_2^3}{c_1^3}\right) - p_1 \log \frac{c_1}{c_2}\right]$$

$$\Rightarrow \quad \frac{1}{2}\upsilon^2 = \frac{c_1^3}{c_2^3}\left[\frac{1}{3}\left(1 - \frac{c_2^3}{c_1^3}\right)\frac{\Pi}{\rho} - \frac{p_1}{\rho}\log\frac{c_1}{c_2}\right] \Big/ 1 - \frac{c_2}{R_2}$$

22. A centre of force attracting inversely as the square of the distance is at the centre of a spherical cavity within an infinite mass of incompressible fluid, the pressure which at an infinite distance is Π, and is such that the work done by this pressure on a unit of area through a unit of length is one half the work done by the attractive force on a unit volume of the fluid from infinity to the initial boundary of the cavity, prove that the time of filling up the cavity will be

$$\Pi a \left(\frac{\rho}{\Pi}\right)^{\frac{1}{2}} \left[2 - \left(\frac{3}{2}\right)^{\frac{3}{4}}\right]$$

a being the initial radius of the cavity and ρ the density of the fluid.

Solution: Let υ be the velocity at a distance x at any time t and p be the pressure. Then equations of continuity and motion are

$$x^2\upsilon = F(t) \qquad \dots (1)$$

and

$$\frac{\partial \upsilon}{\partial t} + \upsilon \frac{\partial \upsilon}{\partial x} = -\frac{\mu}{x^2} - \frac{1}{\rho}\frac{\partial p}{\partial x} \qquad \dots (2)$$

From (1) $\dfrac{\partial \upsilon}{\partial t} = \dfrac{F^1(t)}{x^2}$

So from (2)

$$\frac{F^1(t)}{x^2} + \upsilon \frac{\partial \upsilon}{\partial x} = \frac{-\mu}{x^2} - \frac{1}{\rho}\frac{\partial p}{\partial x}$$

Integrating, we get

$$\frac{-F^1(t)}{x} + \frac{1}{2}\upsilon^2 = \frac{\mu}{x} - \frac{p}{\rho} + c \qquad \text{... (3)}$$

When $x = \infty$, $\upsilon = 0$, $p = \Pi$

Hence from (3)

$$c = \frac{\Pi}{\rho}$$

Hence (3) becomes

$$\frac{-F^1(t)}{x} + \frac{1}{2}\upsilon^2 = \frac{\mu}{x} - \frac{p}{\rho} + \frac{\Pi}{\rho} \qquad \text{... (4)}$$

Let r be the radius of cavity at any time t. Then when $x = r$, $\upsilon = r = u$, $p = 0$, as pressure vanishes on the surface of cavity.

Thus from (1)

$$\frac{-F^1(t)}{r} + \frac{1}{2}u^2 = \frac{\mu}{r} + \frac{\Pi}{\rho} \qquad \text{... (5)}$$

From (1) $r^2u = F(t)$

Hence (5) becomes

$$\frac{-F^1(t)}{r} + \frac{1}{2}\frac{F^2(t)}{r^4} = \frac{\mu}{r} + \frac{\Pi}{\rho}$$

Multiplying by $2r^2u = 2F(t)$

$\Rightarrow$ $2r^2dr = 2F(t)dt$, we get

$$\frac{-F^1(t)F(t)dt}{r} + \frac{F^2(t)}{r^2} = 2\,\mu r dr + \frac{\Pi}{\rho}2r^2dr$$

$$\Rightarrow \; d\left(-\frac{F^2}{r}\right) = 2\mu r dr + \frac{\Pi}{\rho}2r^2dr$$

Integrating, we get

$$\frac{-F^2}{r} = \mu r^2 + \frac{2\Pi}{3\rho}r^3 + C_1 \qquad \text{... (6)}$$

Also when $r = a$, $\upsilon = u = 0$, $F(t) = 0$

Thus from (6)

$$0 = \mu a^2 + \frac{2\Pi}{3\rho} a^3 + C_1 \qquad \text{... (7)}$$

(6) – (7) gives

$$\frac{-F^2}{r} = \mu\left(r^2 - a^2\right) + \frac{2\Pi}{2\rho}\left(r^3 - a^3\right)$$

$$\Rightarrow \quad r^3u^2 = \mu\left(a^2 - r^2\right) + \frac{2\Pi}{3\rho}\left(a^3 - r^3\right) \qquad \text{... (8)}$$

Also it is given that

Work done by Π on unit area through a unit length

$$= \frac{1}{2} \text{ work done by } \frac{-\mu}{x^2} \text{ on a unit volume of fluid}$$

from $x = \infty$ to $x = a$

$$\text{Hence } \Pi \cdot 1 \cdot 1 = \frac{1}{2}\int_{\infty}^{a} \frac{-\mu}{x^2}\rho dx$$

$$\Rightarrow \quad \Pi = \frac{\mu\rho}{2a}$$

$$\Rightarrow \quad \mu = \frac{2a\Pi}{\rho}$$

Now (8) becomes

$$r^3u^2 = \frac{2a\Pi}{\rho}\left(a^2 - r^2\right) + \frac{2\Pi}{3\rho}\left(a^3 - r^3\right)$$

$$\Rightarrow \quad r^3u^2 = \frac{3\Pi}{3\rho}\left[3a\left(a^2 - r^2\right) + \left(a^3 - r^3\right)\right]$$

$$\Rightarrow \quad u^2 = \frac{2\Pi}{3\rho r^3}\left[3a\left(a^2 - r^2\right) + a^3 - r^3\right]$$

$$\Rightarrow \quad \frac{dr}{dt} = -\left(\frac{2\Pi}{3\rho}\right)^{\frac{1}{2}}\left[\frac{3a\left(a^2 - r^2\right) + a^3 - r^3}{r^3}\right]^{\frac{1}{2}}$$

Here negative sign is taken since r decreases when t increases.

Let T be the required time, then

$$\int_0^T dt = -\left(\frac{3\rho}{2\Pi}\right)^{\frac{1}{2}} \int_a^0 \frac{r^{3/2}dr}{\left[3a\left(a^2 - r^2\right) + \left(a^3 - r^3\right)\right]^{\frac{1}{2}}}$$

$$\Rightarrow \quad T = \left(\frac{3\rho}{3\Pi}\right)^{\frac{1}{2}} \int_0^a \frac{r^{3/2}dr}{\left[3a\left(a^2 - r^2\right) + \left(a^3 - r^3\right)\right]^{\frac{1}{2}}}$$

$$= \left(\frac{3\rho}{2\Pi}\right)^{\frac{1}{2}} \int_0^a \frac{r^{3/2}dr}{\left[(a-r)(2a+r)^2\right]^{\frac{1}{2}}}$$

$$= \left(\frac{3\rho}{2\Pi}\right)^{\frac{1}{2}} \int_0^a \frac{r^{3/2}dr}{(a-r)^{\frac{1}{2}}(2a+r)}$$

Putting $r = a \sin^2 \theta$

$$dr = 2a \sin\theta \cos\theta \, d\theta$$

Hence

$$T = \left(\frac{3\rho}{2\Pi}\right)^{\frac{1}{2}} \int_0^{\frac{\pi}{2}} \frac{a^{3/2} \sin^3\theta \, 2a \sin\theta \cos\theta d\theta}{a^{1/2} \cos\theta \, a\left(2 + \sin^2\theta\right)}$$

$$= 2a\left(\frac{3\rho}{2\Pi}\right)^{\frac{1}{2}} \int_0^{\frac{\pi}{2}} \frac{\sin^4\theta \, d\theta}{\left(2 + \sin^2\theta\right)}$$

$$= 2a\left(\frac{3\rho}{2\Pi}\right)^{\frac{1}{2}} \int_0^{\frac{\pi}{2}} \left(\sin^2\theta - 2 + \frac{4}{2 + \sin^2\theta}\right) d\theta$$

$$T = 2a\left(\frac{3\rho}{2\Pi}\right)^{\frac{1}{2}} \left[\frac{\Pi}{4} - 2 \cdot \frac{\Pi}{2} + 4 \int_0^{\frac{\pi}{2}} \frac{\sec^2\theta d\theta}{2 + 3\tan^2\theta}\right]$$

Let $y = \tan\theta$

$$dy = \sec^2\theta \, d\theta$$

Hence

$$T = 2a\left(\frac{3\rho}{2\Pi}\right)^{\frac{1}{2}}\left[-\frac{3\pi}{4} + 4\int_0^{\frac{\pi}{2}}\frac{dy}{2+3y^2}\right]$$

$$= 2a\left(\frac{3\rho}{2\Pi}\right)^{\frac{1}{2}}\left[-\frac{3\pi}{4} + \frac{4}{3}\int_0^{\frac{\pi}{2}}\frac{dy}{y^2+\frac{2}{3}}\right]$$

$$= 2a\left(\frac{3\rho}{2\Pi}\right)^{\frac{1}{2}}\left[-\frac{3\pi}{4} + \frac{4}{3}\sqrt{\frac{3}{2}}\left\{\tan^{-1}\left(y\sqrt{\frac{3}{2}}\right)\right\}_0^{\infty}\right]$$

$$= 2a\left(\frac{3\rho}{2\Pi}\right)^{\frac{1}{2}}\left[-\frac{3\pi}{4} + \frac{4}{3}\sqrt{\frac{3}{2}}\,\frac{\pi}{2}\right]$$

$$= \pi a\left(\frac{\rho}{\Pi}\right)^{\frac{1}{2}}\left[-\frac{3}{4}\sqrt{6} + \frac{4}{3}\sqrt{\frac{3}{2}}\,\frac{1}{2}\sqrt{6}\right]$$

$$= \pi a\left(\frac{\rho}{\Pi}\right)^{\frac{1}{2}}\left[2-\left(\frac{3}{2}\right)^{\frac{3}{2}}\right]$$

23. A spherical hollow of radius a initially exists in an infinite fluid subject to a constant pressure at infinity. Show that the pressure at distance r from the centre when the radius of the cavity is x, is to the pressure at infinity as

$$3x^2r^4 + \left(a^3 - 4x^3\right)r - \left(a^3 - x^3\right)x^3 : 3x^2r^4$$

Solution: Let at time t, υ be the velocity at a distance x and p the pressure thereat, then equations of continuity and motion are

$$x^2\upsilon = F(t) \qquad \text{... (1)}$$

and

$$\frac{\partial \upsilon}{\partial t} + \upsilon\frac{\partial \upsilon}{\partial x} = -\frac{1}{\rho}\frac{\partial p}{\partial x} \qquad \text{... (2)}$$

From (1)

$$x^2 \frac{\partial \upsilon}{\partial \rho} = F^1(t)$$

$$\Rightarrow \quad \frac{\partial \upsilon}{\partial \rho} = \frac{F^1(t)}{x^2}$$

So from (2), we get

$$\frac{F^1(t)}{x^2} + \frac{\partial}{\partial x}\left(\frac{1}{2}\upsilon^2\right) = -\frac{\partial}{\partial x}\left(\frac{p}{\rho}\right)$$

Integrating, we get

$$\frac{-F^1(t)}{x} + \frac{1}{2}\upsilon^2 = -\frac{p}{\rho} + c \qquad \text{... (3)}$$

Let r be the radius of cavity at any time t.

When $x = \infty$, $\upsilon = 0$, $p = \Pi$

Thus from (3)

$$C = \frac{\Pi}{\rho}$$

Hence from (3), we get

$$\frac{-F^1(t)}{x} + \frac{1}{2}\upsilon^2 = \frac{\Pi - p}{\rho} \qquad \text{... (4)}$$

Also when $x = r$, $\upsilon = u = r$, $p = 0$

as pressure vanishes on the surface of cavity.

Therefore from (3), we get

$$\frac{-F^1(t)}{r} + \frac{1}{2}u^2 = 0 + C = \frac{\Pi}{\rho} \qquad \text{... (5)}$$

Also from (1)

$$r^2 u = F(t) = x^2 \upsilon$$

Hence (5) becomes

$$\frac{-F^1(t)}{r} + \frac{1}{2}\frac{F^2}{r^4} = \frac{\Pi}{\rho} \qquad \text{... (6)}$$

Multiplying both sides by $2r^2u = 2F(t)$ or $2r^2 dr = 2F(t)dt$

$$\frac{-F^1(t)2F(t)}{r}+\frac{F^2 dr}{r^2}=\frac{\Pi}{\rho}2r^2 dr$$

$$\Rightarrow \quad d\left(\frac{-F^2}{r}\right)=\frac{2\Pi}{\rho}r^2 dr$$

Integrating, we get

$$\frac{-F^2}{r}=\frac{2\Pi}{3\rho}r^3+C_1 \qquad \text{... (7)}$$

When $r = a$, $\upsilon = u = 0$, so that $F(t) = 0$

Hence from (7)

$$C_1=\frac{-2\Pi}{3\rho}a^3$$

Therefore from (7)

$$\frac{-F^2}{r}=\frac{2\Pi}{3\rho}\left(r^3-a^3\right)$$

$$\Rightarrow \quad F^2(t)=\frac{2}{3}\frac{\Pi r}{\rho}\left(a^3-r^3\right)$$

Hence from (6), we have

$$\frac{-F^1(t)}{r}=\frac{\Pi}{\rho}-\frac{1}{2}\frac{F^2}{r^4}$$

$$=\frac{\Pi}{\rho}-\frac{1}{2r^4}\frac{2}{3}\frac{\Pi r}{\rho}\left(a^3-r^3\right)$$

$$=\frac{\Pi}{\rho}\left[1-\frac{a^3-r^3}{3r^3}\right]$$

$$=\frac{\Pi}{3\rho r^3}\left(4r^3-a^3\right)$$

$$\Rightarrow \quad F^1(t)=\frac{\Pi\left(a^3-4r^3\right)}{3\rho r^2} \qquad \text{... (8)}$$

Hence from (3), we have

$$-\frac{\Pi}{3\rho r^2}\frac{a^3-4r^3}{x}+\frac{1}{2}\frac{F^2(t)}{x^4}=\frac{\Pi-p}{\rho}$$

$$\Rightarrow \quad -\frac{\Pi\left(a^3-4r^3\right)}{3\rho r^2 x}+\frac{1}{2x^4}\frac{2}{3}\frac{\Pi r}{\rho}\left(a^3-r^3\right)=\frac{\Pi-p}{\rho}$$

$$\Rightarrow \quad -\frac{\Pi\left(a^3-4r^3\right)}{3r^2 x}+\frac{\Pi r\left(a^3-r^3\right)}{3x^4}=1-\frac{p}{\Pi}$$

$$\Rightarrow \quad \frac{p}{\Pi}=1+\frac{a^3-4r^3}{3r^2 x}-\frac{r\left(a^3-r^3\right)}{3x^4}$$

$$=\frac{3x^4r^2+\left(a^3-4r^3\right)x^3-r^3\left(a^3-r^3\right)}{3x^4r^2}$$

This gives the pressure at a distance x, when r is the radius of cavity. In order to get the pressure at a distance r when the radius of cavity is x, we replace r by x and x by r. Thus we have

$$\frac{p}{\Pi}=\frac{3r^4x^2+\left(a^3-4x^3\right)r^3-x^3\left(a^3-x^3\right)}{3x^2r^4}$$

24. A mass of uniform liquid is in the form of a thick spherical shell bounded by concentric spheres of radii a and b $(a < b)$. The cavity is filled with gas the pressure of which varies according to Boyle's law and is initially equal to atmospheric pressure Π and the mass of which may be neglected. The outer surface of the shell is exposed to atmospheric pressure. Prove that if the system is symmetrically distributed, so that particle moves along a line joining it to the centre, the time of small oscillation is $2\Pi a\left[\dfrac{\rho\,(b-a)}{3\Pi b}\right]^{\frac{1}{2}}$, where ρ is the density of the liquid.

Solution: Let υ be the velocity at a distance x and p be the pressure there, then equations of continuity and motion are

$$x^2\upsilon = F(t) \qquad \text{... (1)}$$

and

$$\frac{\partial \upsilon}{\partial t}+\upsilon\frac{\partial \upsilon}{\partial x}=-\frac{1}{\rho}\frac{\partial p}{\partial x} \qquad \text{... (2)}$$

From (1)

$$\frac{\partial \upsilon}{\partial t} = \frac{F^1(t)}{x^2}$$

Thus (2) becomes

$$\frac{F^1(t)}{x^2} + \upsilon \frac{\partial \upsilon}{\partial x} = -\frac{\partial}{\partial x}\left(\frac{p}{\rho}\right)$$

Integrating, we get

$$\frac{-F^1(t)}{x} + \frac{\upsilon^2}{2} = -\frac{p}{\rho} + c \qquad \text{... (3)}$$

When $x = R$, $\upsilon = R = U$, $p = \Pi$

Since the outer surface is exposed to atmospheric pressure Π.

Hence from (3), we get

$$\frac{-F^1(t)}{R} + \frac{1}{2}U^2 = -\frac{\Pi}{\rho} + c \qquad \text{... (4)}$$

Again when $x = r$, $\upsilon = r = u$, $p = p_1 = \dfrac{a^3\Pi}{r^3}$

Thus from (3), we have

$$\frac{-F^1(t)}{r} + \frac{1}{2}u^2 = -\frac{a^3\Pi}{\rho r^3} + c \qquad \text{... (5)}$$

(4) – (5) gives

$$\left\{\frac{1}{r} - \frac{1}{R}\right\} F^1(t) + \frac{1}{2}\left(U^2 - u^2\right) = \frac{\Pi}{\rho}\left(\frac{a^3}{r^3} - 1\right)$$

For small oscillations, U^2 and u^2 are small quantities and hence neglected.

Thus

$$F^1(t) = \frac{\Pi}{\rho} \frac{a^3 - r^3}{r^2} \frac{R}{R - r} \qquad \text{... (6)}$$

Also from (2)

$$r^2 u = F(t)$$

$$\Rightarrow \; 2ru^2 + r^2\dot{u} = F^1(t)$$

$\Rightarrow \quad r^2\dot{u} = F^1(t)$ as u^2 is small

$$\Rightarrow \quad r^2\dot{u} = r^2\ddot{r} = \frac{\Pi}{\rho}\frac{a^3 - r^3}{r^2}\frac{R}{R-r} \qquad \text{From (6)}$$

Since displacement is small. So let $r = a + x$, $R = b + x^1$. Then

$$(a+x)^2\ddot{x} = \frac{\Pi}{\rho}\frac{\left[a^3 - (a+x)^3\right]}{(a+x)^2}\frac{b+x^1}{b+x^1-a-x}$$

$$\text{or} \quad \ddot{x} = \frac{\Pi}{\rho}\frac{\left[a^3 - (a+x)^3\right]}{(a+x)^4}\frac{b+x^1}{b-a+x^1-x}$$

$$= \frac{\Pi}{a\rho}\frac{\left[1 - \left(1+\frac{x}{a}\right)^3\right]\left(b+x^1\right)}{\left(1+\frac{x}{a}\right)^4\left(b-a+x^1-x\right)}$$

$$= \frac{\Pi\left[1 - \left(1+\frac{3x}{a}\right)\right]\left(b+x^1\right)}{a\rho\left(1+\frac{4x}{a}\right)\left(x^1-x+b-a\right)}$$

(neglecting higher powers of $\frac{x}{a}$)

$$= \frac{\Pi}{a\rho}\frac{\left(-\frac{3x}{a}\right)\left(b+x^1\right)}{\left(1+\frac{4x}{a}\right)\left(x^1-x+b-a\right)} \qquad \dots (7)$$

Let r and R be the internal and external radii of the shell at any time t. Since the shell contains gas and hence there will be pressure on the inner surface.

Also mass of the liquid remains constant.

Hence

$$\frac{4}{3}\pi\left(R^3 - r^3\right)\rho = \frac{4}{3}\pi\left(b^3 - a^3\right)\rho$$

$$\Rightarrow \quad R^3 - r^3 = b^3 - a^3 \qquad \dots (8)$$

Let $p = p_1$, when $x = r$.

By Boyle's law $\frac{4}{3}\pi r^3 p_1 = \frac{4}{3}\pi a^3 \Pi$

From (8)

$$(b + x^1)^3 - (a + x)^3 = b^3 - a^3$$

$$\Rightarrow \quad b^3\left(1 + \frac{3x^1}{b}\right) - a^3\left(1 + \frac{3x}{a}\right) = b^3 - a^3$$

$$\Rightarrow \quad x^1 = \frac{a^2 x}{b^2}$$

Now Nr of (7) $= \left(-\frac{3x}{a}\right)(b + x^1)$

$$= -\frac{3x}{a}\left(b + \frac{a^2 x}{b^2}\right)$$

$$= -\frac{3xb}{a} \text{ as } x \text{ is very small}$$

Also Dr of (7) $= \left(1 + \frac{4x}{a}\right)(x^1 - x + b - a)$

$$= \left(1 + \frac{4x}{a}\right)\left(\frac{a^2 x}{b^2} - x + b - a\right)$$

$$= \frac{a^2 x}{b^2} - x + b - a + \frac{4xb}{a} - 4x$$

$$= x\left(\frac{a^2}{b^2} - 5 + \frac{4b}{a}\right) + b - a$$

Hence from (7), wc get

$$= \frac{\Pi}{a\,\rho} \frac{\left(\frac{-3xb}{a}\right)}{x\left(\frac{a^2}{b^2} - 5 + \frac{4b}{a}\right) + b - a}$$

$$= \frac{-3xb\Pi}{a\,a\rho\,(b - a)}\left[1 + \left(\frac{a^2}{b^2} - 5 + \frac{4b}{a}\right)\frac{x}{b - a}\right]^{-1}$$

$$= \frac{-3xb\Pi}{\rho a^2 (b-a)}$$

$$= -\mu x \quad \text{Where } \mu = \frac{3b\Pi}{\rho a^2 (b-a)}$$

Hence time of small oscillation

$$= \frac{2\Pi}{\sqrt{\mu}} = 2\Pi a \left[\frac{\rho (b-a)}{3b\Pi}\right]^{\frac{1}{2}}$$

25. Two equal closed cylinders of height c, with their bases in the same horizontal plane, are filled; one with water, and the other with air of such a density as to support a column h of water, h being less than c. If a communication be opened between them at their bases, the height x to which the water rises, is given by the equation

$$cx - x^2 + ch \log \frac{c-x}{x} = 0$$

Solution: Let A and B be two cylinders of cross section K, containing water and air respectively. Initially they are at rest. After the communication is opened, water has risen to a height x in the cylinder B and after that water and air are again at rest. Thus the initial as well as the final K.E. are zero. That is gain in K.E. = 0.

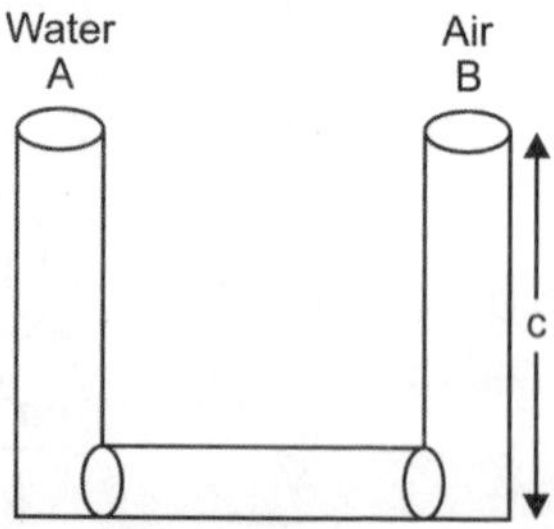

Fig. 2.10

Hence total work done is zero.

Now, initial P.E. due to water of height c in A

$$= \int_0^c \rho gkzdz = \frac{1}{2}\rho gkc^2$$

Final P.E. due to water of height x in B and $c - x$ in A

$$= \int_0^x \rho gkzdz + \int_0^{c-x} \rho gkzdz$$

$$= \frac{\rho g k}{2}\left[x^2 + (c - x)^2\right]$$

Hence loss in P.E. $= \frac{\rho g k}{2}\left[c^2 - x^2 - (c - x)^2\right]$

$$= \frac{\rho g k}{2}\left[c^2 - x^2 - \left(c^2 + x^2 - 2x\right)\right]$$

$$= \rho g k\ (cx - x^2)$$

$$= \rho g k x\ (c - x)$$

Also some work is done against the compression of air in the cylinder B. This force of compression is equal to the pressure of the air inside B. Let p be the force of compression when water level is at a depth y. Then

$$\rho g h c k = p\ (c - y)\ k$$

$$\Rightarrow \quad p = \frac{\rho g h c}{c - y}$$

Work done by this force $p = -\rho g h c k \int_0^x \frac{dy}{c - y}$

$$= \rho g h c k \log \frac{c - x}{c}$$

Since the total work done is zero, we have

$$\Rightarrow \quad \rho g k x\,(c - x) + \rho g h c k \log \frac{c - x}{c} = 0$$

$$\Rightarrow \quad cx - x^2 + ch \log \frac{c - x}{c} = 0$$

26. Show that the rate per unit of time at which work is done by the internal pressure between the parts of a compressible fluid obeying Boyle's law is

$$\iiint p\left(\frac{\partial u}{\partial x} + \frac{\partial v}{\partial y} + \frac{\partial w}{\partial z}\right) dxdydz$$

where p is the pressure and (u, v, w) the velocity at any point, and the integration extends through the volume of the fluid.

Solution: Let W denote work done, then rate of work done

$$= \frac{dW}{dt}$$

Let $\vec{q} = u\vec{i} + v\vec{j} + w\vec{k}$ and $dv = dxdydz$

Now

$$W = \int -p dv$$

$$\therefore \quad \frac{dW}{dt} = \int -\frac{dp}{dt} dv$$

$$= -\int \frac{d}{dt}(k\rho)\, dv$$

$$\Rightarrow \quad \frac{dW}{dt} = -k \int \frac{d\rho}{dt} dv \qquad \dots (1)$$

Also equation of continuity is

$$\frac{d\rho}{dt} + \rho\left(\vec{\nabla}\cdot\vec{q}\right) = 0$$

$$\frac{d\rho}{dt} = -\rho\left(\vec{\nabla}\cdot\vec{q}\right)$$

Hence (1) becomes

$$\frac{dW}{dt} = -k \int -\left(\vec{\nabla}\cdot\vec{q}\right)\rho dv$$

$$= \int p\left(\vec{\nabla}\cdot\vec{q}\right) dv$$

EXERCISE 2

1. An elastic fluid, the weight of which is neglected obeying Boyle's law is in motion in a uniform straight tube, show that on the hypothesis of parallel sections the velocity at any time t at a distance r from a fixed point in the tube is defined by the equation

$$\frac{\partial^2 \upsilon}{\partial t^2} + \frac{\partial}{\partial r}\left(2\upsilon \frac{\partial \upsilon}{\partial t} + \upsilon^2 \frac{\partial \upsilon}{\partial r}\right) = k \frac{\partial^2 \upsilon}{\partial r^2}$$

2. Prove that the velocity components $q_r\,(r, 0) = -U\left(1 - \frac{a^2}{r^2}\right)$

$\cos\theta$, $q_s\ (r,\ 0) = U\left(1+\dfrac{a^2}{r^2}\right)\sin\theta$ satisfy the equation of motion for a two dimensional inviscid incompressible flow. Find the pressure associated with this velocity field, U and a are constants.

$$\left[\textbf{Ans. } p = -\rho U^2 a^2\left[\frac{3}{r^2}\cos 2\theta + \frac{a^4}{2r^4}\right] + c\right]$$

3. A spherical mass of liquid of radius a has a concentric sphere cavity of radius a, which contains gas at pressure p whose mass may be neglected, at every point of the external boundary of the liquid an impulsive pressure $\bar{w}$ per unit area is applied. Assuming that the gas obeys Boyle's law, show that when the liquid first comes to rest, the radius of the internal spherical surface will be a exp $\left\{\dfrac{-W^2 b}{2p\rho a^2(b-a)}\right\}$ where ρ is the density of the liquid.

4. A given quantity of liquid moves, under no forces, in a smooth conical tube having a small vertical angle, and the distances of its nearer and farther extremities for the vertex at time t are r and r^1, show that

$$2r\frac{d^2r}{dt^2} + \left(\frac{dr}{dt}\right)^2\left[3 - \frac{r}{r^1} - \frac{r^2}{r^{1^2}} - \frac{r^3}{r^{1^3}}\right] = 0$$

the pressure at the two surfaces being equal. Show also that the preceding equation results from supposing the vis-viva of the mass of the liquid to be constant and that the velocity of inner surface is given by the equation

$$V^2 = \frac{Cr^1}{r^3\left(r^1 - r\right)},\ r^{1^3} - r^3 = c^3$$

C and c being constants.

5. A sphere whose radius at time t is $b + a\cos nt$ is surrounded by liquid extending to infinity under no forces. Prove that the pressure at a distance r from the centre is less than the pressure at an infinite distance by

$$\frac{na^2}{r}(b + a\cos nt)\left\{a\left(1 - 3\sin^2 nt\right) + b\cos nt + \frac{a}{2r^3}(b + a\cos nt)^3\sin^2 nt\right\}$$

6. Liquid is contained between two parallel planes, the free surface is a circular cylinder of radius a whose axis is perpendicular to the planes. All the liquid within a concentric circular cylinder of radius b is suddenly annihilated. Prove that if Π be the pressure at the outer surface, the initial pressure at any point of the liquid, distant r from the centre is

$$\Pi\left[\frac{\log r - \log b}{\log a - \log b}\right]$$

7. A sphere of radius a is alone in an unbounded liquid which is at rest at a great distance from the sphere and is subject to no external force. The sphere is forced to vibrate radially keeping its spherical shape, the radius r at any time t being given by $r = a + b\cos nt$. Show that if Π is the pressure in the liquid at a great distance from the sphere, the last pressure (assumed positive at the surface if the sphere during the motion) is $\Pi - n^2\rho b\,(a + b)$.

8. A sphere is at rest in an infinite mass of homogeneous liquid of density ρ, the pressure at infinity being P. If the radius R of the sphere varies in such a way that $R = a + b\cos nt$, where $b < a$, show that the pressure at the surface of the sphere at any time is $P + \dfrac{bn^2\rho}{4}(b - 4a\cos nt - 5\,b\cos nt)$.

9. A homogeneous liquid is contained between two concentric spherical surfaces, the radius of the inner being a and that of the outer indefinitely great. The fluid is attracted to the centre of these surfaces by a force $\phi(r)$, and constant pressure Π is exerted at the outer surface.

Suppose $\int \phi(r)\,dr = \psi(r)$

and that $\psi(r)$ vanishes when r is infinite, show that if the inner surface is suddenly removed, the pressure at the distance r is suddenly diminished by $\Pi \frac{a}{r} - \frac{a\rho}{r} \psi(a)$.

Find $\phi(r)$, so that the pressure immediately after the inner surface is removed may be the same as it would be if no attractive force existed. Also with this value of $\phi(r)$, find the velocity of the inner, boundary of the fluid at any period of the motion.

[**Ans.** $\phi(r) = \frac{k}{r^2}$

where k is a constant

$$\upsilon^2 r^3 = k\left(a^2 - r^2\right) + \frac{2\Pi}{3\rho}\left(a^3 - r^3\right)$$]

10. A fine tube whose cross section k is a function of its lengths, in the form of a closed plane curve of area A, filled with ice, is moved in any manner. When the component angular velocity of the tube about a normal to its plane is Ω, the ice melts without change of the volume. Prove that the velocity of the fluid relatively to the tube at a point where the section is k at any subsequent time when w is the angular velocity is $\dfrac{2A(\Omega - w)}{k\int \frac{ds}{k}}$, the integral being taken once round the table.

3 MOTION IN TWO DIMENSIONS

3.1. Motion in Two Dimensions

If the lines of motion are parallel to a fixed plane (say *xy* plane) and if the velocity at corresponding points of all planes has the same magnitude and direction, then the motion is said to be two dimensional. Evidently in this case $w = 0$ and $u = u\ (x, y, t)$, $v = v\ (x, y, t)$.

In the figure, let *xoy* be the fixed plane and $x^1o^1y^1$ a plane parallel to it. If from a point *P* in the *xy* plane a normal is drawn to meet x^1y^1 in P^1. P^1 is called the point corresponding to *P*. Let $\vec{q}$ be the velocity at *P* making an angle θ with *ox*, then the velocity at P^1 will also be $\vec{q}$ making and angle θ with ox^1. This velocity $\vec{q}$ shall be a function of *x*, *y* and time, but not of *z*.

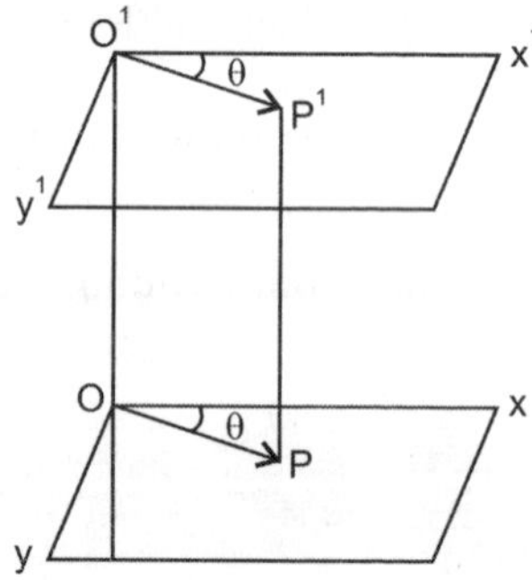

Fig. 3.1

For convenience, let us consider the fluid in two dimensional motion to be confined between two hypothetical parallel planes at unit distance apart. The plane of reference, this is, the plane *xoy* is taken midway and parallel to these planes. Therefore any closed curve drawn in the plane *xoy* will be a cross section of a cylinderical surface of unit length. Accordingly, when we speak of the flow across a curve in the *xoy* plane, we really mean the flow across unit length of the cylinder whose cross section on the xoy plane is that curve.

3.2. Lagrange's Stream Function or Current Function

When the motion is two dimensional $w = 0$

The differential equation of the streamlines are given by $\frac{dx}{u} = \frac{dy}{\upsilon}$

i.e. $\upsilon dx - u dy = 0$ ... (1)

Also for an incompressible fluid, the equation of continuity in two dimensions is

$$\frac{\partial u}{\partial x} + \frac{\partial \upsilon}{\partial y} = 0 \qquad \text{... (2)}$$

The condition for equation (1) to be exact is

$$\frac{\partial \upsilon}{\partial y} = \frac{\partial}{\partial x}(-u)$$

which is the same as (2).

Hence $\upsilon dx - u dy$ is an exact differential say $d\psi$.

i.e. $\upsilon dx - u dy = d\psi = \frac{\partial \psi}{\partial x} dx + \frac{\partial \psi}{\partial y} dy$

So that $u = -\frac{\partial \psi}{\partial y}, \upsilon = \frac{\partial \psi}{\partial x}$

Now (1) is expressible as $d\psi = 0$

Integrating it we get $\psi = \text{const.}$

The function ψ is called the Lagrange's stream function or current function.

The streamlines are given by $\psi = \text{const.}$

It follows that stream function is constant along a streamline.

Remark. (1) It is clear that the existence of a stream function is a consequence of streamlines and equation of continuity for incompressible fluid.

(2) Stream function exists for all types of two dimensional motion—rotational or irrotational.

(3) The necessary conditions for the existence of ψ are:

(i) the flow must be continuous

(ii) the flow must be incompressible.

3.3. Physical Significance of Stream Function

The difference of the values of ψ at the two points represents the flux of a fluid across any curve joining the two points.

Let ds be a line element at a point $P(x, y)$ of a curve AB. Let PT and PN be the tangent and normal at P to the curve. Let (u, v) be the velocity components of the fluid at P.

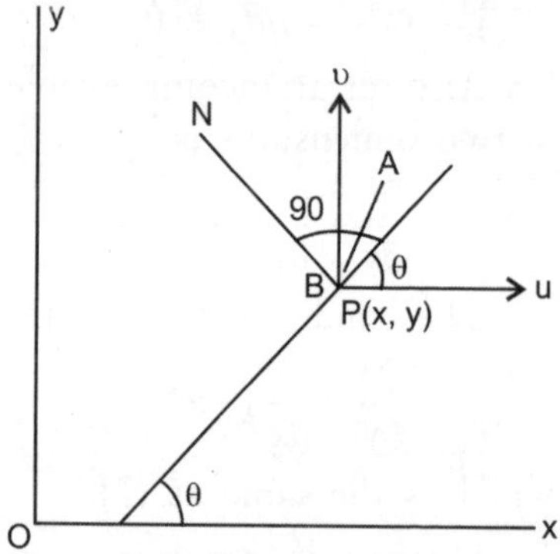

Fig. 3.2

Direction cosines of the normal PN are

$\cos(90 + \theta), \cos\theta, \cos 90$

or $-\sin\theta, \cos\theta, 0$

Inward normal velocity

$$= \hat{n} \cdot \vec{q}$$

$$= u(-\sin\theta) + v(\cos\theta)$$

$$= -u\sin\theta + v\cos\theta$$

Flux across the curve AB from right to left

= density · normal velocity · area of the cross section

$$= \int_{AB} \rho\,(\hat{n} \cdot \vec{q})\, ds$$

$$= \int_{AB} \rho\,(-u\sin\theta + v\cos\theta)\, ds$$

$$= \rho \int_{AB} \left(-u\frac{dy}{ds} + v\frac{dx}{ds}\right) ds$$

$$= \rho \int_{AB} (-u\,dy + v\,dx)$$

$$= \rho \int_{AB} \left\{ \left(\frac{\partial \psi}{\partial y} \right) dy + \left(\frac{\partial \psi}{\partial x} \right) dx \right\}$$

$$= \rho \int_{AB} d\psi$$

$$= \rho (\psi_2 - \psi_1)$$

Where ψ_1 and ψ_2 are the values of ψ at A and B respectively.

Hence flux across $AB = \rho(\psi_2 - \psi_1)$.

3.4. Irrotational Motion in Two Dimensions

To show that in steady two dimensional irrotational motion, stream function satisfies Laplace's equation.

Let the fluid motion be irrotational so that there exists a velocity potential ϕ such that

$$\vec{q} = -\vec{\nabla}\phi$$

Hence

$$u = -\frac{\partial \phi}{\partial x}, v = -\frac{\partial \phi}{\partial y} \qquad \text{... (1)}$$

If ψ is a stream function, then

$$u = -\frac{\partial \psi}{\partial y}, v = \frac{\partial \psi}{\partial x} \qquad \text{... (2)}$$

From (1) and (2), we have

$$-\frac{\partial \phi}{\partial x} = -\frac{\partial \psi}{\partial y}, -\frac{\partial \phi}{\partial y} = \frac{\partial \psi}{\partial x}$$

Now

$$\frac{\partial^2 \psi}{\partial x^2} + \frac{\partial^2 \psi}{\partial y^2} = \frac{\partial}{\partial x}\left(\frac{\partial \psi}{\partial x}\right) + \frac{\partial}{\partial y}\left(\frac{\partial \psi}{\partial y}\right)$$

$$= \frac{\partial}{\partial x}\left(-\frac{\partial \phi}{\partial y}\right) + \frac{\partial}{\partial y}\left(\frac{\partial \phi}{\partial x}\right) = 0$$

Hence ψ satisfies Laplace's equation.

Also from (1) and (2), we have

$$\frac{\partial \phi}{\partial x}\frac{\partial \psi}{\partial x} + \frac{\partial \phi}{\partial y}\frac{\partial \psi}{\partial y} = (-u)\, v + (-v)(-u) = 0$$

Hence the family of curves $\phi(x, y)$ = const. and $\psi(x, y)$ = const. cut orthogonally at their points of intersection.

3.5. Complex Potential and Velocity

Complex Potential. The relation $w = \phi + i\psi$, where ϕ and ψ are velocity potential and stream function of the two dimensional irrotational motion of a perfect liquid, is defined as the complex potential of the fluid motion.

Since $\dfrac{\partial\phi}{\partial x} = \dfrac{\partial\psi}{\partial y}, \dfrac{\partial\phi}{\partial y} = -\dfrac{\partial\psi}{\partial x}$

which are Cauchy Riemann equations, it follows that w is an analytic function of z in a region where ϕ and ψ are single valued functions.

Conversely, *if w is analytic, its real and imaginary parts give the velocity potential and stream function of a possible irrotational two dimensional fluid motion.*

Complex Velocity. We have

$$w = \phi + i\psi = f(z) \qquad \ldots (1)$$

Differentiating (1) partially w.r.t. x, we get

$$\frac{\partial w}{\partial x} = \frac{\partial\phi}{\partial x} + i\frac{\partial\psi}{\partial x} = f^1(z)\frac{\partial z}{\partial x}$$

or $\dfrac{dw}{dz}\dfrac{\partial z}{\partial x} = -u + iv = f^1(z)$

Since $\dfrac{\partial z}{\partial x} = 1, \dfrac{\partial\phi}{\partial x} = -u, \dfrac{\partial\psi}{\partial x} = v$

Hence $\dfrac{dw}{dz} = -u + iv$

Thus $\left|\dfrac{dw}{dz}\right| = \sqrt{u^2 + v^2}$

Therefore $\left|\dfrac{dw}{dz}\right|$ represents the complex velocity at any point.

The points where velocity is zero are called stagnation points. Thus, for stagnation points, $\dfrac{dw}{dz} = 0$.

3.6. Two Dimensional Sources, Sinks, Strength

1. Sources. A two dimensional source is a point from which the liquid is emitted radially and symmetrically in all directions in xy plane.

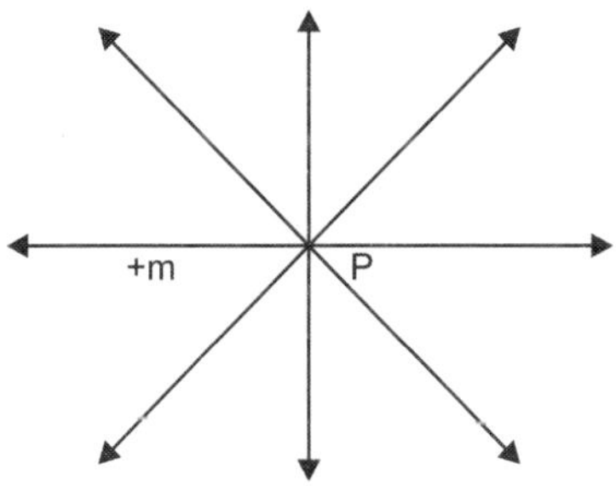

Fig. 3.3

2. Sink. A point to which fluid is flowing in symmetrically and radially in all directions is called sink. Thus sink is a negative of source.

Difference between Source and Sink

Sink is a point which liquid is continuously created and source is a point at which liquid is continuously annihilated. In fact, source and sink are purely abstract conceptions which do not occur in nature.

The sources and sinks are the points at which the velocity potential and stream functions becomes infinite.

3. Strength. Strength of a source is defined as total volume of flow per unit time from it.

Thus if $2\Pi m$ is the total volume of flow across any small circle surrounding the source, then m is called strength of the source. Sink is a source of strength $(-m)$.

3.7. Complex Potential Due to a Source

Let a source of strength m be situated at origin. We are required to determine complex potential w at a point $P(r, \theta)$ due to this source. The velocity at P due to the source is purely radial, let this velocity be q_r. Flux across a circle of radius r surrounding the source at O is $2\Pi r q_r$.

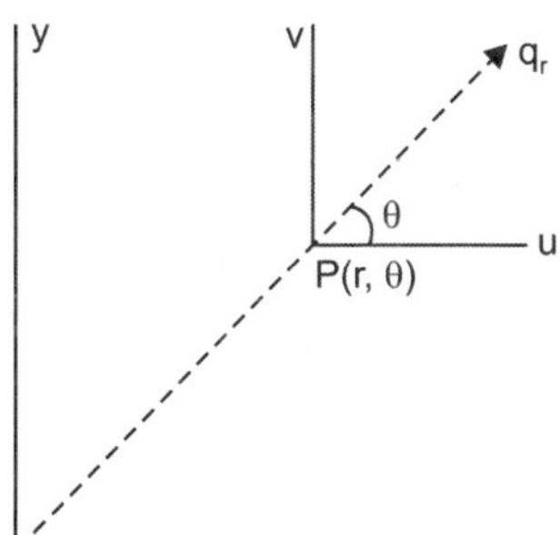

Fig. 3.4

By definition of strength $2\Pi rq_r = 2\Pi m$.

Hence $q_r = \dfrac{m}{r}$.

If u and v are the compound velocity at $P(r, \theta)$ then

$$u = q_r \cos\theta = \frac{m}{r}\cos\theta$$

$$v = q_r \sin\theta = \frac{m}{r}\sin\theta$$

Also

$$\frac{dw}{dz} = -u + iv$$

$$= \frac{m}{r}(-\cos\theta + i\sin\theta)$$

$$= -\frac{m}{r}e^{-i\theta} = \frac{-m}{re^{i\theta}}$$

$$= -\frac{m}{z} \text{ as } z = re^{i\theta}$$

$$\Rightarrow \quad dw = -m\frac{dz}{z}$$

Integrating, we get $w = -m \log z$ neglecting constant of integration.

Deductions. (1) If the source $+m$ is at a point $z = z_1$ in plane of $z = 0$, then by shifting the origin we have

$$w = -m \log(z - z_1)$$

This is the required expression for w in this case.

(2) To find the complex potential w at any point z due to sources of strengths m_1, m_2, m_3 ..... situated at the points z_1, z_2, z_3 .....

In this case, we have

$$w = -m_1 \log(z - z_1) - m_2 \log(z - z_2) \text{ .....}$$

Now $w = \phi + i\psi$ and $z - z_n = r_n e^{i\theta_n}$

$$\phi + i\psi = -m_1 \log(r_1 e^{i\theta_1}) - m_2 \log(r_2 e^{i\theta_2}) \text{ .....}$$

$$= -m_1 \log r_1 - m_1 i\theta_1$$

$$= m_2 \log r_2 - m_2 i\theta_2$$

Equating real and imaginary parts, we get

$$\phi = -m_1 \log r_1 - m_2 \log r_2 \ .....$$

and

$$\psi = -m_1\theta_1 - m_2\theta_2 \ .....$$

3.8. Two Dimensional Doublets

A doublet is a combination of a source of strength m and a sink of strength $-m$ at a small distance δs apart such that $m\delta s$ is finite.

Strength of a doublet. If $m\delta s = \mu$, a finite quantity, where m is taken infinitely large and δs infinitely small, then μ is called the strength of the doublet and line δs is called the axis of the doublet and its direction is taken from sink to source.

3.9. Complex Potential for a Doublet

Let a doublet be formed of a sink $-m$ at $z = a$ and a source $+m$ at $z = a + \delta a$, AB making an angle α with x-axis. Let μ be the strength of the doublet, then $\mu = mAB$ and $\delta a = ABe^{i\alpha}$ for $z = re^{i\theta}$.

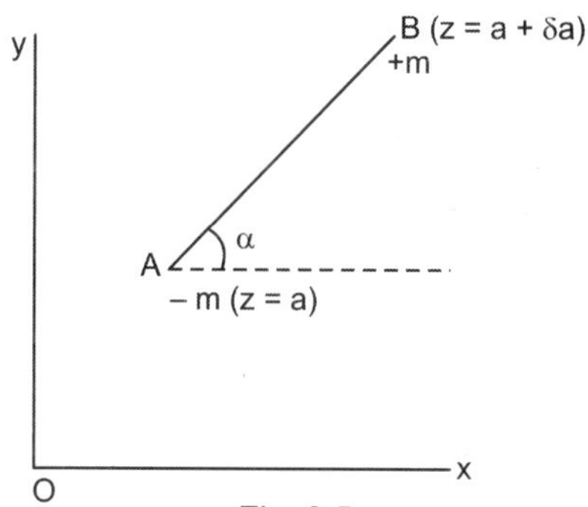

Fig. 3.5

The complex potential w due to this doublet at any point P (z) is given by

$$w = +m \log (z - a) - m \log [z - (a + \delta a)]$$

$$= -m \log \frac{z - a - \delta a}{z - a}$$

$$= -m \log \left(1 - \frac{\delta a}{z - a}\right)$$

$$= \frac{m\delta a}{z - a} \text{ upto first approximation}$$

$$= \frac{m\, ABe^{i\alpha}}{z - a} = \frac{\mu e^{i\alpha}}{z - a}$$

Deductions

(1) If the axis of doublet is along x-axis, then $\alpha = 0$, so that

$$w = \frac{\mu}{z - a}$$

(2) If the axis of a doublet is along x-axis and the doublet is at origin, then $\alpha = 0$, $a = 0$ so that $w = \dfrac{\mu}{z}$.

(3) If a system consists of doublets of strenghts $\mu_1, \mu_2, \ldots.$ placed at $z = a_1, a_2, \ldots..$ then the complex potential due to the above system is given by

$$W = \frac{\mu_1 e^{i\alpha_1}}{z - a_1} + \frac{\mu_2 e^{i\alpha_2}}{z - a_2} + \ldots..$$

3.10. Image

If there exists a curve C in xy plane in a fluid such that there is no flow across it, then the system of sources, sinks and doublets on one side of C is said to be the image of the sources, sinks and doublets on the other side of C.

3.11. To Find the Image of a Source with Respect to a Straight Line

Let there be two sources each of strength m at A and B on opposite sides of an equidistant from the line OP.

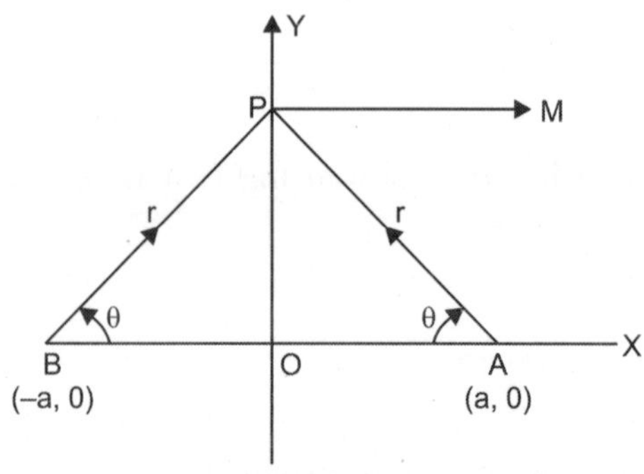

Fig. 3.6

Since $\phi = -m \log r$

$$\frac{\partial \phi}{\partial r} = -\frac{m}{r}$$

$$\Rightarrow \quad \frac{-\partial\phi}{\partial r} = \frac{m}{r}$$

Thus velocity at P due to m at $A = \frac{m}{r}$ along AP and the velocity at P due to m at $B = \frac{m}{r}$ along BP.

Hence the normal velocity at P

$$= -\frac{m}{r}\cos\theta + \frac{m}{r}\cos\theta$$

$$= 0$$

Thus there is no flow across the straight line OP. Hence *the image of a simple source in a straight line is an equal source equidistant from the line.*

3.12. Image of a Doublet in a Straight Line

If AA^1 is a doublet with its axis inclined at angle α to OX, then by considering the image of the source and sink with respect to the line OY, the image of the doublet AA^1 is a doublet BB^1 of equal strength symmetrically situated opposit to AA^1 with its axis antiparallel to that of AA^1.

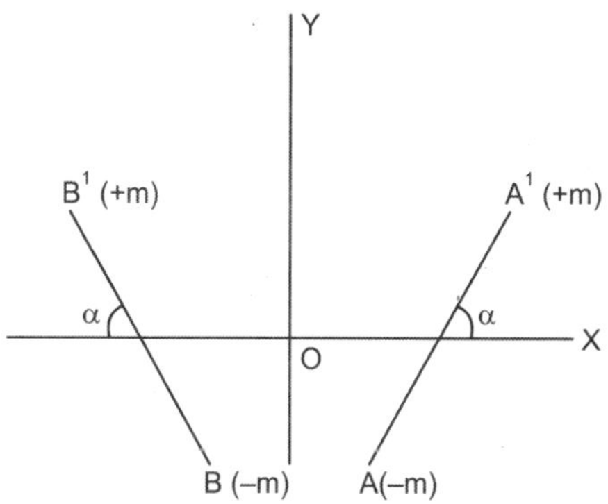

Fig. 3.7

3.13. Image of a Source with Respect to a Circle in Two Dimensions

Let there be a source of strength m at a point A where $OA = f$ outside the circle. Let B be the inverse point of A, so that if a be the radius of the circle then

$$OB \cdot OA = a^2$$

$$OB = \frac{a^2}{f}$$

and we place an equal source at B.

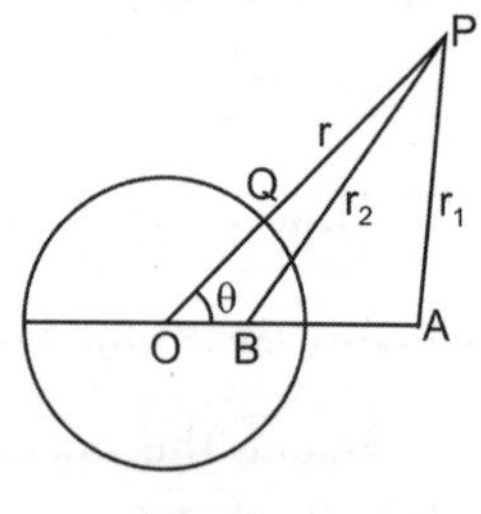

Fig. 3.8

Due to the sources at A and B, velocity potential at any point P is

$$\phi = -m \log r_1 r_2$$

Where

$$AP = r_1,\ BP = r_2$$

$$= -\frac{m}{2} \log r_1^2 r_2^2$$

$$= -\frac{m}{2} \log \left(r^2 + f^2 - 2\,fr \cos\theta\right) \left(r^2 + \frac{a^4}{f^2} - 2r\frac{a^2}{f}\cos\theta\right)$$

Normal velocity at any point Q on the circle is given

$$= \left(-\frac{\partial\phi}{\partial r}\right)_{r=a}$$

Now

$$\frac{-\partial\phi}{\partial r} = m\left[\frac{r - f\cos\theta}{a^2 + f^2 - 2fr\cos\theta} + \frac{r - \dfrac{a^2}{f}\cos\theta}{r^2 + \dfrac{a^4}{f^2} - \dfrac{2ra^2}{f}\cos\theta}\right]$$

At $r = a$

$$\frac{-\partial\phi}{\partial r} = m\left[\frac{a - f\cos\theta}{a^2 + f^2 - 2fa\cos\theta} + \frac{a - \dfrac{a^2}{f}\cos\theta}{a^2 + \dfrac{a^4}{f^2} - \dfrac{2a^3}{f}\cos\theta}\right]$$

$$= m\left[\frac{a - f\cos\theta}{a^2 + f^2 - 2fa\cos\theta} + \frac{a - \dfrac{a^2}{f}\cos\theta}{\dfrac{a^2}{f^2}\left(f^2 + a^2 - 2af\cos\theta\right)}\right]$$

$$= m \left[\frac{a - f \cos \theta + \frac{f^2}{a} - f \cos \theta}{a^2 + f^2 - 2af \cos \theta} \right]$$

$$= \frac{m}{a}$$

Hence if we put a sink at O of strength $-m$ then velocity along OQ due to the two sources at A and B and a sink at O is zero, so that circle is a streamline. Therefore, the image of a source at A is an equal source at B and an equal sink at O.

3.14. Image of a Doublet with Regard to a Circle

Let us determine the image of a doublet AA^1 with its axis making an angle α with OA, outside the circle, there being a sink $-m$ at A and a source m at A^1. Join OA and OA^1. Let B and B^1 be the inverse point of A and A^1 with respect to the circle with O as centre.

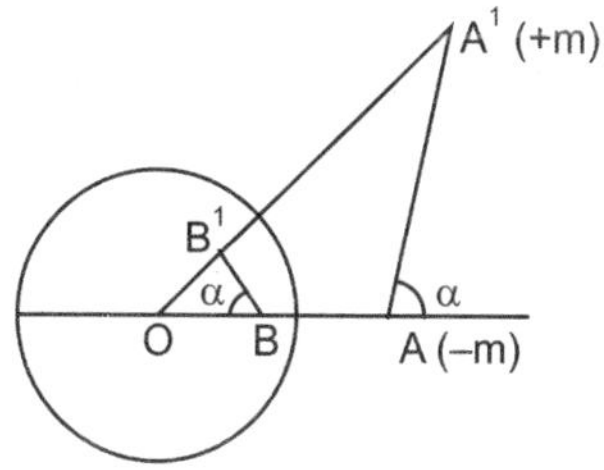

Fig. 3.9

Then

$$OAOB = OA^1OB^1 = a^2 \qquad \text{... (1)}$$

where a is the radius of the circle.

Now the image of source m at A^1 consists of a source m at B^1 and a sink $-m$ at O. Similarly the image of sink $-m$ at A consists of a sink at B and a source m at O. Compounding these we see that source m and sink $-m$ at O cancel each other and hence the image of the given doublet AA^1 is another doublet B^1B.

Let the strength of the given doublet AA^1 be μ.

Then

$$\mu = \underset{A \to A^1}{Lt} (m \cdot AA^1) \qquad \text{... (2)}$$

From (1)

$$\frac{OA}{OA^1} = \frac{OB^1}{OB} \qquad \text{... (3)}$$

and $\angle BOB^1 = \angle AOA^1$

which shows that $\Delta\ OAA^1$ and $\Delta\ OBB^1$ are similar.

Hence

$$\frac{BB^1}{AA^1}=\frac{OB^1}{OA}=\frac{OB^1}{OA}\frac{OA^1}{OA^1}=\frac{a^2}{OAOA^1} \quad \text{... (4)}$$

$\therefore\ \mu^1 =$ strength of the doublet B^1B

$$= \underset{B^1 \to B}{Lt}\ (mB^1B)$$

$$= \underset{A^1 \to A}{Lt}\ \frac{a^2}{OAOA^1}(mAA^1) \qquad \text{From (4)}$$

$$= \frac{\mu a^2}{f^2} \qquad \text{From (2)}$$

and taking $OA = OA^1 = f$

Thus the image of a two dimensional doublet at A with respect to a circle is another doublet at the inverse point B, the axes of the doublets making supplementary angles with the radius OBA.

3.15. The Milni Thomson Circle Theorem or Circle Theorem

Statement. Let $f(z)$ be the complex potential for a flow having no rigid boundaries and such that there are no singularities of flow within the circle $|z| = a$. Then if a circular cylinder $|z| = a$ is inserted in the flow field, the complex potential of the resulting motion is given by

$$w = f(z) + \overline{f}\left(-\frac{a^2}{\overline{z}}\right) \text{ for } |z| \geq a$$

Proof. Let C be the cross section of the circular cylinder $|z| = a$. Then on C, $z\overline{z} = a^2$ or $\overline{z} = \dfrac{a^2}{z}$ so that $f(\overline{z}) = f\left(\dfrac{a^2}{z}\right)$ and so $\overline{f}(\overline{z}) = \overline{f}\left(\dfrac{a^2}{z}\right)$

Hence, for points on C, complex potential is

$$w = f(z) + \overline{f}\left(\frac{a^2}{z}\right)$$

$$= f(z) + \bar{f}(\bar{z})$$

= a real quantity

But $w = \phi + i\psi$ = a real quantity

Equating imaginary part on both the sides, we get

$\psi = 0$ or ψ = const, for 0 is also a constant.

Also we know that if z lies outside C, then the point $\frac{a^2}{z}$ is inside C. Also it is given that $f(z)$ has singularities outside C. Consequently $f\left(\frac{a^2}{z}\right)$ and therefore $\bar{f}\left(\frac{a^2}{z}\right)$ has singularities inside C. It means that the additional term $\bar{f}\left(\frac{a^2}{z}\right)$ introduces no new singularity outside C.

Hence $|z| = a$ is a possible boundary for the new flow and $w = f(z) + \bar{f}\left(\frac{a^2}{z}\right)$ is the appropriate complex potential for the new flow.

Remark. The Milni Thomson circle theorem provides a conventional method for finding the image system of a given two dimensional system which lies outside a circular boundary. For if $w = f(z)$ represents the given system in presence of the circular boundary $|z| = a$ then $w = \bar{f}\left(\frac{a^2}{z}\right)$ represents the image system.

3.16. To Find Image of a Source with Respect to a Circle of Radius *a* with the Help of Circle Theorem

Let Q be an inverse point of P $(z = f)$ with regard to the circle such that

$$OP \cdot OQ = a^2$$

$$\Rightarrow \quad OQ = \frac{a^2}{f}$$

When the source is alone in the fluid the complex potential at a point $P(z)$ is given by

$$f(z) = -m \log (z - f)$$

Let a circular cylinder $|z| = a$ (where $a < f$) be inserted, then by circle theorem the complex potential is given by

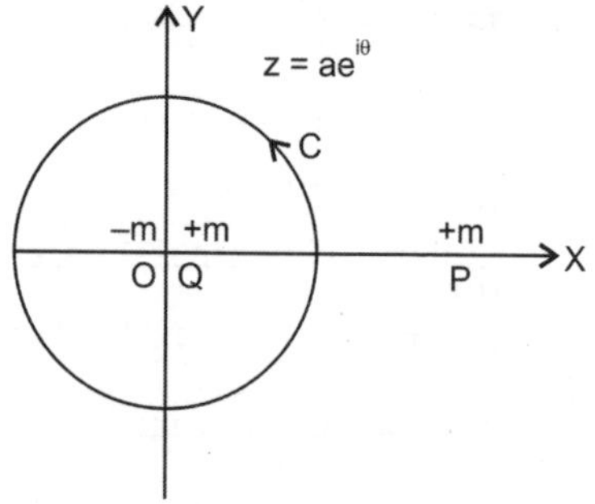

Fig. 3.10

$$w = f(z) + \bar{f}\left(\frac{a^2}{z}\right)$$

$$= -m \log (z - f) - m \log \left(\frac{a^2}{z} - f\right)$$

$$= -m \log (z - f) - m \log \left(\frac{a^2 - fz}{z}\right)$$

$$= -m \log (z - f) - m \log \left(a^2 - fz\right) + m \log z$$

$$= -m \log (z - f) - m \log \left[(-f)\left(z - \frac{a^2}{f}\right)\right] + m \log z$$

$$= -m \log (z - f) - m \left[\log (-f) + \log \left(z - \frac{a^2}{f}\right)\right] + m \log z$$

$$= -m \log (z - f) - m \log \left(z - \frac{a^2}{f}\right) + m \log z - m \log (-f)$$

$$\Rightarrow \quad w = -m \log (z - f) - m \log \left(z - \frac{a^2}{f}\right) + m \log z + \text{constant}$$

Ignoring the constant term, we get

$$w = -m \log (z - f) - m \log \left(z - \frac{a^2}{f}\right) + m \log z \qquad \dots (1)$$

Hence (1) shows that w is the complex potential of

(i) a source of strength $+m$ at P, $z = f$

(ii) a source of strength $+m$ at an inverse point Q, $z = \dfrac{a^2}{f}$

(iii) a sink of strength $-m$ at the origin O, $z = 0$

Let C ($z = ae^{i\theta}$) be any point on the circle, then from (1), we get

$$w = m \log\left\{\left(ae^{i\theta} - f\right)\left(ae^{i\theta} - \frac{a^2}{f}\right)\right\} + m \log\left(ae^{i\theta}\right)$$

Equating the imaginary parts, we get

$$\psi = -m \tan^{-1}\left(\frac{a \sin\theta}{a \cos\theta - f}\right) - m \tan^{-1}\left(\frac{a \sin\theta}{a\cos\theta - \dfrac{a^2}{f}}\right) + m\theta$$

$$= -m \tan^{-1}\left(\tan\theta\right) + m\theta = -m\theta + m\theta = 0$$

which shows that the circle is a streamline. Thus the image system of a source of strength $+m$ at an inverse point and a sink $-m$ at the origin, the centre of the circle.

3.17. To Find the Image of a Doublet with Respect to a Circle by Circle Theorem

Let a two-dimensional doublet of strength μ be place at $z = f$ and making an angle α with the axis of X.

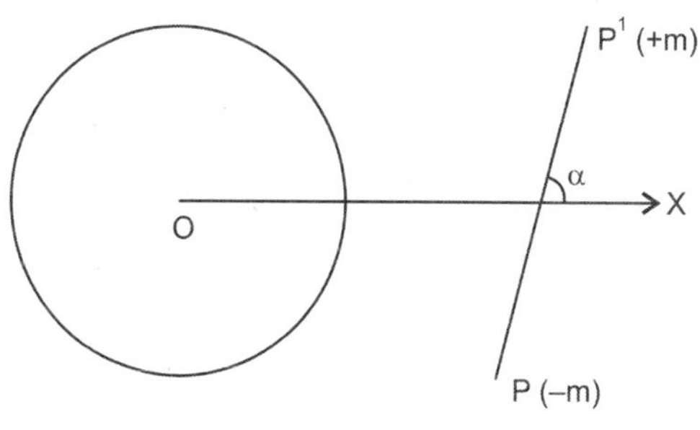

Fig. 3.11

Then complex potential due to this doublet at any point z is $f(z) = \dfrac{\mu e^{i\alpha}}{z - f}$.

If a circular cylinder $|z| = a$ $(a < b)$ is inserted, then by the circle theorem, the complex potential $w(z)$ of the flow field in the region $|z| \geq a$ is

$$w = f(z) + \bar{f}\left(\frac{a^2}{z}\right)$$

$$= \frac{\mu e^{i\alpha}}{z - f} + \frac{\mu e^{i\alpha}}{\dfrac{a^2}{z} - f}$$

$$= \frac{\mu e^{i\alpha}}{z - f} + \frac{\mu e^{-i\alpha}}{\dfrac{a^2}{z} - f}$$

$$= \frac{\mu e^{i\alpha}}{z - f} - \frac{\mu e^{i(\pi - \alpha)}}{\dfrac{a^2}{z} - f}$$

$$= \frac{\mu e^{i\alpha}}{z - f} + \frac{\mu z\, e^{i(\pi - \alpha)}}{fz - a^2}$$

$$= \frac{\mu e^{i\alpha}}{z - f} + \frac{\mu e^{i(\pi - \alpha)} z}{f\left(z - \dfrac{a^2}{f}\right)}$$

$$= \frac{\mu e^{i\alpha}}{z - f} + \frac{\mu e^{i(\pi - \alpha)}\left[z - \dfrac{a^2}{f} + \dfrac{a^2}{f}\right]}{f\left(z - \dfrac{a^2}{f}\right)}$$

$$w = \frac{\mu e^{i\alpha}}{z - f} + \frac{\mu a^2}{f^2}\frac{e^{i(\pi - \alpha)}}{z - \dfrac{a^2}{f}}\ \frac{\mu}{f} e^{i(\pi - \alpha)}$$

Neglecting the constant term $\frac{\mu e^{i(\pi-\alpha)}}{f}$, we get

$$w = \frac{\mu e^{i\alpha}}{z-f} + \frac{\mu a^2}{f^2}\frac{e^{i(\pi-\alpha)}}{z-\frac{a^2}{f}}$$

It follows that the image system consists of—

(i) a doublet of strength μ at $z = f$ inclined at an angle α with the x-axis.

(ii) a doublet of strength $\frac{\mu a^2}{f^2}$ at an inverse point $z = \frac{a^2}{f}$ inclined at an angle $\pi - \alpha$ with the x-axis.

For this complex potential, circle is a streamline and hence the image system for a doublet of strength μ at $z = f$ (outside the circle) is a doublet of strength $\mu^1 = \frac{\mu a^2}{f^2}$ and its axis inclined at an angle $\pi - \alpha$.

3.18. Theorem of Blasius (I.A.S. 85)

Statement. In a steady two dimensional irrotational motion given by the complex potential $w = f(z)$, if the pressure thrusts on the fixed cylinder of any shape are represented by a force (X, Y) and a couple of moment M about the origin of coordinates then, neglecting the external forces,

$$X - iY = \frac{1}{2} i\rho \int_C \left(\frac{dw}{dz}\right)^2 dz$$

$$M = \text{Real part of} \left[-\frac{1}{2}\rho \int_C \left(\frac{dw}{dz}\right)^2 z dz\right]$$

where ρ is the density and integrals are taken round the contour C of the cylinder.

Proof. We consider a cylinder of any cross section C, placed in steady, irrotational flow of a fluid.

Let $P(x, y)$ and Q $(x + \delta x, y + \delta y)$ be two adjacent points on C such that arc $PQ = \delta s$. Let the tangent at P makes an

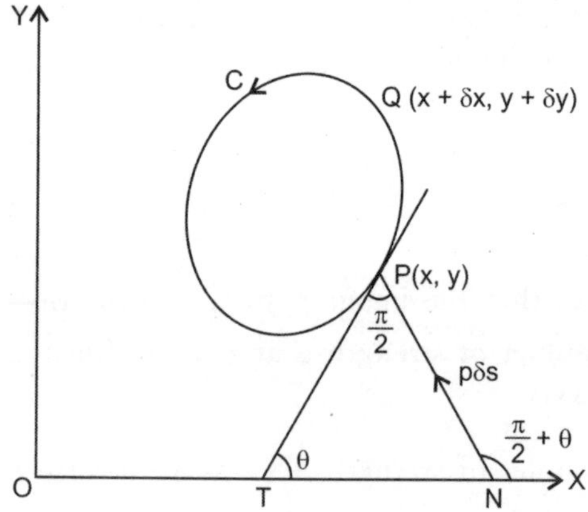

Fig. 3.12

angle θ with x-axis, so that the inward normal at P will make angle 90° + θ with x-axis.

Now if p denote the pressure at P, the thrust $p\delta s$ at P acts along inward normal. Its component along x and y axes are $pds \cos(90 + \theta)$, $pds \sin(90 + \theta)$.

or $-pds \sin\theta$, $pds \cos\theta$

Hence $X = \int_C -pds \sin\theta$, $Y = \int_C pds \cos\theta$

Thus $X - iY = \int_C p(-\sin\theta - i\cos\theta)ds$

$$= -i \int_C p(\cos\theta - i\sin\theta)ds$$

Bernoulli's equation for steady motion gives

$$\frac{p}{\rho} + \frac{1}{2}q^2 = A = \text{Const.}$$

$$\Rightarrow \quad p = \left(A - \frac{1}{2}q^2\right)\rho \qquad \text{... (1)}$$

So from (1)

$$X - iY = -i\rho \int_C \left(A - \frac{1}{2}q^2\right)(\cos\theta - i\sin\theta)\, ds$$

$$= \frac{i\rho}{2}\int_C q^2 e^{-i\theta} ds - i\rho A \int (\cos\theta - i\sin\theta)ds$$

$$= \frac{i\rho}{2} \int_C q^2 e^{-i\theta} ds - i\rho A \int (dx - idy)$$

$$\text{Since } \cos\theta = \frac{dx}{ds}, \sin\theta = \frac{dy}{ds}$$

But $\int_C (dx - idy) = \int_C d\bar{z} = 0$, by Cauchy's theorem

Hence $X - iY = \frac{i\rho}{2} \int_C q^2 e^{-i\theta} ds$ ... (2)

Let u and v be the velocity components. Then

$$\frac{dw}{dz} = -u + iv = -q\cos\theta + iq\sin\theta$$

$$= -q(\cos\theta - i\sin\theta)$$

$$= -q\,\bar{e}^{i\theta}$$

$$\Rightarrow \left(\frac{dw}{dz}\right)^2 dz = q^2 e^{-2i\theta}(dx + idy)$$

$$= q^2 e^{-2i\theta}(\cos\theta + i\sin\theta)\, ds$$

$$= q^2 e^{-i\theta} ds \qquad \text{... (3)}$$

Hence from (2), we have

$$X - iY = \frac{i\rho}{2} \int \left(\frac{dw}{dz}\right)^2 dz$$

The moment of the thrust pds about the origin is

$$M = \int_C \left[-(-pds\sin\theta)y + (pds\cos\theta)x\right]$$

$$= \int_C p(y\sin\theta + x\cos\theta)ds$$

$$= \int_C \left(A - \frac{1}{2}q^2\right)\rho(y\sin\theta + x\cos\theta)\, ds \qquad \text{[From (1)]}$$

$$= A\rho \int_C (y\sin\theta + x\cos\theta)\, ds$$

$$- \frac{\rho}{2} \int_C q^2 (y\sin\theta + x\cos\theta) ds$$

$$M = A\rho \int_C (ydy + xdx) + \frac{\rho}{2} \int_C q^2 (y \sin\theta + x \cos\theta)\, ds$$

But $A\rho \int_C (ydy + xdx) = A\rho \int_C ydy + A\rho \int_C xdx$

$$= 0 + 0 = 0 \quad \text{(By Cauchy's Theorem)}$$

Hence

$$M = -\frac{\rho}{2} \int q^2 (y \sin\theta + x \cos\theta)\, ds$$

$$= \text{Real part of } -\frac{\rho}{2} \int q^2 (x + iy)(\cos\theta - i \sin\theta) ds$$

$$= \text{Real part of } -\frac{\rho}{2} \int z\, q^2 e^{-i\theta} ds$$

$$= \text{Real part of } -\frac{\rho}{2} \int z \left(\frac{dw}{dz}\right)^2 dz \quad \text{From (3).}$$

3.19. Conformal Representation or Transformation of a Mapping

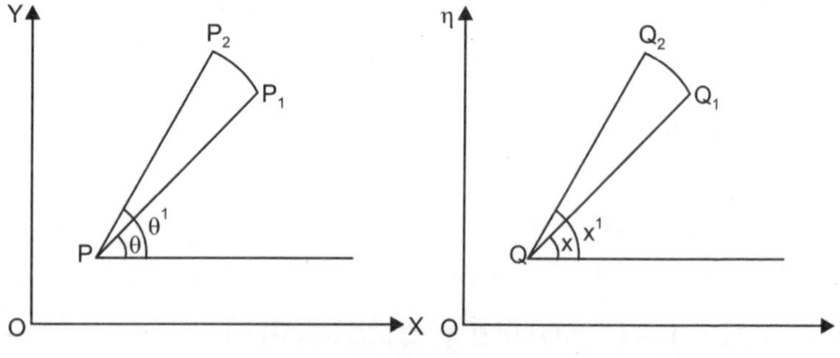

Fig. 3.13

Let $f(z)$ be a function of the complex variable $z = x + iy$ and let $f(z)$ be single valued and differentiable within a closed contour C in the z plane (i.e. xy plane). Let $\zeta = \xi + i\eta$ be another complex variable in ζ plane (i.e. $\xi\eta$ plane) and let there be a relation $\zeta = f(z)$. Then corresponding to each point in the z plane within or on C, there will be a point ζ in ξ plane and points on C or within C will lie on or within a certain contour C^1 in the ζ plane. The necessary condition for existence of such a mapping of z plane into ζ plane is that $f^1(z)$ should never vanish at any point on or within C, or in other words $\frac{d\zeta}{dz}$ must exist independent of the directions of δz. Thus let P, P_1, P_2 be neighbouring points Z, Z_1, Z_2 and Q, Q_1, Q_2 the corresponding points ξ, ζ_1, ζ_2, ζ_3.

Then we have

$$\frac{\zeta_1 - \zeta}{z_1 - z} = \frac{f(z_1) - f(z)}{z_1 - z}$$

and $$\frac{\zeta_2 - \zeta}{z_2 - z} = \frac{f(z_2) - f(z)}{z_2 - z}$$

In the limit when P_1, P_2 approach P, we have

$$\frac{\zeta_1 - \zeta}{z_1 - z} = f^1(z), \frac{\zeta_2 - \zeta}{z_2 - z} = f^1(z) \qquad \text{very nearly}$$

$$\therefore \quad \frac{\zeta_1 - \zeta}{z_1 - z} = \frac{\zeta_2 - \zeta}{z_2 - z} = f^1(z) = \frac{d\zeta}{dz} \qquad \ldots (1)$$

$$\Rightarrow \quad \frac{QQ_1 e^{ix}}{PP_1 e^{i\theta}} = \frac{QQ_2 e^{ix^1}}{PP_2 e^{i\theta^1}}$$

$$\Rightarrow \quad \frac{QQ_1}{PP_1} e^{i(x-\theta)} = \frac{QQ_2}{PP_2} e^{i(x^1 - \theta^1)}$$

$$\Rightarrow \quad \frac{QQ_1}{PP_1} e^{i(x-\theta)} = \frac{QQ_2}{PP_2} e^{i(x^1 - \theta^1)}$$

$$\Rightarrow \quad x - \theta = x^1 - \theta^1$$

$$\Rightarrow \quad x^1 - x = \theta^1 - \theta$$

$$\Rightarrow \quad \angle Q_1QQ_2 = \angle P_1PP_2$$

$$\Rightarrow \quad \frac{QQ_1}{PP_1} = \frac{QQ_2}{PP_2} = |f^1(z)| = \left|\frac{d\zeta}{dz}\right|$$

Thus the triangles P_1PP_2 and Q_1QQ_2 are similar. This establishes the similarity of the corresponding infinitesimal elements of the two planes. Such a relation between the two planes is called the conformal representation of plane on the other.

Again

$$\frac{\Delta Q_1 Q Q_2}{\Delta P_1 P P_2} = \frac{\frac{1}{2} QQ_1 \cdot QQ_2 \sin Q_1 Q Q_2}{\frac{1}{2} PP_1 \cdot PP_2 \sin P_1 P P_2}$$

$$= \frac{QQ_1}{PP_1} \cdot \frac{QQ_2}{PP_2} = | f^1(z) |^2 \qquad \text{... (2)}$$

Also $\zeta = f(z)$

i.e. $\xi + i\eta = f(x + iy)$, we have

$$\frac{\delta\zeta}{\delta z} = \frac{\delta(\xi + i\eta)}{\delta(x + iy)} = \frac{\frac{\partial \xi}{\partial x}\delta x + \frac{\partial \xi}{\partial y}\delta y + i\left(\frac{\partial \eta}{\partial x}\delta x + \frac{\partial \eta}{\partial y}\delta y\right)}{\delta x + i\delta y}$$

$$\Rightarrow \quad \frac{\partial \zeta}{\partial z} = \frac{\left(\frac{\partial \xi}{\partial x} + i\frac{\partial \eta}{\partial x}\right)\delta x + \left(\frac{\partial \xi}{\partial y} + i\frac{\partial \eta}{\partial y}\right)\partial y}{\delta x + i\delta y}$$

Since $\frac{\delta\zeta}{\delta z}$ is independent of $\frac{\delta x}{\delta y}$, we must have

$$\frac{\partial \xi}{\partial y} + i\frac{\partial \eta}{\partial y} = i\left(\frac{\partial \xi}{\partial x} + i\frac{\partial \eta}{\partial x}\right)$$

So $\frac{\partial \xi}{\partial x} = \frac{\partial \eta}{\partial y}$ and $\frac{\partial \xi}{\partial y} = \frac{-\partial \eta}{\partial x}$ ... (3)

Also

$$\frac{d\zeta}{dz} = \frac{\partial \xi}{\partial y} + i\frac{\partial \eta}{\partial y} \qquad \text{... (4)}$$

Hence

$$| f^1(z) |^2 = \left| \frac{d\zeta}{dz} \right|^2 = \left(\frac{\partial \xi}{\partial y}\right)^2 + \left(\frac{\partial \eta}{\partial y}\right)^2 \qquad \text{From (1) and (4)}$$

$$= \left(\frac{\partial \xi}{\partial x}\right)^2 + \left(\frac{\partial \xi}{\partial y}\right)^2 \qquad \text{From (3)}$$

$$= h^2 \text{ (say)}$$

Now from (2)

$$\frac{\Delta Q_1QQ_2}{\Delta P_1PP_2} = h^2$$

So that $d\xi d\eta = h^2\ dxdy$

Showing that the corresponding areas in the ζ and z planes are in the ratio h^2:1.

3.20. Application of Conformal Representation to Hydrodynamics

(I) Transformation of a Doublet

Let there be a doublet at a point P in the z plane then there must be an equal doublet at the corresponding point Q in ζ plane.

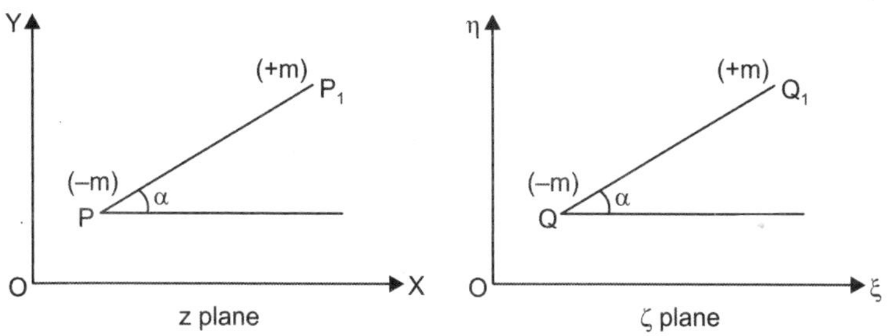

Fig. 3.14

Also $QQ_1 = hPP_1$

Again $\mu^1 = mQQ_1$ and $\mu = mPP_1$

Hence $\mu^1 = mhPP_1 = h\mu$

It follows that if the axis PP_1 of the doublet at P makes an angle α with a given direction in z plane then the axis QQ_1 will also make an angle α with the corresponding direction in ζ plane.

(II) Transformation of a Source

Assuming the motion $\zeta = z^{\frac{1}{n}}$

We consider a source of strength $+m$ at P. Let (r, θ) be the polar coordinates of the branch point P in the z plane and (R, ϕ) be the coordinates of the corresponding point Q in ζ plane.

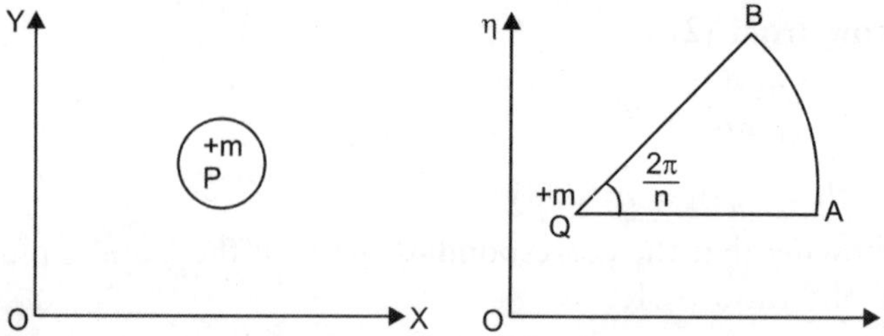

Fig. 3.15

From (1), we have

$$Re^{i\phi} = \left(re^{i\theta}\right)^{\frac{1}{n}} = r^{\frac{1}{n}} e^{\frac{\theta}{n} i}$$

Hence

$$R = r^{\frac{1}{n}} \text{ and } \phi = \frac{\theta}{n}$$

It follows that a complete circle round P in the z plane corresponds to an arc $AB = \dfrac{2\pi}{n}$ in the ζ plane. Since the flow across the circle is equal to that across the arc AB.

Hence

$$2\pi m = \left(\frac{2\pi}{n}\right) m^1$$

$$\Rightarrow \; m^1 = nm$$

Therefore corresponding to a source of strength $+m$ at P, there will be a source of strength $+m^1 = nm$ at Q.

3.21. A Fundamental Theorem

To show that by means of conformal representation as applied to the two-dimensional motion of a liquid, that a source is transformed into an equal source, while a doublet is transformed into a doublet of different strength.

Proof. *Case I:* Let there be a source of strength $+m$ at $A(z = z_1)$ in z plane. Let A^1 $(t = t_1)$ be the point in t plane corresponding to A by means of a conformal map $t = f(z)$. We enclose the point A by means of a small closed curve C_1. Let the curve C_2 in t plane be the image C_1 under the map $t = f(z)$ then C_2 must enclose P_2.

Since the stream function ψ is the same at corresponding points of C_1 and C_2.

Hence

$$\int_{C_2} d\psi = \int_{C_1} d\psi$$

But $\int_{C_1} d\psi = \int_{C_1} \left(\frac{\partial \psi}{\partial x} dx + \frac{\partial \psi}{\partial y} dy \right)$

$$= \int_{C_1} (udy - vdx)$$

= Total flux across C_1

$= -\ 2\pi m$

From $\int_{C_2} d\psi = 2\pi m$, it follows that A^1 is occupied by a source of equal strength m. Thus we can say that in a *conformal transformation a source is transformed into an equal source*. If C_2 encircles A^1 only once, the source will be of strength m. If C_2 encircles A^1 n times when C_1 encircles A once, the strength of the source at A will be $\frac{m}{n}$.

Case II: Let a doublet AB of strength μ_1 in z plane at $A(z)$ be obtained by the combination of source $-m$ at A and $+m$ at B, so that if $mAB = \mu_1$.

Let $AB = \delta z$. Let $A_1B_1 = \delta t$ in t plane be the image of AB by the conformal map $t = f(z)$.

By case I, images of $-m$ and $+m$ in z plane are $-m$ at A_1 and $+m_1$ at B_1 in t plane. Let μ_2 be the strength of image doublet.

Then

$$\mu_2 = \underset{\delta t \to 0}{Lt}\ m|\ \delta t\ |,\ \mu_1 = \underset{\delta z \to 0}{Lt}\ m\ |\ \delta z\ |$$

$$\therefore \quad \mu_2 = \underset{\delta t \to 0}{Lt}\ m \left| \frac{\delta t}{\delta z} \cdot \delta z \right|$$

$$= \underset{\delta z \to 0}{Lt} (m|\ \delta z\ |) \left| \frac{\delta t}{\delta z} \right|$$

$$= \mu_1 \left| \frac{dt}{dz} \right|$$

$$= \mu_1 h \quad \text{where } h = \left| \frac{dt}{dz} \right|$$

$Ash \neq 1$ and So $\mu_2 \neq \mu_1$

Also the axes of the doublets will make equal angles with the corresponding directions.

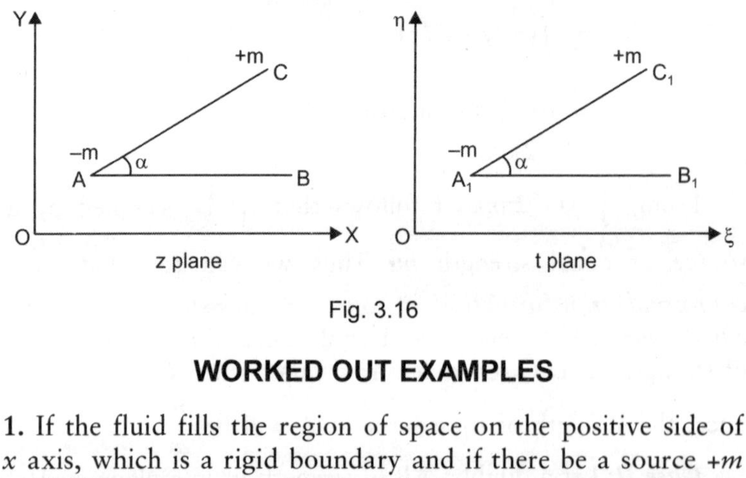

Fig. 3.16

WORKED OUT EXAMPLES

1. If the fluid fills the region of space on the positive side of x axis, which is a rigid boundary and if there be a source $+m$ at the point $(0, a)$ and an equal sink at $(0, b)$, and if the pressure on the negative side of the boundary be the same as the pressure of the fluid at infinity, show that the resultant pressure on the boundary is $\dfrac{\pi \rho m^2 (a-b)^2}{ab(a+b)}$, where ρ is the density of the fluid. [I.A.S. 95]

Solution: The object system consists of source $+m$ at $A\ (0, a)$ i.e. at $z = ia$ and sink $-m$ at $z = ib$.

The image system consists of source $+m$ at A^1 $(z = -\ ia)$ and sink $-m$ at B^1 $(z = -\ ib)$ with respect to the positive line OX which is rigid boundary. The complex potential due to

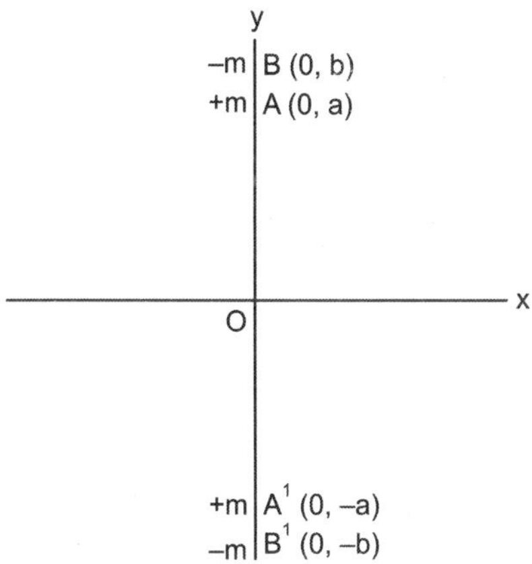

Fig. 3.17

object system with rigid boundary is equivalent to the object system and its image system with no rigid boundary.

$$\therefore \quad w = -m\log(z - ia) + m\log(z - ib) - m\log(z + ia) + m\log(z + ib)$$

$$w = -m\log(z^2 + a^2) + m\log(z^2 + b^2)$$

$$\therefore \quad \frac{dw}{dz} = -2mz\left[\frac{1}{z^2 + a^2} - \frac{1}{z^2 + b^2}\right]$$

$$= \frac{2mz(a^2 - b^2)}{(z^2 + a^2)(z^2 + b^2)}$$

Hence

$$q = \left|\frac{dw}{dz}\right| = \frac{2m(a^2 - b^2)|z|}{|z^2 + a^2||z^2 + b^2|}$$

For any point on x axis we have $z = x$, so that

$$q = \frac{2mx(a^2 - b^2)}{(x^2 + a^2)(x^2 + b^2)}$$

This is an expression for velocity at any point on x axis. Let p_0 be the pressure at $x = \infty$. By Bernoulli's equation for steady motion

$$\frac{p}{\rho}+\frac{1}{2}q^2=c \qquad \ldots (1)$$

$p = p_0$, $q = 0$ when $x = \infty$

Hence from (1), we get

$$c=\frac{p_0}{\rho}$$

Therefore, from (1)

$$\frac{p_0-p}{\rho}=\frac{1}{2}q^2$$

Required pressure P on the boundary is given by

$$P=\int_{-\infty}^{\infty}(p_0-p)\,dx=\int_{-\infty}^{\infty}\frac{1}{2}q^2\rho\,dx$$

$$=\frac{1}{2}\rho\int_{-\infty}^{\infty}\frac{4m^2x^2\left(a^2-b^2\right)^2}{\left(x^2+a^2\right)^2\left(x^2+b^2\right)^2}dx$$

$$=4m^2\rho\left(a^2-b^2\right)^2\int_0^{\infty}\frac{x^2dx}{\left(x^2+a^2\right)^2\left(x^2+b^2\right)^2}$$

$$=4m^2\rho\int_0^{\infty}\left[\frac{a^2+b^2}{a^2-b^2}\left\{\frac{1}{x^2+b^2}-\frac{1}{x^2+a^2}\right\}-\frac{a^2}{\left(x^2+a^2\right)^2}-\frac{b^2}{\left(x^2+b^2\right)^2}\right]dx$$

[By Partial Fraction]

$$=4m^2\rho\left[\frac{a^2+b^2}{a^2-b^2}\left\{\frac{\pi}{2b}-\frac{\pi}{2a}\right\}-\frac{\pi}{4a}-\frac{\pi}{4b}\right]$$

$$=\pi 4m^2\rho\left[\frac{\left(a^2+b^2\right)(a-b)}{\left(a^2-b^2\right)2ab}-\frac{a+b}{4ab}\right]$$

$$=4\pi m^2\rho\left[\frac{a^2+b^2}{(a+b)2ab}-\frac{a+b}{4ab}\right]$$

$$= \frac{\pi m^2 \rho \left[2a^2 + 2b^2 - a^2 - b^2 - 2ab\right]}{ab\,(a+b)}$$

$$= \frac{\pi m^2 \rho\,(a-b)^2}{ab\,(a+b)}$$

Since $\displaystyle\int_0^\infty \frac{dx}{x^2+a^2} = \frac{1}{a}\left[\tan^{-1}\frac{x}{a}\right]_0^\infty = \frac{\pi}{2a}$

and

$$\int_0^\infty \frac{dx}{\left(x^2+a^2\right)^2} = \frac{1}{a^3}\int_0^\infty \cos^2\theta\, d\theta,\ x = a\tan\theta$$

$$= \frac{1}{2a^3}\int_0^{\frac{\pi}{2}} (1+\cos 2\theta)\, d\theta$$

$$= \frac{1}{2a^3}\cdot\frac{\pi}{2} = \frac{\pi}{4a^3}$$

2. Between the fixed boundaries $\theta = \frac{\pi}{4}$ and $\theta = -\frac{\pi}{4}$, there is a two-dimensional fluid motion due to a source of strength m at the point $r = a$, $\theta = 0$ and an equal sink at the point $r = b$, $\theta = 0$. Use the method of images to show that the stream function is

$$-m\tan^{-1}\left[\frac{r^4\left(a^4-b^4\right)\sin 4\theta}{r^8 - r^4\left(a^4+b^4\right)\cos 4\theta + a^4b^4}\right]$$

Show also that the velocity at (r, θ) is

$$\frac{4m\left(a^4-b^4\right)r^2}{\left(r^8 - 2a^4r^4\cos 4\,\theta + a^8\right)^{\frac{1}{2}}\left(r^8 - 2b^4r^4\cos 4\,\theta + b^8\right)^{\frac{1}{2}}}$$

(I.A.S. 94, 91]

Solution: We consider the following conformal transformation from z plane (xy plane) to ζ plane ($\xi\eta$ plane) $\zeta = z^2$.

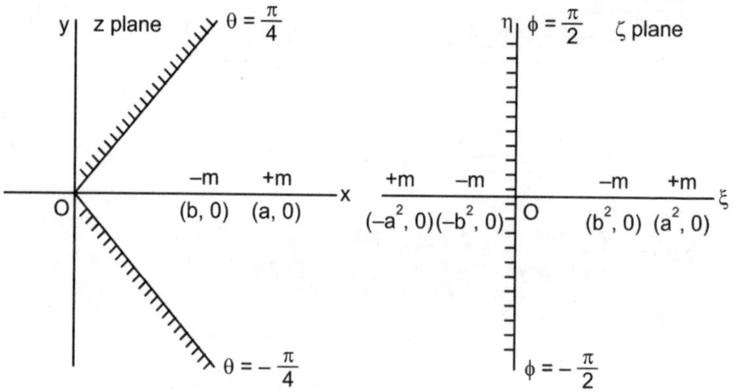

Fig. 3.18

Let $z = re^{i\theta}$, $\zeta = Re^{i\phi}$

Then $\zeta = z^2$

$\Rightarrow \; Re^{i\phi} = r^2 e^{2i\theta}$

$\Rightarrow \; R = r^2, \; \phi = 2\theta$

The boundaries $\theta = \frac{\pi}{4}$ and $\theta = -\frac{\pi}{4}$ in z plane transform to an imaginary axis $\phi = \frac{\pi}{2}$ and $\phi = -\frac{\pi}{2}$ in η axis.

By this transformation points $(a, 0)$ and $(b, 0)$ in z plane are mapped on $(a^2, 0)$ and $(b^2, 0)$ in ζ plane.

Thus the image system consists of—

(i) a source of strength m at $(a^2, 0)$

(ii) a source of strength m at $(-a^2, 0)$

(iii) a sink of strength $-m$ at $(b^2, 0)$

(iv) a sink of strength $-m$ at $(-b^2, 0)$

Hence the complex potential is given by

$$w = -m \log\left(\zeta - a^2\right) - m \log\left(\zeta + a^2\right) + m \log\left(\zeta - b^2\right) + m \log\left(\zeta + b^2\right)$$

$$= -m \log\left(\zeta^2 - a^4\right) + m \log\left(\zeta^2 - b^4\right)$$

$$\Rightarrow \quad w = -m \log \frac{\zeta^2 - a^4}{\zeta^2 - b^4}$$

$$\Rightarrow \quad \phi + i\psi = -m \log \frac{z^4 - a^4}{z^4 - b^4}$$

Equating imaginary parts, we get

$$\psi = -m\left[\tan^{-1}\left(\frac{r^4 \sin 4\theta}{r^4 \cos 4\theta - a^4}\right) - \tan^{-1}\left(\frac{r^4 \sin 4\theta}{r^4 \cos 4\theta - b^4}\right)\right]$$

$$= -m \tan^{-1}\left[\frac{r^4\left(a^4 - b^4\right)\sin 4\theta}{r^8 - r^4\left(a^4 + b^4\right)\cos 4\theta + a^4 b^4}\right]$$

Also

$$\frac{dw}{dz} = -m\left[\frac{4z^3}{z^4 - a^4} - \frac{4z^3}{z^4 - b^4}\right]$$

$$= \frac{-4m\left(a^4 - b^4\right)z^3}{\left(z^4 - a^4\right)\left(z^4 - b^4\right)}$$

$$\therefore \quad q = \left|\frac{dw}{dz}\right| = \frac{4m\left(a^4 - b^4\right)|z^3|}{|z^4 - a^4|\,|z^4 - b^4|}$$

$$= \frac{4m\left(a^4 - b^4\right)|r^3 e^{3i\theta}|}{|r^4 e^{4i\theta} - a^4|\,|r^4 e^{4i\theta} - b^4|}$$

$$= \frac{4m\left(a^4 - b^4\right)r^3}{|r^4 \cos 4\theta - a^4 + 4ir^4 \sin 4\theta|\,|r^4 \cos 4\theta - b^4 + 4ir^4 \sin 4\theta|}$$

$$= \frac{4mr^3\left(a^4 - b^4\right)}{\left(r^8 - 2a^4 r^4 \cos 4\theta + a^8\right)^{\frac{1}{2}}\left(r^8 - 2b^4 r^4 \sin 4\theta + b^8\right)^{\frac{1}{2}}}$$

3. Prove that for the liquid circulating irrotationally in part of the plane between two non-intersecting circles, the curves of constant velocity are Cassini's Ovals. [I.A.S. 93]

Solution: Let CC^1 be the line of centres. Let us take two points A and B such that they are inverse points with respect to both the circles. We consider a point P on one of the circles.

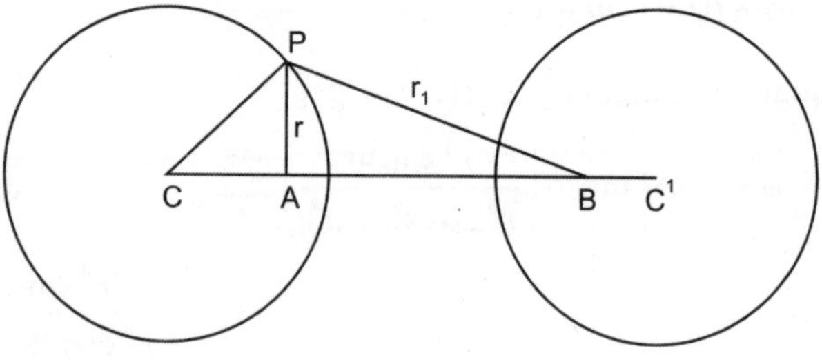

Fig. 3.19

Let $PA = r$, $PB = r_1$

Since $CA, CB = CP^2$

Hence ΔCPA and ΔCPB are similar, so we have

$$\frac{CP}{CB} = \frac{PA}{PB} = \frac{r}{r_1}$$

$$\Rightarrow \quad \frac{r}{r_1} = \text{const.}$$

Therefore the equations of the two circles can be written as

$$\frac{r}{r_1} = K_1, \frac{r}{r_2} = K_2$$

Also these two circles are streamlines, hence ψ must be of the form $\psi = f\left(\frac{r}{r_1}\right)$ but $f\left(\frac{r}{r_1}\right)$ is plane harmonic.

Consequently $\psi = f\left(\frac{r}{r_1}\right) = A \log\left(\frac{r}{r_1}\right)$ as $\log r$ is the only function of r which is plane harmonic. Hence

$$\phi = -A(\theta - \theta_1) \text{ as } \frac{1}{r}\frac{\partial\phi}{\partial\theta} = \frac{-\partial\psi}{\partial r}$$

$$w = \phi + i\psi$$

$$= -A(\theta - \theta_1) + iA \log\left(\frac{r}{r_1}\right)$$

$$= iA\left[\log \frac{r}{r_1} + i(\theta - \theta_1)\right]$$

$$= iA \log\left[\frac{r}{r_1} e^{i(\theta - \theta_1)}\right]$$

$$= iA \log\left[\frac{re^{i\theta}}{r_1 e^{i\theta_1}}\right]$$

Taking A to be $(-a, 0)$ and B to be $(a, 0)$ then

$$w = iA \log\left(\frac{z+a}{z-a}\right)$$

Thus

$$q = \left|\frac{dw}{dz}\right| = |iA| \left|\frac{1}{z+a} - \frac{1}{z-a}\right|$$

$$= \frac{2Aa}{|z+a||z-a|} = \frac{2Aa}{rr_1}$$

Curves of constant velocity are given by

$$q = \text{Const. i.e. } \frac{2Aa}{rr_1} = \text{Const. i.e. } rr_1 = \text{Const.}$$

Which are clearly Cassine's Ovals.

4. Find the lines of flow in two-dimensional fluid motion given by $\phi + i\psi = -\frac{1}{2} n (x+iy)^2 e^{2int}$. [I.A.S. 92]

Prove or verify that the paths of the particles of the fluid (in polar coordinates) may be obtained by eliminating t from the equations.

$$r\cos(nt + \theta) - x_0 = r\sin(nt + \theta) - y_0 = nt(x_0 - y_0)$$

Solution: Since

$$\phi + i\psi = -\frac{1}{2} n (x+iy)^2 e^{2int} \qquad \ldots (1)$$

Putting $x = r\cos\theta$, $y = r\sin\theta$, we have

$$\phi + i\psi = -\frac{1}{2}n\left(r\cos\theta + ir\sin\theta\right)^2 e^{2int}$$

$$= -\frac{1}{2}nr^2 e^{2i(\theta + nt)}$$

$$\phi + i\psi = -\frac{1}{2}nr^2\left[\cos 2(\theta + nt) + i\sin 2(\theta + nt)\right]$$

Equating real and imaginary parts on both the sides, we get

$$\phi = -\frac{1}{2}nr^2\cos 2(\theta + nt)$$

and

$$\psi = -\frac{1}{2}nr^2\sin 2(\theta + nt)$$

The lines of flow are given by

ψ = constant

$$\Rightarrow -\frac{1}{2}nr^2\sin 2(\theta + nt) = \text{const}$$

$\Rightarrow r^2 \sin 2(\theta + nt) = \text{const}$

Now we shall determine path of the particles of fluid.

$$\frac{dr}{dt} = \frac{-\partial\phi}{\partial r} = nr\cos 2(\theta + nt) = nr\cos 2\lambda \qquad \text{... (2)}$$

Where $\lambda = \theta + nt$

and

$$r\frac{d\theta}{dt} = -\frac{1}{r}\frac{\partial\phi}{\partial\theta} = -nr\sin 2(\theta + nt)$$

$$= -nr\sin 2\lambda \qquad \text{... (3)}$$

From (2), we have

$$nr\cos 2\lambda = \frac{dr}{dt} = \frac{dr}{d\lambda}\frac{d\lambda}{dt} = \frac{dr}{d\lambda}\left(\frac{d\theta}{dt} + n\right)$$

$$= \frac{dr}{d\lambda}(n - n\sin 2\lambda) \qquad \text{From (3)}$$

$$\Rightarrow \quad r\cos 2\lambda = \frac{dr}{d\lambda}(1-\sin 2\lambda)$$

$$\Rightarrow \quad \frac{2dr}{r} = \frac{2\cos 2\lambda\, d\lambda}{1-\sin 2\lambda}$$

Integrating, we get

$$2\log r = -\log(1-\sin 2\lambda) + \log c$$

$$\Rightarrow \quad \log r^2(1-\sin 2\lambda) - \log c$$

$$\Rightarrow \quad r(\cos\lambda - \sin\lambda) = c_1 \qquad \text{... (4)}$$

At $t = 0$, $\lambda = \theta$, $r = r_0$ ... (5)

Hence from (4)

$$r_0\cos\theta_0 - r_0\sin\theta_0 = c_1$$

$$\Rightarrow \quad c_1 = x_0 - y_0$$

Hence from (4), we have

$$r(\cos\lambda - \sin\lambda) = x_0 - y_0$$

$$\Rightarrow \quad r\cos(\theta + nt) - x_0 = r\sin(\theta + nt) - y_0 \qquad \text{... (6)}$$

Also

$$\lambda = \theta + nt$$

$$\therefore \quad \frac{d\lambda}{dt} = \frac{d\theta}{dt} + n = n - n\sin 2\lambda$$

$$\Rightarrow \quad \frac{d\lambda}{dt} = n(1-\sin 2\lambda)$$

$$\Rightarrow \quad \int\frac{d\lambda}{1-\sin 2\lambda} = n\int dt$$

$$\Rightarrow \quad \int\frac{d\lambda}{(\cos\lambda - \sin\lambda)^2} = n\int dt$$

$$\Rightarrow \quad \int\frac{\sec^2\lambda d\lambda}{(1-\tan\lambda)^2} = n\int dt$$

$$\Rightarrow \quad \frac{1}{1-\tan\lambda} = nt + c_2$$

$$\Rightarrow \quad \frac{\cos\lambda}{\cos\lambda - \sin\lambda} = nt + c_2 \qquad \text{... (7)}$$

From (5)

$$\frac{\cos\theta_0}{\cos\theta_0 - \sin\theta_0} = c_2$$

$$\Rightarrow \quad c_2 = \frac{x_0}{x_0 - y_0}$$

Hence from (7), we get

$$\frac{r\cos\lambda}{r\cos\lambda - r\sin\lambda} = nt + \frac{x_0}{x_0 - y_0}$$

$$\Rightarrow \quad r\cos\lambda = nt\left(x_0 - y_0\right) + x_0 \qquad \text{From (6)}$$

$$\Rightarrow \quad r\cos\left(\theta + nt\right) = nt\left(x_0 - y_0\right) + x_0 \qquad \ldots (8)$$

From (6) and (8), we get

$$r\cos\left(\theta + nt\right) - x_0 = r\sin\left(\theta + nt\right) - y_0 = \left(x_0 - y_0\right)nt$$

5. Two sources each of strength m are placed at the points $(a, 0)$ and $(-a, 0)$ and a sink of stength $2m$ is placed at the origin. Show that the streamlines are the curves

$$\left(x^2 + y^2\right)^2 = a^2\left(x^2 - y^2 + \lambda xy\right)$$

λ being a variable parameter. Show also that the fluid speed at any point is $\dfrac{2ma^2}{r_1 r_2 r_3}$ where r_1, r_2, r_3 are the distances of the point from the sources and sink respectively. [I.A.S. 90]

Solution: Let the two sources of strength m be placed at the

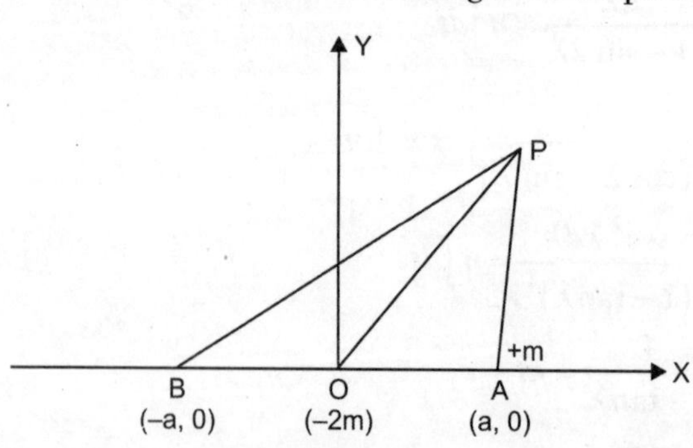

Fig. 3.20

points A $(a, 0)$ and B $(-a, 0)$ and the sink of strength $2m$ be placed at O, the mid-point of AB, which is taken as origin.

If P is any point such that

$$AP = r_1,\ BP = r_2,\ CP = r_3$$

The complex potential w at any point whose complex coordinate is z, is given by

$$w = -m\log(z-a) - m\log(z+a) + 2m\log z$$

$$= -m\log\left(z^2 - a^2\right) + m\log z^2$$

$$\Rightarrow\quad \phi + i\psi = -m\log\left[\left(x^2 - y^2 - a^2\right) + 2xyi\right] + m\log\left[\left(x^2 - y^2\right) + 2xyi\right]$$

Equating imaginary parts, we get

$$\psi = -m\tan^{-1}\frac{2xy}{x^2 - y^2 - a^2} + m\tan^{-1}\frac{2xy}{x^2 - y^2}$$

$$= -m\left[\tan^{-1}\frac{2xy}{x^2 - y^2 - a^2} - \tan^{-1}\frac{2xy}{x^2 - y^2}\right]$$

$$= -m\tan^{-1}\frac{2a^2xy}{\left(x^2 + y^2\right)^2 - a^2\left(x^2 - y^2\right)}$$

Now the streamlines are given by

$$\psi = \text{constant}$$

$$\Rightarrow\quad -m\tan^{-1}\frac{2a^2xy}{\left(x^2 + y^2\right)^2 - a^2\left(x^2 - y^2\right)} = \text{Const.}$$

$$\Rightarrow\quad \frac{2a^2xy}{\left(x^2 + y^2\right)^2 - a^2\left(x^2 - y^2\right)} = \frac{2}{\lambda} \qquad \text{(Say)}$$

$$\Rightarrow\quad \left(x^2 + y^2\right)^2 - a^2\left(x^2 - y^2\right) = \lambda a^2 xy$$

$$\Rightarrow\quad \left(x^2 + y^2\right)^2 = a^2\left(x^2 - y^2 + \lambda xy\right)$$

where λ is a variable parameter.

Again the fluid speed at any point

$$= \left| \frac{dw}{dz} \right|$$

$$= \left| \frac{-2mz}{z^2 - a^2} + \frac{2m}{z} \right|$$

$$= \left| \frac{(z^2 - a^2 - z^2)\,2m}{z(z^2 - a^2)} \right|$$

$$= \frac{2ma^2}{|z||z-a||z+a|}$$

$$= \frac{2ma^2}{r_3 r_2 r_1}$$

6. Show that the velocity potential

$$\phi = \frac{1}{2} \log \frac{(x+a)^2 + y^2}{(x-a)^2 + y^2}$$

gives a possible motion. Determine the streamlines and prove that the curves of equal speed are the Ovals of Cassini given by rr^1 = constant. [I.A.S. 90]

Solution:

Here

$$\phi = \frac{1}{2} \log \frac{(x+a)^2 + y^2}{(x-a)^2 + y^2} \qquad \text{... (1)}$$

$$\therefore \quad \frac{\partial \phi}{\partial x} = \frac{x+a}{(x-a)^2 + y^2} - \frac{x-a}{(x-a)^2 + y^2}$$

and

$$\frac{\partial \phi}{\partial y} = \frac{y}{(x+a)^2 + y^2} - \frac{y}{(x-a)^2 + y^2}$$

Also

$$\frac{\partial^2\phi}{\partial x^2} = \frac{\left[(x+a)^2 + y^2\right] - 2(x+a)^2}{\left[(x+a)^2 + y^2\right]^2} - \frac{\left[(x-a)^2 + y^2\right] - 2(x-a)^2}{\left[(x-a)^2 + y^2\right]^2}$$

$$\Rightarrow \quad \frac{\partial^2\phi}{\partial x^2} = \frac{y^2 - (x+a)^2}{\left[(x+a)^2 + y^2\right]^2} - \frac{y^2 - (x-a)^2}{\left[(x-a)^2 + y^2\right]^2} \qquad \ldots (2)$$

Again

$$\frac{\partial^2\phi}{\partial y^2} = \frac{(x+a)^2 + y^2 - 2y^2}{\left[(x+a)^2 + y^2\right]^2} - \frac{(x-a)^2 + y^2 - 2y^2}{\left[(x-a)^2 + y^2\right]^2}$$

$$\Rightarrow \quad \frac{\partial^2\phi}{\partial y^2} = \frac{(x+a)^2 - y^2}{\left[(x+a)^2 + y^2\right]^2} - \frac{(x-a)^2 - y^2}{\left[(x-a)^2 + y^2\right]^2} \qquad \ldots (3)$$

Adding (2) and (3), we get

$$\frac{\partial^2\phi}{\partial x^2} + \frac{\partial^2\phi}{\partial y^2} = 0$$

Thus the equation of continuity is satisfied and hence (1) gives a possible liquid motion.

To determine streamlines, we have

$$\frac{-\partial\phi}{\partial x} = u = \frac{-\partial\psi}{\partial y}, \frac{-\partial\phi}{\partial y} = v = \frac{\partial\psi}{\partial x}$$

Hence

$$\frac{\partial\phi}{\partial x} = \frac{\partial\psi}{\partial y}, \frac{\partial\phi}{\partial y} = \frac{-\partial\psi}{\partial x}$$

Now

$$\frac{\partial\psi}{\partial y} = \frac{x+a}{(x+a)^2 + y^2} - \frac{x-a}{(x-a)^2 + y^2}$$

Integrating w.r.t. y, we get

$$\psi = \tan^{-1}\frac{y}{x+a} - \tan^{-1}\frac{y}{x-a} + F(x) \qquad \ldots (4)$$

Where $F(x)$ is constant of integration

To determine $F(x)$, we have

$$\frac{\partial\psi}{\partial x} = \frac{-\partial\phi}{\partial y} = \frac{-y}{(x+a)^2+y^2} + \frac{y}{(x-a)^2+y^2} \qquad \ldots (5)$$

From (4)

$$\frac{\partial\psi}{\partial x} = \frac{-y}{(x+a)^2+y^2} + \frac{y}{(x-a)^2+y^2} + F^1(x) \qquad \ldots (6)$$

(6) – (5) gives

$$F^1(x) = 0$$

Integrating, we get $F(x)$ = const. and hence neglected, as it has no effect on the fluid motion.

Now from (4), we have

$$\psi = \tan^{-1}\frac{y}{x+a} - \tan^{-1}\frac{y}{x-a}$$

$$= \tan^{-1}\frac{-2ay}{x^2-a^2+y^2}$$

Streamlines are given by ψ = const.

$$\Rightarrow \quad \tan^{-1}\left(\frac{-2ay}{x^2-a^2+y^2}\right) = \text{const.}$$

$$\Rightarrow \quad \frac{y}{x^2-a^2+y^2} = \text{const.} = A$$

The constant $A = 0$ gives the streamline $y = 0$ i.e. a real axis and $A = \infty$ gives the streamline $x^2 + y^2 = a^2$ i.e. a circle.

To determine curves of equal speed, we obtain

$$w = \phi + i\psi = \frac{1}{2}\log\left[\left(x+a^2\right)+y^2\right]$$

$$-\frac{1}{2}\log\left[(x-a)^2+y^2\right] + i\tan^{-1}\frac{y}{x+a} - i\tan^{-1}\frac{y}{x-a}$$

$$= \log\left[(x + a) + iy\right] - \log\left[(x - a) + iy\right]$$

$$w = \log(z + a) - \log(z - a)$$

Hence

$$\frac{dw}{dz} = \frac{1}{z+a} - \frac{1}{z-a}$$

$$= \frac{-2a}{(z-a)(z+a)}$$

$$\therefore \quad q = \left|\frac{dw}{dz}\right| = \frac{2a}{|z-a||z+a|}$$

$$= \frac{2a}{rr^1}$$

Where $r = |z - a| = PA$

$r^1 = |z + a| = PB$

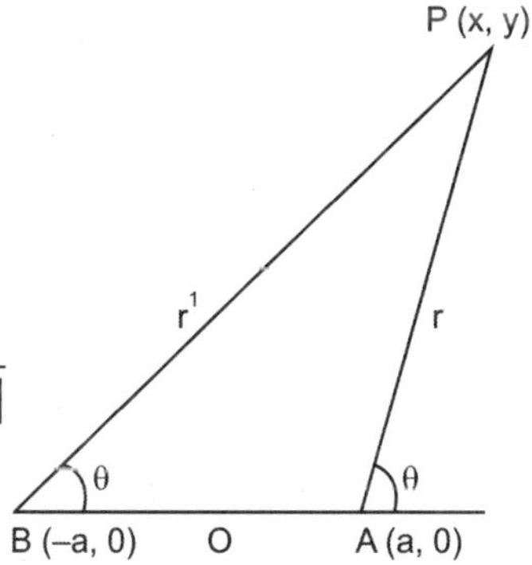

Fig. 3.21

Hence the curves of equal speed are given by

$$\frac{2a}{rr^1} = \text{const.}$$

i.e. rr^1 = const; which are Cassini Ovals.

7. A region is bounded by a fixed quadrantal arc and its radii, with a source and an equal sink at the ends of one of the bounding radii. Show that motion is given by

$$w = -m \log\left(\frac{z^2 - a^2}{z}\right)$$

and prove that the streamline leaving either the source or the sink at an angle α with the radius is $r^2 \sin(\alpha + \theta) = a^2 \sin(\alpha - \theta)$.

Solution: Let AOB be the circular quadrant of radius a with OA and OB as bounding radii. Let there be a source of strength m at A and a sink $-m$ at O.

The image system consists of—

(i) a source m at A $(a, 0)$

(ii) a source m at A^1 $(-a, 0)$

(iii) a sink $-m$ at O $(0, 0)$

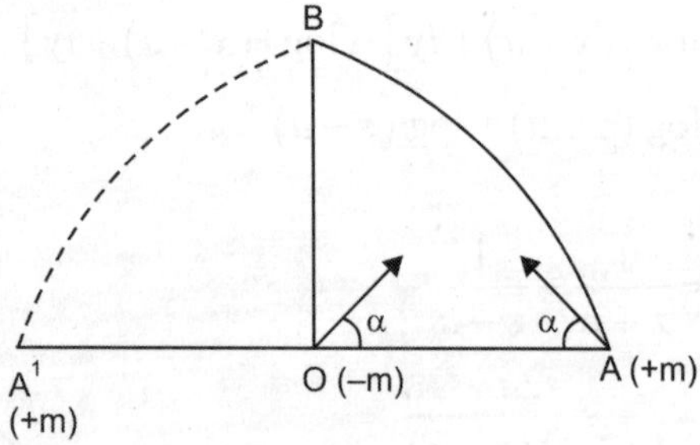

Fig. 3.22

Hence complex potential w is given by

$$w = -m \log (z - a) - m \log (z + a) + m \log z$$

$$= -m \log \frac{z^2 - a^2}{z}$$

$$\Rightarrow \quad \phi + i\psi = -m \log \frac{r^2 e^{2i\theta} - a^2}{re^{i\theta}}$$

$$\Rightarrow \quad \phi + i\psi = -m \log \left(r^2 e^{2i\theta} - a^2\right) + m \log re^{i\theta}$$

$$= -m \log \left[r^2 \cos 2\theta - a^2 + ir^2 \sin 2\theta\right]$$

$$+ m \log (r \cos \theta + ir \sin \theta)$$

Equating imaginary parts, we get

$$\psi = -m \tan^{-1}\left(\frac{r^2 \sin 2\theta}{r^2 \cos 2\theta - a^2}\right) + m \tan^{-1}\left(\frac{\sin \theta}{\cos \theta}\right)$$

$$= -m \left[\tan^{-1}\left(\frac{r^2 \sin 2\theta}{r^2 \cos 2\theta - a^2}\right) - \tan^{-1}\left(\frac{\sin \theta}{\cos \theta}\right)\right]$$

$$= -m \tan^{-1}\left[\frac{r^2 (\sin 2\theta \cos \theta - \sin \theta \cos 2\theta) + a^2 \sin \theta}{\left(r^2 \cos 2\theta - a^2\right) \cos \theta + r^2 \sin 2\theta \sin \theta}\right]$$

$$= -m \tan^{-1} \frac{r^2 \sin \theta + a^2 \sin \theta}{r^2 \cos \theta - a^2 \cos \theta}$$

$$\Rightarrow \quad \psi = -m \tan^{-1} \frac{(r^2 + a^2)\sin\theta}{(r^2 - a^2)\cos\theta} \qquad \text{... (1)}$$

If $r = a$ or $\theta = \frac{\pi}{2}$, then

$$\psi = \frac{-m\pi}{2}$$

If $\theta = 0$, then $\psi = 0$

It means that OA is the streamline $\psi = 0$ and arc AB, line OB are the streamlines $\psi = \frac{-m\pi}{2}$.

It means that the relation (1) determines the motion within the quadrant.

Again $\psi = -m(\theta_1 + \theta_2 - \theta)$

If P is a point on the circle $r = a$, then

$$\theta_2 = \pi - \theta_1 \text{ i.e. } \pi = \theta_1 + \theta_2, \text{ so that}$$

$$\psi = -m\left(\pi - \frac{\pi}{2}\right) = \frac{-m\pi}{2}$$

For a point very close to O

$$\theta_2 = 0,\ \theta_1 = \pi,\ \theta = \alpha.$$

Hence

$$\psi = -m(\pi - \alpha) \qquad \text{... (2)}$$

For a point very close to A,

$$\theta_1 = \pi - \alpha,\ \theta_2 = 0,\ \theta = 0$$

$$\psi = -m(\pi - \alpha)$$

Thus $\psi = -m(\pi - \alpha)$ gives the streamline which makes angle α at O and at A.

From (1) and (2), we have

$$-m(\pi - \alpha) = -m \tan^{-1}\left[\frac{(r^2 + a^2)\sin\theta}{(r^2 - a^2)\cos\theta}\right]$$

$$= -m \tan^{-1}\left[\frac{r^2 + a^2}{r^2 - a^2}\tan\theta\right]$$

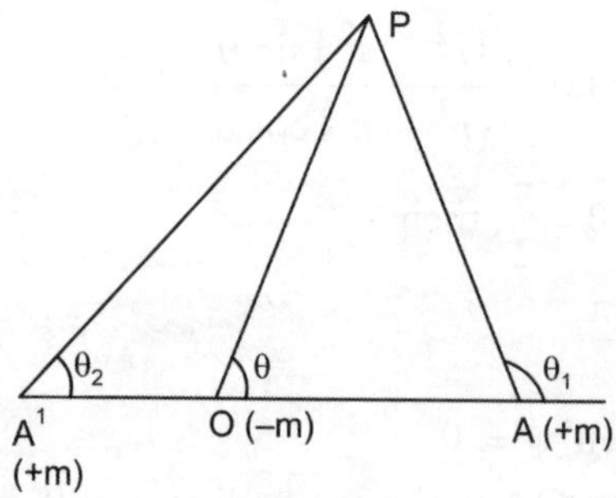

Fig. 3.23

$$\Rightarrow \quad \pi - \alpha = \tan^{-1}\left[\frac{r^2 + a^2}{r^2 - a^2} \tan \theta\right]$$

$$\Rightarrow \quad -\tan \alpha = \frac{r^2 + a^2}{r^2 - a^2} \tan \theta$$

$$\Rightarrow \quad \frac{-\sin \alpha}{\cos \alpha} = \frac{r^2 + a^2}{r^2 - a^2} \frac{\sin \theta}{\cos \theta}$$

$$\Rightarrow \quad \sin \alpha \cos \theta \left(-r^2 + a^2\right) = \left(r^2 + a^2\right) \sin \theta \cos \alpha$$

$$\Rightarrow \quad a^2 (\sin \alpha \cos \theta - \sin \theta \cos \alpha) = r^2 (\sin \theta \cos \alpha + \sin \alpha \cos \theta)$$

$$\text{or} \quad a^2 \sin(\alpha - \theta) = r^2 \sin(\alpha + \theta)$$

8. Within a circular boundary of radius a there is two-dimensional liquid motion due to a source producing liquid at the rate m, at a distance f from the centre and an equal sink at the centre. Find the velocity potential and show that the resultant of the pressure on the boundary is $\dfrac{\rho m^2 f^3}{2a^2\pi\left(a^2 - f^2\right)}$.

Deduce as a limit, the velocity potential due to a doublet at the centre.

Solution: Let K be the strength of the source, then by definition $m = 2\pi K$

$$K = \frac{m}{2\pi}$$

Let O be the centre of the circle and the source be at the point A where $OA = f\ (< a)$. A^1 is the inverse point of the point with respect to the circle. Then $OA^1 = \frac{a^2}{f}$, a be the radius of the circle.

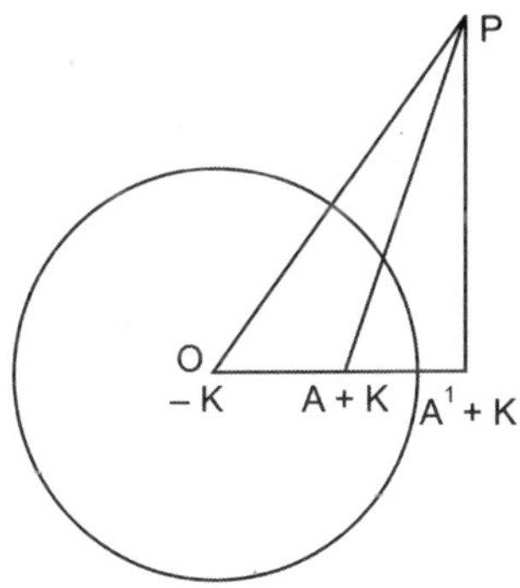

Fig. 3.24

The image of a source at A is

(i) a source at A^1; and

(ii) a sink at O

The image of the sink at O is

(i) a sink at ∞ (the inverse point of O); and

(ii) a source at the point O.

Hence, one source and one sink at the point O cancel out and we are left with a source at A, a source at A^1 and a sink at O. The velocity potential ϕ is therefore given by

$$W = -K \log(z - f) - K \log\left(z - f^1\right) + K \log z \quad \text{... (1)}$$

Equating real parts from both sides, we get

$$\phi = -K \log|z - f| - K \log\left|z - f^1\right| + K \log|z|$$

$$= -K \log AP - K \log A^1P + K \log OP$$

$$= K \log \frac{AP \cdot A^1P}{OP} \quad ,$$

From (1), we have

$$\frac{dw}{dz} = \frac{-K}{z - f} - \frac{K}{z - f^1} + \frac{K}{z}$$

$$\therefore \quad \frac{1}{K^2}\left(\frac{dw}{dz}\right)^2 = \frac{1}{(z - f)^2} + \frac{1}{\left(z - f^1\right)^2} + \frac{1}{z^2}$$

$$+ 2\left[\frac{1}{(z - f)\left(z - f^1\right)} - \frac{1}{z(z - f)} - \frac{1}{z\left(z - f^1\right)}\right]$$

$$=\left[\frac{1}{(z-f)^2}+\frac{1}{(z-f^1)^2}+\frac{1}{z^2}+\frac{2}{f-f^1}\left(\frac{1}{z-f}-\frac{1}{z-f^1}\right)\right.$$
$$\left.+\frac{2}{f^1}\left(\frac{1}{z}-\frac{1}{z-f}\right)+\frac{2}{f}\left(\frac{1}{z}-\frac{1}{z-f}\right)\right]$$

The poles inside the boundary C of the circle are $z = 0$ and $z = f$. Hence the sum of the residues i.e. the sum of the coefficients of z^{-1} and $(z - f)^{-1}$

$$=\left[\frac{2}{f-f^1}+\frac{2}{f^1}+\frac{2}{f}-\frac{2}{f}\right]$$

$$=\frac{2f}{(f-f^1)f^1}$$

By Cauchy's residue theorem

$$\int_C \frac{1}{K^2}\left(\frac{dw}{dz}\right)^2 = 2\pi i \quad \text{(Sum of the residues of the function at the poles)}$$

$$=\frac{2\pi i\, f 2K^2}{(f-f^1)f^1}$$

$$=\frac{2\pi\, fK^2 i}{(f-f^1)f^1}$$

Hence by Blasius theorem

$$X-iY=\frac{i\rho}{2}\int_C\left(\frac{dw}{dz}\right)^2 dz$$

$$=\frac{i\rho}{2}\,\frac{4\pi fK^2 i}{(f-f^1)f^1}$$

$$=\frac{2\pi\rho fK^2}{f^1(f^1-f)}=\frac{2\pi\rho fK^2}{\frac{a^2}{f}\left(\frac{a^2}{f}-f\right)}$$

$$=\frac{2\pi\rho f^3}{a^2(a^2-f^2)}\,\frac{m^2}{4\pi^2}=\frac{\rho f^3 m^2}{2a^2\pi(a^2-f^2)}$$

Equating real and imaginary parts on both the sides, we get

$$X = \frac{\rho f^3 m^2}{2\pi a^2 (a^2 - f^2)},\ Y = 0$$

Hence resultant pressure on the boundary

$$= \sqrt{X^2 + Y^2} = \frac{\rho f^3 m^2}{2\pi a^2 (a^2 - f^2)}$$

To deduce velocity potential due to a doublet at O as a limit.

If we take limit as $f \to \infty$, so that the source at A is outside the circular boundary and the source at A^1 comes near the centre, $OA^1 = \frac{a^2}{f}$. There was already a sink at O and we have brought a source of the same strength near it. These, therefore, form a doublet of strength μ, where $\mu = \frac{ma^2}{2\pi f}$.

In order to deduce the velocity potential due to this doublet at the centre, we have to consider the complex potential due to the images alone and so

$$w = -\frac{m}{2\pi}\log\left(z - \frac{a^2}{f}\right) + \frac{m}{2\pi}\log z$$

$$= -\frac{m}{2\pi}\log\frac{z - \frac{a^2}{f}}{z}$$

$$= -\frac{m}{2\pi}\log\left(1 - \frac{a^2}{fz}\right)$$

$$= \frac{m}{2\pi}\left[\frac{a^2}{fz} + \frac{1}{2}\frac{a^4}{f^2 z^2} + ...\right]$$

$$= \frac{m}{2\pi}\frac{a^2}{fz}, \text{ neglecting higher order terms}$$

$$= \frac{ma^2}{2\pi f r e^{i\theta}}$$

$$\phi + i\psi = \frac{ma^2}{2\pi fr}(\cos\theta - i\sin\theta)$$

Hence $\phi = \dfrac{ma^2 \cos\theta}{2\pi fr}$

This is the required velocity potential.

9. A source of fluid situated in space of two dimensions of such strength that $2\pi\rho m$ represents the mass of fluid of density ρ emitted per unit of time. Show that the force necessary to hold a circular disc at rest in the plane of source is $\dfrac{2\pi\rho a^2 m}{r(r^2 - a^2)}$. Where a is the radius of the disc and r the distance of the source from its centre. In what direction is the disc urged by the pressure.

Solution: Let A^1 be an inverse point of A situated at a distance r from the centre with respect to the circular disc, such that

$$OA \cdot AO^1 = a^2$$

$$OA^1 = \frac{a^2}{r}$$

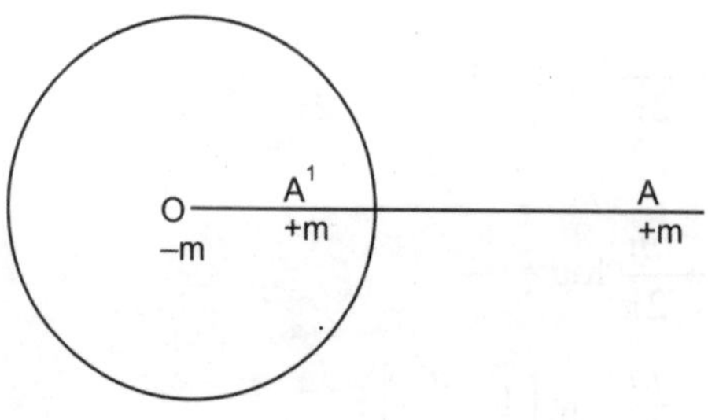

Fig. 3.25

Since a source of strength $+m$ is situated at the point A, so the image system gives rise to a source of strength $+m$ at an inverse point A^1 and a sink of strength $-m$ at the origin O.

Thus the complex potential due to the above system, is given by

$$w = -m\log(z - r) - m\log\left(z - \frac{a^2}{r}\right) + m\log z$$

$$\therefore \quad \frac{dw}{dz} = \frac{-m}{z-r} - \frac{m}{z - \dfrac{a^2}{r}} + \frac{m}{z} \qquad \ldots (1)$$

Let X and Y be the components of the forces that act on the disc, then by Blasius theorem, we have

$$X - iY = \frac{1}{2} i\rho \int_C \left(\frac{dw}{dz}\right)^2 dz \qquad \ldots (2)$$

From (1), we have

$$\left(\frac{dw}{dz}\right)^2 = m^2 \left[\frac{1}{(z-r)^2} + \frac{1}{\left(z - \dfrac{a^2}{r}\right)^2} + \frac{1}{z^2} + \frac{2}{(z-r)\left(z - \dfrac{a^2}{r}\right)} - \frac{2}{z(z-r)} - \frac{2}{z\left(z - \dfrac{a^2}{r}\right)}\right]$$

$$\Rightarrow \left(\frac{dw}{dz}\right)^2 = m^2 \left[\frac{1}{(z-r)^2} + \frac{1}{\left(z - \dfrac{a^2}{r}\right)^2} + \frac{1}{z^2} + \frac{2}{\left(r - \dfrac{a^2}{r}\right)(z-r)} + \frac{2}{\left(\dfrac{a^2}{r} - r\right)\left(z - \dfrac{a^2}{r}\right)} - \frac{2}{r(z-r)} + \frac{2}{rz} - \frac{2}{\dfrac{a^2}{r}\left(z - \dfrac{a^2}{r}\right)} + \frac{2}{\dfrac{a^2}{r} z}\right] \qquad \ldots (3)$$

[Resolving into partial fractions]

From (3), we find that the poles inside the circular contour C are

$$z = 0 \text{ and } z = \frac{a^2}{r}$$

Hence the required sum of the residues

= the sum of coefficients of z^{-1} and

$$\left(z - \frac{a^2}{r}\right)^{-1} \text{ in R.H.S. of (3)}$$

$$= 2m^2\left[\frac{1}{r} + \frac{1}{\frac{a^2}{r}} + \frac{1}{\frac{a^2}{r} - r} - \frac{1}{\frac{a^2}{r}}\right]$$

$$= 2m^2\left[\frac{1}{r} + \frac{r}{a^2} + \frac{r}{a^2 - r^2} - \frac{r}{a^2}\right]$$

$$= 2m^2\left[\frac{a^2}{r(a^2 - r^2)}\right]$$

By Cauchy's residue theorem

$$\int_C \frac{1}{m^2}\left(\frac{dw}{dz}\right)^2 dz = 2\pi i \cdot \text{Sum of residues within C}$$

$$= 2\pi i \frac{2a^2}{r(a^2 - r^2)}$$

Therefore from (2), we have

$$X - iY = \frac{1}{2} i\rho \cdot \frac{2\pi i \, 2m^2 a^2}{r(a^2 - r^2)}$$

$$= \frac{2a^2 m^2 \pi\rho}{r(r^2 - a^2)}$$

Thus

$$X = \frac{2a^2 m^2 \pi\rho}{r(r^2 - a^2)}, \ Y = O$$

Hence

$$\sqrt{X^2 + Y^2} = \frac{2\pi a^2 m^2 \rho}{r(r^2 - a^2)}$$

Thus the disc is attracted towards the source along OA. Hence the pressure will be greater than on the opposite side of the disc to that of the source.

10. In the case of two-dimensional fluid motion produced by a source of strength m placed at a point S outside a rigid circular disc of radius a whose centre is O, show that velocity of slip of the fluid in contact with the disc is greatest at the points where the lines joining S to the ends of the diameter right angles to OS cut the circle, and prove that its magnitude at these points is $\dfrac{2mr}{r^2 - a^2}$ where $OS = r$.

Solution: Let S^1 be the inverse point of S with respect to the circle so that

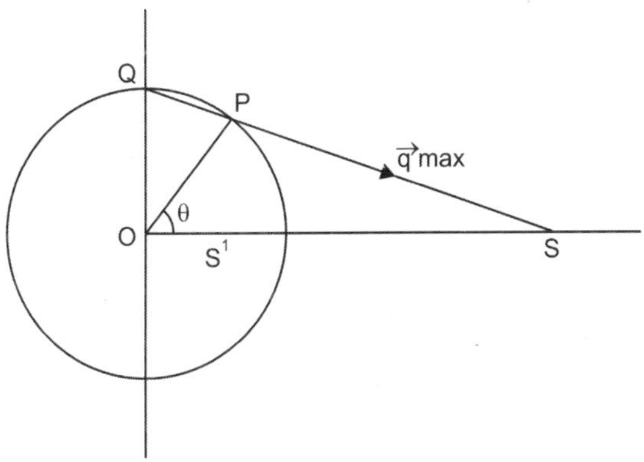

Fig. 3.26

$$OS \cdot OS^1 = a^2$$

$$\Rightarrow \quad OS^1 = \frac{a^2}{r} = r^1$$

The image system of a source of strength $+m$ placed at the point S $(a, 0)$ outside the circular disc consists of

(i) a source of strength $+m$ at S $(a, 0)$

(ii) a source of strength $+m$ at $S^1\left(\dfrac{a^2}{r}, 0\right)$

(iii) a sink of strength $-m$ at O $(0, 0)$

The complex potential w for the motion of the fluid element at any point z is given by

$$w = -m \log (z - r) - m \log (z - r^1) + m \log z$$

$$\frac{dw}{dz} = -\frac{m}{z-r} - \frac{m}{z-r^1} + \frac{m}{z}$$

$$\therefore \quad q = \left|\frac{dw}{dz}\right| = m\left|\frac{1}{z-r} + \frac{1}{z-r^1} - \frac{1}{z}\right|$$

$$= m\left|\frac{z^2 - rr^1}{(z-r)(z-r^1)z}\right|$$

$$= m\left|\frac{z^2 - a^2}{(z-r)\left(z - \frac{a^2}{r}\right)z}\right|$$

In order to determine velocity at any point on the boundary of the disc, we shall put $z = ae^{i\theta}$

$$q = m\left|\frac{a^2\left(e^{2i\theta} - 1\right)}{ae^{i\theta}\left(ae^{i\theta} - r\right)\left(ae^{i\theta} - \frac{a^2}{r}\right)}\right|$$

$$= ma\left|\frac{\left(e^{2i\theta} - 1\right) r e^{-i\theta}}{\left(ae^{i\theta} - r\right)\left(re^{i\theta} - a\right)}\right|$$

$$= mar\left|\frac{2i \sin\theta}{are^{2i\theta} - \left(r^2 + a^2\right)e^{i\theta} + ar}\right|$$

$$= \frac{2mr \sin\theta}{\left|e^{i\theta}\left(2ar\cos\theta - a^2 - r^2\right)\right|}$$

$$q = \frac{2mr\sin\theta}{a^2 + r^2 - 2ar\cos\theta} \quad \text{... (1)}$$

For maximum q, $\frac{dq}{d\theta} = 0$

$$\Rightarrow \quad \frac{2mr\left[\cos\theta\left(a^2 + r^2 - 2ar\cos\theta\right) - 2ar\sin^2\theta\right]}{\left(a^2 + r^2 - 2ar\cos\theta\right)^2} = 0$$

$$\Rightarrow \quad \left(a^2 + r^2 - 2ar\cos\theta\right)\cos\theta - 2ar\sin^2\theta = 0$$

$$\Rightarrow \quad \cos\theta = \frac{2ar}{a^2 + r^2} \quad \text{... (2)}$$

At $\theta = 0$, $q = 0$ which gives minimum velocity.

Hence value of θ given by (2), gives the maximum velocity.

From (2), we get

$$\sin\theta = \frac{r^2 - a^2}{r^2 + a^2}$$

Hence

$$q_{\max} - \frac{2mr\left(r^2 - a^2\right)}{\left(r^2 + a^2\right)\left[a^2 + r^2 - \dfrac{4a^2r^2}{a^2 + r^2}\right]}$$

$$= \frac{2mr\left(r^2 - a^2\right)}{\left(r^2 - a^2\right)^2}$$

$$= \frac{2mr}{r^2 - a^2}$$

The velocity will be along the direction of the tangent to the boundary and is equal to the velocity of slip as the boundary of the circular disc is a streamline.

11. Between the fixed boundaries $\theta = \dfrac{\pi}{\sigma}$ and $\theta = -\dfrac{\pi}{\sigma}$ and there is a two-dimensional liquid motion due to a source at the point $r = e$, $\theta = \alpha$ and a sink at the origin, absorbing water at the same rate as the source produces it. Find the stream function and show that one of the streamlines is a part of the curve $r^3 \sin 3\alpha = c^3 \sin 3\theta$.

Solution: Let the transformation be $\zeta = z^3$ from z plane to ζ plane.

Let $z = re^{i\theta}$, $\zeta = Re^{i\phi}$

Hence $Re^{i\phi} = r^3 e^{3i\theta}$

$$R = r^3 \qquad \text{... (1)}$$

$$\text{and } \phi = 3\theta \qquad \text{... (2)}$$

By this map the boundaries $\theta = \pm\dfrac{\pi}{\sigma}$ are mapped on the boundaries $\phi = \pm\dfrac{\pi}{2}$.

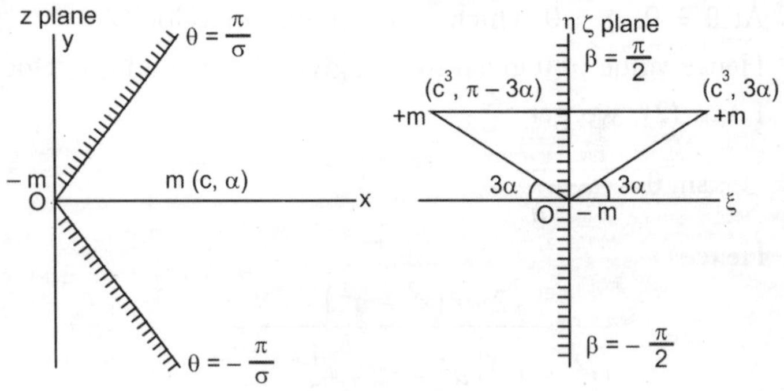

Fig. 3.27

i.e. η axis is the new boundary in ζ plane.

The points (c, α) and $(0, 0)$ in z plane are mapped respectively on the points $(c^3, 3\alpha)$ and $(0, 0)$ by virtue of (1) and (2).

The image system with respect to imaginary axis $\phi = \pm \dfrac{\pi}{2}$ in ζ plane consists of—

(i) a source of strength m at $(c^3, 3\alpha)$

(ii) a sink of strength $-m$ at $(0, 0)$

(iii) a source of strength m at $(c^3, \pi - 3\alpha)$

(iv) a sink of strength $-m$ at $(0, 0)$

Therefore the complex potential is given by

$$w = -m \log\left(\zeta - c^3 e^{3i\alpha}\right) + m \log(\zeta - 0)$$

$$- m \log\left\{\zeta - c^3 e^{i(\pi - 3\alpha)}\right\} + m \log(\zeta - 0)$$

$$= 2\, m \log \zeta - m \log\left(\zeta - c^3 e^{3i\alpha}\right) - m \log\left(\zeta + c^3 e^{-3i\alpha}\right)$$

$$= 2m \log z^3 - m \log\left[\left(z^3 - c^3 e^{3i\alpha}\right)\left(z^3 + c^3 e^{-3i\alpha}\right)\right]$$

$$= m \log z^6 - m \log\left[z^6 - c^6 - 2iz^3 c^3 \sin 3\alpha\right]$$

$$= -\, m \log \frac{z^6 - c^6 - 2iz^3 c^3 \sin 3\alpha}{z^6}$$

$$= -\, m \log \left[1 - \frac{c^6}{r^6 e^{6i\theta}} - \frac{2i \sin 3\alpha, c^3}{r^3 e^{3i\theta}}\right]$$

$$\phi + i\psi = -m\log\left[1 - \frac{2c^3}{r^3}\sin 3\alpha \sin\theta - \frac{c^6}{r^6}\cos 6\theta + i\left(\frac{c^6}{r^6}\sin 6\theta - \frac{2c^3}{r^3}\cos 3\theta \sin 3\alpha\right)\right]$$

Equating imaginary parts, we get

$$\psi = -m\tan^{-1}\frac{\frac{c^6}{r^6}\sin 6\theta - \frac{2c^3}{r^3}\cos 3\theta\sin\alpha}{1 - \frac{2c^3}{r^3}\sin 3\alpha\sin\theta - \frac{c^6}{r^6}\cos 6\theta}$$

Streamlines are given by ψ = Const.

It is easy to see that, for $\psi = 0$, one of the streamlines is given by

$$\frac{c^6}{r^6}\sin 6\theta = \frac{2c^3}{r^3}\sin 3\alpha\cos 3\theta$$

$$\Rightarrow \quad c^3 \sin 3\theta = r^3 \sin 3\alpha$$

12. In the case of motion of liquid in a part of a plane bounded by a straight line due to a source in the plane, prove that if $m\rho$ is the mass of the liquid (of density ρ) generated at the source per unit time, the pressure on the length $2l$ of the boundary immediately opposite to the source is less than that on an equal length at a great distance by

$$\frac{1}{2}\frac{m^4\rho}{\pi^2}\left(\frac{1}{c}\tan^{-1}\frac{l}{c} - \frac{l}{l^2+c^2}\right)$$

where c is the distance of the source from the boundary.

Solution: Let Y axis be the bounding line and μ is the strength of the source at P where $OP = c$. Then by definition of strength

$$2\pi\mu\rho = m\rho$$

$$\Rightarrow \quad \frac{m}{2\pi} = \rho$$

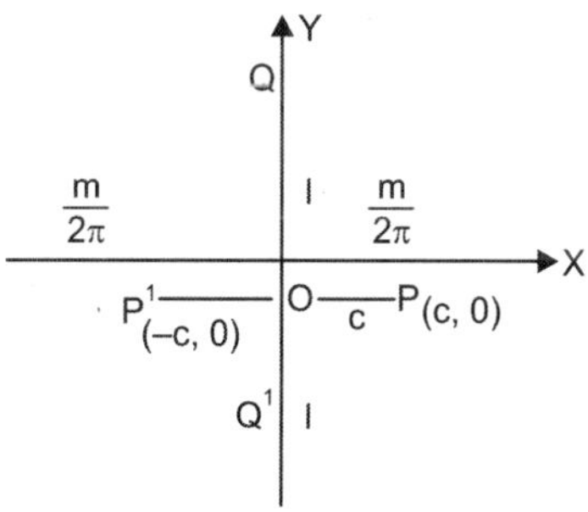

Fig. 3.28

The image system consists of—

(i) a source of strength $+\dfrac{m}{2\pi}$ at $(c, 0)$

(ii) a source of strength $-\dfrac{m}{2\pi}$ at $(-c, 0)$

The complex potential w is given by

$$w = -\frac{m}{2\pi}\log(z-c) - \frac{m}{2\pi}\log(z+c)$$

$$= -\frac{m}{2\pi}\log\left(z^2 - c^2\right)$$

The velocity q is given by

$$q = \left|\frac{dw}{dz}\right| = \frac{m}{2\pi}\left|\frac{2z}{z^2 - c^2}\right|$$

For any point on Y axis, $z = iy$, so that

$$q = \frac{m}{\pi}\left|\frac{iy}{-y^2 - c^2}\right| = \frac{my}{\pi\left(y^2 + c^2\right)}$$

Let p_0 be the pressure on Y axis at a great distance from O then by Bernaulli's theorem, we have

$$\frac{p_0 - p}{\rho} = \frac{1}{2}q^2 = \frac{1}{2}\frac{m^2}{\pi^2}\frac{y^2}{\left(y^2 + c^2\right)^2}$$

Let P be the pressure on the length $2l$ of the boundary, then

$$P = \int_{-l}^{l}(p_0 - p)\,y = \int_{-l}^{l}\frac{1}{2}\frac{m^2\rho}{\pi^2}\frac{y^2}{\left(y^2 + c^2\right)^2}dy$$

$$= \frac{1}{2}\frac{m^2\rho}{\pi^2}\int_{-l}^{l}\frac{y^2 dy}{\left(y^2 + c^2\right)^2}$$

Putting $y = c\tan\theta$, $dy = c\sec^2\theta\, d\theta$

$$= \frac{m^2}{\pi^2}\rho\int_{0}^{l}\frac{c^2\tan^2\theta c\sec^2\theta d\theta}{c^4\sec^4\theta}$$

$$= \frac{m^2\rho}{\pi^2 c}\int_{0}^{l}\sin^2\theta d\theta$$

$$= \frac{m^2\rho}{2\pi^2 c}\int_0^l (1 - \cos 2\theta)\, d\theta$$

$$= \frac{m^2\rho}{2\pi^2 c}\left[\theta = \frac{\sin 2\theta}{2}\right]_0^l$$

$$= \frac{m^2\rho}{2\pi^2 c}\left[\tan^{-1}\frac{y}{c} - \frac{\tan\theta}{1+\tan^2\theta}\right]_0^l$$

$$= \frac{m^2\rho}{2\pi^2 c}\left[\tan^{-1}\frac{y}{c} - \frac{y}{c\left(1+\frac{y^2}{c^2}\right)}\right]_0^l$$

$$= \frac{m^2\rho}{2\pi^2 c}\left[\tan^{-1}\frac{y}{c} - \frac{yc}{c^2+y^2}\right]_0^l$$

$$= \frac{m^2\rho}{2\pi^2 c}\left[\tan^{-1}\frac{l}{c} - \frac{lc}{c^2+l^2}\right]$$

$$= \frac{m^2\rho}{2\pi^2}\left[\frac{1}{c}\tan^{-1}\frac{l}{c} - \frac{l}{l^2+c^2}\right]$$

13. Within a rigid boundary in the form of the circle $(x + \alpha)^2 + (y - 4\alpha)^2 = 8\alpha^2$, there is liquid motion due of strength μ at the point $(0, 3\alpha)$ with its axis along the axis of y. Show that velocity potential is

$$\mu\left[\frac{4(x-3\alpha)}{(x-3\alpha)^2+y^2} + \frac{y-3\alpha}{x^2+(y-3\alpha)^2}\right]$$

Solution: The equation to the rigid boundary is given by

$$(x + \alpha)^2 + (y - 4\alpha)^2 = 8\alpha^2$$

which is a circle whose centre is $(-\alpha, 4\alpha)$ and radius $2\sqrt{2}\alpha$.

Object doublet is at $P(0, 3\alpha)$.

From C we draw CM perpendicular to x axis. From P we draw PN perpendicular to CM. Join PC and produce it to meet OX at Q.

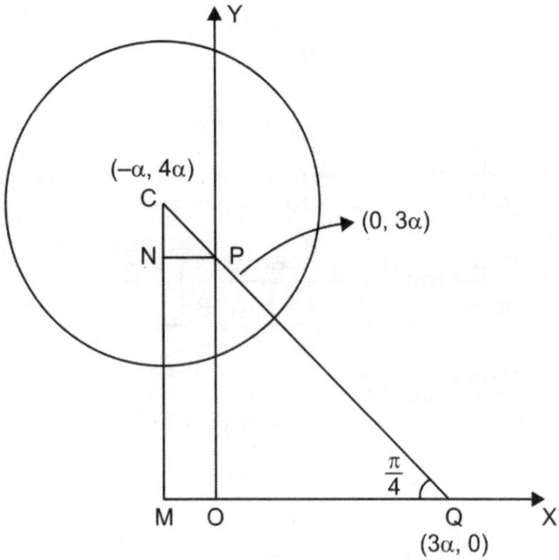

Fig. 3.29

In ΔPCN

$$PN = CN = \alpha$$

Hence

$$\angle NPC = \angle NCP = \frac{\pi}{4}$$

Therefore,

$$CM = MQ = 4\alpha, \ CP = \alpha\sqrt{2}$$

and

$$CQ = \sqrt{CM^2 + MQ^2} = 4\alpha\sqrt{2}$$

Now since $CPCQ = \alpha\sqrt{2} \cdot 4\alpha\sqrt{2} = 8\alpha^2 = (\text{radius})^2$

Hence Q is the inverse point of P with respect to the circle. It is at this point Q that the image doublet is formed.

$$\frac{\text{Strength of the doublet at } Q}{\text{Strength of the doublet at } P} = \frac{(\text{radius})^2}{(CP)^2} = \frac{8\alpha^2}{2\alpha^2} = 4$$

But it is given that the strength of the doublet at P is μ, hence the strength of the doublet at Q is 4μ. The given doublet is situated at $z = 3i\alpha$ and its axis makes an angle $\frac{\pi}{4}$ with the line CPQ.

Hence the image doublet, situated at $z = 3\alpha$, will make the supplementary angle 135° with the line CPQ. Therefore, the axis of the image doublet is along the x axis.

Now, the complex potential w is given by

$$w = \frac{\mu e^{i\pi/2}}{z - 3i\alpha} + \frac{\mu^1 e^{i\theta}}{z - 3\alpha}$$

$$= \frac{\mu i}{x + iy - 3i\alpha} + \frac{4\mu}{x + iy - 3\alpha}$$

$$= \mu\left[\frac{i}{x + i(y - 3\alpha)} + \frac{4}{(x - 3\alpha) + iy}\right]$$

$$\mu = \left[\frac{i\{x - i(y - 3\alpha)\}}{x^2 + (y - 3a)^2} + \frac{4\{(x - 3\alpha) - iy\}}{(x - 3\alpha)^2 + y^2}\right]$$

Hence the velocity potential ϕ

$$= \frac{\mu(y - 3\alpha)}{x^2 + (y - 3\alpha)^2} + \frac{4\mu(x - 3\alpha)}{(x - 3\alpha)^2 + y^2}$$

14. In the part of an infinite plane bounded by a circular quadrant AB and the production of the radii OA, OB there is a two-dimensional motion due to the productions of liquid at A, and its absorption at B, at the uniform rate m. Find the velocity potential of the motion, and show that the fluids which issues from A in the direction making an angle μ with OA follows the path whose polar equation is

$$r = a \sin^{1/2} 2\theta\left[\cot\mu + \sqrt{\{\cot^2\mu + \operatorname{cosec}^2 2\theta\}}\right]^{\frac{1}{2}}$$

the positive sign being taken for all the square roots.

Solution: The strength of the source is $\dfrac{m}{2\pi}$.

With respect to the circular boundary

(i) the image of the source $\dfrac{m}{2\pi}$ at A is $\dfrac{m}{2\pi}$ at A itself and a sink $-\dfrac{m}{2\pi}$ at O.

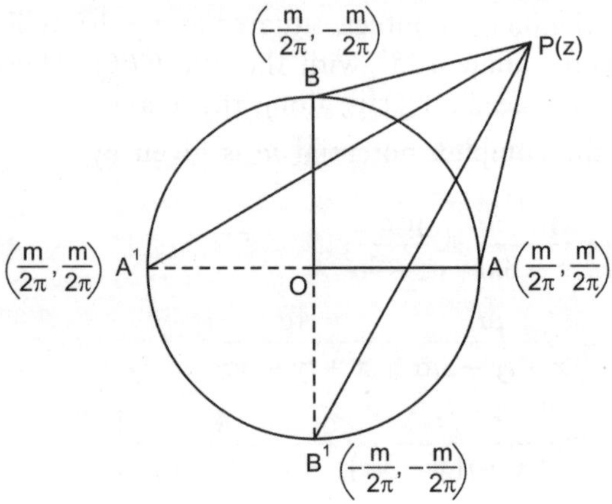

Fig. 3.30

(ii) the image of the sink $-\frac{m}{2\pi}$ at B is a sink $-\frac{m}{2\pi}$ at B itself and a source $\frac{m}{2\pi}$ at O.

With respect to the bounding planes

(i) the image of the source $\frac{m}{2\pi}$ at A is a source $\frac{m}{2\pi}$ at A^1.

(ii) the image of the sink $-\frac{m}{2\pi}$ at B is a sink $-\frac{m}{2\pi}$ at B^1.

Also the images at A and B have their images $\frac{m}{2\pi}$ at A^1 and $-\frac{m}{2\pi}$ at B^1.

The image system, therefore, consists of sources $\frac{m}{2\pi}, \frac{m}{2\pi}$ at A, $\frac{m}{2\pi}, \frac{m}{2\pi}$ at A^1

sink $-\frac{m}{2\pi}, -\frac{m}{2\pi}$ at B, $-\frac{m}{2\pi}, -\frac{m}{2\pi}$ at B^1

(the source $\frac{m}{2\pi}$ and sink $-\frac{m}{2\pi}$ at O conceal eath other).

Taking OA as the x axis, the complex velocity potential at any point P ($z = x + iy = re^{i\theta}$) is given by

$$\phi = -\frac{2m}{2\pi}\log AP - \frac{2m}{2\pi}\log A^1P + \frac{2m}{2\pi}\log BP + \frac{2m}{2\pi}\log B^1P$$

$$= -\frac{m}{\pi}\log\frac{APA^1P}{BPB^1P}$$

The complex potential of the motion w is given by

$$w = -\frac{2m}{2\pi}\log(z-a) - \frac{2m}{2\pi}\log(z+a) + \frac{2m}{2\pi}\log(z-ia) + \frac{2m}{2\pi}\log(z+ia)$$

$$= -\frac{m}{\pi}\log\left(z^2 - a^2\right) + \frac{m}{\pi}\log\left(z^2 + a^2\right)$$

$$= -\frac{m}{\pi}\log\frac{z^2 - a^2}{z^2 + a^2}$$

Putting $z = re^{i\theta}$, we get

$$w = -\frac{m}{\pi}\log\frac{r^2e^{2i\theta} - a^2}{r^2e^{2i\theta} + a^2}$$

$$\phi + i\psi = -\frac{m}{\pi}\left[\log\left\{\left(r^2\cos 2\theta - a^2\right) + ir^2\sin 2\theta\right\} - \log\left\{\left(r^2\cos 2\theta + a^2\right) + ir^2\sin 2\theta\right\}\right]$$

Equating imaginary parts, we get

$$\psi = -\frac{m}{\pi}\left[\tan^{-1}\frac{r^2\sin 2\theta}{r^2\cos 2\theta - a^2} - \tan^{-1}\frac{r^2\sin 2\theta}{r^2\cos 2\theta + a^2}\right]$$

$$= -\frac{m}{\pi}\tan^{-1}\frac{\dfrac{r^2\sin 2\theta}{r^2\cos 2\theta - a^2} - \dfrac{r^2\sin 2\theta}{r^2\cos 2\theta + a^2}}{1 + \dfrac{r^4\sin^2 2\theta}{r^4\cos^2 2\theta - a^4}}$$

$$\psi = -\frac{m}{\pi}\tan^{-1}\frac{2a^2r^2\sin 2\theta}{r^4 - a^4} \qquad \text{... (1)}$$

For a particular streamline which leaves A at an angle μ

$$\mu = -\frac{m}{\pi}\mu \qquad \text{... (2)}$$

From (1) and (2), we have

$$-\frac{m}{\pi}\mu = -\frac{m}{\pi}\tan^{-1}\frac{2a^2r^2\sin 2\theta}{r^4 - a^4}$$

$$\Rightarrow \quad \tan\mu = \frac{2a^2r^2\sin^2\theta}{r^4 - a^4}$$

$$\Rightarrow \quad r^4 - 2a^2r^2\sin 2\theta\cot\mu - a^4 = 0$$

This is a quadratic equation in r^2

Hence

$$r^2 = \frac{2a^2\sin 2\theta\cot\mu \pm \sqrt{4a^4\sin^2 2\theta\cot^2\mu + 4a^4}}{2}$$

Here negative sign has been neglected as r^2 is a non negative quantity.

Thus we have

$$r = a\sin^{1/2}2\theta\left[\cot\mu + \sqrt{\left(\cot^2\mu + \operatorname{cosec}^2 2\theta\right)}\right]^{\frac{1}{2}}$$

15. The space on one side of an infinite plane wall $y = 0$ is filled with inviscid, incompressible fluid, moving at infinity with velocity U in the direction of x axis. The motion of the fluid is wholly two-dimensional in xy plane. A doublet of strength μ is at a distance a from the wall and the points in the negative direction of x axis. Show that if $\mu < 4a^2U$, the pressure of the fluid on the wall is maximum at points distant $a\sqrt{3}$ from O, the foot of the perpendicular from the doublet on the wall and is minimum also.

If $\mu = 4a^2U$, find points where the velocity of the fluid is zero and show that streamlines include the circle $x^2 + (y - a)^2 = 4a^2$.

Solution: Since the points of the doublet are in the negative direction of x axis so that the doublet makes an angle π with the x axis. Image of the given doublet is an equal doublet, similarly oriented at $z = -ia$.

The system consists object doublet, image doublet and stream with velocity U parallel to x axis. Hence

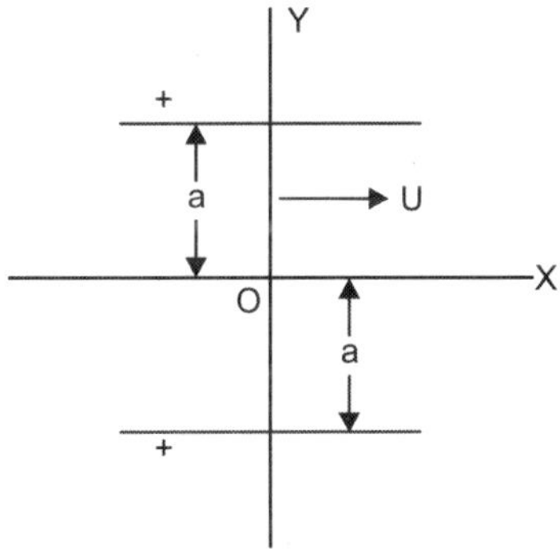

Fig. 3.31

$$w = \frac{\mu e^{i\pi}}{z - ia} + \frac{\mu e^{i\pi}}{z + ia} - Uz$$

$$= \frac{-\mu}{z - ia} - \frac{\mu}{z + ia} - Uz$$

$$\Rightarrow \quad w = \frac{-2\mu z}{z^2 + a^2} - Uz$$

$$\therefore \quad -\frac{dw}{dz} = U + \frac{2\mu}{\left(z^2 + a^2\right)^2}\left[z^2 + a^2 - 2z^2\right]$$

$$= U + \frac{2\mu\left(a^2 - z^2\right)}{\left(z^2 + a^2\right)^2} \quad \ldots (1)$$

$$\left|\frac{-dw}{dz}\right| = q = \left|U + \frac{2\mu\left(a^2 - z^2\right)}{\left(z^2 + a^2\right)^2}\right|$$

For any point on the wall $z = x$ so that

$$q = U + \frac{2\mu\left(a^2 - x^2\right)}{\left(x^2 + a^2\right)^2} \quad \ldots (2)$$

$$\Rightarrow \quad q^2 - U^2 = \frac{4\mu^2\left(a^2 - x^2\right)^2}{\left(x^2 + a^2\right)^4} + \frac{4\mu U\left(a^2 - x^2\right)}{\left(x^2 + a^2\right)^2} \quad \ldots (3)$$

By Bernoulli's equation for steady motion

$$\frac{p}{\rho} + \frac{1}{2}q^2 = C \quad \ldots (4)$$

Subjecting this to the condition $p = \Pi$, $q = U$, where $z = \infty$, so that

$$\frac{\Pi}{\rho} + \frac{1}{2}U^2 = C \qquad \text{... (5)}$$

(5) – (4) gives

$$\Rightarrow \quad \frac{\Pi - p}{\rho} + \frac{1}{2}\left(U^2 - q^2\right) = 0$$

$$\Rightarrow \quad \frac{\Pi - p}{\rho} = \frac{1}{2}\left(q^2 - U^2\right)$$

$$\Rightarrow \quad \frac{\Pi - p}{\rho} = \frac{2\mu^2\left(a^2 - x^2\right)^2}{\left(x^2 + a^2\right)^4} + \frac{2\mu U\left(a^2 - x^2\right)}{\left(a^2 + x^2\right)^2}$$

$$\Rightarrow \quad -\frac{1}{\rho}\frac{dp}{dx} = \frac{-8\mu^2 x\left(a^2 - x^2\right)}{\left(a^2 + x^2\right)^4} - \frac{16\mu^2\left(a^2 - x^2\right)^2 x}{\left(a^2 + x^2\right)^5} + 2\mu U\left[\frac{-2x}{\left(a^2 + x^2\right)^2} - \frac{4x\left(a^2 - x^2\right)}{\left(a^2 + x^2\right)^3}\right]$$

$$= \frac{-8\mu^2 x\left(a^2 - x^2\right)}{\left(a^2 + x^2\right)^5}\left[a^2 + x^2 + 2\left(a^2 - x^2\right)\right] - \frac{4\mu U x}{\left(a^2 + x^2\right)^2}\left[a^2 + x^2 + 2\left(a^2 - x^2\right)\right]$$

$$\Rightarrow \quad \frac{1}{\rho}\frac{dp}{dx} = \frac{4\mu x\left(3a^2 - x^2\right)}{\left(a^2 + x^2\right)^5}\left[2\mu\left(a^2 - x^2\right) + U\left(a^2 + x^2\right)^2\right]$$

For extremum value of p, $\dfrac{dp}{dx} = 0$

which gives $x\ (3a^2 - x^2) = 0$

$$\Rightarrow \quad x = 0, \pm a\sqrt{3}$$

so that on the wall ($y = 0$) at $x = \pm\sqrt{3}a$, we have from (2)

$$\left[\frac{dw}{dz}\right]_{x = \sqrt{3}a} = U - \frac{\mu}{4a^2}$$

If $\mu < 4a^2U$, $\dfrac{d^2p}{dx^2}$ is negative at $x = \sqrt{3}a$ so that p is maximum at the wall.

Again if $\mu < 4a^2U$, then $\dfrac{d^2p}{dx^2} > 0$ where $z = 0$ so that p is minimum.

In the case when $\mu = 4a^2U$, we find

$$\frac{dw}{dz} = 0$$

Hence from (1), we have

$$U + \frac{8a^2U(a^2 - z^2)}{(a^2 + z^2)^2} = 0$$

$$\Rightarrow \quad (z^2 + a^2)^2 + 8a^2(a^2 - z^2) = 0$$

On the wall, $z = x$

$$\therefore \quad (x^2 + a^2)^2 + 8a^2(a^2 - x^2) = 0$$

$$\Rightarrow \quad x^4 + a^4 + 2a^2x^2 + 8a^4 - 8a^2x^2 = 0$$

$$\Rightarrow \quad x^4 - 6a^2x^2 + 9a^4 = 0$$

$$\Rightarrow \quad (x^2 - 3a^2)^2 = 0$$

$$x = \pm a\sqrt{3}$$

Hence $(\pm a\sqrt{3}, 0)$ are the points where the velocity vanishes.

Again we have

$$w = \frac{-2\mu z}{z^2 + a^2} - Uz$$

$$\Rightarrow \quad \phi + i\psi = \frac{-2\mu(x + iy)}{(x + iy)^2 + a^2} - U(x + iy)$$

$$= -U(x + iy) - \frac{2\mu(x + iy)}{(x^2 - y^2 + a^2) + 2xiy}$$

$$= -U(x + iy) - \frac{2\mu(x + iy)\left[(x^2 - y^2 + a^2) - 2ixy\right]}{(x^2 - y^2 + a^2)^2 + 4x^2y^2}$$

Equating imaginary parts on both the sides, we get

$$\psi = -Uy - \frac{2\mu\left[y\left(x^2 - y^2 + a^2\right) - 2x^2y\right]}{\left(x^2 - y^2 + a^2\right)^2 + 4x^2y^2}$$

Streamlines are given by

ψ = const. Taking const. = 0

Hence the streamlines are given by

$$YU = \frac{24a^2U\left[2x^2y - Y\left(x^2 - y^2 + a^2\right)\right]}{\left(x^2 - y^2 + a^2\right)^2 + 4x^2y^2}$$

$$\Rightarrow \; Y\left[\left(x^2 + y^2\right)^2 + 2a^2\left(x^2 - y^2\right) + a^4\right] = 8a^2y\left[x^2 + y^2 - a^2\right]$$

$$\left(x^2 + y^2\right)^2 + 2a^2\left(x^2 - y^2\right) + a^4 = 8a^2\left(x^2 + y^2 - a^2\right)$$

$$\Rightarrow \; \left(x^2 + y^2\right)^2 - 6a^2x^2 - 10a^2y^2 + 9a^4 = 0$$

$$\Rightarrow \; \left(x^2 + y^2 - 3a^2\right)^2 - 4a^2y^2 = 0$$

$$\Rightarrow \; \left(x^2 + y^2 - 3a^2 - 2ay\right)\left(x^2 + y^2 - 3a^2 + 2ay\right) = 0$$

which includes the circle

$$x^2 + y^2 - 2ay - 3a^2 = 0$$

i.e. $x^2 + (y - a)^2 = 4a^2$

16. A source S and sink T of equal strength m are situated within the space bounded by a circle whose centre is O. If S and T are at equal distance from O on opposite sides of it and on the same diameter AOB. Show that velocity of the liquid at any point P is

$$2m\frac{OS^2 + OA^2}{OS} \cdot \frac{PA \cdot PB}{PS \cdot PS^1 PT \cdot PT^1}$$

where S^1 and T^1 are inverse points of S and T with respect to the circle.

Solution: We take O as origin and OA as x axis.

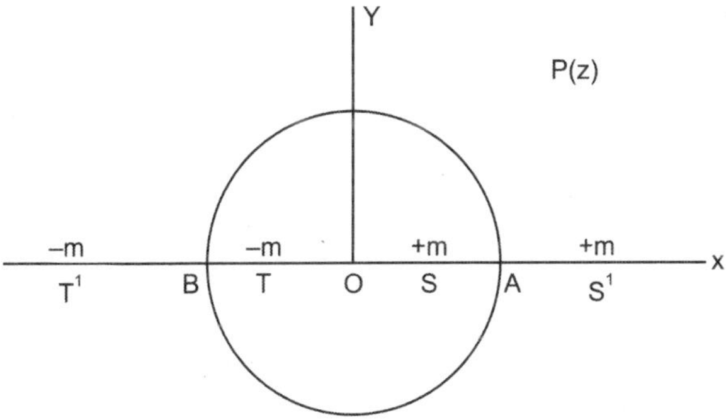

Fig. 3.32

Let $OS = OT = c$, $OA = OB = a$

Then $OS \cdot OS^1 = a^2$, $OT \cdot OT^1 = a^2$

Hence $OS^1 = OT^1 = \dfrac{a^2}{c}$

The image system of a source of strength $+m$ and a sink of strength $-m$ placed at the points S and T within the circle consist of

(i) a source of strength $+m$ at $S\ (c,\ 0)$

(ii) a source of strength $+m$ at $S\left(\dfrac{a^2}{c}, 0\right)$

(iii) a sink of strength $-m$ at the origin O

(iv) a sink of strength $-m$ at $T\ (-c,\ 0)$

(v) a sink of strength $-m$ at $T^1\left(-\dfrac{a^2}{c}, 0\right)$

(vi) a source of strength $+m$ at the origin O.

The source and sink of same strength canceals each other at the origin. Thus the complex potential w becomes

$$w = -m\log(z-c) + m\log(z+c) - m\log\left(z - \frac{a^2}{c}\right) + m\log\left(z + \frac{a^2}{c}\right)$$

$$\frac{-dw}{dz} = m\left[\frac{1}{z-c} - \frac{1}{z+c} + \frac{1}{z-c^1} - \frac{1}{z+c^1}\right] \quad \text{where } c^1 = \frac{a^2}{c}$$

$$= m\left[\frac{2c}{z^2-c^2} + \frac{2c^1}{z^2-c^{1^2}}\right]$$

$$= 2m\left[\frac{c\left(z^2 - \frac{a^4}{c^2}\right) + \frac{a^2}{c}\left(z^2-c^2\right)}{\left(z^2-c^2\right)\left(z^2-c^{1^2}\right)}\right]$$

$$= 2m\left[\frac{cz^2 - \frac{a^4}{c} + \frac{a^2z^2}{c} - a^2c}{(z-c)(z+c)\left(z-c^1\right)\left(z+c^1\right)}\right]$$

$$= \frac{2m\left[\frac{z^2}{c}\left(c^2+a^2\right) - \frac{a^2}{c}\left(a^2+c^2\right)\right]}{(z-c)(z+c)\left(z-c^1\right)\left(z+c^1\right)}$$

$$\frac{-dw}{dz} = \frac{2m\left(a^2+c^2\right)\left[z^2-a^2\right]}{c(z-c)(z+c)\left(z-\frac{a^2}{c}\right)\left(z+\frac{a^2}{c}\right)}$$

$$= \frac{2m\left(a^2+c^2\right)(z-a)(z+a)}{c(z-c)(z+c)\left(z-\frac{a^2}{c}\right)\left(z+\frac{a^2}{c}\right)}$$

The velocity q at any point P is given by

$$q = \left|\frac{-dw}{dz}\right| = 2m\frac{\left(OA^2+OS^2\right)PA\cdot PB}{OS\,PS\cdot PT\,PS^1\cdot PT^1}$$

17. Parallel line sources (perpendicular to the xy plane) of equal strength m are placed at the points $z = nia$, where $n = \ldots.\ -2, -1, 0, 1, \ldots.$ Prove that the complex potential for two dimensional doublets (line doublets) with their axes parallel to the x-axis, of strength μ at the same points, is given by

$$w = \mu \cot h\left(\frac{\Pi z}{a}\right)$$

Solution: Sources of equal strength m are at $z = \pm\, nia$. Where $n = 0, 1, 2, 3, \ldots$. The complex potential due to this system at any point z is given by

$$w = -m \log (z - 0) - \sum_{n=1}^{\infty} m \log (z - nia)$$

$$- \sum_{n=1}^{\infty} m \log (z + nia)$$

$$= -m \log z - \sum_{n=1}^{\infty} m \log \left(z^2 + n^2a^2\right)$$

$$= -\sum_{n=1}^{\infty} m \log \left(1 + \frac{z^2}{n^2a^2}\right) n^2a^2z$$

$$= -\sum_{n=1}^{\infty} m \log \left\{\left(1 + \frac{z^2}{n^2a^2}\right) n^2a^2 \cdot \frac{\pi z}{a} \frac{a}{\pi}\right\}$$

$$w = -\sum_{n=1}^{\infty} m \log \left(1 + \frac{z^2}{n^2a^2}\right) \frac{\pi z}{a} - \sum_{n=1}^{\infty} m \log \left(n^2a^2 \frac{a}{\pi}\right)$$

Neglecting constant

$$w = -\sum_{n=1}^{\infty} m \log \frac{\pi z}{a} \left(1 + \frac{z^2}{n^2a^2}\right)$$

Putting $\frac{\theta}{\pi} = \frac{z}{a}$, we get

$$w = -\sum_{n=1}^{\infty} m \log \theta \left(1 + \frac{\theta^2}{n^2\pi^2}\right)$$

$$= -m \log \theta \left(1 + \frac{\theta^2}{\pi^2}\right)\left(1 + \frac{\theta^2}{2^2\pi^2}\right)$$

$$= -m \log \sin h\theta$$

$$= -m \log \sin h \left(\frac{\pi z}{a}\right) \qquad \ldots (1)$$

Since $w = -m \log (z - a)$ due to source $+m$ at $z = a$ and $w = \dfrac{m}{z-a}$ due to doublet $+m$ at $z = a$ with its axis along x axis

i.e. $w = \dfrac{d}{dz}\left[m \log (z - a)\right]$ for a

doublet $+m$ at $z = a$ with its axis along x axis.

Therefore the complex potential for the doublets of strength m at these points is negative derivative of (1), so that

$$w = \frac{d}{dz}\left[m \log \sin h\left(\frac{\pi z}{a}\right)\right]$$

$$= \frac{m\pi}{a} \cot h\left|\frac{z\pi}{a}\right| = \mu \cot h\left(\frac{\pi z}{a}\right)$$

18. Prove that in the two-dimensional liquid motion due to any number of sources at different points on a circle, the circle is a streamline provided that there is no boundary and that the algebraic sum of strengths of the sources is zero.
Show that the same is true if the region of flow is bounded by a circle in which cuts orthogonally the circle in question.

Solution: Let A_1, A_2, A_3 be the positions of the sources of strengths m_1, m_2, m_3, ...

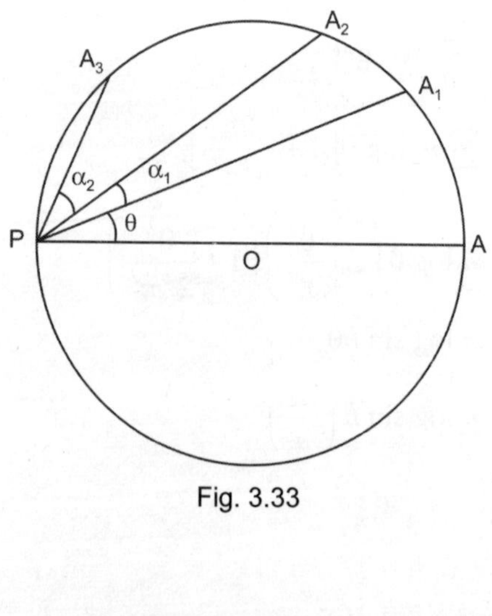

Fig. 3.33

We take any point P on the circle and the diameter through it as the initial line.

Let $\angle A_1PA = \theta$, $\angle A_2PA_1 = \alpha_1$, $\angle A_3PA_2 = \alpha_2$ .... etc.

The streamline is given by

$$\psi = -m_1\theta - m_2(\theta + \alpha_1) - m_2(\alpha_2 + \alpha_1 + \theta).....$$

$$= -(m_1 + m_2 + ...)\theta - [m_2\alpha_1 + m_3(\alpha_1 + \alpha_2) + ...]$$

$$= -\theta\sum m - \text{const.}$$

For whatever be the position of P, α_1, α_2, ... etc. do not change. Since the angle subtended at the circumference by an arc is always the same.

If the algebraic sum of the strengths is zero, i.e. if $\sum m = 0$, then ψ = const, meaning thereby circle is a streamline. Hence the result.

Second Part

Let O^1 be the centre of a circle which cuts the given circle orthogonally join O^1A_1. If O^1A_1 cuts the original at A^1, then A^1 is the inverse point of A_1 w.r.t. circle whose centre is O^1.

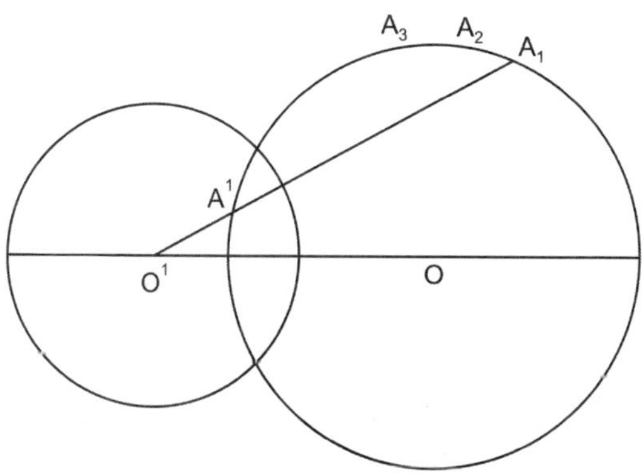

Fig. 3.34

Relative to the circle whose centre is O^1, the image of source + m_1 at A_1 is a source + m_1 at A^1 and a sink $-m_1$ at O^1. If the barries are omitted, we are left with system $2\sum m$

on the original circle and $-\sum m$ at O^1 and as $\sum m = 0$, we again get the same result i.e. ψ = constant.

19. Find the velocity potential when there is a source and an equal sink inside a circular cavity and show that one of the streamlines is an arc of the circle which passes through the source and sink and cuts orthogonally the boundary of the cavity.

Solution: We consider a source $+m$ at A and a sink $-m$ at B respectively inside a circular cavity whose centre is O and radius a. Let $OA = b$, $OB = c$, $\angle BOA = \alpha$. Let A^1 and B^1 be respectively inverse points of A and B respectively.

Then $OA \cdot OA^1 = a^2 = OB \cdot OB^1$

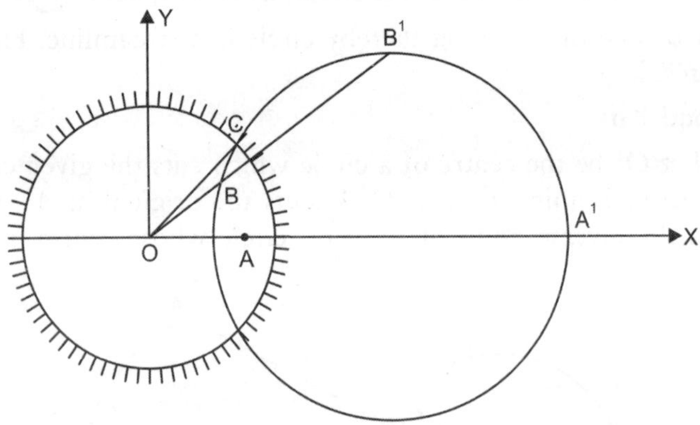

Fig. 3.35

The image of source $+m$ at A is a source $+m$ at A^1 and a sink $-m$ at O. The image of sink $-m$ at B is a sink $-m$ at B^1 and a source $+m$ at O. The source $+m$ and sink $-m$ both at O conceal each other. Then complex potential w is given by

$$w = -m \log(z-b) - m \log\left(z - \frac{a^2}{b}\right)$$

$$+ m \log\left(z - ce^{i\alpha}\right) + m \log\left(z - \frac{a^2}{c} e^{i\alpha}\right)$$

Equating real and imaginary part on both the sides, we can easily get velocity potential and stream function respectively.

Since $OA \cdot OA^1 = OB \cdot OB^1 = a^2$

Hence points A, A^1, B, B^1 are concyclic. Let the circle through these points meet the cavity in C and C^1. Then $OA \cdot OA^1 = OC^2$. Hence OC is a tangent at C to the circle through B, B^1, A, A^1. Thus the two circles intersect orthogonally. Also the circle through A, A^1, B, B^1 passes through A and B are the same source and sink, hence it must be a streamline.

20. If a homogeneous liquid is acted on by a repulsive force from the origin, the magnitude of which at distance r from the origin is μr per unit mass, show that it is possible for the liquid to move steadily, without being constrained by any boundaries, in the space between one branch of the hyperbola $x^2 - y^2 = a^2$ and the asymptotes and find the velocity potential.

Solution: The liquid moves steadily between the space given by one branch of

$$x^2 - y^2 = a^2 \qquad \text{... (1)}$$

and its asymptote is given by

$$x^2 - y^2 = 0 \qquad \text{... (2)}$$

Clearly (1) and (2) are streamlines. For $x^2 - y^2$ is a harmonic function as it satisfies Laplace's equation.

Thus

$$\psi = A\,(x^2 - y^2) = Ar^2\,(\cos^2\theta - \sin^2\theta) = Ar^2 \cos 2\theta$$

$$\Rightarrow \quad \psi = Ar^2 \sin\left(\frac{\pi}{2} + 2\theta\right), \ A \text{ being a constant}$$

Also

$$\frac{\partial \phi}{\partial r} = \frac{1}{r}\frac{\partial \psi}{\partial \theta} = -\frac{1}{r}\,Ar^2 \cos 2\theta$$

$$\Rightarrow \quad \phi = -\,Ar^2 \sin 2\theta = Ar^2 \cos\left(\frac{\pi}{2} + 2\theta\right)$$

$$w = \phi + i\psi = Ar^2\left[\cos\left(\frac{\pi}{2} + 2\theta\right) + i \sin\left(\frac{\pi}{2} + 2\theta\right)\right]$$

$$= Ar^2 e^{i\left(\frac{\pi}{2}+2\theta\right)}$$

$$= Ar^2 e^{2i\theta} i = Aiz^2 \text{ as } z = re^{i\theta}$$

$$\therefore \quad q = \left|\frac{dw}{dz}\right| = |\,2Aiz\,| = 2A|\,z\,| = 2Ar$$

In case of steady motion, the equation of motion is

$$\frac{p}{\rho} + \frac{1}{2}q^2 + \Omega = \text{const.} \qquad \dots (1)$$

Given $\frac{-\partial\Omega}{\partial r} = \mu r \left(\because\ \vec{F} = -\vec{\nabla}\Omega\right)$

$$\Omega = \frac{-\mu r^2}{2}, \text{ neglecting constant}$$

Hence (1) becomes $\frac{p}{\rho} + 2A^2r^2 - \frac{pr^2}{2} = \text{const.}$

On the free surface p = const.

$$\therefore \quad 2A^2r^2 - \frac{\mu r^2}{2} = 0$$

$$\Rightarrow \quad A = \frac{\sqrt{\mu}}{2}$$

Hence

$$q = 2Ar = \frac{2\sqrt{\mu}}{2} r = r\sqrt{\mu}$$

Thus velocity potential $= \phi = Ar^2 \cos\left(\frac{\pi}{2} + 2\theta\right)$

$$= -Ar^2 \sin 2\theta$$

$$= \frac{-\sqrt{\mu}}{2} r^2 \sin 2\theta$$

21. In two-dimensional irrotational fluid motion show that if the streamlines are confocal ellipses

$$\frac{x^2}{a^2+\lambda} + \frac{y^2}{b^2+\lambda} = 1$$

$$\psi = A \log\left[\sqrt{a^2+\lambda}+\sqrt{b^2+\lambda}\right]+B$$

and the velocity at any point is inversely proportional to the square root of the rectangle under the focal radii of the point.

Solution: We consider the relation $z = c \cos w$

$$\Rightarrow \quad x + iy = c \cos (\phi + i\psi)$$

$$\Rightarrow \quad x = c \cos \phi \cos h\psi$$

and

$$y = c \sin \phi \sin h\psi$$

Eliminating ϕ, we get

$$\frac{x^2}{c^2 \cos h^2\psi}+\frac{y^2}{c^2 \sin h^2\psi}=1 \qquad \dots (1)$$

Streamlines are given by ψ = const. Hence (1) shows that streamlines are confocal ellipses.

Comparing (1) with $\dfrac{x^2}{a^2+\lambda}+\dfrac{y^2}{b^2+\lambda}=1$

We get

$$\left.\begin{aligned} c \cos h\psi &= \sqrt{a^2+\lambda} \\ \text{and } c \sin h\psi &= \sqrt{b^2+\lambda} \end{aligned}\right\} \qquad \dots (2)$$

Thus

$$c(\cos h\psi + \sin h\psi) = \sqrt{a^2+\lambda}+\sqrt{b^2+\lambda}$$

$$\Rightarrow \quad ce^{\psi} = \sqrt{a^2+\lambda}+\sqrt{b^2+\lambda}$$

$$\Rightarrow \quad \psi = \log\left[\sqrt{a^2+\lambda}+\sqrt{b^2+\lambda}\right] = \log c \qquad \dots (3)$$

If ϕ and ψ are velocity potential and stream function, $A\phi$ and $A\psi$ are also velocity potential and stream function.

Hence (3) gives

$$\psi = A \log\left[\sqrt{a^2+\lambda}+\sqrt{b^2+\lambda}\right]-B$$

Also

$$z = c \cos w$$

$$\frac{dz}{dw} = -c \sin w$$

$$= -c\sqrt{1-\cos^2 w} = -c\sqrt{1-\frac{z^2}{c^2}} = -\sqrt{c^2 - z^2}$$

$$q^{-1} = \left|\frac{-dz}{dw}\right| = \sqrt{|c^2 - z^2|} = \sqrt{|c-z||c+z|} \qquad \dots (4)$$

From (2)

$$c^2\left(\cos h^2\psi - \sin h^2\psi\right) = a^2 + \lambda - \left(b^2 + \lambda\right) = a^2 - b^2$$

$$\Rightarrow \; c^2 = a^2 - a^2\left(1-e^2\right) = a^2e^2$$

$$\Rightarrow \; c = ae$$

Thus from (4), we get

$$q^{-1} = \sqrt{|z-ae||z+ae|} \qquad \dots (5)$$

($\pm$ *ae*, 0) are coordinates of focii, denoted by S and S^1. P is a point z.

Then

$$r_1 = SP = |z - ae|$$

$$r_2 = S^1P = |z + ae|$$

Hence from (5), we have

$$q^{-1} = \sqrt{r_1 r_2}$$

$$q = \frac{1}{\sqrt{r_1 r_2}}$$

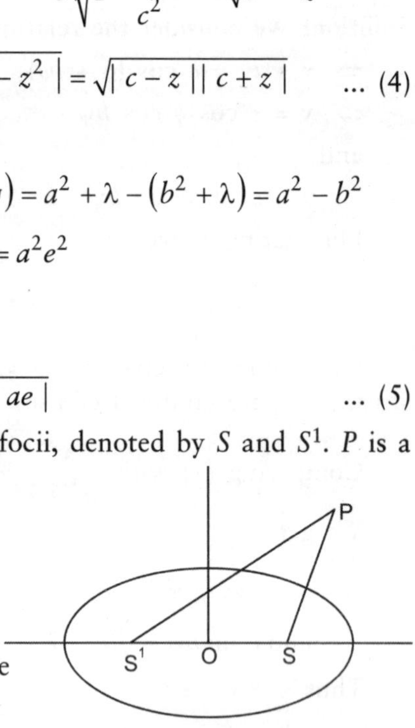

Fig. 3.36

22. An area A is bounded by that part of the x axis for which $x > a$ and by that branch of $x^2 - y^2 = a^2$ which is in the positive quadrant. There is a two-dimensional unit source at $(a, 0)$ which sends out liquid uniformly in all directions. Show by means of the transformation $w = \log(z^2 - a^2)$ that in steady motion the streamlines of the liquid within the area A are portions of rectangular hyperbolas. Determine the streamlines corresponding to $\psi = 0, \frac{\pi}{4}$, and $\frac{\pi}{2}$. If ρ_1 and ρ_2 are the distances of a point P within the fluid from the points $(\pm a, 0)$ show that velocity of the fluid at P is measured by $\frac{2OP}{\rho_1\rho_2}$, O being the origin.

Solution: Given

$$w = \log (z^2 - a^2) \qquad \dots (1)$$

$$\Rightarrow \quad \phi + i\psi = \log\left[(x+iy)^2 - a^2\right]$$

$$= \log\left[(x^2 - y^2 - a^2) + 2xyi\right]$$

Equating imaginary part on both the sides, we get

$$\psi = \tan^{-1}\frac{2xy}{x^2 - y^2 - a^2}$$

The streamlines are given by

$$\psi = \text{const.}$$

$$\Rightarrow \quad \frac{2xy}{x^2 - y^2 - a^2} = \text{const.} = K \text{ (Let)}$$

If $K = 0$, the streamlines are $x = 0$, $y = 0$

If K is infinite, the streamlines are $x^2 = y^2 = a^2$.

Thus the liquid flows in the area A bounded by $x = 0$, $y = 0$ and $x^2 - y^2 = a^2$ i.e. the portion of a rectangular hyperbola in the positive quadrant.

Again the complex potential (1) can be written as

$$w = \log (z - a) + \log (z + a)$$

which shows that the image of a unit source at the point $(a, 0)$ consists a unit source at the point $(-a, 0)$ with respect to y axis.

The velocity of fluid at any point is

$$q = \left|\frac{dw}{dz}\right| = \left|\frac{1}{z-a} + \frac{1}{z+a}\right| = \frac{|2z|}{|z-a||z+a|} = \frac{2OP}{\rho_1\rho_2}$$

The streamlines corresponding to $\psi = 0$ and $\psi = \frac{\pi}{2}$ are $x = 0$, $y = 0$ and $x^2 - y^2 = a^2$.

Also the streamlines corresponding to $\psi = \frac{\pi}{4}$ are

$$\frac{2xy}{x^2 - y^2 - a^2} = \tan\frac{\pi}{4} = 1$$

$$\Rightarrow \; x^2 - y^2 - a^2 = 2xy$$

23. What arrangement of source S and sink S will give rise to the function $w = \log\left(z - \frac{a^2}{z}\right)$? Draw a rough sketch of the streamlines in this case and prove that two of them subdivide into the circle $r = a$ and axis of y.

Solution: Given

$$w = \log\left(z - \frac{a^2}{z}\right)$$

$$\Rightarrow \; w = \log\frac{z^2 - a^2}{z}$$

$$\Rightarrow \; w = \log(z - a) + \log(z + a) - \log z \qquad \dots (1)$$

This shows that the given arrangement consists of two sinks each of strength -1 at $z = a$ and $z = -a$ and a source of strength $+1$ at the origin.

To determine streamlines, from (1)

$$\phi + i\psi = \log(x - a + iy) + \log(x + a + iy) - \log(x + iy)$$

Equating imaginary parts, we get

$$\psi = \tan^{-1}\frac{y}{x-a} + \tan^{-1}\frac{y}{x+a} - \tan^{-1}\frac{y}{x}$$

$$= \tan^{-1}\frac{\dfrac{y}{x-a} + \dfrac{y}{x+a}}{1 - \dfrac{y^2}{x^2 - a^2}} - \tan^{-1}\frac{y}{x}$$

$$= \tan^{-1}\frac{2xy}{x^2 - y^2 - a^2} - \tan^{-1}\frac{y}{x}$$

$$= \tan^{-1} \frac{\dfrac{2xy}{x^2 - a^2 - y^2} - \dfrac{y}{x}}{1 + \dfrac{y}{x}\dfrac{2xy}{x^2 - y^2 - a^2}}$$

$$\psi = \tan^{-1} \frac{y(x^2 + y^2 + a^2)}{x(x^2 + y^2 - a^2)}$$

Streamlines are given by ψ = const.

$$\Rightarrow \quad \tan^{-1} \frac{y(x^2 + y^2 + a^2)}{x(x^2 + y^2 - a^2)} = \text{const.} \qquad \text{... (2)}$$

$$\Rightarrow \quad \frac{y(x^2 + y^2 + a^2)}{x(x^2 + y^2 - a^2)} = \text{const.}$$

If const. = 0, then $y\ (x^2 + y^2 + a^2) = 0$

$y = 0$, for $x^2 + y^2 + a^2 \neq 0$

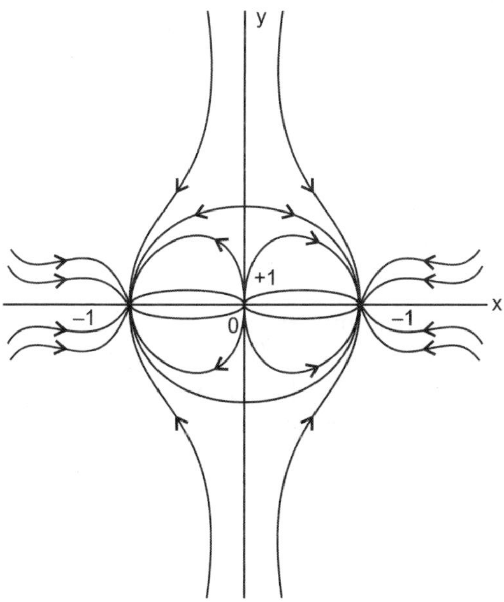

Fig. 3.37

If const. = ∞, then from (2)

$$x\,(x^2 + y^2 - a^2) = 0$$

$$\Rightarrow \quad x = 0,\ r^2 = a^2$$

i.e. $x = 0,\ r = a$

But $x = 0$ represent y axis and $r = a$ represents circle with radius a and centre at the origin. Thus we see that particular steamlines are x axis and the circle $r = a$. A rough sketch of the streamlines is given in the figure.

EXERCISE 3

1. Use the method of images to prove that if there be a source m at the point z_0 in a fluid bounded by the lines $\theta = 0$ and $\theta = \dfrac{\pi}{3}$, the solution is

   $$\phi + i\psi = -\,m \log\left[\left(z^3 - z_0^3\right)\left(z^3 - z_0^{1^3}\right)\right]$$

2. Prove that $u = -\,wy$, $v = wx$, $w = 0$ represents a possible motion of inviscid fluid. Find the stream function and sketch the streamlines. What is the basic difference between this motion and one represented by the potential $\phi = A \log r$, $r = (x^2 + y^2)^{1/2}$

   [**Ans.** Streamlines are $x^2 + y^2 = c$. The basic difference between the two motions is that velocity potential does not exist in first case. Whereas in the second case it exists.]

3. An infinite mass of liquid is moving irrotationally and steadily under the influence of a source of strength μ and an equal sink at a distance $2a$ from it. Prove that the kinetic energy of the liquid which passes in unit time across the plane which bisects at right angles the line joining the source and sink is $\dfrac{8\pi}{7a^4}\rho\mu^3$, ρ being the density of the liquid.

4. A single source is placed in an infinite perfectly elastic fluid, which is also a perfect conductor of head, show that if the motion be steady, the velocity at a distance r from

the source satisfies the equation $\left(V-\frac{K}{V}\right)\frac{\partial V}{\partial r}=\frac{2K}{r}\,V^2\big/4K$

and hence that $r=\frac{1}{\sqrt{V}}\,e.$

5. λ denoting a variable parameter, and f a given function, find the condition that $f(x, y, \lambda) = 0$ should be a possible system of streamlines for steady irrotational motion in two dimensions.

$$\left[\textbf{Ans. }\frac{d^2\psi}{d\lambda^2}\left[\left(\frac{\partial\lambda}{\partial x}\right)^2+\left(\frac{\partial\lambda}{\partial y}\right)^2\right]+\frac{d\psi}{d\lambda}\left[\frac{\partial^2\lambda}{\partial x^2}+\frac{\partial^2\lambda}{\partial y^2}\right]=0\right]$$

6. A line source is in the presence of an infinite plane on which is placed a semi-circular cylinder boss, the direction of the source is parallel to the axis of the boss, the source is at distance c from the plane and the axis of the boss, whose radius is a. Show that the radius to the point on the boss at which velocity is a maximum makes an angle θ with the radius to the source, where

$$\theta=\cos^{-1}\left\{\frac{a^2+c^2}{\sqrt{2\left(a^4+c^4\right)}}\right\}$$

7. Show that $u = 2Axy$, $V = (a^2 + x^2 - y^2)$ are the velocity components of a possible fluid motion. Determine the stream function.

$$\left[\textbf{Ans. }\psi=A\left(a^2x-x^2y+\frac{1}{3}x^3\right)+g(t)\right]$$

8. A velocity field is given by $\vec{q}=-x\vec{i}+(y+t)\vec{j}$. Find the stream function and the streamlines for this field at $t = 0$.

$$\left[\begin{array}{l}\textbf{Ans. }\psi=x(y+2)+g(2)\\ \text{Streamlines are } x\,(y+2)=\text{con.}\end{array}\right]$$

4 MOTION OF CYLINDERS

PART I – MOTION OF CIRCULAR CYLINDER

4.1. Introduction

This chapter deals with two-dimensional irrotational motion produced by the motion of cylinders in an infinite mass of liquid at rest at infinity or when a cylinder is inserted in a stream. For the sake of simplicity, we take the cylinders of unit length.

4.2. Boundary Conditions for the Stream Function

The stream function ψ is determined in the light of the following conditions:

1. ψ satisfied Laplace's equation i.e. $\vec{\nabla}^2 \psi = 0$ at every point of the liquid.

2. The liquid is at rest at infinity so that $\frac{\partial \psi}{\partial x} = 0$ and $\frac{\partial \psi}{\partial y} = 0$ at infinity.

3. Along any fixed boundary, the normal component of velocity must be zero so that $-\frac{\partial \psi}{\partial s} = 0$ and hence ψ = constant.

It means that the boundary must coincide with a streamline ψ = const.

4. Along the boundary of moving cylinder, the normal component of the velocity of the liquid must be equal to normal component of velocity of cylinder.

4.3. General Motion of a Cylinder in Two Dimensions

Let a cylinder of any shape and size be moving in a fluid at rest at infinity. Let the cylinder be moving at right angles to its generators and let the reference axes be fixed in the cylinder which are moving with linear velocities U and V respectively and rotational velocity w as shown in the figure. We consider any point $P(x, y)$ on the boundary. Let PT and PW be the tangent and normal at P to the curve.

We have $x = r \cos \theta$, $y = r \sin \theta$.

$$\dot{x} = \frac{d}{dt}(r \cos \theta) = -r \sin \theta \, \dot{\theta} = -yw$$

$$\dot{y} = \frac{d}{dt}(r \sin \theta) = r \cos \theta \, \theta = xw$$

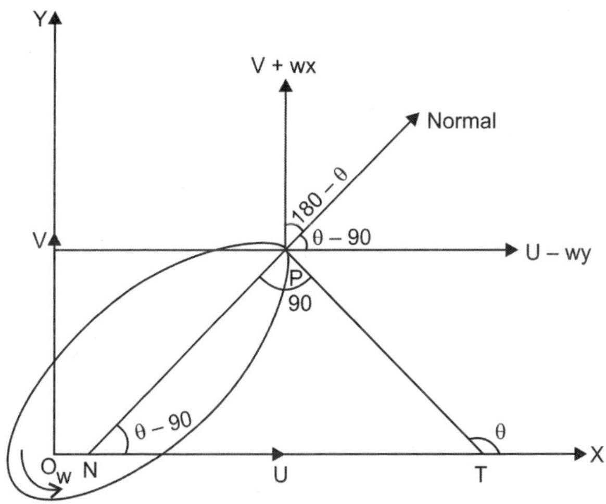

Fig. 4.1

Velocity components at P along x and y axes are $U + \dot{x}$, $V + \dot{y}$; i.e. $U - yw$, $V + xw$.

But normal component of velocity of boundary

= normal component of velocity of liquid

i.e. $\dfrac{-\partial\psi}{\partial s}$ = $(U - yw) \cos (\theta - 90) + (V + xw) \cos (180 - \theta)$

$$\Rightarrow \quad \frac{-\partial\psi}{\partial s} = (U - yw)\frac{dy}{ds} - (V + xw)\frac{dx}{ds}$$

$$\Rightarrow \quad \partial\psi = (U + yw)dy + (V + xw)dx$$

Integrating, we get

$$\psi = -Uy + \frac{y^2}{2}w + Vx + \frac{x^2}{2}w + C$$

$$\Rightarrow \quad \psi = (Vx - Uy) + \frac{w}{2}(x^2 + y^2) + C$$

This is the required expression for the general motion of the cylinder.

Deductions. (1) If there is a pure rotation, then $U = 0 = V$, so that $\psi = \frac{w}{2}(x^2 + y^2) + C$.

(2) If the cylinder has only linear velocity U along x axis then $w = 0$, $V = 0$, so that

$$\psi = -Uy + C$$

4.4. Motion of a Circular Cylinder in a Uniform Stream

To discuss the motion of a circular cylinder moving with velocity U along x axis in an infinite mass of liquid at rest at infinity.

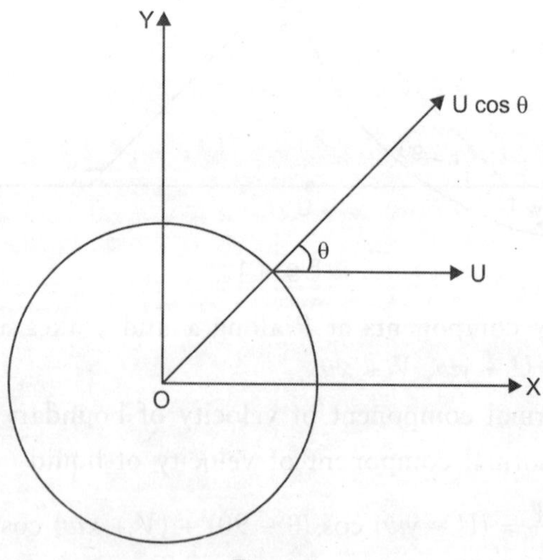

Fig. 4.2

We consider the irrotational motion, originated from rest. Let the origin be taken in the axes of the circular cylinder and the coordinate axes be in a plane perpendicular to its length.

Since the viscous terms are neglected in the equation of fluid motion, a velocity potential ϕ may be defined so that

$$\vec{q} = -\vec{\nabla}\phi, \vec{\nabla}^2\phi = 0 \quad \dots (1)$$

Also ϕ satisfies the following conditions

$$\left(\frac{-\partial\phi}{\partial r}\right)_{r=a} = U\cos\theta \quad \dots (2)$$

$$\left(\frac{-\partial\phi}{\partial r}\right)_{r\to\infty} = 0 \quad \dots (3)$$

and $$\left(-\frac{1}{r}\frac{\partial\phi}{\partial\theta}\right)_{r\to\infty} = 0 \quad \dots (4)$$

The condition (2) suggests that

$$\phi = Ar\cos\theta + \frac{B}{r}\cos\theta$$

Then $$-\frac{\partial\phi}{\partial r} = -\left(A - \frac{B}{r^2}\right)\cos\theta \quad \dots (5)$$

Hence from (2)

$$U\cos\theta = -\left(A - \frac{B}{a^2}\right)\cos\theta$$

$$\Rightarrow \quad U = -\left(A - \frac{B}{a^2}\right) \quad \dots (6)$$

Again from (3) and (5), we get

$$O = -A\cos\theta$$

$$\therefore \quad A = O$$

Thus from (6), we get $B = a^2U$

Hence $$\phi = \frac{a^2U}{r}\cos\theta \quad \dots (7)$$

It evidently satisfies the condition (7).

Since $\frac{\partial\phi}{\partial r} = \frac{1}{r}\frac{\partial\psi}{\partial\theta}$

Therefore

$$\frac{\partial\psi}{\partial\theta} = r\frac{(-a^2U)}{r^2}\cos\theta$$

$$\Rightarrow \qquad \frac{\partial\psi}{\partial\theta} = \frac{-a^2U}{r}\cos\theta$$

Integrating, we get $\psi = \frac{-a^2U}{r}\sin\theta$, neglecting constant of integration.

$$w = \phi + i\psi = \frac{Ua^2}{r}(\cos\partial - i\sin\theta)$$

$$= \frac{Ua^2}{re^{i\theta}} = \frac{Ua^2}{z}$$

This is the required complex potential.

4.5. To Discuss the Motion in the Case of a Liquid Streaming Past a Fixed Circular Cylinder of Radius *a* with Velocity *U*.

This theorem is divided into two parts.

1. To discuss the motion of a circular cylinder moving with U along x axis in an infinite mass of liquid at rest at infinity, which the previous theorem 4.4.

2. To discuss the motion in case of a liquid streaming past a fixed circular cylinder of radius a with velocity U along x axis.

We now suppose that the cylinder is at rest and the liquid is flowing past the cylinder with velocity U in the negative direction of x axis.

This can be obtained from the theorem 4.4 by imposing a velocity U in the negative direction of x axis to both the liquid and the cylinder.

This velocity is given by $\frac{-\partial\phi}{\partial x} = -U$

So that $\phi = Ux$

Thus we have to add a term xU in the velocity potential in theorem 4.4.

Hence
$$\phi = \frac{Ua^2}{r}\cos\theta + Ux$$
$$= \frac{Ua^2}{r}\cos\theta + Ur\cos\theta$$
$$\phi = U\left(r + \frac{a^2}{r}\right)\cos\theta$$

Now
$$\frac{\partial\phi}{\partial r} = \frac{1}{r}\frac{\partial\psi}{\partial\theta}$$

Hence
$$U\left(1 - \frac{a^2}{r^2}\right)\cos\theta = \frac{1}{r}\frac{\partial\psi}{\partial\theta}$$

$$\Rightarrow \qquad \phi = Ur\left(1 - \frac{a^2}{r^2}\right)\sin\theta$$
$$= U\left(r - \frac{a^2}{r}\right)\sin\theta$$

neglecting constant of integration.

Hence
$$w = \phi + i\psi$$
$$= U\left(r + \frac{a^2}{r}\right)\cos\theta + iU\left(r - \frac{a^2}{r}\right)\sin\theta$$
$$= U\left[r(\cos\theta + i\sin\theta) + \frac{a^2}{r}(\cos\theta - i\sin\theta)\right]$$
$$= U\left[re^{i\theta} + \frac{a^2}{re^{i\theta}}\right]$$
$$= U\left(z + \frac{a^2}{z}\right)$$

This is the required complex potential.

4.6. Application of Circle Theorem

Suppose $w = f(z)$ is the complex potential for a two-dimensional motion having no rigid boundaries. If a circular cylinder of radius a is inserted in the fluid and held fixed, then the new complex potential is given by

$$w = f(z) + f\left(\frac{a^2}{z}\right)$$

1. If $w = Uz$ in the absence of cylinder, then on inserting a cylinder $|z| = a$, $w = Uz + \dfrac{Ua^2}{z}$

and so $$\phi = U\left(r + \frac{a^2}{r}\right)\cos\theta$$

and $$\psi = U\left(r - \frac{a^2}{r}\right)\sin\theta$$

2. If the stream makes an angle α with Ox, its complex potential is $Uze^{-i\alpha}$, then on inserting the cylinder $|z| = a$, we have

$$w = U\left(ze^{-i\alpha} + \frac{a^2}{z}e^{i\alpha}\right)$$

But for a change of origin, if the centre of the cylinder is at the point z_0,

$$w = U\left(ze^{-i\alpha} + \frac{a^2 e^{i\alpha}}{z - z_0}\right)$$

4.7. Initial Motion Between Two Coaxial Circular Cylinders

To find the velocity potential and stream function at any point of a liquid contained between coaxial cylinders of radii a and b ($a < b$) when the cylinders are moved suddenly parallel to themselves in directions at right angles with velocities U and V respectively.

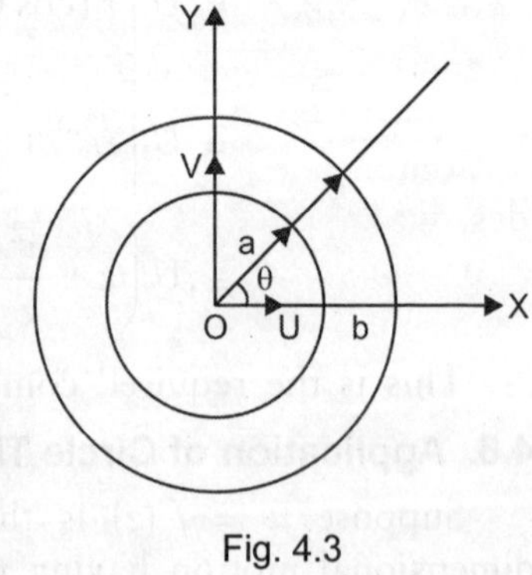

Fig. 4.3

We consider two coaxial cylinders of radii a and b ($a < b$), the space between the cylinders being filled with liquid of density ρ. Let the cylinders move parallel to themselves in perpendicular directions with velocities U and V respectively.

The velocity potential ϕ must satisfy the following conditions:

(i) $\vec{\nabla}^2 \phi = 0$, the solutions of this equation are $r^{\pm n} \cos n\theta$ and $r^{\pm n} \sin n\theta$.

(ii) $\left(\dfrac{-\partial\phi}{\partial r}\right)_{r=a} = U \cos\theta$

and $\left(\dfrac{-\partial\phi}{\partial r}\right)_{r=b} = V \sin\theta$

(iii) Liquid is at rest at infinity so that

$$\left(\frac{-\partial\phi}{\partial r}\right)_{r\to\infty} = 0 \text{ and } \left(\frac{-\partial\phi}{r\partial\theta}\right)_{r\to\infty} = 0$$

To satisfy these conditions, let us assume velocity potential ϕ where

$$\phi = \left(Ar + \frac{B}{r}\right)\cos\theta + \left(Cr + \frac{D}{r}\right)\sin\theta \qquad \text{... (1)}$$

Hence $\dfrac{\partial\phi}{\partial r} = \left(A - \dfrac{B}{r^2}\right)\cos\theta + \left(C - \dfrac{D}{r^2}\right)\sin\theta$

By virtue of (ii), we have

$$-U\cos\theta = \left(A - \frac{B}{a^2}\right)\cos\theta + \left(C - \frac{D}{a^2}\right)\sin\theta$$

and $-V\sin\theta = \left(A - \dfrac{B}{b^2}\right)\cos\theta + \left(C - \dfrac{D}{b^2}\right)\sin\theta$

Comparing the coefficients of $\cos\theta$, $\sin\theta$ on both the sides, we get

$$A - \frac{B}{a^2} = -U \qquad \text{... (2)}$$

$$C - \frac{D}{a^2} = 0 \qquad \text{... (3)}$$

$$A - \frac{B}{b^2} = 0 \qquad \dots (4)$$

$$C - \frac{D}{b^2} = -V \qquad \dots (5)$$

Solving (2) and (4), we get

From (4) $A = \dfrac{B}{b^2}$

Hence from (2)

$$B\left(\frac{1}{b^2} - \frac{1}{a^2}\right) = -U$$

$$\Rightarrow \quad B = -\frac{Ua^2b^2}{a^2 - b^2}$$

Hence $A = \dfrac{-Ua^2}{a^2 - b^2}$

From (3) $C = \dfrac{D}{a^2}$

From (5) $D\left(\dfrac{1}{a^2} - \dfrac{1}{b^2}\right) = -V$

$$D = \frac{Vb^2a^2}{a^2 - b^2}$$

$$C = \frac{Vb^2}{a^2 - b^2}$$

Now (1) becomes

$$\phi = \left(\frac{Ua^2r}{b^2 - a^2} + \frac{Ua^2b^2}{b^2 - a^2}\frac{1}{r}\right)\cos\theta - \left(\frac{Vb^2r}{b^2 - a^2} + \frac{Va^2b^2}{b^2 - a^2}\frac{1}{r}\right)\sin\theta$$

$$\Rightarrow \quad \phi = \frac{Ua^2}{b^2 - a^2}\left(r + \frac{b^2}{r}\right)\cos\theta - \frac{Vb^2}{b^2 - a^2}\left(r + \frac{a^2}{r}\right)\sin\theta \qquad \dots (6)$$

Since $\dfrac{\partial\phi}{\partial r}=\dfrac{1}{r}\dfrac{\partial\psi}{\partial\theta}$ and so

$$\frac{Ua^2}{b^2-a^2}\left(1-\frac{b^2}{r^2}\right)\cos\theta-\frac{Vb^2}{b^2-a^2}\left(1-\frac{a^2}{r^2}\right)\sin\theta=\frac{1}{r}\frac{\partial\psi}{\partial\theta}$$

$$\Rightarrow\quad \frac{\partial\psi}{\partial\theta}=\frac{Ua^2}{b^2-a^2}\left(r-\frac{b^2}{r}\right)\cos\theta-\frac{Vb^2}{b^2-a^2}\left(r-\frac{a^2}{r}\right)\sin\theta$$

Integrating, we get

$$\psi=\frac{Ua^2}{b^2-a^2}\left(r-\frac{b^2}{r}\right)\sin\theta+\frac{Vb^2}{b^2-a^2}\left(r-\frac{a^2}{r}\right)\cos\theta \quad \text{... (7)}$$

neglecting constant of integration.

The values of ϕ and ψ given by (6) and (7) hold only at the instant when the cylinders are coaxial.

4.8. Kinetic Energy of a Finite Liquid

The kinetic energy T is given by

$$T=\int_V\frac{1}{2}\rho q^2dv$$

taken throughout the volume occupied by the fluid.

For irrotational motion $\vec{q}=-\vec{\nabla}\phi$

$$T=\frac{1}{2}\rho\int\left(\vec{\nabla}\phi:\vec{\nabla}\phi\right)dv$$

Using Green's theorem

$$T=\frac{1}{2}\rho\int\left(-\phi\vec{\nabla}^2\phi-\phi\frac{\partial\phi}{\partial n}\right)ds$$

$$=-\frac{1}{2}\rho\int\phi\frac{\partial\phi}{\partial n}ds$$

Since $\vec{\nabla}^2\phi=0$. Here dn denotes an element of inward drawn normal.

4.9. Equation of Motion of a Circular Cylinder without Circulation

A circular cylinder is moving in a liquid at rest at infinity, to calculate the forces on the cylinder owing to the pressure of the liquid.

Let a circular cylinder of radius a, and of unit length, move with velocity U in an infinite mass of liquid at rest at infinity then

$$w = \frac{Ua^2}{Z} = \frac{Ua^2}{r} e^{-i\theta}$$

$$\Rightarrow \quad \phi + i\psi = \frac{Ua^2}{r}(\cos\theta - i \sin\theta)$$

Equating real and imaginary parts on both the sides, we get

$$\phi = \frac{Ua^2}{r}\cos\theta, \ \psi = \frac{-Ua^2}{r}\sin\theta$$

Hence $\dfrac{\partial\phi}{\partial r} = \dfrac{-Ua^2}{r^2}\cos\theta$

$$\left(\frac{\partial\phi}{\partial r}\right)_{r=a} = \left(\frac{-Ua^2}{a^2}\cos\theta\right) = -U\cos\theta$$

If T_1 be the Kinetic energy of the liquid on the boundary of the cylinder and T_2 that of the cylinder. Let σ and ρ be the densities of cylinder and liquid respectively

$$T_1 = -\frac{1}{2}\rho \int \phi \frac{\partial\phi}{\partial n} ds \qquad \text{or } r = a$$

Here $\dfrac{\partial\phi}{\partial n}$ denotes derivative of ϕ along normal and normal is along radius.

For circle $|z| = a$, $s = a\theta$, so that $ds = ad\theta$.

Hence $\quad T_1 = \dfrac{1}{2}\rho \displaystyle\int_0^{2\pi} \left(\phi \frac{\partial\phi}{\partial r}\right)_{r=a} ad\theta$

$$= -\frac{1}{2}\rho \int_0^{2\pi} \left(\frac{Ua^2}{a}\cos\theta\right)(-U\cos\theta)\, ad\theta$$

$$= \frac{1}{4}\rho U^2 a^2 \int (1 + \cos 2\theta)\, d\theta$$

$$= \frac{1}{4}\rho U^2 a^2 \left[\theta + \frac{1}{2}\sin 2\theta\right]_0^{2\pi}$$

$$= \frac{1}{2}\rho U^2 a^2 \pi = \left(\pi a^2\right)\frac{\rho U^2}{2} = \frac{1}{2} M^1 U^2$$

Where M^1 = mass of the liquid displaced by the cylinder.

K.E. of the cylinder = $T_2 = \frac{1}{2} MU^2$

Total K.E. of the cylinder and liquid

$$= T_1 + T_2 = \frac{1}{2}\left(M + M^1\right) U^2$$

Let R be the external force in the direction of motion of the cylinder.

Hence by energy equation, rate of change of total energy is the rate at which work is being done by external forces at the boundary.

$$\text{Hence } \frac{d}{dt}(\text{Total K.E.}) = \frac{\text{Work done}}{\text{Time}} = \frac{\text{Force Distance}}{\text{Time}}$$

$$= \text{Force Velocity}$$

$$\Rightarrow \quad \frac{d}{dt}\left[\frac{1}{2}\left(M + M^1\right)U^2\right] = RU$$

$$\Rightarrow \quad (M + M^1)\, U\dot{U} = RU$$

$$\Rightarrow \quad (M + M^1)\, \dot{U} = R \qquad \text{... (1)}$$

$$\Rightarrow \quad M\dot{U} = R - M^1\dot{U} \qquad \text{... (2)}$$

This shows that the presence of the liquid offers resistance (or drag force) of amount M^1U in the direction of motion of the cylinder.

If R^1/M is a constant conservative external force per unit mass acting on the cylinder and liquid alike, then the energy equation at time t provides

$$\frac{1}{2}\left(M + M^1\right)U^2 - \left(M - M^1\right)\left(\frac{R^1}{M}\right) r = \text{const.} \qquad \text{... (3)}$$

Where r is the distance moved by the cylinder on the direction of R^1.

Differentiating (3) with respect to t and cancelling the common factor $U = \frac{dr}{dt}$, we get

$$(M + M^1)\, U\dot{U} = \left(M - M^1\right)\left(\frac{R^1}{M}\right)U$$

$$\Rightarrow \quad (M + M^1)\, \dot{U} = \left(M - M^1\right)\frac{R^1}{M}$$

$$\Rightarrow \quad M\dot{U} = \left(\frac{M - M^1}{M + M^1}\right)R^1$$

$$\Rightarrow \quad M\dot{U} = \frac{\pi a^2 (\sigma - \rho)}{\pi a^2 (\sigma - \rho)} R^1$$

$$\Rightarrow \quad M\dot{U} = \left(\frac{\sigma - \rho}{\sigma + \rho}\right)R^1$$

If U, V be the components of velocity Q of the cylinder and X^1, Y^1 the components of the external forces R^1 then

$$Q = U + iV,\ R^1 = X^1 + iY^1$$

and the two Cartesian equations of motion are

$$M\frac{dU}{dt} = \frac{\sigma - \rho}{\sigma + \rho}X^1,\ M\frac{dV}{dt} = \frac{\sigma - \rho}{\sigma - \rho}Y^1$$

From this it follows that the effect of the presence of the liquid is to reduce the external forces in the ratio $(\sigma - \rho) : (\sigma + \rho)$.

4.10. Equation of Motion of the Circular Cylinder with Circulations

Let Q be the velocity of a circular cylinder moving in an infinite liquid at rest at infinity and let K be the circulation. Then we have seen that

(i) the presence of the liquid offers resistance of amount M^1Q to the motion of the cylinder.

(ii) the effect of the circulation is to give an upward lift of the magnitude. $X + iY = i\rho KQ$.

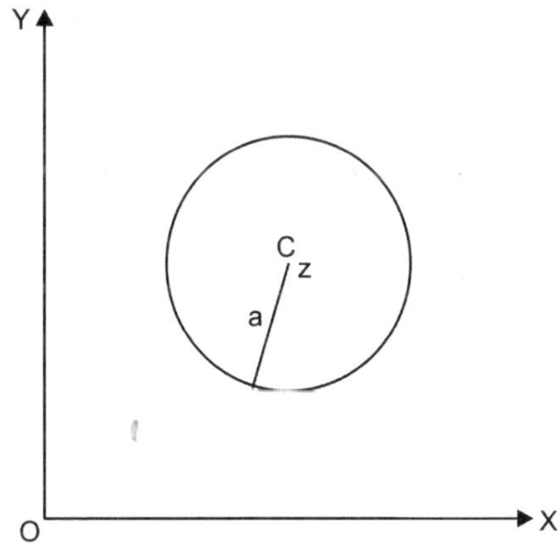

Fig. 4.4

The above figure shows the cross section of the circular cylinder, radius a, centre C, which at time t is at the point z. Let $Q = U + iV$ be the velocity of the centre.

Then $Q = \dfrac{dz}{dt} = U + iV$

$$Q = \frac{d^2z}{dt^2}$$

Therefore, the force exerted by the liquid on account of (i) and (ii) above, is

$$X + iY = -M^1 \frac{d^2z}{dt^2} + i\rho K \frac{dz}{dt}$$

When there is no external force, the equation of motion of the cylinder of mass M is

$$M\frac{d^2z}{dt^2} = -M^2 \frac{d^2z}{dt^2} + i\rho K \frac{dz}{dt}$$

$$\Rightarrow \quad \left(M + M^1\right)\frac{d^2z}{dt^2} = i\rho K \frac{dz}{dt}$$

$$\Rightarrow \quad \frac{\dfrac{d^2z}{dt^2}}{\dfrac{dz}{dt}} = \frac{i\rho K}{M + M^1} = i\lambda \qquad \text{where } \lambda = \frac{\rho K}{M + M^1}$$

Integrating, we get

$$\log \dot{z} = i\lambda t + \log A$$

$$\Rightarrow \quad \dot{z} = Ae^{i\lambda t}$$

$$\Rightarrow \quad \frac{dz}{dt} = Ae^{i\lambda t} \qquad \text{... (1)}$$

Integrating, again we get

$$z = \frac{A}{i\lambda} e^{i\lambda t} + Z_0$$

$$\Rightarrow \quad z - z_0 = \frac{A}{i\lambda} e^{i\lambda t}$$

$$\Rightarrow \quad |z - z_0| = \left|\frac{A}{i\lambda}\right| |e^{i\lambda t}| = B \quad \text{where } B \text{ is real.}$$

Thus the centre of the circle i.e. section of the given cylinder describes a circle with radius B whose centre is z_0.

Also from (1)

$$U + iV = \frac{dz}{dt} = Ae^{i\lambda t} = \lambda B e^{i\lambda t}$$

$$\Rightarrow \quad |U + iV| = \lambda B \left|e^{i\lambda t}\right|$$

$$\Rightarrow \quad \sqrt{U^2 + V^2} = \lambda B$$

$$\Rightarrow \quad U^2 + V^2 = \lambda^2 B^2$$

$$\Rightarrow \quad B = \frac{\sqrt{U^2 + V^2}}{\lambda} = \frac{\sqrt{U^2 + V^2}\left(M + M^1\right)}{\rho K} \qquad \text{... (2)}$$

Therefore

$$|Z - Z_0| = \frac{\sqrt{U^2 + V^2}\left(M + M^1\right)}{\rho K}$$

Hence the cylinder describes a circle of radius B, where B is given by (2) and centre at z_0 with constant velocity. $\sqrt{U^2 + V^2}$ in the sense of cyclic motion.

4.11. Circulation about a Circular Cylinder

Let K be a constant circulation round a circular cylinder of radius a. So we can suppose that K is a circulation round a coaxial circular cylinder of radius r ($<a$).

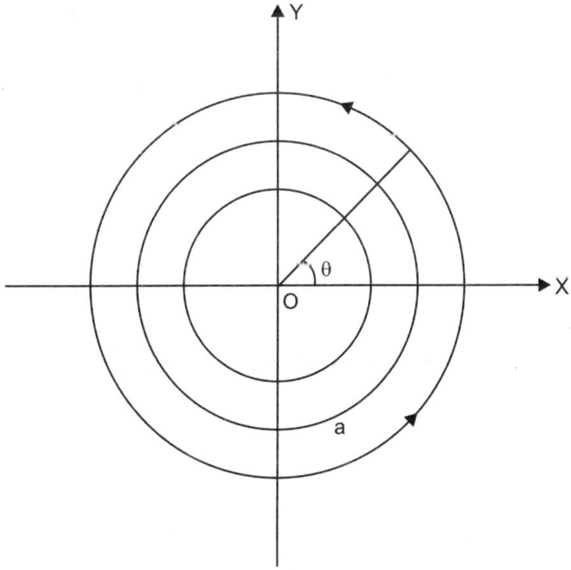

Fig. 4.5

$$\text{Flow along such a circle} = -\frac{1}{r}\frac{\partial\phi}{\partial\theta}2\pi r = K$$

$$\Rightarrow \quad \frac{\partial\phi}{\partial\theta} = \frac{-K}{2\pi} \text{ as circulation} = \int \vec{q}\,d\vec{r}$$

Integrating, we get

$$\phi = \frac{-K\theta}{2\pi}$$

But $\dfrac{\partial\phi}{\partial r} = \dfrac{1}{r}\dfrac{\partial\psi}{\partial\theta}$

and $\dfrac{1}{r}\dfrac{\partial\phi}{\partial\theta} = \dfrac{-\partial\psi}{\partial r}$

Hence $-\dfrac{\partial\psi}{\partial r} = \dfrac{1}{r}\left(-\dfrac{K}{2\pi}\right)$

Integrating, we get $\psi = \frac{K}{2\pi} \log r$

Hence $w = \phi + i\psi$

$$= -\frac{K\theta}{2\pi} + \frac{iK}{2\pi} \log r$$

$$= \frac{iK}{2\pi}(\log r + i\theta)$$

$$= \frac{iK}{2\pi} \log re^{i\theta}$$

$$= \frac{iK}{2\pi} \log z$$

Remark. Some authors write $w = iK \log z$ due to the fact that they take $2\pi K$ instead of K.

4.12. Streaming and Circulation about a Fixed Circular Cylinder

If a liquid is flowing with velocity U in the negative direction of x axis in presence of a circular cylinder of radius a, then the complex potential for this motion is $Uz + \frac{Ua^2}{z}$ and the complex potential due to a constant circulation K is $\frac{iK}{2\pi} \log z$.

Hence the complex potential due to combined motion is

$$w = Uz + \frac{Ua^2}{z} + \frac{iK}{2\pi} \log z$$

Equating real parts on both the sides, we get

$$\phi = Ur \cos\theta + \frac{Ua^2}{r} \cos\theta - \frac{K\theta}{2\pi}$$

$$= U\left(r + \frac{a^2}{r}\right) \cos\theta - \frac{K\theta}{2\pi}$$

To determine stagnation points

We have $\frac{\partial \phi}{\partial r} = U\left(1 - \frac{a^2}{r^2}\right) \cos\theta$

$$\Rightarrow \quad \frac{1}{r}\frac{\partial\phi}{\partial\theta} = \frac{U}{r}\left(r+\frac{a^2}{r}\right)(-\sin\theta) - \frac{K}{2\pi r}$$

$$= -U\left(1+\frac{a^2}{r}\right)\sin\theta - \frac{K}{2\pi r}$$

Now $\quad q^2 = \left(-\frac{\partial\phi}{\partial r}\right)^2_{r=a} + \left(-\frac{1}{r}\frac{\partial\phi}{\partial\theta}\right)^2_{r=a}$

$$\Rightarrow \quad q^2 = \theta + \left(2U\sin\theta + \frac{K}{2\pi a}\right)^2$$

$$\Rightarrow \quad q = 2U\sin\theta + \frac{K}{2\pi a}$$

For stagnation points, $q = 0$, so that

$$2U\sin\theta + \frac{K}{2\pi a} = 0 \qquad \text{... (1)}$$

If there is no circulation i.e. if $K = 0$, the stagnation points are given by $\theta = 0$ and $\theta = \pi$.

For $q = 0$, $K = 0$, $\Rightarrow \sin\theta = 0$

Hence $\theta = 0, \pi$.

The stagnation points in the presence of circulation are given by

$$\sin\theta = \frac{-K}{4\pi aU} \qquad \text{... (2) From (1)}$$

The equation (2) will hold only if

$$|\sin\theta| \le 1$$

i.e. $|K| \le 4\pi aU$

Case I. When $|K| < 4\pi aU$

The stagnation points lie on the cylinder and on a line below the centre parallel to x axis.

Case II. When $|K| = 4\pi aU$

In this case the stagnation points coincide at the bottom of the cylinder.

Case III. When $|K| > 4\pi aU$

In this case there exist no stagnation points on the cylinder but there is one in the fluid on y axis below or above the cylinder according as $K > 0$ or $K < 0$.

We note that any point on the cylinder can be made critical (stagnation) point by a suitable choice of K/U.

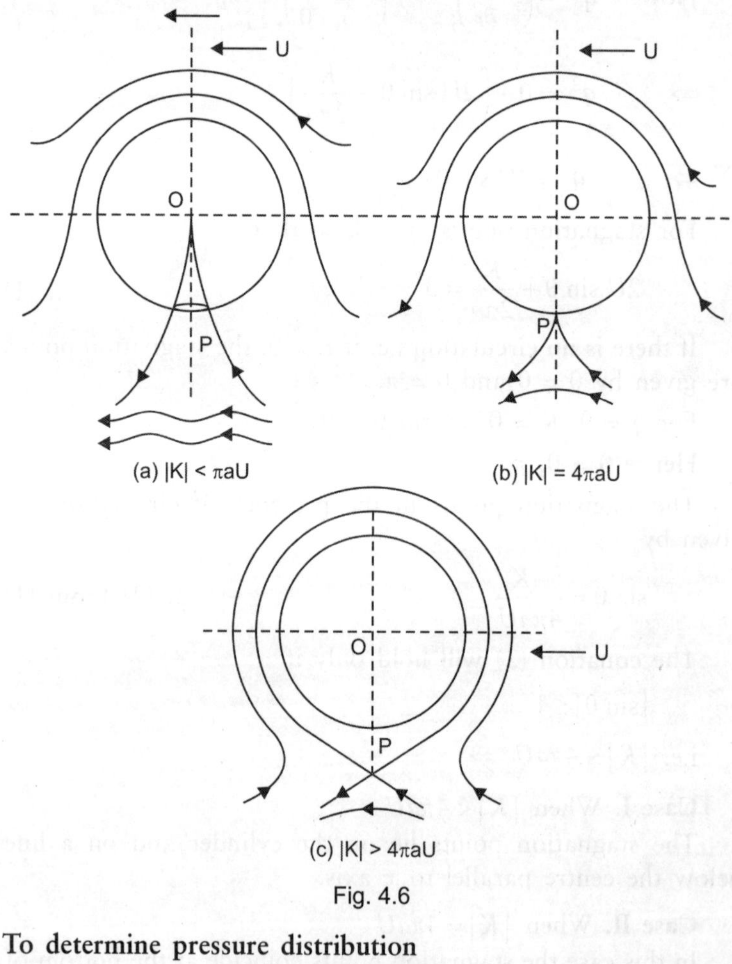

(a) $|K| < \pi aU$ (b) $|K| = 4\pi aU$ (c) $|K| > 4\pi aU$

Fig. 4.6

To determine pressure distribution

We have $w = U\left(z + \dfrac{a^2}{z}\right) + \dfrac{iK}{2\pi}\log z$

Then $\dfrac{dw}{dz} = U\left(1 - \dfrac{a^2}{z^2}\right) + \dfrac{iK}{2\pi z}$

$$\left(\frac{dw}{dz}\right)^2 = U^2\left(1 - \frac{a^2}{z^2}\right)^2 - \frac{K^2}{4\pi^2 z^2} + \frac{iKU}{\pi z}\left(1 - \frac{a^2}{z^2}\right)$$

$$= U^2\left(1 - \frac{a^2}{z^2}\right)^2 - \frac{K^2}{4\pi^2 z^2} + \frac{iKU}{\pi}\left(\frac{1}{z} - \frac{a^2}{z^3}\right)$$

Here coefficient of $\dfrac{1}{z}$ is $\dfrac{iKU}{\pi}$, so that residue at $z = 0$ is $\dfrac{iKU}{\pi}$.

Evidently there is only one pole namely $z = 0$ inside the boundary $|z| = a$. Hence by Cauchy's residue theorem

$$\int_{|z|=a} \left(\frac{dw}{dz}\right)^2 dz = 2\pi i \quad \text{[sum of residues at poles within the domain } |z| = a\text{]}$$

$$= 2\pi i\left(\frac{iKU}{\pi}\right)$$

$$= -2KU$$

By Blasius theorem, if X and Y be the components of the thrust, then

$$X - iY = \frac{i\rho}{2} \int_{|z|=a} \left(\frac{dw}{dz}\right)^2 dz$$

$$= \frac{i\rho}{2}(-2KU) = -iK\rho U$$

Thus $X = 0$, $Y = K\rho U$.

It means that the cylinder experiences a force of magnitude $K\rho U$ which is called an upward lift.

Note. The above article may also be restated as follows:

Show that the complex potential

$$w = U\left(z + \frac{a^2}{z}\right) + iK \log z$$

represents a possible flow past a circular cylinder. Sketch the streamlines, find the stagnation points and calculate the force on the cylinder.

WORKED OUT EXAMPLES

1. An infinite cylinder of radius a and density σ is surrounded by a fixed concentric cylinder of radius b, and the intervening space is filled with liquid of density ρ. Prove that the impulse per unit length necessary to start the inner cylinder with velocity V is $\dfrac{\pi a^2}{b^2 - a^2}\left[(a+\rho)\,b^2 - (\sigma - \rho)\,a^2\right]V$. [I.A.S. 85]

Solution: Let ϕ be the velocity potential, then the boundary conditions are

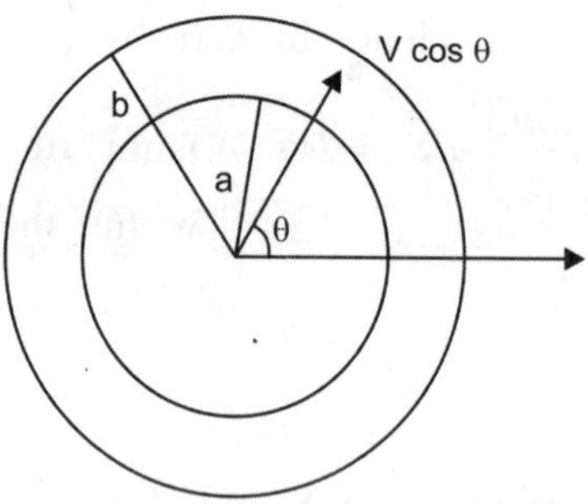

Fig. 4.7

(i) $\vec{\nabla}^2 \phi = 0$

(ii) $\left(-\dfrac{\partial \phi}{\partial r}\right)_{r=a} = V\cos\theta$

(iii) $\left(-\dfrac{\partial \phi}{\partial r}\right)_{r=b} = 0$

(iv) $\left(-\dfrac{\partial \phi}{\partial r}\right)_{r\to\infty} = 0, \left(-\dfrac{1}{r}\dfrac{\partial \phi}{\partial \theta}\right)_{r\to\infty} = 0$

To satisfy these conditions, let us assume

$$\phi = \left(Ar + \frac{B}{r}\right)\cos\theta + \left(Cr + \frac{D}{r}\right)\sin\theta$$

So that $\dfrac{\partial \phi}{\partial r} = \left(A - \dfrac{B}{r^2}\right)\cos\theta + \left(C - \dfrac{D}{r^2}\right)\sin\theta$ ... (1)

Subjecting (1) to (ii) and (iii), we get

$$- V\cos\theta = \left(A - \frac{B}{a^2}\right)\cos\theta + \left(C - \frac{D}{a^2}\right)\sin\theta$$

and $$0 = \left(A - \frac{B}{b^2}\right)\cos\theta + \left(C - \frac{D}{b^2}\right)\sin\theta$$

which gives $A - \dfrac{B}{a^2} = -V,\ A - \dfrac{B}{b^2} = 0$

$$A = C - \frac{D}{a^2} = 0 = C - \frac{D}{b^2}$$

Hence $B = Ab^2$

$$\Rightarrow\ A - \frac{Ab^2}{a^2} = -V$$

$$\Rightarrow\ A = \frac{Va^2}{b^2 - a^2},\ B = \frac{Va^2b^2}{b^2 - a^2}$$

$$C = D = 0$$

Therefore $\phi = \dfrac{Va^2}{b^2 - a^2}\left(r + \dfrac{b^2}{r}\right)\cos\theta$

The impulsive pressure p at point (a, θ) on the inner cylinder is given by

$$p = \rho\phi = \frac{Va^2\rho}{b^2 - a^2}\left(a + \frac{b^2}{a}\right)\cos\theta$$

$$= \frac{Va\rho}{b^2 - a^2}\left(a^2 + b^2\right)\cos\theta$$

The total impulsive force p due to pressure on the cylinder in X-direction is

$$P = -\int_0^{2\pi} p\cos\theta\, ds$$

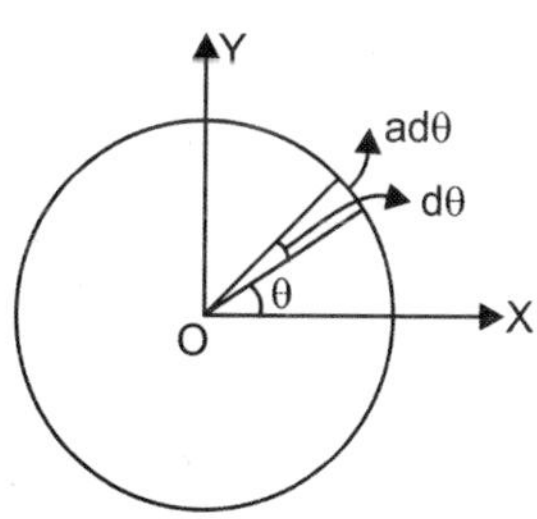

Fig. 4.8

$$= -\int_0^{2\pi} \frac{Va\rho\left(a^2+b^2\right)}{b^2-a^2}\cos^2\theta\, a d\theta$$

$$= -\frac{Va^2\rho\left(a^2+b^2\right)}{b^2-a^2}\int_0^{2\pi}\frac{(1+\cos 2\theta)}{2}d\theta$$

$$= -\pi a^2\rho\,\frac{b^2+a^2}{b^2-a^2}V$$

The impulse required to move the inner cylinder with a velocity V = mass of cylinder × $V = \pi a^2\sigma V$.

Now change in momentum = the sum of impulsive forces.

$\Rightarrow \quad \pi a^2\sigma\ (V - 0) = I + P$

where I is the total impulsive force

$$\Rightarrow \quad I = \pi a^2\sigma V + \frac{\pi a^2\rho\left(b^2+a^2\right)}{b^2-a^2}V$$

$$= \frac{\pi a^2 V}{b^2-a^2}\left[\sigma\left(b^2-a^2\right)+\rho\left(b^2+a^2\right)\right]$$

$$= \frac{\pi a^2 V}{b^2-a^2}\left[(\sigma+\rho)\,b^2-(\sigma-\rho)\,a^2\right]$$

2. A circular cylinder is fixed across a stream of velocity U with circulation K round the cylinder. Show that the maximum velocity in the liquid is $2U + \dfrac{K}{2\pi a}$, where a is the radius of the cylinder.

Solution: The complex potential for streaming past a fixed cylinder with velocity U is

$$w = U\left(z+\frac{a^2}{z}\right)+\frac{iK}{2\pi}\log z$$

$$\Rightarrow \quad w = U\left(re^{i\theta}+\frac{a^2}{r}\overline{e}^{i\theta}\right)+\frac{iK}{2\pi}\log re^{i\theta}$$

K being the circulation about the cylinder.

Hence $\phi = U\left(r + \frac{a^2}{r}\right)\cos\theta - \frac{K\theta}{2\pi}$

$$\therefore \quad \frac{\partial\phi}{\partial r} = U\left(1 - \frac{a^2}{r^2}\right)\cos\theta$$

Also $\frac{\partial\phi}{\partial\theta} = -U\left(r + \frac{a^2}{r}\right)\sin\theta - \frac{K}{2\pi}$

$$\frac{1}{r}\frac{\partial\phi}{\partial\theta} = -U\left(1 + \frac{a^2}{r^2}\right)\sin\theta - \frac{K}{2\pi r}$$

Since $q^2 = \left(-\frac{\partial\phi}{\partial r}\right)^2 + \left(-\frac{1}{r}\frac{\partial\phi}{\partial\theta}\right)^2$

$$\therefore \quad q^2 = \left[U\left(1 - \frac{a^2}{r^2}\right)\cos\theta\right]^2 + \left[U\left(1 + \frac{a^2}{r^2}\right)\sin\theta + \frac{K}{2\pi r}\right]^2$$

We want to find the maximum value of q with respect to r and θ. q will be maximum when r is minimum i.e. when $r = a$. Thus

$$\Rightarrow \quad q^2 = 0 + \left(2U\sin\theta + \frac{K}{2\pi a}\right)^2$$

$$\Rightarrow \quad q = 2U\sin\theta + \frac{K}{2\pi a}$$

Now it is clear that q will be maximum when $\sin\theta$ will be maximum i.e. when $\sin\theta = 1$, max $q = 2U + \frac{K}{2\pi a}$.

3. A stream of water of great depth is flowing with uniform velocity V over a plane level bottom. An infinite cylinder of which the cross section is a semi circle of radius a lies on its flat side with its generating lines making an angle α with the undisturbed streamlines. Prove that the resultant fluid pressure per unit length on the curved surface is $2a\Pi - \frac{5}{3}\rho a V^2 \sin^2\alpha$, where Π is the fluid pressure at a great distance from the cylinder.

Solution: The components of the velocity V are $V \cos \alpha$ along the generator and $V \sin \alpha$ perpendicular to the generator of the cylinder. The component $V \cos \alpha$ will exert no pressure being parallel to the generator of the cylinder, and the component $V \sin \alpha$ give rise a velocity potential ϕ, such that

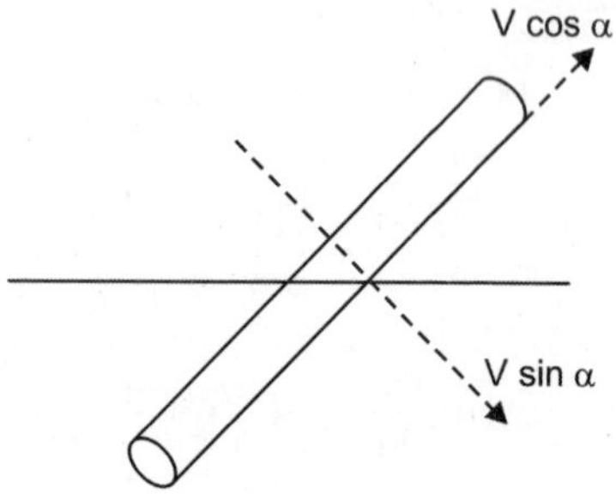

Fig. 4.9

$$\phi = V \sin \alpha \left(r + \frac{a^2}{r} \right) \cos \theta$$

or $$\phi = U \left(r + \frac{a^2}{r} \right) \cos \theta, \text{ where } U = V \sin \alpha$$

$$\therefore \quad \frac{\partial \phi}{\partial r} = U \left(1 - \frac{a^2}{r^2} \right) \cos \theta$$

and $$\frac{1}{\partial} \frac{\partial \phi}{\partial \theta} = -U \left(1 + \frac{a^2}{r^2} \right) \sin \theta$$

Hence $$q^2 = \left(-\frac{\partial \phi}{\partial r} \right)^2 + \left(-\frac{1}{r} \frac{\partial \phi}{\partial \theta} \right)^2$$

$$= U^2 \left[\left(1 - \frac{a^2}{r^2} \right)^2 \cos^2 \theta + \left(1 + \frac{a^2}{r^2} \right)^2 \sin^2 \theta \right]$$

$$= U^2 \left[1 + \frac{a^4}{r^4} - \frac{2a^2}{r^2} \cos 2\theta \right] \quad \ldots (1)$$

Since U is steady and so by Bernoulli's equation for steady motion, we have

$$\frac{p}{\rho} = C - \frac{1}{2} q^2 = C - \frac{U^2}{2} \left[1 + \frac{a^4}{r^4} - \frac{2a^2}{r^2} \cos 2\theta \right] \quad \ldots (2)$$

When $r = \infty$, $p = \Pi$

$$\therefore \quad \frac{\Pi}{\rho} = C - \frac{U^2}{2} \Rightarrow C = \frac{U^2}{2} + \frac{\Pi}{\rho}$$

So from (2)

$$\frac{p}{\rho} = \frac{U^2}{2} + \frac{\Pi}{\rho} - \frac{U^2}{2}\left[1 + \frac{a^4}{r^4} - \frac{2a^2}{r^2}\cos 2\theta\right]$$

$$\Rightarrow \quad \frac{p}{\rho} = \frac{\Pi}{\rho} - \frac{U^2}{2}\left(\frac{a^4}{r^4} - \frac{2a^2}{r^2}\cos 2\theta\right)$$

$$\Rightarrow \quad p = \Pi - \frac{\rho U^2}{2}\left(\frac{a^4}{r^4} - \frac{2a^2}{r^2}\cos 2\theta\right)$$

The pressure at a point (a, θ) on the cylinder is given by

$$p = \Pi - \frac{\rho U^2}{2}(1 - 2\cos 2\theta)$$

$$= \Pi - \frac{\rho U^2}{2}\left(4\sin^2\theta - 1\right) \qquad \text{... (3)}$$

The component $p \cos \theta$ neutralizes each other, as the pressure exerted on the curved surface of semicircular cylinder is equal.

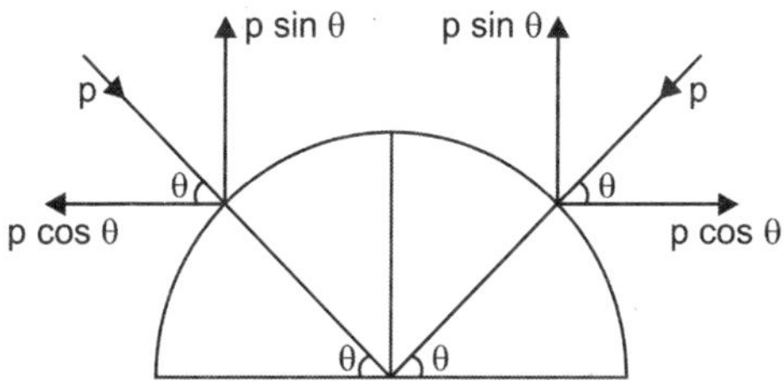

Fig. 4.10

The total resultant pressure P on the semicircular cylinder per unit length is

$$P = \int_0^{\pi} p \sin\theta \, ds$$

$$= \int_0^{\pi} p \sin\theta \, a d\theta \qquad \text{where } s = a\theta$$

$$= \int_0^{\pi}\left[\pi + \frac{\rho U^2}{2}\left(1 - 4\sin^2\theta\right)\right] a \sin\theta d\theta \qquad \text{From (3)}$$

$$= a\left[\Pi\,(-\cos\theta)_0^{\Pi} + \frac{\rho U^2}{2}\left\{(-\cos\theta)_0^{\Pi} - \frac{4.2}{3}\right\}\right]$$

$$= a\left[2\Pi + \rho\left\{\frac{U^2}{2}\cdot 2 - \frac{8}{3}U^2\right\}\right]$$

$$= 2a\Pi - \frac{5}{3}\rho U^2$$

$$= 2a\Pi - \frac{5}{3}\rho V^2 \sin^2\alpha \qquad \text{Since } U = V\sin\alpha$$

4. The space between two infinitely long coaxial cylinders of radii a and b respectively is filled with homogeneous liquid of density ρ and the inner cylinder is suddenly moved with velocity U perpendicular to the axis, the outer one being kept fixed. Show that the resultant impulsive pressure on a length l of the inner cylinder is $\pi\rho a^2 l\left(\dfrac{b^2+a^2}{b^2-a^2}\right)U$.

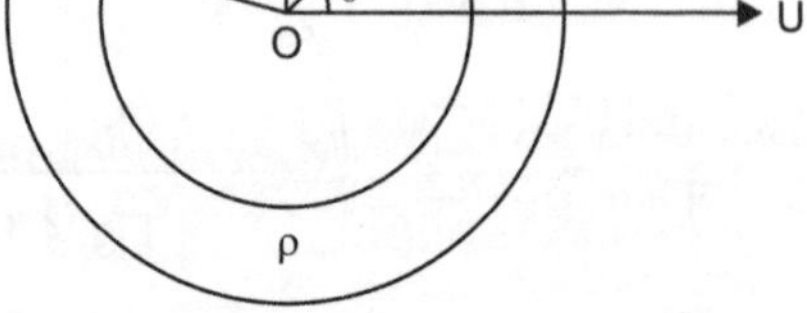

Fig. 4.11

Solution: Liquid of density ρ is filled between the intervening space of two coaxial cylinders. The inner cylinder suddenly moves with velocity U perpendicular to the axis and the outer cylinder being kept fixed. The velocity potential ϕ must satisfy the Laplace's equation and the boundary conditions

(i) $\left(-\dfrac{\partial\phi}{\partial r}\right)_{r=a} = U\cos\theta$

(ii) $\left(-\dfrac{\partial\phi}{\partial r}\right)_{r=b} = 0$

To satisfy these conditions, let us assume that

$$\phi = \left(Ar + \frac{B}{r}\right)\cos\theta + \left(Cr + \frac{D}{r}\right)\sin\theta \qquad \text{... (1)}$$

$$\therefore \quad \frac{\partial\phi}{\partial r} = \left(A - \frac{B}{r^2}\right)\cos\theta + \left(C - \frac{D}{r^2}\right)\sin\theta \qquad \text{... (2)}$$

Subjecting (2) to (i) and (ii), we have

$$U\cos\theta = \left(A - \frac{B}{a^2}\right)\cos\theta + \left(C - \frac{D}{a^2}\right)\sin\theta \qquad \text{... (3)}$$

$$\text{and } O = \left(A - \frac{B}{b^2}\right)\cos\theta + \left(C - \frac{D}{b^2}\right)\sin\theta \qquad \text{... (4)}$$

Which gives $A - \frac{B}{a^2} = -U,\ A - \frac{B}{b^2} = O,\ C - \frac{D}{a^2} = O,$ $C - \frac{D}{b^2} = O$

Hence $B = Ab^2$

$$\therefore \quad A - \frac{Ab^2}{a^2} = -U$$

$$\Rightarrow \quad \frac{A}{a^2}\left(a^2 - b^2\right) = -U$$

$$\Rightarrow \quad A = \frac{Ua^2}{b^2 - a^2},\ B = \frac{Ua^2b^2}{b^2 - a^2}$$

and $C = D = O$

Therefore from (1)

$$\phi = \frac{Ua^2}{b^2 - a^2}\left(r + \frac{b^2}{r}\right)\cos\theta$$

Let F be the impulsive pressure at a point $P(a, \theta)$ of the inner cylinder

$$F = (\rho\phi)_{r=a}$$

$$= \frac{\rho Ua^2}{b^2 - a^2}\left(a + \frac{b^2}{a}\right)\cos\theta$$

$$= \frac{\rho Ua\left(a^2 + b^2\right)}{b^2 - a^2} \cos\theta$$

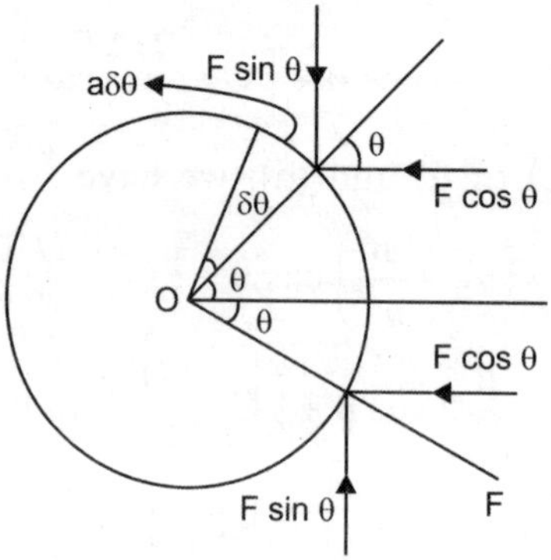

Fig. 4.12

The total impulsive pressure on the cylinder of length l is

$$= \int_0^{2\pi} l\, F \cos\theta\, a d\theta$$

$$= \frac{\rho Ua^2\left(a^2 + b^2\right) l}{b^2 - a^2} \int_0^{2\pi} \cos^2\theta d\theta$$

$$= \frac{\rho Ua^2 l\left(a^2 + b^2\right)}{b^2 - a^2} \int_0^{2\pi} \frac{1 + \cos 2\theta}{2} d\theta$$

$$= \frac{\rho Ua^2 l\left(a^2 + b^2\right)}{2\left(b^2 - a^2\right)} \left[\theta + \frac{\sin 2\theta}{2}\right]_0^{2\pi}$$

$$= \frac{\pi\rho Ua^2 lU\left(b^2 + a^2\right)}{b^2 - a^2}$$

5. A circular cylinder of radius a is moving with velocity U along the axis of x, show that the motion produced by the cylinder in a mass of fluid at rest, is given by the complex potential $w = \phi + i\psi = \dfrac{a^2 U}{z - ut}$ where $z = x + iy$.

Find the magnitude and direction of velocity in the fluid and deduce that for a marked particle of fluid, whose polar coordinates are r, θ referred to the centre of the cylinder as origin

$$\frac{1}{r}\frac{dr}{dt}+\frac{i\,d\theta}{dt}=\frac{U}{r}\left(\frac{a^2}{r^2}e^{i\theta}-e^{-i\theta}\right)$$

and $\left(r-\dfrac{a^2}{r}\right)\sin\theta=b$

Hence prove that the path of such particle is the elastic curve given by $\rho\left(y-\dfrac{1}{2}b\right)=\dfrac{a^2}{4}$

Where ρ is the radius of curvature of the path.

[I.A.S. 86]

Solution: Let the line of the cylinder be taken as x axis, then in time t, it moves a distance Ut along the line of motion. The complex potential, taking the origin at $z = Ut$, for the motion produced by the cylinder in a mass of fluid is, therefore

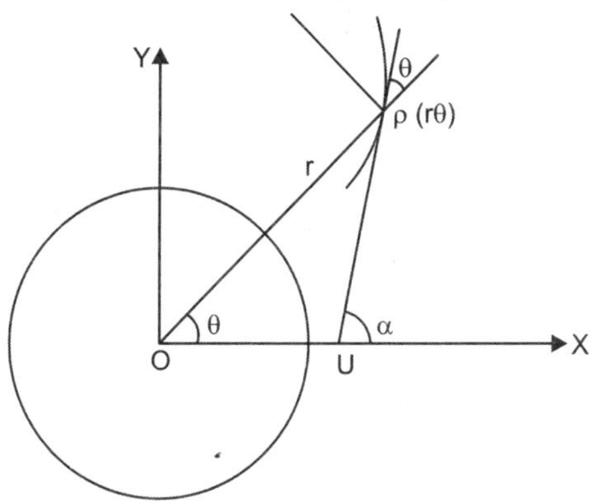

Fig. 4.13

$$w=\phi+i\psi=\frac{Ua^2}{z-Ut} \qquad \text{... (1)}$$

Differentiating (1) w.r.t. z, we get

$$-\frac{dw}{dz}=\frac{Ua^2}{(z-Ut)^2}$$

$$=\frac{Ua^2}{r^2}e^{-2i\theta}$$

where $z - Ut = re^{i\theta}$

$$\Rightarrow \quad V - iV = \frac{Ua^2}{r^2}(\cos 2\theta - i \sin 2\theta)$$

$$\Rightarrow \quad U = \frac{a^2}{r^2} U \cos 2\theta, \ V = \frac{Ua^2}{r^2} \sin 2\theta$$

Let α be the angle made by the direction of velocity with x axis. Then

$$q = \sqrt{U^2 + V^2} = \frac{Ua^2}{r^2}$$

$$\tan \alpha = \frac{V}{U} = \frac{\sin 2\theta}{\cos 2\theta} = \tan 2\theta$$

$$\Rightarrow \alpha = 2\theta$$

The complex potential of the motion at $P\,(r, \theta)$ relative to the cylinder becomes

$$w = Uz + \frac{Ua^2}{z}$$

$$\therefore \quad \phi = Ux + \frac{Ua^2 x}{x^2 + y^2} = Ur \cos \theta + \frac{Ua^2 \cos \theta}{r}$$

and $\psi = Uy - \dfrac{Ua^2 y}{x^2 + y^2} = Ur \sin \theta - \dfrac{Ua^2 \sin \theta}{r}$

The lines of flow at P are given by

$$\psi = \text{constant}$$

$$\Rightarrow \quad U\left(r - \frac{a^2}{r}\right) \sin \theta = \text{constant}$$

$$\Rightarrow \quad \left(r - \frac{a^2}{r}\right) \sin \theta = b \qquad \text{... (2)}$$

Also $\dfrac{dr}{dt} = -\dfrac{\partial \phi}{\partial r} = -U \cos \theta + \dfrac{Ua^2}{r^2} \cos \theta$

and $r\dfrac{d\theta}{dt} = -\dfrac{1}{r}\dfrac{\partial \phi}{\partial \theta}$

$$= -\frac{1}{r}\left[-Ur \sin \theta - \frac{Ua^2 \sin \theta}{r}\right]$$

$$= U \sin\theta + \frac{Ua^2}{r^2} \sin\theta$$

Hence $\frac{1}{r}\frac{dr}{dt} + \frac{i\,d\theta}{dt}$

$$= -\frac{U\cos\theta}{r} + \frac{Ua^2}{r^3}\cos\theta + \frac{iU}{r}\left(\sin\theta + \frac{a^2}{r^2}\sin\theta\right)$$

$$= \frac{U}{r}(-\cos\theta + i\sin\theta) + \frac{Ua^2}{r^3}(\cos\theta + i\sin\theta)$$

$$= \frac{U}{r}\left[-e^{-i\theta} + \frac{a^2}{r^2}e^{i\theta}\right]$$

$$\Rightarrow \quad \frac{1}{r}\frac{dr}{dt} + \frac{i\,d\theta}{dt} = \frac{U}{r}\left[\frac{a^2}{r^2}e^{i\theta} - e^{-i\theta}\right]$$

Let (x, y) be the coordinates of P referred to O as origin, which is at $z = Ut$. The coordinates of the fluid particle are

$$x = r\cos\theta + Ut,\; y = r\sin\theta$$

From (2) $\left(r - \frac{a^2}{r}\right)\sin\theta = b$

$$\Rightarrow \quad y - \frac{a^2\sin^2\theta}{y} = b$$

$$\Rightarrow \quad y^2 - a^2\sin^2\theta = by$$

Differentiating w.r.t. θ, we get

$$2y\frac{dy}{d\theta} - a^2\sin 2\theta = b\frac{dy}{d\theta}$$

$$\Rightarrow \quad \frac{dy}{d\theta}(2y - b) = a^2\sin 2\theta$$

$$\Rightarrow \quad \frac{dy}{d\theta} = \frac{a^2\sin 2\theta}{2y - b}$$

Also $\rho = \frac{ds}{d\psi}$

In this case $\rho = \dfrac{ds}{d\alpha} = \dfrac{ds}{dy}\dfrac{dy}{d\alpha}$

$$= \text{cosec}\,\alpha\,\frac{dy}{d\alpha}$$

$$= \text{cosec}\,2\theta\,\frac{dy}{2d\theta}$$

$$= \frac{1}{2\sin 2\theta}\frac{dy}{d\theta}$$

$$\Rightarrow \quad \rho = \frac{1}{2\sin 2\theta}\frac{a^2 \sin 2\theta}{2y - b}$$

$$\Rightarrow \quad 2\rho\,(2y - b) = a^2$$

$$\Rightarrow \quad \rho\left(y - \frac{b}{2}\right) = \frac{a^2}{4}$$

6. An infinite circular cylinder of radius a is in motion in homogeneous fluid which extends to infinity and is at rest there. Show that at any moment, the pressure at a point of the fluid at a distance r from the axis of the cylinder exceeds the hydrostatic pressure by

$$\rho\left[\frac{a^2}{r}f_1 + \frac{a^2}{r^2}\left\{\left(1 - \frac{a^2}{2r^2}\right)u_1^2 - \left(1 + \frac{a^2}{2r^2}\right)v_1^2\right\}\right]$$

where f_1 is the component acceleration of the centre of the cylinder in the direction of r, and u_1 and v_1 are the component velocities in and perpendicular to the direction.

Solution: Let (x_0, y_0) be the coordinates of the centre of the cylinder. Let U and V be the velocity components of the cylinder, then

$$x_0 = \frac{dx_0}{dt} = U,\ y_0 = \frac{dy_0}{dt} = V$$

and hence $\dot{z}_0 = U + iV$

Also the complex potential w is given by

$$w = \frac{Qa^2}{z - z_0} \quad \text{where } Q = U + iV,\ z - z_0 = re^{i\theta}$$

i.e. $\quad |z - z_0| = r$

$$q = \left| -\frac{dw}{dz} \right| = \left| \frac{Qa^2}{(z - z_0)^2} \right|$$

$$= \frac{a^2}{r^2} \sqrt{U^2 + V^2}$$

$$\Rightarrow \quad q^2 = \frac{a^4}{r^4} \left(U^2 + V^2 \right) \qquad \text{... (1)}$$

Now $w = \phi + i\psi = \dfrac{a^2 (U + iV)}{z - z_0}$

Differentiating w.r.t. t, we get

$$\frac{\partial w}{\partial t} = \frac{\partial \phi}{\partial t} + \frac{i\,\partial \psi}{\partial t}$$

$$= \frac{a^2 (U + iV)}{z - z_0} - \frac{a^2 (U + iV)}{(z - z_0)^2} \left(-\frac{\partial z_0}{\partial t} \right)$$

$$= \frac{a^2}{r} (u + iV)\, e^{-i\theta} + \frac{a^2 (U + iV)^2}{r^2} e^{-i\theta}$$

as $z_0 = U + iV$

Equating real parts on both the sides, we get

$$\frac{\partial \phi}{\partial t} = \frac{a^2}{r} (U \cos \theta + V \sin \theta)$$

$$+ \frac{a^2}{r^2} \left[\left(U^2 - V^2 \right) \cos 2\theta + 2UV \sin 2\theta \right] \qquad \text{... (2)}$$

By pressure equation

$$\frac{p}{\rho} + \frac{1}{2} q^2 - \frac{\partial \phi}{\partial t} = F(t) \qquad \text{... (3)}$$

Using (1) and (2), (3) becomes

$$\frac{p}{\rho} + \frac{a^4}{2r^4} \left(U^2 + V^2 \right) - \frac{a^2}{r} \left(\dot{U} \cos \theta + \dot{V} \sin \theta \right)$$

$$- \frac{a^2}{r^2} \left[\left(U^2 - V^2 \right) \cos 2\theta + 2UV \sin 2\theta \right] = F(t) \qquad \text{... (4)}$$

Let p_0 be the hydrostatic pressure, then when $r = \infty$, $p = p_0$, $q = 0$, so that $U = 0 = V$. Hence from (4)

$$\frac{p_0}{\rho} = F(t)$$

Hence from (4), we get

$$\frac{p - \rho_0}{\rho} = \frac{a^4}{2r^4}\left(U^2 + V^2\right) + \frac{a^2}{r}\left(U \cos\theta + V \sin\theta\right)$$

$$+ \frac{a^2}{r^2}\left[\left(U^2 - V^2\right)\cos 2\theta + 2UV \sin 2\theta\right] \quad \dots (5)$$

f_1 = component of acceleration in the direction of r

Hence $f_1 = \dot{U}\cos\theta + \dot{V}\sin\theta$ ... (6)

U_1 = component of velocity in the direction of r

$\therefore\ U_1 = U\cos\theta + V\sin\theta$ ... (7)

V_1 = component of velocity perpendicular to the direction of r

$= U\cos(90 + \theta) + V\sin(90 + \theta)$

$= -U\sin\theta + V\cos\theta$ ... (8)

From (7) and (8), we get

$$U_1^2 + V_1^2 = U^2 + V^2$$

and $$U_1^2 - V_1^2 = U^2\cos 2\theta - V^2\cos 2\theta + 4UV\sin\theta\cos\theta$$

$$= \left(U^2 - V^2\right)\cos 2\theta + 2UV\sin 2\theta$$

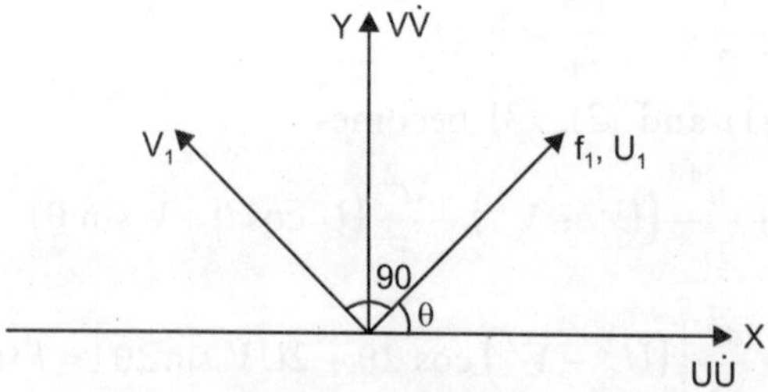

Fig. 4.14

Hence from (5), we get

$$\frac{p-p_0}{\rho} = -\frac{a^4}{2r^4}\left(U_1^2+V_1^2\right)+\frac{a^2}{r}f_1+\frac{a^2}{r^2}\left(U_1^2-V_1^2\right)$$

$$\Rightarrow \quad p-p_0 = \rho\left[\frac{a^2}{r}f_1+\frac{a^2}{r^2}\left\{U_1^2\left(1-\frac{a^2}{2r^2}\right)\right.\right.$$
$$\left.\left.-\left(1+\frac{a^2}{2r^2}\right)V_1^2\right\}\right]$$

7. In the case of two-dimensional motion of a liquid streaming past a fixed circular disc, the velocity at infinity is u in a fixed direction where u is variable. Show that the maximum value of velocity at any point of the fluid is $2u$. Prove that the necessary to hold the disc at rest is $2mu$ where m is the mass of the liquid, d is placed by the disc.

Solution: The velocity potential of a liquid streaming past a fixed circular disc is given by

$$\phi = u\left(r+\frac{a^2}{r}\right)\cos\theta \qquad \text{... (1)}$$

$$\therefore \quad \frac{\partial\phi}{\partial r} = u\left(1-\frac{a^2}{r^2}\right)\cos\theta$$

and $$\frac{1}{r}\frac{\partial\phi}{\partial\theta} = u\left(1+\frac{a^2}{r^2}\right)(-\sin\theta)$$

Let q be the velocity at the point $P\ (r, \theta)$, then

$$q^2 = \left(-\frac{\partial\phi}{\partial r}\right)^2+\left(-\frac{1}{r}\frac{\partial\phi}{\partial\theta}\right)^2$$

$$= u^2\left[\left(1-\frac{a^2}{r^2}\right)^2\cos^2\theta+\left(1+\frac{a^2}{r^2}\right)^2\sin^2\theta\right]$$

or $$q^2 = u^2\left[1+\frac{a^4}{r^4}-\frac{2a^2}{r^2}\cos 2\theta\right] \qquad \text{... (2)}$$

The velocity q will be maximum with respect to θ.

When $\cos 2\theta = -1$, i.e. $2\theta = \pi$.

Then $$q^2 = u^2\left(1+\frac{a^4}{r^4}+\frac{2a^2}{r^2}\right)$$

$$= u^2\left(1+\frac{a^2}{r^2}\right)^2$$

$$\therefore \quad q = u\left(1+\frac{a^2}{r^2}\right) \qquad \dots (3)$$

Again q is maximum when r is minimum i.e. $r = a$

Then $$q_{\max}^2 = u^2(1+1+2) = 4u^2$$

$$q = 2u$$

Thus maximum velocity is $2u$ and it occurs at the point $\left(a, \frac{\pi}{2}\right)$.

Now we suppose that the disc is at rest so that $\dot{r} = \theta$ and $r\dot{\theta} = 0$.

Now differentiating (1) with respect to t, we get

$$\frac{\partial \phi}{\partial t} = \dot{u}\left(r+\frac{a^2}{r}\right)\cos\theta \qquad \dots (4)$$

By pressure equation

$$\frac{p}{\rho}+\frac{1}{2}q^2-\frac{\partial\phi}{\partial t} = F(t)$$

$$\Rightarrow \frac{p}{\rho} = -\frac{1}{2}q^2+\frac{\partial\phi}{\partial t}+F(t)$$

$$= F(t)+\dot{u}\left(r+\frac{a^2}{r}\right)\cos\theta$$

$$-\frac{u^2}{2}\left[1+\frac{a^4}{r^4}-\frac{2a^2}{r^2}\cos 2\theta\right]$$

From (2) and (4)

$$\Rightarrow \left(\frac{p}{\rho}\right)_{r=a} = F(t) + 2a\dot{u}\cos\theta - u^2(1-\cos 2\theta)$$

$$= F(t) + 2a\dot{u}\cos\theta - 2u^2\sin^2\theta \qquad \dots (5)$$

Let $a\delta\theta$ be an element on the surface of the disc. The resultant pressure on the disc is

$$= \int_0^{2\pi} -p\cos\theta ds = -\int_0^{2\pi} p\cos\theta\, a d\theta$$

$$= -a\int_0^{2\pi} \rho\left[F(t) + 2a\dot{u}\cos\theta - 2u^2\sin^2\theta\right]\cos\theta\, d\theta$$

From (5)

$$= -a\rho\left[\left\{F(t)\sin\theta - \frac{2u^2}{3}\sin^3\theta\right\}_0^{2\pi} + a\dot{u}\left\{\theta + \frac{\sin 2\theta}{2}\right\}_0^{2\pi}\right]$$

$$= -a\rho\left[0 + a\dot{u}(2\pi - 0)\right] = -2\pi a^2\rho\dot{u}$$

$$= -2m\dot{u}, \text{ since } m = \pi a^2\rho$$

Hence the force necessary to hold the disc at rest is $2m\dot{u}$.

8. If a long circular cylinder of radius a moves on a straight line at right angles to its length in liquid at rest at infinity, show that when a particle of liquid in the plane of symmetry, initially at distance b in advance of the axis of the cylinder, has moved through a distance $c + \dfrac{b^2 - a^2}{b + wth\dfrac{a}{c}}$.

Solution: Let the line of motion and a line perpendicular to it be taken as the coordinate axes. Let the cylinder moves with velocity V and $P(x, y)$ be any point on it.

After any time t it has described a distance $x = Vt$, $y = 0$.

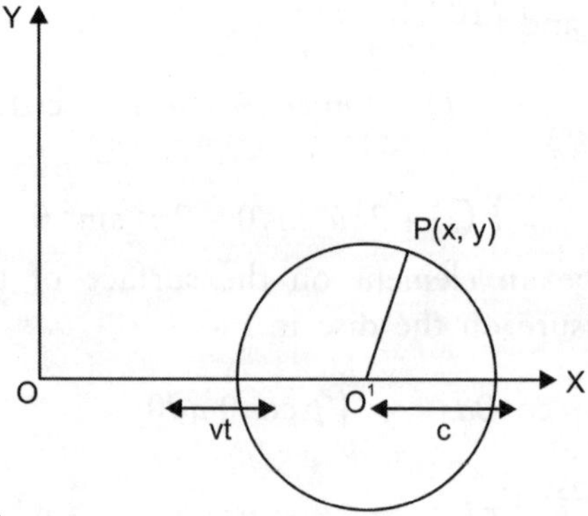

Fig. 4.15

The complex potential at $z = Vt$ is

$$w = \frac{Va^2}{x - Vt}$$

$$q = -\frac{dw}{dz} = \frac{Va^2}{(z - Vt)^2}$$

Along the real axis $y = 0$ and velocity is $\frac{dx}{dt}$, so that

$$\frac{dx}{dt} = \frac{Va^2}{(x - Vt)^2} \qquad \text{... (1)}$$

Putting $x - Vt = \lambda$

$$\frac{dx}{dt} = \frac{d\lambda}{dt} + V$$

Hence from (1), we have

$$\frac{d\lambda}{dt} + V = \frac{Va^2}{\lambda^2}$$

$$\Rightarrow \quad \frac{d\lambda}{dt} = \frac{V(a^2 - \lambda^2)}{\lambda^2}$$

$$\Rightarrow \quad Vdt = \frac{\lambda^2}{a^2 - \lambda^2} d\lambda$$

$$\Rightarrow \quad Vdt = -\left(1 + \frac{a^2}{\lambda^2 - a^2}\right) d\lambda$$

Inegrating, we get

$$Vt + A = -\lambda - \frac{a}{2}\log\frac{\lambda - a}{\lambda + a}$$

$$= -x + Vt - \frac{a}{2}\log\frac{x - Vt - a}{x - Vt + a}$$

$$\Rightarrow \quad x + A = -\frac{a}{2}\log\frac{x - Vt - a}{x - Vt + a} \qquad \text{... (2)}$$

When $x = b$, $t = 0$, so from (2), we have

$$b + A = -\frac{a}{2}\log\frac{b - a}{b + a} \qquad \text{... (3)}$$

(2) – (3) gives

$$x - b = -\frac{a}{2}\log\frac{x - Vt - a}{x - Vt + a} + \frac{a}{2}\log\frac{b - a}{b + a} \qquad \text{... (4)}$$

Also when $t = T$, $x = b + c$

So from (4)

$$c = -\frac{a}{2}\log\frac{b + c - Vt - a}{b + c - Vt + a} + \frac{a}{2}\log\frac{b - a}{b + a}$$

$$\Rightarrow \quad -\frac{2c}{a} = \log\frac{b + a}{b - a}\left[\frac{(b - a) + (c - Vt)}{(b + a) + (c - Vt)}\right]$$

$$\Rightarrow \quad e^{-\frac{2c}{a}} = \frac{b^2 - a^2 + (b + a)(c - Vt)}{b^2 - a^2 + (b - a)(c - Vt)}$$

By componendo and dividendo.

$$\frac{1 + e^{-\frac{2c}{a}}}{1 - e^{-\frac{2c}{a}}} = \frac{2\left[\left(b^2 - a^2\right) + b(c - Vt)\right]}{-2a(c - Vt)}$$

$$\Rightarrow \quad -a\coth\frac{c}{a} = \frac{b^2 - a^2}{c - Vt} + b$$

$$\Rightarrow \quad b + a \cot h\frac{c}{a} = \frac{b^2 - a^2}{Vt - c}$$

$$\Rightarrow \quad Vt - c = \frac{b^2 - a^2}{b + a \cot h\dfrac{c}{a}}$$

$$\Rightarrow \quad Vt = c + \frac{b^2 - a^2}{b + a \cot h\dfrac{c}{a}}$$

9. The circle $(x + a)^2 + y^2 = a^2$ is placed in an on coming wind of velocity U and there is a circulation $2\Pi K$. Find the complex potential and show that moment about the origin is $2\Pi K\rho aU$.

Solution: The given circle is $|z + a| = a$.

The complex potential for the said motion is

$$w = Uz + \frac{Ua^2}{z} + iK \log (z - a)$$

$$\therefore \quad \frac{dw}{dz} = U\left(1 - \frac{a^2}{z^2}\right) + \frac{iK}{z - a}$$

$$\Rightarrow \quad \left(\frac{dw}{dz}\right)^2 = U^2\left(1 - \frac{a^2}{z^2}\right)^2 - \frac{K^2}{(z - a)^2} + \frac{2iKU}{z - a}\left(1 - \frac{a^2}{z^2}\right)$$

$$\Rightarrow \quad \left(\frac{dw}{dz}\right)^2 = U^2\left(1 - \frac{a^2}{z^2}\right)^2 - \frac{K^2}{z^2}\left(1 - \frac{a}{z}\right)^{-2} + \frac{2iKU}{z}\left(1 - \frac{a}{z}\right)^{-1}\left(1 - \frac{a^2}{z^2}\right)$$

$$\Rightarrow \quad z\left(\frac{dw}{dz}\right)^2 = zU^2\left(1 - \frac{a^2}{z^2}\right)^2 - \frac{K^2}{z}\left(1 + \frac{2a}{z} + ...\right) + 2iKU\left(1 + \frac{a}{z} + ...\right)\left(1 - \frac{a^2}{z^2} + ...\right)$$

Coefficient of $\frac{1}{z}$ in this expansion

$$= -2U^2a^2 - K^2 + 2iKUa$$

Consequently residue at $z = 0$ is

$$-2U^2a^2 - K^2 + 2iKUa$$

$$\therefore \quad \int z\left(\frac{dw}{dz}\right)^2 dz = 2\Pi i\left[-2U^2a^2 - K^2 + 2iKUa\right]$$

For $z = 0$ is the only pole inside $|z + a| = a$

By Blasius theorem

$$N = R.P \text{ of } -\frac{1}{2}\rho \int z\left(\frac{dw}{dz}\right)^2 dz$$

$$= R.P \text{ of } -\frac{1}{2}\rho\left[2\Pi i\left(-2U^2a^2 - K^2 + 2iKUa\right)\right]$$

$$= 2\Pi K\rho aU$$

Hence moment of the couple = $2\Pi K\rho aU$

10. A circular cylinder of radius a and infinite length lies on a plane in an infinite depth of liquid. The velocity of liquid of a great distance from the cylinder is U perpendicular to generators, and the motion is irrotational and two-dimensional. Verify that the stream function is the imaginary part of $w = \Pi aU \cot h\left(\frac{\Pi a}{z}\right)$ where $z = 0$ on the line of contact. Prove that the pressure at the two ends of the diameter of the cylinder normal to the plane differs by $\frac{1}{32}\Pi^4\rho U^2$.

Solution: We have

$$\frac{1}{z} = \frac{1}{x + iy} = \frac{x - iy}{x^2 + y^2}$$

Also

$$w = \phi + i\psi = \Pi aU \cot h\frac{\Pi a}{z}$$

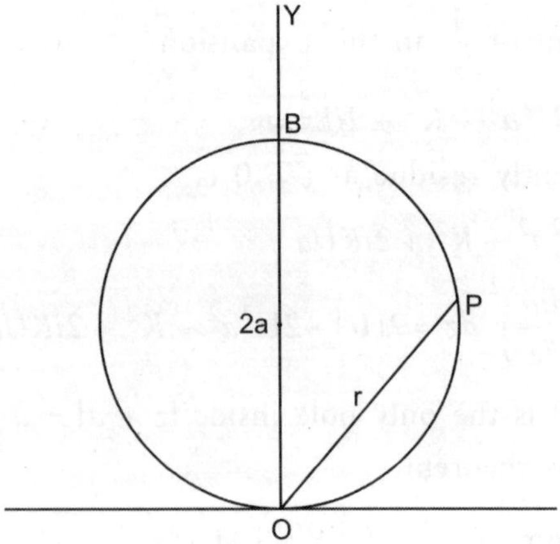

Fig. 4.16

$$= \frac{\Pi a\, U \cos \dfrac{h\Pi a}{r^2}(x - iy)}{\sin h \dfrac{\Pi a}{r}(x - iy)}$$

$$= \frac{2\Pi a U \cos \dfrac{h\Pi a}{r^2}(x - iy) \sin h \dfrac{\Pi a}{r^2}(x + iy)}{2 \sin h \dfrac{\Pi a}{r^2}(x - iy) \sin h \dfrac{\Pi a}{r^2}(x + iy)}$$

$$= \Pi a U \frac{\sin h\left(\dfrac{2\Pi ax}{r^2}\right) + \sin h\left(\dfrac{2\Pi aiy}{r^2}\right)}{\cos h\left(\dfrac{2\Pi ax}{r^2}\right) - \cos h\left(\dfrac{i2\Pi ay}{r^2}\right)}$$

$$\phi + i\psi = \Pi\, aU \frac{\sin h\left(\dfrac{2\Pi ax}{r^2}\right) + i \sin\left(\dfrac{2\Pi ay}{r^2}\right)}{\cos h\left(\dfrac{2\Pi ax}{r^2}\right) - \cos\left(\dfrac{2\Pi ay}{r^2}\right)}$$

Equating imaginary parts, we get

$$\psi = \frac{\Pi a U \sin\left(\dfrac{2\Pi ay}{r^2}\right)}{\cos h\left(\dfrac{2\Pi ax}{r^2}\right) - \cos\left(\dfrac{2\Pi ay}{r^2}\right)}$$

Streamlines are given by ψ = const.

When $\psi = 0$ then $\sin\left(\frac{2\Pi ay}{r^2}\right) = 0$

$$\Rightarrow \quad \frac{2\Pi ay}{r^2} = 0 \text{ or } \pi$$

$$\Rightarrow \quad y = 0,\ r^2 = 2ay$$

$$\Rightarrow \quad r = 2a \sin\theta$$

But $r = 2a \sin\theta$ is the equation of the circle OBP. Thus the circle is also a streamline. Thus

$w = \Pi aU \cot h\left(\frac{\Pi a}{z}\right)$ gives correct motion.

Also it shows that velocity U is at great distance.

Also pressure p is given by $\frac{p}{\rho} + \frac{1}{2}q^2 = C$ ... (1)

When $q = U$, $p = \Pi$, we get from (1)

$$C = \frac{\Pi}{\rho} + \frac{U^2}{2}$$

Hence from (1)

$$\frac{p}{\rho} + \frac{1}{2}q^2 = \frac{\Pi}{\rho} + \frac{U^2}{2}$$

$$\Rightarrow \quad \frac{p - \Pi}{\rho} = \frac{1}{2}\left(U^2 - q^2\right)$$

$$\Rightarrow \quad p - \Pi = \frac{\rho}{2}\left(U^2 - q^2\right)$$

$$\Rightarrow \quad p = \Pi + \frac{\rho}{2}\left(U^2 - q^2\right) \qquad \text{... (2)}$$

Also $q = \left|\frac{dw}{dz}\right|$

$$= \Pi aU \left|\operatorname{cosec} h^2\left(\frac{\Pi a}{z}\right)\right|\left|-\frac{\Pi a}{z^2}\right| \qquad \text{... (3)}$$

At O, let $p = p_0$

Here $q = 0$

At B, $q = q_1$ (say) and $p = p_1$

From (2)

$$p_0 = \Pi + \frac{\rho}{2}\left(U^2 - 0\right)$$

$$p_1 = \Pi \frac{\rho}{2}\left(U^2 - q_1^2\right)$$

$$\therefore \quad p_0 - p_1 = \frac{\rho}{2} q_1^2 \quad \text{... (4)}$$

From (3)

$$q_1 = \Pi^2 a^2 U \left| \frac{1}{\sin h^2\left(\frac{\Pi a}{0 + i2a}\right)} \right| \frac{1}{r^2}$$

$$= \frac{\Pi^2 a^2 U}{4a^2} = \frac{\Pi^2 U}{4}$$

as $x = 0$, $y = 2a$, $r = 2a$ at B.

and $\left|\sin h^2\left(-\frac{i\Pi}{2}\right)\right| = \left|\sin^2 \frac{\Pi}{2}\right| = 1$

Hence from (4)

$$p_0 - p_1 = \frac{1}{2}\rho \frac{\Pi^4 U^2}{16} = \frac{\Pi^4 \rho U^2}{32}$$

EXERCISE
(PART I)

1. The space between two fixed coaxial circular cylinders of radii a and b, and between two planes perpendicular to the axis and distance c apart, is occupied by the liquid of density ρ. Show that the velocity potential of a motion whose kinetic energy shall equal a given quantity is given by $A\theta$ where

$$\Pi \rho A^2 C^2 \log\left(\frac{b}{a}\right) = T$$

2. Show that when a cylinder moves uniformly in a given straight line in an infinite liquid, the path of any point in the fluid is given by the equations

$$\frac{dz}{dt} = \frac{Va^2}{\left(z^1 - Vt\right)^2}, \frac{dz^1}{dt} = \frac{Va^2}{\left(z - Vt\right)^2}$$

where $z = x + iy$, $z^1 = x - iy$, V = velocity of cylinder. The coordinates (x, y) of a point are measured from the starting point of the axis, along and perpendicular to its direction of motion.

3. A circular cylinder is placed in a uniform stream, find the forces acting on the cylinder.

[**Ans.** No force and no couple act on the cylinder.]

PART II – ELLIPTIC CYLINDER

4.13. Elliptic Coordinates

Let $z = c\cos h\zeta$ where $z = x + iy$, $\zeta = \xi + i\eta$

Then $x + iy = c\cos h\,(\xi + i\eta)$

$\Rightarrow \quad x + iy = c\,(\cos h\,\xi\cos\eta + i\sin h\,\xi\sin\eta)$

So that $x = c\cos h\,\xi\cos\eta,\ y = c\sin h\,\xi\sin\eta$

Eliminating η, we get

$$\frac{x^2}{c^2\cos h^2\xi} + \frac{y^2}{c^2\sin h^2\xi} = 1$$

and eliminating ξ, we have

$$\frac{x^2}{c^2\cos^2\eta} - \frac{y^2}{c^2\sin^2\eta} = 1$$

Thus ξ = constant and η = constant represent confocal ellipses and hyperbolas respectively. The distance between the focii in each case is $2c$. The parameters ξ and η are called the elliptic coordinates when $\xi = \alpha$, we get the elliptic cylinder $\frac{x^2}{a^2} + \frac{y^2}{b^2} = 1$

where $a = c\cos h\,\alpha$, $b = c\sin h\,\alpha$.

The distance between the focii

$$= 2ac = 2\sqrt{a^2 - b^2}$$

$$= 2\sqrt{c^2\cos h^2\alpha - c^2\sin h^2\alpha}$$

$$= 2c$$

Also $a + b = c\cos h\,\alpha + c\sin h\,\alpha = ce^{\alpha}$

$a - b = c\cos h\,\alpha - c\sin h\,\alpha = ce^{-\alpha}$

Hence $e^{2\alpha} = \frac{a+b}{a-b}$

$\Rightarrow \quad e^{\alpha} = \sqrt{\frac{a+b}{a-b}}$

4.14. Solution of Laplace's Equation

With the help of elliptic coordinates, $x = c \cos h \xi \cos \eta$, $y = c \sin h \xi \sin \eta$, the Laplace's equation

$$\vec{\nabla}^2 \psi = 0 \text{ i.e. } \frac{\partial^2 \psi}{\partial x^2} + \frac{\partial^2 \psi}{\partial y^2} = 0$$

can be transformed into $\dfrac{\partial^2 \psi}{\partial \xi^2} + \dfrac{\partial^2 \psi}{\partial \eta^2} = 0$

Its solution are of the type

$$\psi = e^{\pm \xi} (A \cos hn\, \eta + B \sin hn\, \eta)$$

When the motion is cyclic, ψ is single valued, i.e. ψ is unaltered. When η is increased by 2Π. This requires that n should be an integer.

The velocity components $\dfrac{\partial \psi}{\partial n}$ and $\dfrac{\partial \psi}{\partial \xi}$ contain $e^{\pm n\xi}$. Further the fluid is at rest at infinity i.e. $\dfrac{\partial \psi}{\partial \eta} = 0 = \dfrac{\partial \psi}{\partial \xi}$ as $\xi \to \infty$.

This shows that $\psi = e^{-n\xi}(A \cos n\eta + B \sin n\eta)$ and a solution of the type $\psi = e^{n\xi}(A \cos n\eta + B \sin n\eta)$ cannot be used.

4.15. The Joukowski Transformation

The two-dimensional transformation

$$z = t + \frac{c^2}{4t} \qquad \dots (1)$$

is called Joukowski transformation.

Obviously when $|z|$ is large, $t = z$ nearly, so that the corresponding distant parts of the two planes are unaltered. Thus a uniform stream at infinity in the z plane corresponds to that, of the same strength and direction in t plane.

From (1) $4t^2 - 4tz + c^2 = 0$

$$\Rightarrow \quad t = \frac{4z \pm \sqrt{16z^2 - 16c^2}}{2.4}$$

$$\Rightarrow \quad t = \frac{1}{2}\left[z \pm \sqrt{z^2 - c^2}\right]$$

Taking positive square root, we get

$$t = \frac{1}{2}\left(z + \sqrt{z^2 - c^2}\right)$$

The inverse transformation maps the domain outside the ellipse $\frac{x^2}{a^2} + \frac{y^2}{b^2} = 1$ on the domain outside the circle $|t| = \frac{1}{2}(a + b)$.

4.16. Streaming Past a Fixed Elliptic Cylinder

We consider a stream whose complex potential in t plane is $Ut\, e^{-i\beta}$. Now if we introduce a circular cylinder $|t| = \frac{a+b}{2}$, the complex potential becomes

$$w = Ute^{-i\beta} + \frac{U(a+b)^2}{4t} e^{i\beta}$$

by Circle theorem.

By Joukowski's transformation

$$t = \frac{1}{2}\left[z + \sqrt{z^2 - c^2}\right] \qquad \text{... (2)}$$

the domain outside the cylinder $|t| = \frac{1}{2}(a + b)$ is mapped on the domain outside $\frac{x^2}{a^2} + \frac{y^2}{b^2} = 1$. Hence the complex potential for a fluid streaming past a fixed elliptic cylinder is obtained by eliminating t between (1) and (2). Hence we have

$$w = \frac{Ue^{-i\beta}}{2}\left[z + \sqrt{z^2 - c^2}\right] + \frac{Ue^{i\beta}(a+b)^2}{2\left[z + \sqrt{z^2 - c^2}\right]}$$

To simplify the form of w, we use elliptic coordinates. Accordingly we put $z = c \cos h\ \xi$ so that $\sqrt{z^2 - c^2} = c \sin h\ \zeta$ and therefore

$$z + \sqrt{z^2 - c^2} = c\left(\cos h\,\zeta + \sin h\,\zeta\right) = ce^{\zeta}$$

and $z - \sqrt{z^2 - c^2} = c\left(\cos h\,\zeta - \sin h\,\zeta\right) = ce^{-\zeta}$

Thus
$$w = \frac{U}{2}e^{-i\beta} + \frac{Ue^{i\beta}}{2}\frac{(a+b)^2}{ce^{\zeta}}$$

$$= \frac{U}{2}e^{\zeta - i\beta}\sqrt{a^2 - b^2} + \frac{U}{2}e^{-\zeta + i\beta}\frac{(a+b)^2}{\sqrt{a^2 - b^2}}$$

$$= \frac{U}{2}\sqrt{\frac{a+b}{a-b}}\left[(a-b)\,e^{\zeta - i\beta} + (a+b)\,e^{-\zeta + i\beta}\right]$$

$$= \frac{U}{2}e^{\alpha}\left[ce^{-\alpha}e^{\zeta - i\beta} + ce^{\alpha}\,e^{-\zeta + i\beta}\right]$$

$$= \frac{U}{2}ce^{\alpha}\left[e^{\zeta - (\alpha + i\beta)} + e^{-\zeta + (\alpha + i\beta)}\right]$$

$$= \frac{U}{2}ce^{\alpha}\,2\cos h\left[\zeta - (\alpha + i\beta)\right]$$

$$= U\,(a+b)\cos h\,(\zeta - \alpha - i\beta)$$

This is the complex potential for the streaming motion past a fixed elliptic cylinder.

Particular case. Let the whole system is given a velocity U inclined at an angle α with the real axes, then the system is reduced to rest and the cylinder moves with velocity U. Then complex potential reduces to

$$w = \frac{U(a+b)^2\,e^{i\beta}}{4t}$$

$$= \frac{e^{i\beta}U\,(a+b)^2\,2}{4\left(z + \sqrt{z^2 - c^2}\right)}$$

$$= \frac{U\,(a+b)^2\,e^{i\beta}}{2ce^{\zeta}} = \frac{U(a+b)^2\,e^{-\xi}\,e^{i\beta}}{2(a+b)\,e^{-\alpha}}$$

$$= \frac{U}{2}(a+b)\,e^{\alpha + i\beta}\,e^{-\xi - i\eta}$$

$$w = \frac{U}{2}(a+b)\, e^{\alpha - \xi} e^{i(\beta - n)}$$

Hence equating real and imaginary parts on both the sides, we get

$$\phi = \frac{U}{2}(a+b)\, e^{\alpha - \xi} \cos(\beta - n)$$

and $\psi = \dfrac{U}{2}(a+b)\, e^{\alpha - \xi} \sin(\beta - n)$

Remark. Also we know that Kinetic energy

$$T = -\frac{1}{2}\rho \int \phi \frac{\partial \phi}{\partial n}\, ds$$

$$= -\frac{1}{2}\rho \int \phi \frac{\partial \psi}{\partial s}\, ds$$

$$= -\frac{1}{2}\rho \int \phi\, d\psi$$

4.17. Motion of a Moving Elliptic Cylinder

To determine the stream function and velocity potential when an elliptic cylinder moves in an infinite liquid with velocity U parallel to major axis of the cross section.

Let the cross section of an elliptic cylinder be given by $\dfrac{x^2}{a^2} + \dfrac{y^2}{b^2} = 1.$

Let us suppose that boundary is given by $\xi = \alpha$, so that

$$a = c \cosh \alpha,\; b = c \sinh \alpha$$

The boundary condition for ψ is

$$\psi = Vx - Uy + \frac{1}{2} w\left(x^2 + y^2\right) + B$$

In the present case $V = 0$, $w = 0$.

$$\psi = -Uy + B = -Uc \sin h\, \xi \sin \eta + B$$

On boundary this becomes

$$\psi = -Uc \sin h\, \alpha \sin \eta + B$$

$$\Rightarrow \quad \psi = -Ub \sin \eta + B \qquad \text{... (1)}$$

Since the effect is to vanish at infinity and $\sin \eta$ is the only variable factor in the boundary condition, we must assume a complex relation which gives to ψ the form $e^{-\xi} \sin \eta$.

So, let ψ at any point be given by

$$\psi = Ae^{-\xi} \sin \eta \qquad \dots (2)$$

At the boundary, we have

$$Ae^{-\alpha} \sin \eta = \psi = -Ub \sin \eta + B$$

Comparing the coefficients, $Ae^{-\alpha} = -Ub$, $B = 0$.

Hence $A = -Ube^{\alpha}$.

Therefore from (2)

$$\psi = -Ube^{\alpha}e^{-\xi} \sin \eta$$

$$\psi = -Ub\, e^{\alpha - \xi} \sin \eta \qquad \dots (3)$$

$$\therefore \quad \frac{\partial \psi}{\partial \eta} = -Ube^{\alpha - \xi} \cos \eta$$

But $\dfrac{\partial \psi}{\partial \eta} = \dfrac{\partial \phi}{\partial \xi}$

Hence $\dfrac{\partial \phi}{\partial \xi} = -Ube^{\alpha - \xi} \cos \eta$

Thus $\phi = Ube^{\alpha - \xi} \cos \eta \qquad \dots (4)$

Where $e^{\alpha} = \sqrt{\dfrac{a+b}{a-b}}$

The required expressions are given by (3) and (4).

Deduction. In this case kinetic energy T of the liquid on the boundary of the cylinder is given by

$$T = -\frac{1}{2} \rho \int (\phi \, d\psi)_{\xi = \alpha}$$

$$= \frac{\rho}{2} \int Ube^{\alpha - \xi} \cos \eta \, Ube^{\alpha - \xi} \cos \eta \, d\eta$$

$$= \frac{\rho}{2} U^2 b^2 \int_0^{2\pi} \cos^2 \eta \, d\eta$$

$$= \frac{\rho}{4} U^2 b^2 \int_0^{2\pi} (1 + \cos 2\eta)\, d\eta$$

$$= \frac{\rho}{4} U^2 b^2 \left[\eta + \frac{\sin 2\eta}{2}\right]_0^{2\pi}$$

$$= \frac{\Pi\rho}{2} U^2 b^2$$

Corollary. If the cylinder moves parallel to the axial plane through the minor axis of its cross section with velocity V, we get in the same manner.

$$\phi = Va \sqrt{\frac{a+b}{a-b}}\, e^{-\xi} \sin \eta$$

and $\psi = Va \sqrt{\dfrac{a+b}{a-b}}\, e^{-\xi} \cos \eta$

Also kinetic energy $= \dfrac{\Pi\rho}{2} V^2 a^2$

4.18. Motion of a Fixed Elliptic Cylinder

To determine ϕ and ψ for a liquid streaming past a fixed elliptic cylinder with velocity U parallel to major axis of the section.

Superimposing a velocity U on the cylinder and liquid both in the sense opposite to the velocity of the liquid. This brings the liquid at rest and the cylinder in motion with velocity U. Accordingly, we have to add a term in the stream function and the velocity potential when the velocity of the stream is from right to left parallel to the major axis, we have

$$\frac{\partial \phi}{\partial x} = \frac{-\partial \psi}{\partial y} = -U \qquad \text{... (1)}$$

Hence we add a term Uy in the expression for ψ and Ux in that of ϕ as obtained in 4.17.

Therefore

$$\phi = Ux + Ub\left(\frac{a+b}{a-b}\right)^{\frac{1}{2}} e^{-\xi}\cos\eta$$

$$= Uc\cosh\xi\cos\eta + Ub\left(\frac{a+b}{a-b}\right)^{\frac{1}{2}} e^{-\xi}\cos\eta$$

$$\phi = U\sqrt{a^2-b^2}\cosh\xi\cos\eta + Ub\left(\frac{a+b}{a-b}\right)^{\frac{1}{2}} e^{-\xi}\cos\eta \quad \ldots (2)$$

$$\text{and}\,\psi = Uy - Ub\left(\frac{a+b}{a-b}\right)^{\frac{1}{2}} e^{-\xi}\sin\eta$$

$$= U\sqrt{a^2-b^2}\sinh\xi\sin\eta - Ub\left(\frac{a+b}{a-b}\right)^{\frac{1}{2}} e^{-\xi}\sin\eta \quad \ldots (3)$$

Thus the complex potential is given by

$$w = \phi + i\psi$$

$$= U(x+iy) + Ub\left(\frac{a+b}{a-b}\right)^{\frac{1}{2}} e^{-\xi}(\cos\eta - i\sin\eta)$$

From (2) and (3)

$$= Uz + Ub\,e^{\alpha}e^{-\xi}e^{-i\eta}$$

$$= Uz + Ub\,e^{\alpha-(\xi+i\eta)}$$

$$= Uz + Ube^{\alpha-\zeta}$$

Another Form for ϕ, ψ and w

Rewriting (2), we have

$$\phi = U\cos\eta\left[be^{-\xi} + (a-b)\cosh\xi\right]\left(\frac{a+b}{a-b}\right)^{\frac{1}{2}}$$

$$= U\sqrt{\frac{a+b}{a-b}}\cos\eta\left[b(\cosh\xi - \sinh\xi) + (a-b)\cosh\xi\right]$$

$$= U\sqrt{\frac{a+b}{a-b}} \cos\eta\,(a\cosh\xi - b\sinh\xi)$$

$$= U\sqrt{\frac{a+b}{a-b}} \cos\eta\,(c\cosh\alpha\cosh\xi - c\sinh\alpha\sinh\xi)$$

$$= Uc\sqrt{\frac{a+b}{a-b}} \cos\eta\cosh(\xi-\alpha)$$

$$\phi = U(a+b)\cos\eta\cosh(\xi-\alpha)$$

as $c = \sqrt{a^2-b^2}$

$$\Rightarrow \phi = Uc\,e^{\alpha}\cos\eta\cosh(\xi-\alpha)$$

Similarly from (3), we have

$$\psi = Uce^{\alpha}\sin\eta\sinh(\xi-\alpha)$$

Hence complex potential is given by

$$w = \phi + i\psi$$

$$= Uce^{\alpha}\left[\cosh(\xi-\alpha)\cos\eta + i\sinh(\xi-\alpha)\sin\eta\right]$$

$$= Uce^{\alpha}\left[\cos i(\xi-\alpha)\cos\eta - i^2\sin i(\xi-\alpha)\sin\eta\right]$$

$$= Uce^{\alpha}\cos\left[i(\xi-\alpha)-\eta\right]$$

$$= Uce^{\alpha}\cos i\,[\xi-\alpha+i\eta]$$

$$= Uce^{\alpha}\cosh(\xi-\alpha)$$

4.19. Motion of a Rotating Elliptic Cylinder

To find the complex potential and stream function in case of an elliptic cylinder rotating in an infinite mass of liquid at rest at infinity.

Let the cross section of the elliptic cylinder be given by $\frac{x^2}{a^2}+\frac{y^2}{b^2}=1$ and the boundary is given by $a = c\cosh\alpha$, $b = c\sinh\alpha$. The elliptic coordinates are $x = c\cosh b\cos\eta$, $y = c\sinh b\sin\eta$. Let w be the angular velocity of the cylinder. The boundary condition for ψ is

$$\psi = Vx - Uy + \frac{1}{2} w \left(x^2 + y^2\right) + B$$

Hence $U = V = 0$

Hence

$$\begin{aligned}
\psi &= \frac{1}{2} w \left(x^2 + y^2\right) + B \\
&= \frac{1}{2} w \left(c^2 \cosh^2 \xi \cos^2 \eta + c^2 \sinh^2 \xi \sin^2 \eta\right) + B \\
&= \frac{1}{2} wc^2 \left[\cosh^2 \xi \left(1 - \sin^2 \eta\right) + \sinh^2 \xi \sin^2 \eta\right] + B \\
&= \frac{1}{2} wc^2 \left[\cosh^2 \xi \left(1 - \sin^2 \eta\right) \right. \\
&\qquad \left. + \left(\cosh^2 \xi - 1\right) \sin^2 \eta\right] + B \\
&= \frac{1}{2} wc^2 \left[\cosh^2 \xi - \sin^2 \eta\right] + B \\
&= \frac{1}{4} wc^2 \left[1 + \cosh 2\xi - (1 - \cos 2\eta)\right] + B \\
&= \frac{1}{4} wc^2 \left(\cosh^2 \xi + \cos 2\eta\right) + B
\end{aligned}$$

At the boundary $\xi = \alpha$, so that

$$\begin{aligned}
\psi &= \frac{1}{4} wc^2 \left(\cosh^2 \alpha + \cos 2\eta\right) + B \\
&= \frac{1}{4} wc^2 \cos 2\eta + C
\end{aligned}$$

where $c = \frac{1}{4} wc^2 \cosh 2\alpha + B$

Since the effect is to vanish at infinity and the only variable term in the boundary conditions is cos 2 η, let us assume that at any point

$$\psi = Ae^{-2\xi} \cos 2\eta$$

At the boundary $\psi = Ae^{-2\alpha} \cos 2\eta$

$$= \frac{1}{4} wc^2 \cos 2\eta + c$$

Hence $A = \frac{1}{4} wc^2 e^{2\alpha}, c = 0$

So that $\psi = \frac{1}{4} wc^2 e^{2\alpha - 2\xi} \cos 2\eta$

and $\phi = \frac{1}{4} wc^2 e^{2\alpha - 2\xi} \sin 2\eta$

But $ce^{\alpha} = a + b$

$\Rightarrow c^2 e^{2\alpha} = (a + b)^2$

Hence $\psi = \frac{w}{4} (a+b)^2 e^{-2\xi} \cos 2\eta$

and $\phi = \frac{w}{4} (a+b)^2 e^{-2\xi} \sin 2\eta$

Kinetic Energy

If T be the Kinetic energy, then

$$T = -\frac{\rho}{2} \int \phi \, d\psi$$

$$= -\frac{\rho}{2} \int \frac{w}{4} (a+b)^2 e^{-2\alpha} \sin 2\eta \, d\psi \quad \text{At } \xi = \alpha$$

$$= -\frac{\rho}{2} \int \frac{w}{4} (a+b)^2 e^{-2\alpha} \sin 2\eta$$

$$\left\{ \frac{w}{4} (a+b)^2 e^{-2\alpha} 2 (\sin 2\eta) \right\} d\eta \quad \text{for } \xi = \alpha$$

$$= \frac{\rho w^2}{16} \int_0^{2\pi} (a+b)^4 e^{-4\alpha} \sin^2 2\eta \, d\eta$$

$$= \frac{\rho w^2}{16} (a+b)^4 e^{-4\alpha} \int_0^{2\pi} \frac{(1 - \cos 4\eta) \, d\eta}{2}$$

$$= \frac{\rho w^2}{16} \frac{(a+b)^4}{2} e^{-4\alpha} \left[\eta - \frac{\sin 4\eta}{4} \right]_0^{2\pi}$$

$$= \frac{\rho w^2}{16} \frac{(a+b)^4}{2} 2\Pi \, e^{-4\alpha}$$

$$= \frac{\rho w^2 (a+b)^4 \Pi e^{-4\alpha}}{16}$$

$$= \frac{\Pi \rho w^2 c^4}{16}$$

Remark. We have seen that if the cylinder has velocity U, kinetic energy $= \frac{\Pi \rho b^2 U^2}{2}$ and if it has a velocity V (in the perpendicular direction) kinetic energy $= \frac{\Pi \rho a^2 V^2}{2}$.

The kinetic energy when the cylinder has both these velocities, will be the sum of the kinetic energies in the two separate cases.

Since at the boundary $\psi = Vx - Uy$, hence at any point ψ is linear in U and V. This statement must be true for ϕ as well.

Thus $T = -\frac{\rho}{2} \int \phi \, d\psi$, must be a quadratic in U and V and will be of the form $AU^2 + 2HUV + BV^2$, where A, H, B are certain constants. If V is kept the same and U is reversed, the kinetic energy must be unaffected. Hence $H = 0$, therefore

$$2T = AU^2 + BV^2$$

or $$T = \frac{1}{2} \rho \Pi \left(b^2 U^2 + a^2 V^2\right)$$

4.20. Kinetic Energy of the Liquid Contained in Rotating Elliptic Cylinder

Let $w = -iAz^2$

Then $\phi + i\psi = -iA\left(x^2 - y^2 + 2ixy\right)$

Hence $\phi = 2Axy$, $\psi = -A(x^2 - y^2)$

The boundary condition for ψ is

$$\psi = \frac{1}{2} w \left(x^2 + y^2\right) + B$$

Hence $-A\left(x^2 - y^2\right) = \frac{1}{2} w \left(x^2 + y^2\right) + B$

$$\Rightarrow \quad -B = x^2\left(\frac{w}{2} + A\right) + y^2\left(\frac{w}{2} - A\right)$$

$$\Rightarrow \quad \frac{x^2}{\dfrac{-B}{\dfrac{w}{2} + A}} + \frac{y^2}{\dfrac{-B}{\dfrac{w}{2} - A}} = 1$$

Hence $a^2 = \dfrac{-B}{A + \dfrac{w}{2}}, b^2 = \dfrac{-B}{\dfrac{w}{2} - A}$

Then $\dfrac{a^2}{b^2} = \dfrac{\dfrac{w}{2} - A}{\dfrac{w}{2} + A}$

$$\Rightarrow \quad a^2\left(\frac{w}{2} + A\right) = b^2\left(\frac{w}{2} - A\right)$$

$$\Rightarrow \quad \frac{w}{2}\left(a^2 - b^2\right) = -A\left(a^2 + b^2\right)$$

$$\Rightarrow \quad -A = \frac{w\left(a^2 - b^2\right)}{2\left(a^2 + b^2\right)}$$

Hence $\phi = -w\left(\dfrac{a^2 - b^2}{a^2 + b^2}\right)xy$

and $\psi = \dfrac{w\left(a^2 - b^2\right)}{2\left(a^2 + b^2\right)}\left(x^2 - y^2\right)$

Let T be the required kinetic energy.

Then

$$\begin{aligned} T &= \frac{1}{2}\rho \iint q^2 dx\, dy \\ &= \frac{1}{2}\rho \iint\left[\left(-\frac{\partial\phi}{\partial x}\right)^2 + \left(-\frac{\partial\phi}{\partial y}\right)^2\right] dx\, dy \\ &= \frac{1}{2}\rho\left(\frac{a^2 - b^2}{a^2 + b^2}\right)^2 w \iint\left(x^2 + y^2\right) dx\, dy \qquad \dots (1) \end{aligned}$$

But $\iint x^2 dx\, dy$ = moment of inertia of the ellipse about y axis

$$= \Pi ab \frac{a^2}{4} = \frac{1}{4} \Pi a^3 b$$

and $\iint y^2 dx\, dy$ = Moment of inertia of the ellipse about x axis

$$= \Pi ab \frac{b^2}{4} = \frac{1}{4} \Pi ab^3$$

Hence from (1)

$$T = \frac{1}{2} \rho w \left(\frac{a^2 - b^2}{a^2 + b^2} \right)^2 \int \frac{\Pi ab}{4} \left(a^2 + b^2\right)$$

$$= \frac{\Pi}{8} \rho w\, ab \frac{\left(a^2 - b^2\right)^2}{a^2 + b^2}$$

4.21. Path of the Particles

Liquid is contained in a rotating elliptic cylinder. By making use of elliptic transformation $z = c \cosh \zeta$. Show directly that the stream function of the motion is

$$\psi = \frac{1}{2} w \left(\frac{a^2 - b^2}{a^2 + b^2} \right) \left(x^2 - y^2\right)$$

Hence prove that the paths of the particles are similar ellipses described in time $\Pi\ (a^2 + b^2)/wab$.

Let (x, y) be the coordinates of a point on the rotating elliptic cylinder. Then

$$-\frac{\partial \phi}{\partial x} = \text{velocity of the fluid particle along } x \text{ axis}$$

$$= \dot{x} - wy$$

and $$-\frac{\partial \phi}{\partial y} = \dot{y} + wx$$

$$\left[\because \dot{x} = \frac{d}{dt}(r \cos \theta) = -r \sin \theta\, \dot{\theta} = -yw \right]$$

Hence $\dot{x} - wy = -\dfrac{\partial \phi}{\partial x}$

$$= w\left(\frac{a^2 - b^2}{a^2 + b^2}\right) y$$

$$\left[\because \phi = -w\left(\frac{a^2 - b^2}{a^2 + b^2}\right) xy\right]$$

$$\dot{x} = \frac{2wya^2}{a^2 + b^2} \qquad \text{... (1)}$$

So that $\ddot{x} = \dfrac{2wa^2\dot{y}}{a^2 + b^2}$ ... (2)

Also $\dot{y} + wx = -\dfrac{\partial \phi}{\partial y} = w\left(\dfrac{a^2 - b^2}{a^2 + b^2}\right) x$

$$\therefore \quad \ddot{y} = -\frac{2b^2wx}{a^2 + b^2}$$

Hence from (2)

$$\ddot{x} = \frac{2wa^2}{a^2 + b^2} \frac{\left(-2b^2wx\right)}{a^2 + b^2}$$

$$\ddot{x} = -\left(\frac{2wab}{a^2 + b^2}\right)^2 x \qquad \text{... (3)}$$

Which is a simple harmonic motion.
Thus time period T is given by

$$T = \frac{2\Pi\left(a^2 + b^2\right)}{2wab} = \frac{\Pi\left(a^2 + b^2\right)}{wab}$$

Also solution of (3) is

$$x = A_1 \cos\left(\frac{2abwt}{a^2 + b^2} + \alpha\right) \qquad \text{... (4)}$$

$$\dot{x} = -A_1 \frac{2abw}{a^2 + b^2} \sin\left(\frac{2abw}{a^2 + b^2} t + \alpha\right)$$

$$\Rightarrow \quad \frac{2wya^2}{a^2+b^2} = -\frac{A_1\, 2abw}{a^2+b^2} \sin\left(\frac{2abwt}{a^2+b^2}+\alpha\right) \qquad \text{[From (1)]}$$

$$\Rightarrow \quad \frac{y}{b} = -\frac{A_1}{a} \sin\left(\frac{2abwt}{a^2+b^2}+\alpha\right)$$

$$\Rightarrow \quad \frac{a}{b} y = -A_1 \sin\left(\frac{2abwt}{a^2+b^2}+\alpha\right) \qquad \dots (5)$$

Squaring and adding (4) and (5), we get

$$x^2 + \frac{a^2y^2}{b^2} = A_1^2$$

$$\Rightarrow \quad \frac{x^2}{a^2} + \frac{y^2}{b^2} = \frac{A_1^2}{a^2}$$

Which represents an ellipse similar to the given ellipse.

4.22. Circulation about an Elliptic Cylinder

We consider the elliptic transformation $z = c \cosh \zeta$.

i.e. $x + iy = c \cosh (\xi + i\eta)$

Hence $x = c \cosh \xi \cos \eta$, $y = \sinh \xi \sin \eta$

$$\text{Therefore} \quad \frac{x^2}{(c \cosh \xi)^2} + \frac{y^2}{(c \sinh \xi)^2} = 1 \qquad \dots (1)$$

If the streamlines are confocal with the given ellipse (1) then

$$\psi = f(\xi)$$

Also ψ satisfies $\vec{\nabla}^2 V = 0$

$$\Rightarrow \quad \frac{\partial^2 \psi}{\partial \xi^2} + \frac{\partial^2 \psi}{\partial \eta^2} = 0$$

$$\Rightarrow \quad \frac{\partial^2 \psi}{\partial \xi^2} = 0 \text{ as } \psi \text{ is independent of } \eta.$$

Integrating this, we get $\dfrac{\partial \psi}{\partial \xi} = A$

Again integrating, we have $\psi = A\xi$

But $\dfrac{\partial\phi}{\partial\eta} = -\dfrac{\partial\psi}{\partial\xi}$

Hence $\phi = -A\eta$

Now $w = \phi + i\psi = A(-\eta + i\xi) = iA\zeta$ ... (1)

Let K be the circulation about the cylinder, then

$$K = \int -\frac{\partial\phi}{\partial s}\,dx = -\int d\phi = A\int_0^{2\pi} d\eta \qquad \text{Since } \phi = -A\eta$$

$\Rightarrow K = 2\Pi A$

$\Rightarrow A = \dfrac{K}{2\Pi}$

From (1) $w = iA\zeta = \dfrac{iK}{2\Pi}(\xi + i\eta)$

This is the complex potential when there is circulation round an elliptic cylinder.

WORKED OUT EXAMPLES

1. Prove that when an infinitely long cylinder of density σ whose cross section is ellipse of semi axes a, b is immersed in an infinite liquid of density ρ, the square of its radius of gyration about its axis is effectively increased by the quantity $\dfrac{\rho}{8\sigma}\dfrac{\left(a^2-b^2\right)^2}{ab}$.

Solution: Let T_1 and T_2 be kinetic energies of cylinder and liquid outside the cylinder which is rotating with angular velocity w in an infinite mass of liquid.

Then $T_1 = \dfrac{1}{2}MK^2w^2$ where K^2 is the radius of gyration of an elliptic plate about a line passing through the centre and perpendicular to its plane and M is the mass of the elliptic plate.

Then $T_1 = \dfrac{1}{2}\Pi ab\sigma\left(\dfrac{a^2+b^2}{4}\right)w^2$

Now, we shall determine the K.E. of the liquid when the elliptic cylinder is immersed in an infinite liquid of density ρ.

The velocity potential and stream function for an elliptic cylinder rotating with angular velocity w, in an infinite liquid becomes

$$\phi = \frac{1}{4} w (a+b)^2 e^{-2\alpha} \sin 2\eta$$

and $\psi = \frac{1}{4} w (a+b)^2 e^{-2\alpha} \cos 2\eta$

Then

$$T_2 = -\frac{1}{2}\rho \int \phi \frac{\partial \phi}{\partial n} ds$$

$$= -\frac{1}{2}\rho \int \phi \, d\psi \quad \left(\because \frac{\partial \phi}{\partial \eta} = d\psi\right)$$

$$= -\frac{1}{2}\rho \int_0^{2\pi} \frac{w}{4}(a+b)^2 e^{-2\alpha} \sin 2\eta \times \left\{-\frac{w}{4}(a+b)^2 e^{-2\alpha} 2 \sin 2\eta\right\} d\eta$$

$$= \frac{\rho w^2}{16} \int_0^{2\pi} (a+b)^4 e^{-4\alpha} \sin^2 2\eta \, d\eta$$

$$= \frac{\rho w^2 (a+b)^4}{16} e^{-4\alpha} \int_0^{2\pi} \frac{(1-\cos 4\eta)}{2} d\eta$$

$$= \frac{\rho w^2 (a+b)^4 e^{-4\alpha} \Pi}{16}$$

$$= \frac{\Pi \rho w^2 c^4}{16} = \frac{1}{16} \Pi \rho w^2 \left(a^2 - b^2\right)^2$$

Hence the total kinetic energy

$$= T_1 + T_2$$

$$= \frac{1}{2} \Pi ab\sigma \left(\frac{a^2+b^2}{4}\right) w^2 + \frac{1}{16} \Pi \rho w^2 \left(a^2 - b^2\right)^2 \quad \ldots (1)$$

Let μ be the effective increase in the radius of gyration then, we have

$$\text{total K.E.} = \frac{1}{2}\Pi ab\sigma\left(\frac{a^2+b^2}{4}+\mu^2\right)w^2 \qquad \dots (2)$$

Equating (1) and (2), we get

$$\frac{1}{2}\Pi ab\sigma\left(\frac{a^2+b^2}{4}+\mu^2\right)w^2$$

$$= \frac{\Pi ab}{8}\sigma\left(a^2+b^2\right)w^2 + \frac{\Pi}{16}\rho w^2\left(a^2-b^2\right)^2$$

$$\Rightarrow \quad \frac{1}{2}\Pi ab\sigma\mu^2 w^2 = \frac{\Pi}{16}\rho w^2\left(a^2-b^2\right)^2$$

$$\Rightarrow \quad \mu^2 = \frac{\rho}{86} - \frac{\left(a^2-b^2\right)^2}{ab}$$

2. A thin shell in the form of an infinitely long elliptic cylinder, semi axes a and b, is rotating about its axis in an infinite liquid otherwise at rest. It is filled with the same liquid. Prove that the ratio of the Kinetic energy of the liquid inside to that of the liquid outside is $2ab : a^2 + b^2$.

Solution: Let the elliptic cylinder be rotating with angular velocity w in an infinite liquid at rest at infinity then the complex potential is given by

$$W = \frac{1}{4}iw\left(a+b\right)^2 e^{-2\zeta} \text{ where } \zeta = \xi + i\eta$$

Equating the real and imaginary parts, we have

$$\phi = \frac{1}{4}w(a+b)^2 e^{-2\xi}\sin 2\eta$$

$$\text{and } \psi = \frac{1}{4}w\left(a+b\right)^2 e^{-2\xi}\cos 2\eta$$

Let T_1 and T_2 be the Kinetic energies of the liquid inside and outside the same elliptic cylinder which is rotating with angular velocity w in an infinite mass of liquid at rest at infinity.

Then

$$T_1 = -\frac{1}{2}\rho \int \phi \, d\psi$$

$$= \frac{1}{16}\rho\, w^2 (a+b)^4 e^{-4\alpha} \int_0^{2\pi} \sin 2\eta \, d\eta$$

$$= \frac{1}{16}\Pi\rho w^2 \left(a^2 - b^2\right)^2 \qquad \text{... (1)}$$

The complex potential of the liquid contained in the rotating elliptic cylinder is given by

$$w = -iAz^2 \text{ where } A \text{ is a constant} \qquad \text{... (2)}$$

$$\Rightarrow \quad \phi + i\psi = -iA\left(x^2 - y^2 + 2xyi\right)$$

Equating imaginary parts, we get

$$\psi = -A\left(x^2 - y^2\right) \qquad \text{... (3)}$$

For rotating cylinder, the stream function is given by

$$\psi = \frac{1}{2}w\left(x^2 + y^2\right) = \text{constant} \qquad \text{... (4)}$$

At the boundary of the ellipse, we must have

$$\frac{1}{2}w\left(x^2 + y^2\right) = \text{constant} = -A\left(x^2 - y^2\right)$$

$$\Rightarrow \quad \left(\frac{1}{2}w + A\right)x^2 + \left(\frac{1}{2}w - A\right)y^2 = \text{constant} \qquad \text{... (5)}$$

Also the point (x, y) on the boundary satisfy

$$\frac{x^2}{a^2} + \frac{y^2}{b^2} = 1 \qquad \text{... (6)}$$

From (5) and (6), we get

$$a^2\left(\frac{1}{2}w + A\right) = b^2\left(\frac{1}{2}w - A\right)$$

$$\Rightarrow \quad A\left(a^2 + b^2\right) = \frac{w}{2}\left(b^2 - a^2\right)$$

$$A = \frac{w}{2}\frac{b^2 - a^2}{b^2 + a^2}$$

Hence from (6)

$$w = -iz^2 \frac{1}{2} w \frac{b^2 - a^2}{b^2 + a^2}$$

$$= iz^2 \frac{1}{2} w \frac{a^2 - b^2}{a^2 + b^2}$$

$$= \frac{1}{2} iw \frac{a^2 - b^2}{a^2 + b^2} c^2 \cosh^2 \zeta, \; z = c \cosh \zeta$$

$$\Rightarrow \quad \phi + i\psi = \frac{1}{2} iw \frac{a^2 - b^2}{a^2 + b^2} c^2 (\cosh \xi \cos \eta + i \sinh \xi \sin \eta)^2$$

Equating real and imaginary parts, on both the sides, we get

$$\phi = -w \frac{(a^2 - b^2)}{a^2 + b^2} c^2 \cosh \xi \sinh \xi \cos \eta \sin \eta$$

$$\text{and } \psi = \frac{1}{2} w \frac{a^2 - b^2}{a^2 + b^2} c^2 \left(\cosh^2 \xi \cos^2 \eta - \sinh^2 \xi \sin^2 \eta\right)$$

Also

$$T_2 = -\frac{1}{2} \rho \int \phi d\psi \quad \text{at } \xi = \alpha$$

$$= -\frac{1}{2} \rho w^2 \left(\frac{a^2 - b^2}{a^2 + b^2}\right)^2 c^4 \cosh \alpha \sinh \alpha$$

$$\left(\cosh^2 \alpha + \sinh^2 \alpha\right) \int_0^{2\pi} \cos^2 \eta \sin^2 \eta \, d\eta$$

$$= \frac{1}{8} \Pi \rho w^2 \left(\frac{a^2 - b^2}{a^2 + b^2}\right)^2 c^4 \cosh \alpha \sinh \alpha$$

$$= \frac{1}{8} \Pi \rho w^2 \left(\frac{a^2 - b^2}{a^2 + b^2}\right)^2 ab \left(a^2 + b^2\right)$$

$$T_2 = \frac{1}{8} \Pi \rho w^2 \frac{(a^2 - b^2)^2}{a^2 + b^2} ab \qquad \ldots (7)$$

Dividing (1) by (7), we get

$$\frac{T_1}{T_2} = \frac{\Pi\rho w^2\left(a^2 - b^2\right)}{16} \times \frac{8\left(a^2 + b^2\right)}{\Pi\rho w^2\left(a^2 - b^2\right)^2 ab}$$

$$= \frac{a^2 + b^2}{2ab}$$

$$\Rightarrow \quad \frac{T_2}{T_1} = \frac{2ab}{a^2 + b^2}$$

3. If an elliptic cylinder of semi axes a, b filled with a liquid, rotates with a uniform angular velocity about its axis, show that the Kinetic energy of the liquid contained is less than if it were moving as a solid in the ratio $(a^2 - b^2)^2 : (a^2 + b^2)^2$.

Solution: Let T_1 be the Kinetic energy of the liquid contained in the rotating elliptic cylinder, then

$$T_1 = \frac{\Pi}{8}\rho ab w^2 \frac{\left(a^2 - b^2\right)^2}{a^2 + b^2}$$

Let T_2 be the Kinetic energy of the liquid contained in the cylinder, when it is moving as a solid with angular velocity w.

Then

$$T_2 = \frac{1}{2} MK^2 w^2$$

$$= \frac{1}{2}\Pi ab\rho\left(\frac{a^2 + b^2}{4}\right)w^2$$

Hence

$$\frac{T_1}{T_2} = \frac{\Pi ab\rho\, w^2}{8}\frac{\left(a^2 - b^2\right)^2}{a^2 + b^2} \times \frac{8}{\Pi ab\rho\left(a^2 + b^2\right)w^2}$$

$$= \frac{\left(a^2 - b^2\right)^2}{\left(a^2 + b^2\right)^2}$$

4. An infinite elliptic cylinder with semi axes a, b is rotating round its axis with angular velocity w, in an infinite liquid of density ρ which is at rest at infinity. Show that if the liquid is under the action of no forces, the moment of the fluid pressure on the cylinder round the centre is $\frac{1}{8}\Pi\rho c^4 \frac{dw}{dt}$ where $c^2 = a^2 - b^2$.

Solution: Let T_1 be the K.E. of the liquid outside the rotating cylinder, then

$$T_1 = \frac{\Pi}{16}\rho w^2 c^4$$

Also the K.E. T_2 of the cylinder, from dynamical considerations is $\frac{1}{2}MK^2w^2$, where MK^2 is the moment of inertia of the cylinder about its axis.

Thus, the total K.E. T of the cylinder composite of the cylinder and the liquid outside it, is

$$T = T_1 + T_2$$

$$\Rightarrow \; T = \frac{\Pi}{16}\rho w^2 c^4 + \frac{1}{2}MK^2 w^2$$

$$\therefore \; \frac{dT}{dt} = \frac{\Pi}{8}\rho w c^4 \frac{dw}{dt} + MK^2 w \frac{dw}{dt}$$

Here $\frac{dT}{dt} = 0$, as external forces are absent

$$\Rightarrow \; \frac{\Pi}{8} + wc^4 \frac{dw}{dt} + MK^2 w \frac{dw}{dt} = 0$$

$$\Rightarrow \; MK^2 \frac{dw}{dt} = -\frac{\Pi}{8}\rho c^4 \frac{dw}{dt}$$

But $MK^2 \frac{dw}{dt} = MK^2 \frac{d^2\theta}{dt^2}$, Since $w = \dot{\theta}$

Hence, the moment of the fluid pressure on the cylinder about the centre is

$$= \frac{\Pi}{8}\rho c^4 \frac{dw}{dt}$$

5. In the two dimensional irrotational motion of a liquid streaming past a fixed elliptic disc $\frac{x^2}{a^2}+\frac{y^2}{b^2}=1$, the velocity at infinity being parallel to the major axes and equal to V, prove that if $x+iy=c\cosh(\xi+i\eta)$, $a^2-b^2=c^2$ and $a = c\cosh\alpha$, $b = c\sinh\alpha$, the velocity at any point is

$$q^2=V^2\left(\frac{a+b}{a-b}\right)\frac{\sinh^2(\xi-\alpha)+\sin^2\eta}{\sinh^2\xi+\sin^2\eta}$$

and that it has its maximum value $\frac{V(a+b)}{a}$ at the end of the minor axis.

Solution: The complex potential for liquid streaming past a fixed elliptic disc is

$$w=V(a+b)\cosh(\zeta-\alpha) \text{ where } z=c\cosh\zeta$$

$$\frac{dw}{dz}=\frac{dw}{dt}\frac{d\zeta}{dz}=V(a+b)\sinh\frac{(\zeta-\alpha)1}{c\sinh\zeta}$$

$$=\frac{Ve^{\alpha}\sinh(\xi+i\eta-\alpha)}{\sinh(\xi+i\eta)} \text{ as } a+b=ce^{\alpha}$$

$$q=\left|\frac{dw}{dz}\right|$$

$$=Ve^{\alpha}\left[\frac{\sinh^2(\xi-\alpha)\cos^2\eta+\cosh^2(\xi-\alpha)\sin^2\eta}{\sinh^2\xi\cos^2\eta+\cosh^2\xi\sin^2\eta}\right]^{\frac{1}{2}} \quad \text{... (1)}$$

But

$$\sinh^2(\xi-\alpha)\cos^2\eta+\cosh^2(\xi-\alpha)\sin^2\eta$$

$$=\sinh^2(\xi-\alpha)\left(1-\sin^2\eta\right)+\cosh^2(\xi-\alpha)\sin^2\eta$$

$$=\sinh^2(\xi-\alpha)\left(1-\sin^2\eta\right)+\left[1+\sinh^2(\xi-\alpha)\right]\sin^2\eta$$

$$=\sinh^2(\xi-\alpha)+\sin^2\eta$$

Also $\sinh^2 \xi \left(1 - \sin^2 \eta\right) + \left(1 + \sinh^2 \xi\right) \sin^2 \eta$

$$= \sinh^2 \xi + \sin^2 \eta$$

Hence from (1)

$$q = Ve^{\alpha} \left[\frac{\sinh^2 (\xi - \alpha) + \sin^2 \eta}{\sinh^2 \xi + \sin^2 \eta} \right]^{\frac{1}{2}}$$

$$q^2 = V^2 e^{2\alpha} \left[\frac{\sinh^2 (\xi - \alpha) + \sin^2 \eta}{\sinh^2 \xi + \sin^2 \eta} \right]$$

$$= V^2 \left(\frac{a+b}{a-b} \right) \left[\frac{\sinh^2 (\xi - \alpha) + \sin^2 \eta}{\sinh^2 \xi + \sin^2 \eta} \right]$$

Now to determine maximum value of q at the boundary. At the boundary $\xi = \alpha$.

So that $q^2 = V^2 \left(\dfrac{a+b}{a-b} \right) \left[\dfrac{\sin^2 \eta}{\sinh^2 \alpha + \sin^2 \eta} \right]$

$$\therefore \quad 2q \frac{dq}{d\eta} = V^2 \left(\frac{a+b}{a-b} \right)$$

$$\left[\frac{2 \sin \eta \cos \eta \left(\sinh^2 \alpha + \sin^2 \eta\right) - 2 \sin^3 \eta \cos \eta}{\left(\sinh^2 \alpha + \sin^2 \eta\right)^2} \right]$$

For extremum values of q,

$$\frac{dq}{d\eta} = 0$$

Hence $\sin 2\eta \left[\sinh^2 \alpha + \sin^2 \eta - \sin^2 \eta\right] = 0$

$\Rightarrow$ $\sin 2\eta = 0$

Hence $2\eta = 0$ or Π

$\Rightarrow$ $\eta = 0 \text{ or } \dfrac{\Pi}{2}$

If $\eta = 0$, then $q = 0$

If $\eta = \dfrac{\Pi}{2}$, then $q^2 = V^2 \left(\dfrac{a+b}{a-b} \right) \left(\dfrac{1}{1 + \sinh^2 \alpha} \right)$

$$\Rightarrow \quad q^2 = V^2\left(\frac{a+b}{a-b}\right)\frac{1}{\cosh^2\alpha}$$

$$q^2 = V^2\left(\frac{a+b}{a-b}\right)\frac{c^2}{a^2}$$

$$= V^2\left(\frac{a+b}{a-b}\right)\frac{a^2-b^2}{a^2}$$

$$\Rightarrow \quad q^2 = V^2\left(\frac{a+b}{a}\right)^2$$

Hence $q = V\left(\dfrac{a+b}{a}\right)$

Thus $\eta = 0$, gives minimum velocity and $\eta = \dfrac{\Pi}{2}$ gives maximum velocity, i.e. $V\left(\dfrac{a+b}{a}\right)$ at the end of minor axis.

6. An elliptic cylinder, the semi axes of whose cross sections are a and b, is moving with velocity U parallel to the major axis of its cross section, through an infinite liquid of density ρ which is at rest at infinity, the pressure there being Π. Prove that in order that the pressure may everywhere be positive.

$$\left(2ab + b^2\right)\rho U^2 < 2a^2\Pi$$

Solution: Let p be the pressure and q be the velocity at any point, then by Bernoulli's theorem for steady motion, we have

$$\frac{p}{q} + \frac{1}{2}q^2 = c \qquad \text{... (1)}$$

The motion will be reduced to a steady motion with cylinder at rest by superimposing a velocity $-U$ both on the cylinder and the liquid.

Then complex potential is given by

$$w = Uz + Ube^{\alpha - \zeta}$$

$$w = U(a+b)\cosh(\zeta - \alpha) \qquad \text{With } z = c\cosh\zeta$$

Hence

$$\frac{dw}{dz} = U - Ube^{\alpha - \zeta}\frac{d\zeta}{dz}$$

$$= U - Ube^{\alpha - \zeta}\frac{1}{c\sinh\zeta}$$

$$\left|\frac{dw}{dz}\right| = q = U\left|1 - \frac{be^{\alpha - \zeta}}{c\sinh\zeta}\right|$$

$$= U\left|\frac{c\sinh\zeta - be^{\alpha-\zeta}}{c\sinh\zeta}\right|$$

$$= \frac{U}{2}\left|\frac{c\left(e^{\zeta} - e^{-\zeta}\right) - 2c\sinh\alpha\, e^{\alpha-\zeta}}{c\sinh\zeta}\right|$$

$$= \frac{U}{2}\left|\frac{e^{\zeta} - e^{-\zeta} - \left(e^{\alpha} - e^{-\alpha}\right)e^{\alpha-\zeta}}{\sinh\zeta}\right|$$

$$= \frac{U}{2}\left|\frac{e^{\zeta} - e^{-\zeta} - e^{2\alpha-\zeta} + e^{-\zeta}}{\sinh\zeta}\right|$$

$$= \frac{U}{2}\left|\frac{e^{\alpha}\left\{e^{\zeta-\alpha} - e^{-(\zeta-\alpha)}\right\}}{\sinh\zeta}\right|$$

$$= \frac{Ue^{\alpha}}{2}\left|\frac{2\sinh(\zeta - \alpha)}{\sinh\zeta}\right|$$

$$= Ue^{\alpha}\left|\frac{\sinh(\alpha - \zeta)}{\sinh\zeta}\right|$$

$$= Ue^{\alpha}\left|\frac{\sinh\{\alpha - (\xi + i\eta)\}}{\sinh(\xi + i\eta)}\right|$$

$$= Ue^{\alpha}\left|\frac{\sinh(\alpha - \xi)\cos\eta - i\cosh(\alpha - \xi)\sin\eta}{\sinh\xi\cos\eta + i\cosh\xi\sin\eta}\right|$$

$$q^2 = U^2e^{2\alpha}\left[\frac{\sinh^2(\xi - \alpha)\cos^2\eta + \cosh^2(\xi - \alpha)\sin^2\eta}{\sinh^2\xi\cos^2\eta + \cosh^2\xi\sin^2\eta}\right]$$

$$= U^2 e^{2\alpha}\left[\frac{\sinh^2(\xi-\alpha)\left(1-\sin^2\eta\right)+\left\{1+\sinh^2(\xi-\alpha)\sin^2\eta\right\}}{\sinh^2\xi\left(1-\sin^2\eta\right)+\left(1+\sinh^2\xi\right)\sin^2\eta}\right]$$

$$q^2 = U^2 e^{2\alpha}\left[\frac{\sinh^2(\xi-\alpha)+\sin^2\eta}{\sinh^2\xi+\sin^2\eta}\right]$$

Velocity on the boundary $\xi = \alpha$ is

$$q^2 = U^2 e^{2\alpha}\left[\frac{\sin^2\eta}{\sinh^2\alpha+\sin^2\eta}\right]$$

It is easy to see that maximum velocity is attained when $\eta = \frac{\pi}{2}$

Thus

$$q^2_{\max} = U^2 e^{2\alpha}\frac{1}{\sinh^2\alpha+1}$$

$$= U^2 e^{2\alpha}\,\text{sech}^2\alpha$$

$$\Rightarrow \quad q^2_{\max} = U^2 e^{2\alpha}\frac{c^2}{a^2} = \frac{U^2(a+b)^2}{a^2}$$

$$\Rightarrow \quad q_{\max} = U\left(\frac{a+b}{a}\right)$$

Subjecting (1) to the boundary condition when $|z| = \infty$, $p = \pi$, $q = -U$, we get

$$C = \frac{\pi}{\rho}+\frac{1}{2}U^2$$

Hence

$$\frac{p}{\rho}+\frac{1}{2}q^2 = \frac{\pi}{\rho}+\frac{1}{2}U^2$$

$$\Rightarrow \quad p = \pi + \frac{\rho}{2}\left(U^2-q^2\right)$$

Now $p > 0$ everywhere if $\pi + \frac{1}{2}\rho\left(U^2-q^2\right) > 0$ for energy q.

Hence in particular, $p > 0$ if $\pi + \frac{\rho}{2}\left(U^2 - q_{\max}^2\right) > 0$

$$\Rightarrow \quad \pi + \frac{\rho}{2}\left[U^2 - U^2\left(\frac{a+b}{a}\right)^2\right] > 0$$

$$\Rightarrow \quad \pi + \frac{\rho U^2}{2}\left[-\frac{\left(b^2 + 2ab\right)}{a^2}\right] > 0$$

$$\Rightarrow \quad \rho U^2 > \frac{2a^2\pi}{b^2 + 2ab}$$

7. An elliptic cylinder of semi axes a, b is filled with incompressible fluid and rotates about its axis with constant angular velocity w. Prove that the velocity components (U, V) parallel to OX, OY the axes of ellipse are given by

$$U = W\left(\frac{a^2 - b^2}{a^2 + b^2}\right)y, \; V = w\left(\frac{a^2 - b^2}{a^2 + b^2}\right)x$$

Show that the coordinates X, Y (relative to axis through O fixed in the space) of a given particle at time t can be written as

$$X = \lambda\left[\left(a^2 + b\right)\cos\left\{\frac{(a-b)^2 wt}{a^2 + b^2}\right\} + (a-b)\cos\left\{\frac{(a+b)^2 wt}{a^2 + b^2}\right\}\right]$$

$$Y = \lambda\left[(a+b)\sin\left\{\frac{(a-b)^2 wt}{a^2 + b^2}\right\} + (a-b)\sin\left\{\frac{(a+b)^2 wt}{a^2 + b^2}\right\}\right]$$

where λ is a constant depending on the particle and $b = 0$, when the particle crosses the axis OX.

Solution: The velocity potential ϕ is given by

$$\phi = -w\left(\frac{a^2 - b^2}{a^2 + b^2}\right)xy \qquad \ldots (1)$$

$$U = -\frac{\partial\phi}{\partial x} = w\left(\frac{a^2 - b^2}{a^2 + b^2}\right)y$$

$$V = -\frac{\partial\phi}{\partial y} = w\left(\frac{a^2 - b^2}{a^2 + b^2}\right)x$$

Also, we know that

$$x = A\cos\left(\frac{2abwt}{a^2 + b^2} + c\right)$$

$$y = -\frac{b}{a}A\sin\left(\frac{2abwt}{a^2 + b^2} + c\right) \qquad \ldots (3)$$

Since $t = 0$, $x = \lambda$, $y = 0$

Hence (3) becomes

$$\lambda = A\cos c \text{ and } 0 = -\frac{b}{a}A\sin c$$

Hence $A = \lambda$, $c = 0$

Therefore,

$$\left.\begin{aligned} x &= \lambda\cos\frac{2abwt}{a^2 + b^2} \\ y &= -\frac{b\lambda}{a}\sin\frac{2abwt}{a^2 + b^2} \end{aligned}\right\} \qquad \ldots (4)$$

which represent the coordinates with regard to the rotating axis fixed in the elliptic cylinder.

In order to determine the coordinates X and Y relative to fixed axes in space, we proceed as follows

$$Z = ze^{iwt}$$

$$\Rightarrow \quad X + iY = (x + iy)(\cos wt + i\sin wt)$$

$$\Rightarrow \quad X = x\cos wt - y\sin wt$$

$$Y = x \sin wt + y \cos wt$$

Substituting the value of x and y from the equation (4), we have

$$X = \lambda \cos \frac{2abwt}{a^2 + b^2} \cos wt + \frac{b\lambda}{a} \sin \frac{2abwt}{a^2 + b^2} \sin wt$$

$$\Rightarrow \quad X = \frac{\lambda}{2a}\left[2a \cos wt \cos \frac{2ab}{a^2 + b^2} wt + 2b \sin wt \sin \frac{2abwt}{a^2 + b^2}\right]$$

$$\Rightarrow \quad X = \lambda^1 \left[a \left\{\cos\left(\frac{a^2 + b^2 + 2ab}{a^2 + b^2}\right) wt + \cos\left(\frac{a^2 + b^2 - 2ab}{a^2 + b^2}\right) wt\right\} + b \left\{\cos\left(\frac{a^2 + b^2 + 2ab}{a^2 + b^2}\right) wt - \cos\left(\frac{a^2 + b^2 - 2ab}{a^2 + b^2}\right) wt\right\}\right]$$

where $\lambda^1 = \dfrac{\lambda}{2a}$

$$\Rightarrow \quad X = \lambda^1 \left[(a + b) \cos\left(\frac{a^2 + b^2 + 2ab}{a^2 + b^2}\right) wt + (a - b) \cos\left(\frac{a^2 + b^2 - 2ab}{a^2 + b^2}\right) wt\right]$$

Similarly

$$Y = \lambda^1 \left[(a - b) \sin\left\{\frac{(a + b)^2 wt}{a^2 + b^2}\right\} + (a + b) \sin\left\{\frac{(a - b)^2 wt}{a^2 + b^2}\right\}\right]$$

8. An elliptic cylinder semi axes a and b is held with its length perpendicular to and its major axis making an angle θ with the direction of the stream of velocity V. Prove that the magnitude of the couple per unit length on the cylinder due to fluid pressure is $\Pi\rho\,(a^2 - b^2)\,V^2 \sin\theta\cos\theta$ and determine its sense.

Solution: The complex potential is given by

$$w = V(a+b)\cosh(\zeta - \zeta_0) \text{ where } \zeta = \xi + i\eta,\ \zeta_0 = \alpha + i\theta$$

$$z = c\cosh\zeta$$

Now

$$\frac{dw}{dz} = \frac{dw}{d\zeta}\frac{d\zeta}{dz} = \frac{V(a+b)\sinh(\zeta - \zeta_0)}{c\sinh\zeta}$$

$$= \frac{V(a+b)}{c}\left[\frac{\sinh\zeta\cosh\zeta_0 - \cosh\zeta\sinh\zeta_0}{\sinh\zeta}\right]$$

$$= \frac{V(a+b)}{c}\left[\cosh\zeta_0 - \frac{z}{(z^2 - c^2)^{\frac{1}{2}}}\sinh\zeta_0\right]$$

$$= \frac{V(a+b)}{c}\left[\cosh\zeta_0 - \left(1 + \frac{c^2}{2z^2}\right)\sinh\zeta_0\right]$$

as z is large

$$\Rightarrow \left(\frac{dw}{dz}\right)^2 = \frac{V(a+b)}{c}\left[e^{-\zeta_0} - \frac{c^2}{2z^2}\sinh\zeta_0\right]$$

$$\Rightarrow z\left(\frac{dw}{dz}\right)^2 = \left[\frac{V(a+b)}{c}\right]^2 z\left[e^{-\zeta_0} - \frac{c^2}{2z^2}\sinh\zeta_0\right]^2$$

This has a pole at the origin. Also residue at $z = 0$.

$$= \text{Coeff. of } \frac{1}{2} \text{ in the expansion of } z = \left(\frac{dw}{dz}\right)^2$$

$$= -\left[\frac{V(a+b)}{c}\right]^2 e^{-\zeta_0}\sinh\zeta_0$$

$$= -\left[V(a+b)\right]^2 e^{-\zeta_0}\sinh\zeta_0$$

Hence by Cauchy's residue theorem

$$\int z\left(\frac{dw}{dz}\right)^2 dz$$

$$= 2\Pi i \text{ (sum of residues)}$$

$$= -2\Pi i\left[V(a+b)\right]^2 e^{-\zeta_0}\sinh\zeta_0$$

$$= -2\Pi i V^2(a+b)^2 e^{-(\alpha+i\theta)}\sinh(\alpha+i\theta)$$

$$= -2\Pi i V^2(a+b)^2 e^{-\alpha}e^{-i\theta}(-i)\sin(i\alpha-\theta)$$

$$= -2\Pi V^2(a+b)^2 e^{-\alpha}e^{-i\theta}(\sin i\alpha\cos\theta-\cos i\alpha\sin\theta)$$

$$= -2\Pi V^2(a+b)^2 e^{-\alpha}(\cos\theta-i\sin\theta)$$

$$(i\sinh\alpha\cos\theta-\cosh\alpha\sin\theta)$$

Hence real part of $\int z\left(\frac{dw}{dz}\right)^2 dz$

$$= -2\Pi V^2(a+b)^2 e^{-\alpha}[\sin\theta\cos\theta\sinh\alpha$$

$$-\cos\theta\cosh\alpha\sin\theta]$$

$$= 2\Pi V^2(a+b)^2 e^{-\alpha}\sin\theta\cos\theta\, e^{-\alpha}$$

$$= 2\Pi V^2(a+b)^2 e^{-2\alpha}\sin\theta\cos\theta$$

If N be the required couple about the origin, then by Blasius theorem

$$N = \text{R.P. of } -\frac{1}{2}\rho\int z\left(\frac{dw}{dz}\right)^2 dz$$

$$= -\frac{1}{2}\rho\, 2\Pi V^2(a+b)^2 e^{-2\alpha}\sin\theta\cos\theta$$

$$= -\rho\Pi V^2(a+b)^2\left(\frac{a-b}{a+b}\right)\sin\theta\cos\theta$$

$$= -\Pi\rho V^2\left(a^2-b^2\right)\sin\theta\cos\theta$$

The negative sign shows that the couple tends to set the cylinder broad side the stream.

9. The space between two confocal elliptic cylinders (a_0, b_0) and (a_1, b_1) and two planes perpendicular to their axes is filled with liquid. If both cylinders be made to rotate about their common axis with angular velocity w, the kinetic energy of the motion setup is

$$\frac{1}{8}\frac{Mw^2c^4(b_1a_0 - b_0a_1)}{(a_1a_0 - b_1b_0)(a_1b_1 - a_0b_0)}$$

M being the mass of the liquid and $2c$ the distance between the focii.

Solution: Mass M of the liquid occupying the space between two confocal elliptic cylinders (a_0, b_0) and (a_1, b_1) and between two planes perpendicular to their axes and unit distance apart is

$$M = \Pi(a_1b_1 - a_0b_0)\rho \qquad \text{... (1)}$$

Let the elliptic cylinders be (a_0, b_0) and (a_1, b_1) be given by $\xi = \alpha_0$ and $\xi = \alpha_1$, respectively such that

$$\left.\begin{aligned} a_0 &= c\cosh\alpha_0, \; b_0 = c\sinh\alpha_0 \\ \text{and} \quad a_1 &= c\cosh\alpha_1, \; b_1 = c\sinh\alpha_1 \end{aligned}\right\} \qquad \text{... (2)}$$

Since

$$\left.\begin{aligned} z &= c\cosh\zeta = c\cosh(\xi + i\eta) \\ \text{or} \quad x &= c\cosh\xi\cos\eta, \; y = \sinh\xi\sin\eta \end{aligned}\right\} \qquad \text{... (3)}$$

The boundary condition for ψ is

$$\psi = \frac{1}{2}w\left(x^2 + y^2\right) + \text{const.}$$

$$\psi = \frac{wc^2}{2}\left(\cosh^2\xi\cos^2\eta + \sinh^2\xi\sin^2\eta\right) + \text{const.}$$

$$= \frac{wc^2}{2}\left[\frac{\cosh 2\xi + 1}{2}\,\frac{1 + \cos 2\eta}{2} + \frac{\cosh 2\xi - 1}{2}\,\frac{1 - \cos 2\eta}{2}\right] + \text{const.}$$

$$= \frac{wc^2}{4}\left[\cosh 2\xi + \cos 2\eta\right] + \text{const.}$$

At the boundaries of the cylinder

$$\psi_0 = \frac{wc^2}{4}(\cosh 2\xi_0 + \cos 2\eta) + \text{const.} \quad \text{... (4)}$$

$$\psi_1 = \frac{wc^2}{4}(\cosh 2\xi_1 + \cos 2\eta) + \text{const.} \quad \text{... (5)}$$

We have to find the solution of ψ which satisfy $\nabla^2\psi = 0$, (4) and (5).

It means that we must take ψ in the form

$$\psi = (A \cosh 2\xi + B \sinh 2\xi) \cos 2\eta \quad \text{... (6)}$$

At $\xi = \xi_0$, equating the coefficients of $\cos 2\eta$ from (4) and (6), we get

$$\frac{wc^2}{4} = A \cosh 2\xi_0 + B \sinh 2\xi_0 \quad \text{... (7)}$$

Similarly (5) and (6) give

$$\frac{wc^2}{4} = A \cosh 2\xi_1 + B \sinh 2\xi_1 \quad \text{... (8)}$$

Hence from (7) and (8)

$$A \cosh 2\xi_0 + B \sinh 2\xi_0 - \frac{wc^2}{4} = 0$$

$$A \cosh 2\xi_1 + B \sinh 2\xi_1 - \frac{wc^2}{4} = 0$$

$$\frac{A}{\sinh 2\xi_1 - \sinh 2\xi_0} = \frac{B}{\cosh 2\xi_0 - \cosh 2\xi_1}$$

$$= \frac{wc^2/4}{\sinh 2(\xi_1 - \xi_0)}$$

$$\Rightarrow \quad \frac{A}{\cosh(\xi_0 + \xi_1)} = \frac{B}{\sinh(\xi_0 + \xi_1)} = \frac{wc^2/4}{\cosh(\xi_1 - \xi_0)}$$

$$A = \frac{wc^2}{4}\frac{\cosh(\xi_1 + \xi_0)}{\cosh(\xi_1 - \xi_0)}$$

$$B = \frac{-wc^2}{4} \frac{\sinh(\xi_0 + \xi_1)}{\cosh(\xi_1 - \xi_0)}$$

Hence from (6)

$$\psi = \frac{wc^2}{4}\left[\frac{\cosh(\xi_1 + \xi_0)}{\cosh(\xi_1 - \xi_0)} \cos 2\xi - \frac{\sinh(\xi_0 + \xi_1)}{\cosh(\xi_1 - \xi_0)} \sinh 2\xi\right] \cos 2\eta$$

$$\Rightarrow \quad \psi = \frac{wc^2}{4} \frac{\cosh(2\xi - \xi_1 - \xi_0)}{\cosh(\xi_1 - \xi_0)} \cos 2\eta$$

Hence

$$\phi = -\frac{1}{4} wc^2 \frac{\sinh(2\xi - \xi_1 - \xi_0)}{\cosh(\xi_1 - \xi_0)} \sin 2\eta \qquad \text{as } \frac{\partial \phi}{\partial \xi} = \frac{\partial \psi}{\partial \eta}$$

Hence

$$(\phi d\psi)\, \xi = \xi_0$$

$$= \frac{1}{16}(wc^2)^2 \frac{2 \sinh(\xi_0 - \xi_1) \cosh(\xi_0 - \xi_1) \sin^2 2\eta}{\cosh^2(\xi_1 - \xi_0)} 2\eta d\psi$$

Also $\int_0^{2\pi} \sin^2 2\eta \, d\eta = \frac{1}{2} \int_0^{2\pi} (1 - \cos 4\eta)\, d\eta = \Pi$

$$\therefore \quad T_0 = -\frac{1}{2}\rho \int_0^{2\pi} (\phi d\psi)\, \xi = \xi_0$$

$$= -\frac{1\rho}{32}(w^2 c^4)\Pi \frac{\sinh 2(\xi_0 - \xi_1)}{\cosh^2(\xi_1 - \xi_0)}$$

Similarly

$$T_1 = -\frac{1}{32} \rho\, w^2 c^4 \frac{\sinh 2(\xi_1 - \xi_0)}{\cosh^2(\xi_1 - \xi_0)} \Pi$$

Total K.E. = T = $T_0 - T_1$

$$= \frac{1}{32} \rho w^2 c^4 \Pi \left[\frac{2 \sinh(\xi_1 - \xi_0)}{\cosh(\xi_1 - \xi_0)} + \frac{2 \sinh(\xi_1 - \xi_0)}{\cosh(\xi_1 - \xi_0)}\right]$$

$$= \frac{\Pi}{8} \rho w^2 c^4 \frac{\sinh(\xi_1 - \xi_0)}{\cosh(\xi_1 - \xi_0)}$$

$$= \frac{\Pi}{8} \rho w^2 c^4 \left[\frac{\sinh \xi_1 \cosh \xi_0 - \cosh \xi_1 \sinh \xi_0}{\cosh \xi_1 \cosh \xi_0 - \sinh \xi_1 \sinh \xi_0} \right]$$

$$= \frac{\Pi}{8} \rho w^2 c^4 \frac{(b_1 a_0 - a_1 b_0)}{(a_1 a_0 - b_1 b_0)}$$

$$= \frac{M}{8} \frac{w^2 c^4 (a_0 b_1 - a_1 b_0)}{(a_1 a_0 - b_1 b_0)(a_1 b_1 - a_0 b_0)}$$

10. Verify that the stream function for uniform streaming parallel to the axis past a solid, bounded by those parts of the circles

$$(x+1)^2 + y^2 = 2, (x-1)^2 + y^2 = 2$$

which are external to each other, are

$$\phi = y \left[1 + \frac{1}{x^2 + y^2} + \frac{2}{(x+1)^2 + y^2} + \frac{2}{(x-1)^2 + y^2} \right]$$

$$\psi = -x + \frac{x}{x^2 + y^2} + \frac{2(x+1)}{(x+1)^2 + y^2} + \frac{2(x-1)}{(x-1)^2 + y^2}$$

Solution: We have

$$|z-1|^2 = (x-1)^2 + y^2$$

and $|z+1|^2 = (x+1)^2 + y^2$

Hence the given circles are

$$|z-1| = 2 \text{ and } |z+1| = 2$$

We consider the transformation

$$t = \frac{2z}{z^2 - 1}$$

The circle $|t| = 1$ in t plane is the same as

$$\left| \frac{2z}{z^2 - 1} \right| = 1 \text{ in z plane}$$

Hence

$$|2z|^2 = |z^2 - 1|^2$$

$$\Rightarrow \quad 4|z|^2 = |z-1|^2 |z+1|^2$$

$$\Rightarrow \quad 4(x^2 + y^2) = \left[(x-1)^2 + y^2\right]\left[(x+1)^2 + y^2\right]$$

$$\Rightarrow \quad \left[(x-1)^2 + y^2 - 2\right]\left[(x+1)^2 + y^2 - 2\right] = 0$$

Hence the boundaries of the given circles $(x - 1)^2 + y^2 = 2$ and $(x + 1)^2 + y^2 = 2$ of plane under the given transformation, are transformed to the circle $|t| = 1$ in t plane.

Let V be the velocity of stream at infinity in t plane along positive direction of imaginary axis. The complex potential for streaming past the cylinder $|t| = 1$ is

$$w = iV\left(t - \frac{1}{t}\right)$$

Transforming into z plane, we have

$$w = iv\left[\frac{2z}{z^2 - 1} - \frac{z^2 - 1}{2z}\right]$$

$$= iV\left[\left(\frac{1}{z+1} + \frac{1}{z-1}\right) - \frac{1}{2}\left(z - \frac{1}{z}\right)\right]$$

$$= \frac{iV}{2}\left[2\left\{\frac{(x-1) - iy}{(x-1)^2 + y^2} + \frac{(x+1) - iy}{(x+1)^2 + y^2}\right\} - \frac{1}{2}\left\{z - \frac{x - iy}{x^2 + y^2}\right\}\right]$$

Equating real and imaginary parts, we get

$$\phi = \frac{Vy}{2}\left[2\left\{\frac{1}{(x-1)^2 + y^2} + \frac{1}{(x+1)^2 + y^2}\right\} + \left\{y + \frac{1}{x^2 + y^2}\right\}\right]$$

$$\text{and } \psi = \frac{V}{2}\left[2\left\{\frac{x-1}{(x-1)^2+y^2}+\frac{x+1}{(x+1)^2+y^2}\right\} - \left\{x - \frac{x}{x^2+y^2}\right\}\right]$$

If we take $V = 2$, then we get the required result.

EXERCISE
(PART II)

1. If the liquid is contained between the elliptic cylinders $\frac{x^2}{a^2}+\frac{y^2}{b^2}=1$ and $\frac{x^2}{a^2}+\frac{y^2}{b^2}=K^2$ and the whole rotates about OZ with angular velocity Ω, prove that the velocity potential ϕ referred to the axes OX, OY is given by

$$\phi = -\Omega\left(\frac{a^2-b^2}{a^2+b^2}\right)xy$$

and that the surfaces of equal pressure are the hyperbolic cylinders $\frac{x^2}{3a^2+b^2}-\frac{y^2}{a^2+3b^2}=c$.

2. Show that the motion of a liquid streaming past the elliptic disc $\frac{x^2}{a^2}+\frac{y^2}{b^2}=1$, the velocity at infinity being parallel to axis of x and equal to U can be expressed by the relation

$$\phi + i\psi = \frac{U}{a-b}\left[az - b\sqrt{z^2-c^2}\right]$$

3. Show that with proper choice of units the motion of an infinite liquid produced by the motion of an elliptic cylinder parallel to one of its principal axes, is given by the complex potential $w = e^{-z}$, where $z = 2\cos\zeta$. Deduce the formulae

$$x = \phi\left(1+\frac{1}{\phi^2+\psi^2}\right),\ y = \psi\left(1-\frac{1}{\phi^2+\psi^2}\right)$$

4. Prove that if $2a$, $2b$ are the axes of the cross section of an elliptic cylinder placed across a stream in which the velocity at infinity is U parallel to the major axis of the cross section, the velocity at a point $(a \cos \eta, b \sin \eta)$ on the surface is

$$U(a+b)\sin\eta\left(b^2\cos\eta + a^2\sin\eta\right)^{-\frac{1}{2}}$$

and that, in consequence of the motion of the liquid, the resultant thrust (per unit length) on that half cylinder on which the stream impringes is diminished by

$$\frac{2b^2\rho u^2}{a-b}\left[1-\left(\frac{a+b}{a-b}\right)^{\frac{1}{2}}\tan^{-1}\left(\frac{a-b}{a+b}\right)^{\frac{1}{4}}\right]$$

where ρ is the density of the liquid.

5 IRROTATIONAL MOTION IN THREE DIMENSIONS (MOTION OF A SPHERE)

5.1. Introduction

We shall now study irrotational motion in three dimensions with a particular reference to the motion of a sphere. The treatment is similar to that of the motion of a circular cylinder. However, unlike in the case of the circular cylinder the complex potential w will not be of any use in the motion of a sphere, the motion being now in three dimensions in which complex quantities fail to play any role.

5.2. Equation of Continuity

Equation of continuity for an incompressible fluid is

$$\frac{\partial u}{\partial x}+\frac{\partial v}{\partial y}+\frac{\partial w}{\partial z}=0$$

In spherical coordinates $(r, \theta, \in)$ the equation of continuity is

$$\frac{1}{r^2}\frac{\partial}{\partial r}(r^2 u)+\frac{1}{r\sin\theta}\frac{\partial}{\partial \theta}(v\sin\theta)+\frac{1}{r\sin\theta}\frac{\partial w}{\partial \in}=0 \qquad \text{... (1)}$$

Let the motion be irrotational, so that velocity potential ϕ exists such that $u=\frac{-\partial\phi}{\partial r}$, $v=-\frac{1}{r}\frac{\partial\phi}{\partial\theta}$ and $w=-\frac{1}{r\sin\theta}\frac{\partial\phi}{\partial w}$

Hence (1) becomes

$$\frac{1}{r^2}\frac{\partial}{\partial r}\left(\frac{r^2\partial\phi}{\partial r}\right)+\frac{1}{r\sin\theta}\frac{\partial}{\partial\theta}\left(\frac{1}{r}\frac{\partial\phi}{\partial\theta}\sin\theta\right)$$
$$+\frac{1}{r\sin\theta}\frac{\partial}{\partial \in}\left(\frac{1}{r\sin\theta}\frac{\partial\phi}{\partial w}\right)=0 \qquad \text{... (2)}$$

When there is a symmetry about a line, ϕ is independent of $\in$ and hence (2) becomes

$$\frac{\partial}{\partial r}\left(\frac{r^2\partial\phi}{\partial r}\right)+\frac{1}{\sin\theta}\frac{\partial}{\partial\theta}\left(\sin\theta\frac{\partial\phi}{\partial\theta}\right)=0 \qquad \dots (3)$$

Whose solutions are of the type $r^n P_n$ $(\cos n\,\theta)$ and $r^{-n-1} P_n$ $(\cos\theta)$ for all integral values of n, where P_n is the Legendre's polynomial of order n.

Thus the general solution of (3) is

$$\phi=\sum_n\left\{Anr^n+\frac{Bn}{r^{n+1}}\right\}P_n(\cos\theta)$$

It is also known that $P_0(\mu) = 1$, $P_1(\mu) = \mu$ and $P_2 = \frac{1}{2}(3\mu^2-1)$ etc.

5.3. To Find the Velocity Potential when a Sphere of Radius *a* is Moving with Velocity *U* in a Liquid at Rest at Infinity

(I.A.S. 81)

Let the origin be at the centre of the sphere and the axis of z in the direction of motion. The boundary conditions for ϕ are

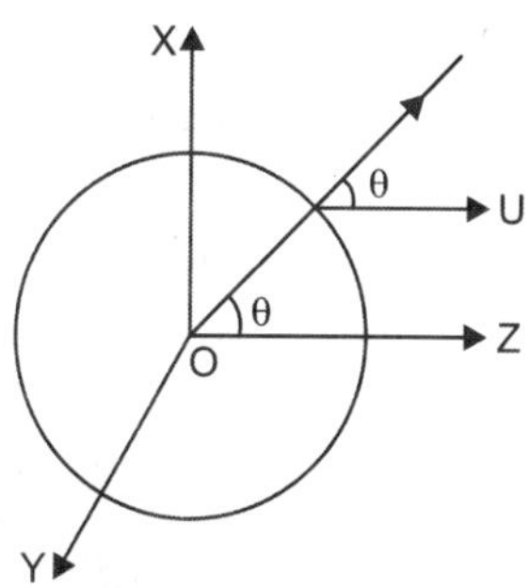

Fig. 5.1

(i) $\vec{\nabla}^2\phi=0$

(ii) $\left(\frac{-\partial\phi}{\partial r}\right)_{r=a}=U\cos\theta$

(iii) $\left(\frac{-\partial\phi}{\partial r}\right)=0=\left(-\frac{1}{r}\frac{\partial\phi}{\partial\theta}\right)$ at $r=\infty$

The conditions (i) and (ii) suggests that ϕ must be of the form

$$\phi=\left(Ar+\frac{B}{r^2}\right)\cos\theta \qquad \dots (1)$$

Hence

$$\frac{\partial\phi}{\partial r}=\left(A-\frac{2B}{r^3}\right)\cos\theta \qquad \dots (2)$$

Putting $r = \infty$ in (2) and using (iii), we get

$$0 = A \cos \theta \text{ so that } A = 0 \qquad \dots (3)$$

Putting $r = a$ in (2) and using (ii) and (3), we get

$$-U \cos \theta = \frac{-2B}{a^3} \cos \theta$$

so that

$$B = \frac{1}{2} Ua^3 \qquad \dots (4)$$

Hence (1) becomes

$$\phi = \left(0 + \frac{a^3 U}{2r^2}\right) \cos \theta$$

$$\Rightarrow \quad \phi = \frac{Ua^3}{2r^2} \cos \theta \qquad \dots (5)$$

We now determine the equations of line (streamlines) of flow. The differential equation of the lines of flow at the instant the centre of sphere is passing through the origin is given by

$$\frac{dr}{\frac{\partial \phi}{\partial r}} = \frac{r d\theta}{\frac{\partial \phi}{\partial \theta}}$$

$$\Rightarrow \quad \frac{dr}{\frac{-Ua^3}{r^3} \cos \theta} = \frac{r^2 d\theta}{\frac{-Ua^3}{2r^2} \sin \theta}$$

$$\Rightarrow \quad \frac{dr}{\cos \theta} = \frac{2r d\theta}{\sin \theta}$$

$$\Rightarrow \quad \frac{dr}{r} = 2 \cot \theta \, d\theta$$

Integrating, we get

$$\log r = \log \sin^2 \theta + \log c$$

$$\Rightarrow r = c \sin^2 \theta \qquad \dots (6)$$

(6) gives the required streamlines.

5.4. Liquid Streaming Past a Fixed Sphere

For a liquid streaming past a fixed sphere, obtain the lines of flow relative to the sphere.

Let the sphere be at rest and let the liquid flow past the cylinder with velocity U in the negative direction of z-axis. We superimpose a velocity $-U$ on the whole system, the velocity potential becomes

$$\phi = Ur\cos\theta + \frac{Ua^3}{2r^2}\cos\theta$$

The streamlines are given by

$$\frac{dr}{\frac{\partial\phi}{\partial r}} = \frac{r\,d\theta}{\frac{1}{r}\frac{\partial\phi}{\partial\theta}}$$

$$\Rightarrow \quad \frac{dr}{\left(1 - \frac{a^3}{r^3}\right)\cos\theta U} = \frac{r^2 d\theta}{-\left(r + \frac{a^3}{2r^2}\right)\sin\theta U}$$

$$\Rightarrow \quad \frac{dr\left(a^3 + 2r^3\right)}{r\left(r^3 - a^3\right)} = -2\cot\theta\, d\theta$$

$$\Rightarrow \quad \left[\frac{3r^2}{r^3 - a^3} - \frac{1}{r}\right] dr = -2\cot\theta\, d\theta$$

Integrating, we get

$$\log r - \log(r^3 - a^3) + \log c = \log \sin^2\theta$$

$$\Rightarrow \quad \log\frac{cr}{r^3 - a^3} = \log\sin^2\theta$$

$$\Rightarrow \quad \sin^2\theta = \frac{cr}{r^3 - a^3}$$

This equation gives the lines of flow relative to the sphere.

5.5. Equations of Motion of a Sphere

Let a sphere of radius a advance with velocity U in an infinite mass of liquid at rest at infinity. We take the origin at the centre of the sphere and the axis of z in the direction of motion.

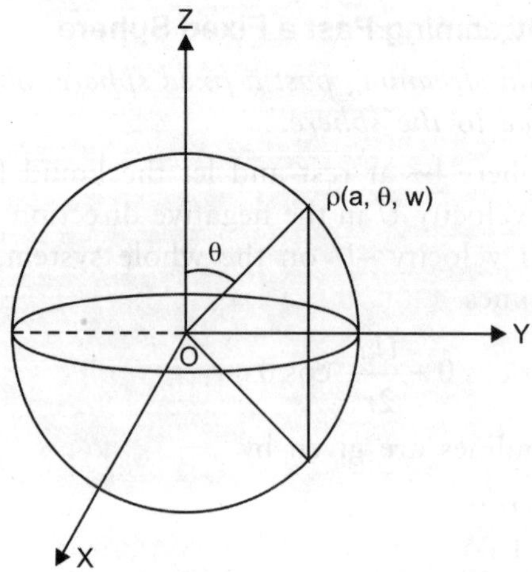

Fig. 5.2

The velocity potential and stream function are given by

$$\phi = \frac{Ua^3}{2r^2} \cos\theta \qquad \text{... (1)}$$

and

$$\psi = -\frac{1}{2}\frac{Ua^3 \sin\theta}{r} \qquad \text{... (2)}$$

Let T_1 be the kinetic energy of the liquid on the surface of the sphere, then

$$T_1 = -\frac{1}{2}\rho \iint_{2\pi} \phi \frac{\partial \phi}{\partial n} ds$$

$$= -\frac{1}{2}\rho \int_{w=0} \int_{\theta=0}^{\pi} \frac{Ua^3}{2r^2} \cos\theta \left(\frac{Ua^3 \cos\theta}{r^3}\right) \times a^3 \sin\theta d\theta dw$$

$$= \frac{U^2}{4} \rho a^3 \int_{\theta=0}^{\pi} \cos^2\theta \sin\theta d\theta \int_{w=0}^{2\pi} dw$$

$$= \frac{U^2}{4} \rho a^3 \int_0^{\pi} -\cos^2\theta d(\cos\theta)\, 2\Pi$$

$$= \frac{U^2}{4} \rho a^3 \frac{2}{3} 2\Pi$$

$$= \frac{\Pi}{3} \rho a^3 U^2$$

$$= \frac{1}{4} M^1 U^2 \qquad \text{... (3)}$$

where $M^1 = \frac{4}{3} \Pi \rho a^3$ is the mass of the liquid displaced by the sphere, ρ is the density of the liquid. Let σ be the density of the sphere and M be the mass of the sphere so that

$$M = \frac{4}{3} \Pi \sigma a^3 \qquad \text{... (4)}$$

and

$$\text{K.E. of the sphere } = T_2 = \frac{1}{2} M U^2 \qquad \text{... (5)}$$

Let T be the total kinetic energy of the liquid and the sphere.

Then

$$T = \frac{1}{2} \left(M + \frac{1}{2} M^1 \right) U^2; \qquad \text{From (3) and (5)}$$

Let R be the external force parallel to the direction of motion.

Since rate of increase of total energy

= Rate of work done

$$= \frac{\text{Force} \times \text{distance}}{\text{time}}$$

= Force × velocity

$$\Rightarrow \quad \frac{d}{dt} \left[\frac{1}{2} \left(M + \frac{M^1}{2} \right) U^2 \right] = RU$$

$$\Rightarrow \quad \left(M + \frac{M^1}{2} \right) U \frac{dU}{dt} = RU$$

$$\Rightarrow \quad \left(M + \frac{M^1}{2}\right)\frac{dU}{dt} + R$$

$$\Rightarrow \quad M\frac{dU}{dt} = R - \frac{M^1}{2}\frac{dU}{dt} \qquad \dots (6)$$

Thus, the liquid exerts a resistance $\frac{1}{2}M^1\frac{dU}{dt}$.

Let R^1 be the external force on the sphere, when the liquid is absent. Then from Hydrostatical considerations, the relation between R and R^1 is found to be

$$R = \frac{\sigma - R}{\sigma} R^1 \qquad \dots (7)$$

From (6) and (7), we get

$$M\frac{dU}{dt} = \frac{\sigma - \rho}{\sigma} R^1 - \frac{M^1}{2}\frac{dU}{dt}$$

$$\Rightarrow \quad \left(M + \frac{1}{2}M^1\right)\frac{dU}{dt} = \frac{\sigma - \rho}{\sigma} R^1$$

$$\Rightarrow \quad \frac{4}{3}\Pi a^3\left(\sigma + \frac{\rho}{2}\right)\frac{dU}{dt} = \frac{\sigma - \rho}{\sigma} R^1$$

$$\Rightarrow \quad \frac{4}{3}\Pi a^3\left(\sigma + \frac{\rho}{2}\right)\frac{dU}{dt} = \frac{4}{3}\Pi a^3 \frac{\sigma - \rho}{\frac{4}{3}\Pi a^3 \sigma} R^1$$

$$\Rightarrow \quad \left(\sigma + \frac{\rho}{2}\right)\frac{dU}{dt} = \frac{\sigma - \rho}{M} R^1$$

$$\Rightarrow \quad M\frac{dU}{dt} = \frac{\sigma - \rho}{\sigma + \frac{\rho}{2}} R^1$$

$$\Rightarrow \quad \frac{MU}{R^1} = \frac{\sigma - \rho}{\sigma + \frac{\rho}{2}}$$

This shows that the effect of presence of liquid is to reduce external forces in the ratio of $\sigma - \rho - \sigma + \frac{\rho}{2}$.

5.6. Motion of a Sphere Under Gravity

In considering the motion of a sphere, we consider its virtual mass equal to $M + \frac{1}{2}M^1$, where M and M^1 are masses of sphere and liquid displaced by sphere. By Newton's second law of motion

$$\left(M + \frac{1}{2}M^1\right)\frac{dU}{dt} = Mg - M^1 g$$

$$\Rightarrow \quad \frac{4}{3}\Pi a^3\left(\sigma + \frac{1}{2}\rho\right)\frac{dU}{dt} = \frac{4}{3}\Pi a^3(\sigma - \rho)g$$

$$\Rightarrow \quad \text{acceleration} = \frac{dU}{dt} = \frac{\sigma - \rho}{\sigma + \frac{1}{2}\rho}g$$

$$= \frac{2(\sigma - \rho)}{2\sigma + \rho}g$$

Hence the centre of the sphere describes a parabola of latus rectum

$$= \frac{2\,(\text{Horizontal velocity})^2}{\text{acceleration}}$$

$$= \frac{2U^2(2\sigma + \rho)}{2(\sigma - \rho)g} = \frac{U^2}{g}\frac{2\sigma + \rho}{\sigma - \rho}$$

5.7. Concentric Spheres (Problem of Initial Motion)

A sphere of radius a is surrounded by a concentric sphere of radius b, the space between being filled with liquid at rest. The inner sphere is given a velocity U and outer sphere a velocity V in the same direction. To determine the initial motion of the liquid.

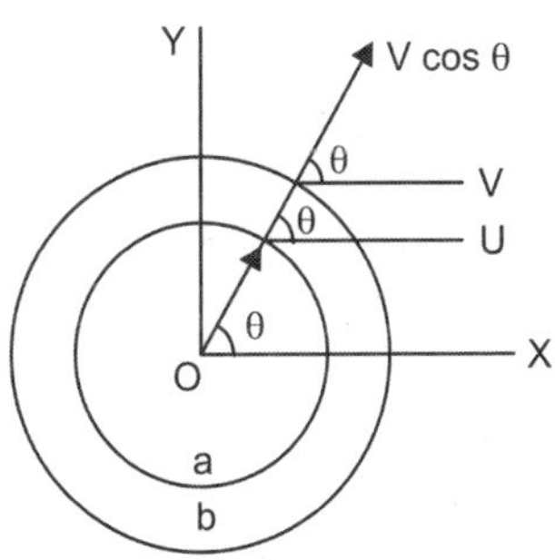

Fig. 5.3

Let O be the common centre and ϕ be the velocity potential of the initial motion. Let U and V be in the direction of initial line OA as shown in the figure 5.3.

The boundary conditions are

(i) $\vec{\nabla}^2\phi = 0$

(ii) $\left(\frac{-\partial\phi}{\partial r}\right)_{r=a} = U\cos\theta$

(iii) $\left(\frac{-\partial\phi}{\partial r}\right)_{r=b} = V\cos\theta$

(iv) $\left(\frac{-\partial\phi}{\partial r}\right)_{r=\infty} = 0 = \left(-\frac{1}{r}\frac{\partial\phi}{\partial\theta}\right)_{r=\infty}$

(i) and (ii) suggests that ϕ must be of the form

$$\phi = \left(Ar + \frac{B}{r^2}\right)\cos\theta \qquad \text{... (1)}$$

when there is a symmetry about a line.

From (1)

$$\frac{\partial\phi}{\partial r} = \left(A - \frac{2B}{r^3}\right)\cos\theta \qquad \text{... (2)}$$

Subjecting (2) to (ii) and (iii), we get

$$-U\cos\theta = \left(A - \frac{2B}{a^3}\right)\cos\theta \qquad \text{... (3)}$$

and

$$-V\cos\theta = \left(A - \frac{2B}{b^3}\right)\cos\theta \qquad \text{... (4)}$$

Hence $A - \frac{2B}{a^3} = -U$ and $A - \frac{2B}{b^3} = -V$

Hence $2B\left(\frac{1}{a^3} - \frac{1}{b^3}\right) = U - V$

$$\Rightarrow \quad B = \frac{(U-V)\,a^3b^3}{2\left(b^3 - a^3\right)}$$

Hence

$$A = \frac{b^3(U - V)}{b^3 - a^3} - U$$

$$= \frac{Ua^3 - Vb^3}{b^3 - a^3}$$

Hence from (1), we get

$$\phi = \frac{1}{b^2 - a^3}\left[\left(Ua^3 - Vb^3\right)r + \frac{(U - V)\,a^3 b^3}{2r^2}\right]\cos\theta$$

This is the required value of velocity potential.

Deductions. Let the outer sphere be at rest, then $V = 0$, then we have

$$\phi = \frac{Ua^3}{b^3 a^3}\left(\frac{b^3}{2r^2} + r\right)\cos\theta$$

Let M be the mass of the inner sphere and I, the impulse necessary to produce the velocity U,

$$I = MU + \iint p\cos\theta ds$$

where $p = (\rho\phi)_{r=a}$

= impulsive pressure of liquid on the sphere $r = a$

$$\therefore \quad I = MU + \int_0^{\pi} \frac{\rho U a^3}{b^3 - a^3}\left(\frac{b^3}{2a^2} + a\right)\cos^2\theta \times (2\Pi a\sin\theta)\, a d\theta$$

$$= MU + \frac{\rho U a}{b^3 - a^3}\left(2a^3 + b^3\right)\int_0^{\pi} \Pi a^2 \sin\theta\cos^2\theta d\theta$$

$$= MU + \frac{\rho U a^3\left(2a^3 + b^3\right)\Pi}{b^3 - a^3}\int_0^{\pi}\left(-\cos^2\theta\right)d(\cos\theta)$$

$$= MU + \frac{\rho U a^3\left(2a^3 + b^3\right)\Pi}{b^3 - a^3}\left(-\frac{1}{3}\cos^3\theta\right)_0^{\pi}$$

$$= MU + \frac{1}{2}\left(\frac{b^3 + 2a^3}{b^3 - a^3}\right)M^1 U$$

where M^1 is the mass of the liquid displaced by the sphere of radius a.

Remark. If the outer sphere becomes infinitely large, then $b \to \infty$, so that $I = MU + \frac{1}{2} M^1 U$.

This shows that the impulse required to impart a velocity U is the same when the sphere is in motion with velocity U in an infinite mass of liquid at rest at infinity.

5.8. Stoke's Stream Function

The equation of continuity in cylindrical coordinates (r, θ, z) is

$$\frac{\partial \rho}{\partial t} + \frac{1}{r}\frac{\partial}{\partial r}(\rho r u) + \frac{1}{r}\frac{\partial}{\partial \theta}(\rho v) + \frac{\partial}{\partial z}(\rho w) = 0$$

For incompressible fluid, it becomes

$$\frac{1}{r}\frac{\partial}{\partial r}(ru) + \frac{1}{r}\frac{\partial v}{\partial \theta} + \frac{\partial w}{\partial z} = 0$$

If the motion be symmetrical about z-axis, the $v = 0$ and the equation of continuity becomes

$$\frac{1}{r}\frac{\partial}{\partial r}(ru) + \frac{\partial w}{\partial z} = 0$$

Instead of z-axis, if we take x-axis as the axis of symmetry, $\bar{w}$ the direction perpendicular to x-axis and u, v velocity components in these directions, then

$$\frac{1}{\bar{w}}\frac{\partial}{\partial \bar{w}}(\bar{w}v) + \frac{\partial u}{\partial x} = 0$$

[Writing $\bar{w}$ for r, v for u, u for w]

$$\Rightarrow \quad \frac{\partial}{\partial x}(u\bar{w}) + \frac{\partial}{\partial \bar{w}}(\bar{w}v) = 0$$

But this is the condition that

$$\begin{aligned} \bar{w}v\,dx - u\bar{w}\,d\bar{w} &= \text{an exact differential} \\ &= d\psi \text{ (say)} \\ &= \frac{\partial \psi}{\partial x}dx + \frac{\partial \psi}{\partial y}dy \end{aligned}$$

So that

$$\bar{w}v = \frac{\partial \psi}{\partial x} \text{ and } -u\bar{w} = \frac{\partial \psi}{\partial \bar{w}}$$

$$\Rightarrow \quad u = -\frac{1}{\bar{w}}\frac{\partial \psi}{\partial \bar{w}} \text{ and } v = \frac{1}{\bar{w}}\frac{\partial \psi}{\partial x}$$

This function ψ is called Stoke's stream function.

The streamlines are given by

$$\Rightarrow \quad \frac{dx}{u} = \frac{d\bar{w}}{v}$$

$$\Rightarrow \quad vdx - ud\bar{w} = 0$$

$$\Rightarrow \quad v\bar{w}dx - u\bar{w}d\bar{w} = 0$$

$$\Rightarrow \quad d\psi = 0$$

$$\Rightarrow \quad \psi = \text{constant}$$

which represents the streamline. Stoke's stream function can exist even when ϕ does not i.e. even if the motion is rotational.

5.9. A Property of Stoke's Stream Function

2Π times the difference of the values of Stoke's stream function at two points in the same meridian plane is equal to the flow across the angular surface obtained by the revolution round the axis of curve joining the points.

Proof. Let AB be an arc of a curve which when rotated about the x-axis will describe an annular surface. Let P be a point in AB, ds an elementary arc at P. Let θ be the inclination of ds to the axis.

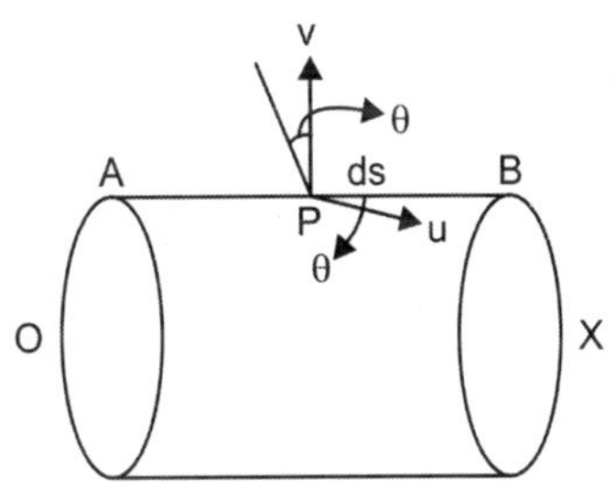

Fig. 5.4

Then velocity across the element ds

$$= v\cos\theta - u\sin\theta$$

$$= \frac{1}{\bar{w}}\frac{\partial \psi}{\partial x}\frac{dx}{ds} + \frac{1}{\bar{w}}\frac{\partial \psi}{\partial \bar{w}}\frac{d\bar{w}}{ds}$$

$$= \frac{1}{\bar{w}} \frac{\partial \psi}{\partial s} \qquad \dots (1)$$

Hence the outward flow across the surface of revolution

$$= \int_A^B (v \cos \theta - u \sin \theta)\, 2\Pi \bar{w} ds$$

$$= 2\Pi \int_A^B \frac{1}{\bar{w}} \frac{\partial \psi}{\partial s} \bar{w} ds \qquad \text{From (1)}$$

$$= 2\Pi \int_A^B d\psi = 2\Pi \left(\psi_B - \psi_A\right)$$

This proves the required result.

5.10. Irrotational Motion

If the motion be irrotational, then ϕ exists.

If ψ be Stoke's function, then

$$u = -\frac{1}{\bar{w}} \frac{\partial \psi}{\partial \bar{w}},\ v = \frac{1}{\bar{w}} \frac{\partial \psi}{\partial x} \qquad \dots (1)$$

Also

$$u = -\frac{\partial \phi}{\partial x},\ v = -\frac{\partial \phi}{\partial \bar{w}} \qquad \dots (2)$$

Hence

$$\frac{\partial \phi}{\partial x} = \frac{1}{\bar{w}} \frac{\partial \psi}{\partial \bar{w}},\ \frac{\partial \phi}{\partial \bar{w}} = -\frac{1}{\bar{w}} \frac{\partial \psi}{\partial x} \qquad \dots (3)$$

From (3)

$$\frac{\partial}{\partial \bar{w}}\left(\frac{\partial \phi}{\partial x}\right) = \frac{\partial}{\partial \bar{w}}\left(\frac{1}{\bar{w}} \frac{\partial \psi}{\partial \bar{w}}\right)$$

and

$$\frac{\partial}{\partial x}\left(\frac{\partial \phi}{\partial \bar{w}}\right) = \frac{\partial}{\partial x}\left(-\frac{1}{\bar{w}} \frac{\partial \psi}{\partial x}\right) = -\frac{\partial}{\partial x}\left(\frac{1}{\bar{w}} \frac{\partial \psi}{\partial x}\right)$$

Hence

$$\frac{\partial}{\partial \bar{w}}\left(\frac{1}{\bar{w}} \frac{\partial \psi}{\partial \bar{w}}\right) = -\frac{\partial}{\partial x}\left(\frac{1}{\bar{w}} \frac{\partial \psi}{\partial x}\right)$$

or $\quad -\dfrac{1}{\bar{w}^2}\dfrac{\partial\psi}{\partial\bar{w}}+\dfrac{1}{\bar{w}}\dfrac{\partial^2\psi}{\partial\bar{w}^2}+\dfrac{1}{\bar{w}}\dfrac{\partial^2\psi}{\partial x^2}=0$

or $\quad \dfrac{\partial^2\psi}{\partial x^2}+\dfrac{\partial^2\psi}{\partial\bar{w}^2}-\dfrac{1}{\bar{w}}\dfrac{\partial\psi}{\partial\bar{w}}=0 \qquad \ldots (4)$

Again from (3)

$$\frac{\partial}{\partial x}\left(\frac{\partial\psi}{\partial\bar{w}}\right)=\frac{\partial}{\partial\bar{w}}\left(\frac{\partial\psi}{\partial x}\right) \text{ gives}$$

$$\Rightarrow \frac{\partial}{\partial x}(\bar{w}u)=\frac{\partial}{\partial\bar{w}}(v\bar{w})$$

$$\Rightarrow \frac{\partial}{\partial x}\left(\bar{w}\frac{\partial\phi}{\partial x}\right)=\frac{\partial}{\partial\bar{w}}\left(-\bar{w}\frac{\partial\phi}{\partial\bar{w}}\right)$$

$$\Rightarrow \bar{w}\frac{\partial^2\phi}{\partial x^2}+\frac{\partial\phi}{\partial\bar{w}}+\bar{w}\frac{\partial^2\phi}{\partial\bar{w}^2}=0$$

$$\Rightarrow \frac{\partial^2\phi}{\partial x^2}+\frac{\partial^2\phi}{\partial\bar{w}^2}+\frac{1}{w}\frac{\partial\phi}{\partial\bar{w}}=0 \qquad \ldots (5)$$

From (4) and (5) it is clear that ϕ and ψ are not interchangeable. This is a fact contrary to two-dimensional irrotational motion. (5) is a Laplace's equation in cylindrical coordinates, when there is a symmetry about x-axis. Hence ψ does not satisfy Laplace's equation.

5.11. Polar Coordinates

Let (r, θ) be polar coordinates of the point $(x, \bar{w})$ so that $x = r\cos\theta,\ \bar{w} = r\sin\theta$.

So if q_r and q_θ be the velocities in the directions of r and θ, and ψ be the Stoke's stream function, then

$$q_r=-\frac{1}{\bar{w}}\frac{\partial\psi}{r\partial\theta}=-\frac{1}{r^2\sin\theta}\frac{\partial\psi}{\partial\theta}$$

$$q_s=v=\frac{1}{\bar{w}}\frac{\partial\psi}{\partial r}=\frac{1}{r\sin\theta}\frac{\partial\psi}{\partial r}$$

Also

$$q_r=-\frac{\partial\phi}{\partial r},\ q_s=-\frac{1}{r}\frac{\partial\phi}{\partial\theta}$$

Hence

$$\frac{\partial \phi}{\partial r} = \frac{1}{r^2 \sin\theta} \frac{\partial \psi}{\partial \theta}$$

and

$$\frac{1}{r \sin \theta} \frac{\partial \psi}{\partial r} = -\frac{1}{r} \frac{\partial \phi}{\partial \theta}$$

5.12. Solution for ψ

We have

$$\frac{\partial \phi}{\partial r} = \frac{1}{r^2 \sin\theta} \frac{\partial \psi}{\partial \theta} \quad \ldots (1)$$

and

$$\frac{1}{r} \frac{\partial \phi}{\partial \theta} = -\frac{1}{r \sin \theta} \frac{\partial \psi}{\partial r} \quad \ldots (2)$$

Since

$$\frac{\partial}{\partial \theta}\left(\frac{\partial \psi}{\partial r}\right) = \frac{\partial}{\partial r}\left(\frac{\partial \psi}{\partial \theta}\right)$$

$$\Rightarrow \quad \frac{\partial}{\partial \theta}\left(-\sin\theta \frac{\partial \phi}{\partial \theta}\right) = \frac{\partial}{\partial r}\left(r^2 \sin\theta \frac{\partial \phi}{\partial r}\right) \quad \ldots (3)$$

Let $\mu = \cos\theta$, so that

$$\sin\theta \frac{\partial}{\partial \mu} = -\frac{\partial}{\partial \theta}$$

Hence from (3), we get

$$-\sin\theta \frac{\partial}{\partial \mu}\left(-\sin\theta \frac{\partial \phi}{\partial \theta}\right) = \frac{\partial}{\partial r}\left(r^2 \frac{\partial \phi}{\partial r}\right) \sin\theta$$

$$\Rightarrow \quad \frac{\partial}{\partial r}\left(r^2 \frac{\partial \phi}{\partial r}\right) + \frac{\partial}{\partial \mu}\left[\left(1-\mu^2\right) \frac{\partial \phi}{\partial \mu}\right] = 0$$

This is Laplace's equation under the present condition and so ϕ has a solution of the form

$$\phi_1 = r^n P_n(\mu) \text{ or } r^{-(n+1)} P_n(\mu) = \phi_2$$

Similarly eliminating ϕ from (1) and (2), we get

$$r^2 \frac{\partial^2 \psi}{\partial r^2} + \left(1-\mu^2\right) \frac{\partial^2 \psi}{\partial \mu^2} = 0 \quad \ldots (4)$$

Let ψ_1 and ψ_2 be the solutions of equation (4) then from (1) and (2), we get

$$\frac{\partial \psi_1}{\partial \mu} = -r^2 \frac{\partial \phi_1}{\partial r} = -r^2 \frac{\partial}{\partial r}\{r^n P_n(\mu)\}$$

$$= -nr^{n+1} P_n(\mu)$$

and

$$\frac{\partial \psi_2}{\partial \mu} = -r^2 \frac{\partial \phi_2}{\partial r} = -r^2 \frac{\partial}{\partial r}\{r^{-n-1} P_n(\mu)\}$$

$$= (n+1)\, r^{-n} P_n(\mu)$$

Also

$$\frac{\partial \psi_1}{\partial r} = (1-\mu^2)\frac{\partial \phi_1}{\partial \mu} = (1-\mu^2)\, r^n \frac{\partial P_n(\mu)}{\partial \mu} \quad \text{... (5)}$$

and

$$\frac{\partial \psi_2}{\partial r} = (1-\mu^2)\frac{\partial \phi_2}{\partial \mu} = (1-\mu^2)\, r^{-n-1} \frac{\partial P_n(\mu)}{\partial \mu} \quad \text{... (6)}$$

By integrating (4) and (5), we have

$$\psi_1 = \frac{1-\mu^2}{n+1}\, r^{n+1} \frac{\partial P_n(\mu)}{\partial \mu},\ \psi_2 = -\frac{1-\mu^2}{n}\frac{1}{r^n}\frac{\partial P_n(\mu)}{\partial \mu}$$

Which are possible solution for the stream function ψ.

5.13. Motion of a Solid of Revolution Along the Axes

Solids of Revolution along their axes in an Infinite Mass of the Liquid

Let the body moves along x-axis with velocity u and ds be an element of the meridian curve. The motion is symmetrical about x-axis and hence Stoke's function ψ exists.

The normal velocity at any point is $u\dfrac{\partial \overline{w}}{\partial s}$.

On boundary,

$$-\frac{1}{\overline{w}}\frac{\partial \psi}{\partial s} = u \cos\theta = u\frac{\partial \overline{w}}{\partial s}$$

$$\Rightarrow \quad \frac{\partial \psi}{\partial s} = -u\overline{w}\frac{\partial \overline{w}}{\partial s}$$

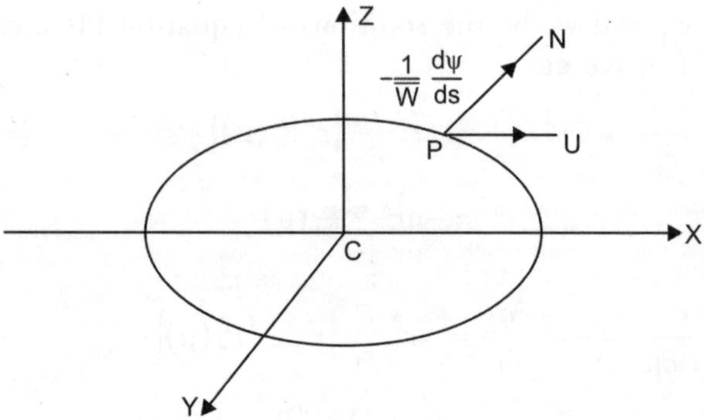

Fig. 5.5

Integrating, we get $\psi = -\frac{1}{2}u\bar{w}^2 + \text{constant}$

$$\Rightarrow \quad \psi = -\frac{1}{2}ur^2\sin^2\theta + \text{constant}\ (\because \bar{w} = r\sin\theta) \qquad \ldots (1)$$

$$\Rightarrow \quad \psi = -\frac{1}{2}ur^2\left(1-\mu^2\right) + C \qquad (\because \mu = \cos\theta)$$

On the surface of the sphere

$$\psi = -\frac{1}{2}ua^2\left(1-\mu^2\right) + C \qquad \ldots (2)$$

But the stream function ψ has the solutions of the form $\frac{1-\mu^2}{n+1}r^{n+1}\frac{\partial P_n}{\partial \mu}$ and $\frac{1-\mu^2}{nr^n}\frac{\partial P_n}{\partial \mu}$

Assuming $\frac{\partial P_n}{\partial \mu} = 1$ or $P_n\ (\mu) = \mu$.

Hence $n = 1$

Also the liquid is at rest at infinity.

Therefore, suitable form of ψ is

$$\psi = A\frac{1-\mu^2}{r} \qquad \ldots (3)$$

On boundary $r = a$

Hence from (2) and (3), we get

$$\frac{A\left(1-\mu^2\right)}{r} = -\frac{1}{2}ua^2\left(1-\mu^2\right) + C$$

So $C = 0,\ A = -\dfrac{1}{2}ua^3$

$$\therefore \quad \psi = -\frac{1}{2}\frac{ua^3}{r}\left(1-\mu^2\right)$$

$$\psi = -\frac{1}{2}\frac{ua^3}{r}\sin^2\theta \qquad \ldots (4)$$

But

$$\frac{\partial\phi}{\partial r} = \frac{1}{r\sin\theta}\frac{1}{r}\frac{\partial\psi}{\partial\theta}$$

$$= \frac{1}{r^2\sin\theta}\left(-\frac{1}{2}\frac{ua^3}{r}2\sin\theta\cos\theta\right)$$

$$= -\frac{ua^3}{r^3}\cos\theta$$

Integrating, we have

$$\phi = \frac{ua^3}{2r^2}\cos\theta \qquad \ldots (5)$$

Neglecting constant of integration (4) and (5) are required expression for ψ and ϕ.

If the liquid flows with velocity u in the negative direction of x-axis in presence of fixed sphere, then we have to add a term

$$\psi = \frac{u}{2}r^2\sin^2\theta$$

to the value of ψ when the sphere moves with velocity u in a liquid at rest at infinity. Hence in this case

$$\psi = \frac{1}{2}ur^2\sin^2\theta - \frac{1}{2}\frac{ua^3}{r}\sin^2\theta$$

$$= \frac{1}{2}ur^2\sin^2\theta\left(1-\frac{a^3}{r^3}\right)$$

In this case $\phi = ur\cos\theta + \dfrac{ua^3}{2r^2}\cos\theta.$

5.14. Three-Dimensional Source and Sink

If the motion of a liquid consists of symmetrical outward radial flow in all directions proceeding from a point, then the point is called a three-dimensional source.

5.15. To Determine ϕ due to a Three-dimensional Source μ at the Origin

If u is the radial velocity at a distance r from the origin due to a source $+m$ at the origin, then by definition of strength,

$$4\Pi r^2 u = 4\Pi m$$

$$\Rightarrow \quad u = \frac{m}{r^2}$$

But

$$u = -\frac{\partial\phi}{\partial r} = \frac{m}{r^2}$$

Hence

$$\phi = \frac{m}{r}$$

Neglecting constant of integration.

5.16. Three-Dimensional Doublet

A three-dimensional doublet is a combination of three-dimensional sink of strength $-m$ and a source of strength $+m$ at a distance δs apart. Now suppose that $m \to \infty$, $\delta s \to 0$ in such a way that the product $m\delta s = \mu$ remains finite and constant. The quantity μ is called the strength of the doublet.

Velocity Potential due to a Doublet

Let AB be a doublet with a sink $-m$ at A and a source $+m$ at B. Let p be any point in fluid such that $AP = r$ and $BP = r + \delta r$.

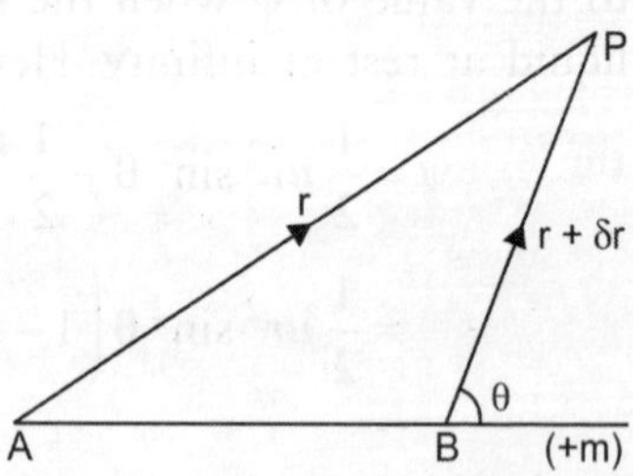

Fig. 5.6

The velocity potential ϕ due to the doublet at P is given by

$$\phi = -\frac{m}{r} + \frac{m}{r + \delta r}$$

$$= -\frac{m}{r}\left[1-\left(1+\frac{\delta r}{r}\right)^{-1}\right]$$

$$= -\frac{m}{r}\left[1-\left\{1-\frac{\partial r}{r}+\left(\frac{\partial r}{r}\right)^{2}.....\right\}\right]$$

$$= -\frac{m\delta r}{r^2} \text{ to first approximation}$$

$$= \frac{m\delta s \cos\theta}{r^2}$$

Where $AB = \delta s$, $\cos\theta = -\dfrac{dr}{ds}$

or $\phi = \dfrac{\mu\cos\theta}{r^2}$ where $\mu = m\delta s$

Let (r, θ, ψ) be the coordinates of the point preferred to the axis of the doublet as initial line. The velocity components (q_r, q_θ, q_ψ) at P are

$$q_r = -\frac{\partial\phi}{\partial r} = \frac{2\mu\cos\theta}{r^2}$$

$$q_\theta = -\frac{1}{r}\frac{\partial\phi}{\partial\theta} = \frac{\mu\sin\theta}{r^3}$$

$$q_\psi = -\frac{1}{r\sin\theta}\frac{\partial\phi}{\partial\psi} = 0$$

5.17. Value of Stoke's Stream Function in Simple Cases

1. To determine ψ for a single source of strength m placed at the origin

Let there be a three-dimensional source of strength m at the origin,

We have $\phi = \dfrac{m}{r}$

but $\dfrac{\partial\phi}{\partial r} = \dfrac{1}{r^2\sin\theta}\dfrac{\partial\psi}{\partial\theta}$

$$\Rightarrow \quad -\frac{m}{r^2} = \frac{1}{r^2\sin\theta}\frac{\partial\psi}{\partial\theta}$$

$$\therefore \quad \frac{\partial \psi}{\partial \theta} = -m \sin \theta$$

Integrating, we get $\psi = m \cos \theta$ or $\dfrac{mx}{r}$ neglecting constant of integration.

2. To determine ψ in case of a uniform line

Let OA be a uniform line source of strength m per unit length along z-axis. Let PN be the perpendicular on the z-axis from any point P and let $PN = p$, $ON = K$, $OB = z$.

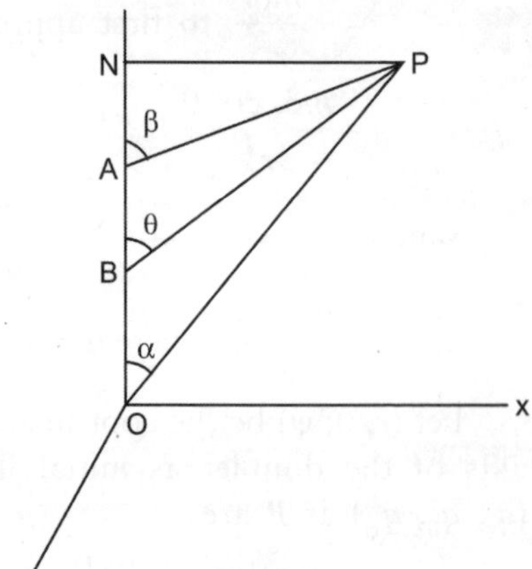

Fig. 5.7

Then
$$\begin{aligned} z &= OB \\ &= ON - NB \\ &= K - p \cot \theta \\ dz &= p \operatorname{cosec}^2\theta \, d\theta, \end{aligned}$$

Strength of the source in dz is mdz.

The value of ψ due to line source OA at P is

$$\psi = \int m \, dz \cos \theta \text{ as } \psi = m \cos \theta \text{ for source } +m$$

$$= \int_{\alpha}^{\beta} mp \operatorname{cosec}^2\theta \cos \theta \, d\theta$$

$$= mp \int_{\alpha}^{\beta} \cot \theta \operatorname{cosec} \theta \, d\theta$$

$$= mp \left[-\operatorname{cosec} \theta \right]_{\alpha}^{\beta}$$

$$= m \left[p \operatorname{cosec} \alpha - p \operatorname{cosec} \beta \right]$$

$$= m \, (OP - AP)$$

5.18. Image of a Source with Regard to a Sphere (I.A.S. 87)

Let O be the centre of the sphere and a be its radius. If Q be the inverse point of P, then

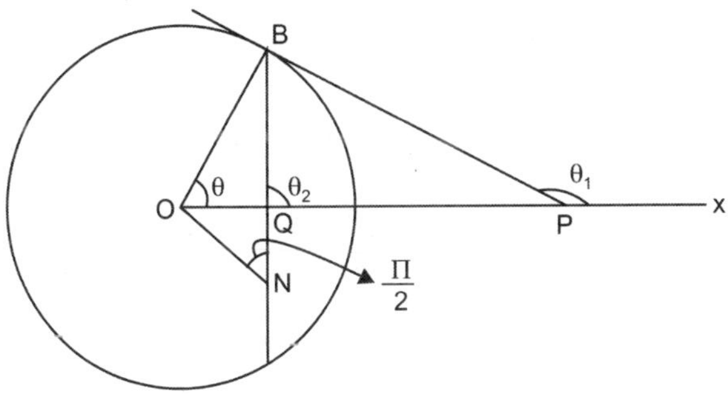

Fig. 5.8

$$OQ \cdot OP = a^2$$

$$\Rightarrow \quad OQ = \frac{a^2}{f}$$

Let there be any point B on the sphere. We draw a perpendicular ON on the produced part of BQ. Since

$$OQ \cdot OP = a^2 = OB^2$$

$$\Rightarrow \quad \frac{OQ}{OB} = \frac{OB}{OP}$$

Also

$$\angle QOB = \angle BOP$$

Hence Δs OQB and BPO are similar.

Hence $\angle OBQ = \angle OPB = \Pi - \theta_1$

In right angle ΔONB

$$\cos(\Pi - \theta_1) = \frac{BN}{OB}$$

$$\Rightarrow \quad BN = -a \cos \theta_1$$

$$BQ = BN - NQ = -a \cos \theta_1 - OQ \cos \theta_2$$

$$= -a \cos \theta_1 - \frac{a^2}{f} \cos \theta_2$$

We now consider the following distribution:

(i) a source of strength $+m$ at P.

(ii) a source of strength $\frac{ma}{f}$ at Q

(iii) a uniform line sink of strength $-\dfrac{m}{a}$ per unit length from O to Q.

The value of Stoke's stream function at B is given by

$$\psi_B = m\cos\theta_1 + \frac{ma}{f}\cos\theta_2 - \frac{m}{a}(OB - QB)$$

$$= m\cos\theta_1 + \frac{ma}{f}\cos\theta_2 + \frac{m}{a}[QB - a]$$

$$= m\cos\theta_1 + \frac{ma}{f}\cos\theta_2 + \frac{m}{a}\left[-a\cos\theta_1 - \frac{a^2}{f}\cos\theta_2 - a\right]$$

$$= m\cos\theta_1 + \frac{ma}{f}\cos\theta_2 - m\cos\theta_1 - \frac{ma}{f}\cos\theta_2 - m$$

$$= -m$$

$$= \text{constant}$$

Hence the image of a source of strength m at P with regard to a sphere is a source $\dfrac{ma}{f}$ at the inverse point Q and a line sink of strength $-\dfrac{m}{a}$ per unit length from O to Q.

5.19. Image of a Doublet Relative to a Sphere or Image of a Radial Doublet in a Sphere

We consider a doublet of strength μ at A where $OA = f$, O is the centre of the sphere of radius a.

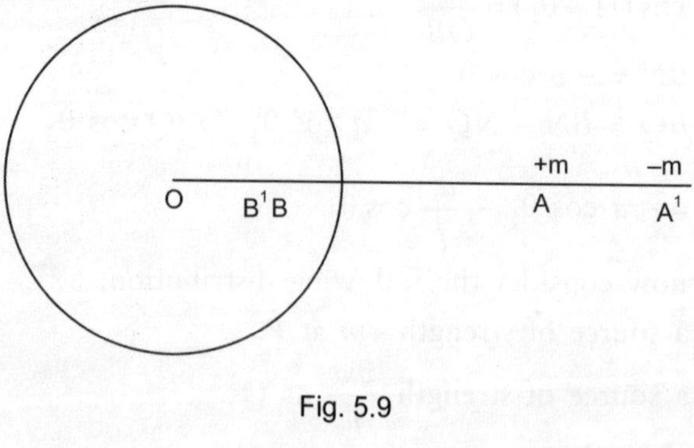

Fig. 5.9

We regard this doublet as a source $+m$ at A and a sink $-m$ at A^1.

Where $OA = f$, $OA^1 = f + \delta f$, $AA^1 = \delta f$ and $\mu = mAA^1 = m\delta f$ as $\delta f \to 0$.

The image of source $+m$ at A is a source $+\dfrac{ma}{f}$ at the inverse point B and line sink $-\dfrac{m}{a}$ per unit length from O to B.

The image of sink $-m$ at A^1 is a sink $-\dfrac{ma}{f+\delta f}$ at the inverse point B^1 and a line source of strength $\dfrac{m}{a}$ per unit length from O to B^1.

Now

$$-\frac{ma}{f+\delta f} = -\frac{ma}{f}\left(1+\frac{\delta f}{f}\right)^{-1}$$

$$= -\frac{ma}{f}\left(1-\frac{\delta f}{f}\right)$$

$$= -\frac{ma}{f} + \frac{ma\,\delta f}{f^2}$$

Compounding these, we get a doublet of strength $\dfrac{ma}{f}BB^1$, a source $\dfrac{ma}{f^2}\delta f$ and a line sink of strength $-\dfrac{m}{a}B^1B$, all ultimately at the inverse point.

Also since $OB = \dfrac{a^2}{f}$

$$BB^1 = OB - OB^1$$

$$= \frac{a^2}{f} - \frac{a^2}{f+\delta f}$$

$$= \frac{a^2}{f}\left[1-\left(1+\frac{\delta f}{f}\right)^{-1}\right]$$

$$= \frac{a^2}{f}\frac{\delta f}{f} \qquad \text{neglecting higher powers of } \frac{\delta f}{f}$$

Hence

$$-\frac{m}{a}B^1B = -\frac{m}{a}\cdot\frac{a^2}{f}\cdot\frac{\delta f}{f} = -\frac{ma}{f^2}\delta f$$

Thus the source and line sink B^1B cancel each other and there remains only a doublet of strength $\frac{ma}{f}B^1B = \frac{ma}{f}\cdot\frac{a^2}{f}\frac{\delta f}{f}$ $= \frac{ma^3}{f^3}\delta f$ at the inverse point in the opposite direction to the given doublet.

Thus the image of a doublet of strength μ with regard to a sphere is a doublet of strength μ^1 at the inverse point in a direction opposite to the given doublet.

Working rules for solving the problem on the motion of a sphere

1. For sphere $ds = (2\Pi\ a \sin\theta)\ (4d\theta)$ and θ varies from 0 to Π.

2. If the sphere is moving with velocity U in an infinite mass of liquid at rest at infinity, then apply the equation

$$\frac{p}{\rho} + \frac{1}{2}q^2 - \frac{\delta\phi}{\delta t} = C$$

At $r = \infty$, $p = p_0$, $q = O_1$

3. If liquid is moving with velocity U parallel to z-axis in presence of fixed sphere, then apply $\frac{p}{\rho} + \frac{1}{2}q^2 = C$ for determining pressure.

WORKED OUT EXAMPLES

1. An infinite ocean of an incompressible liquid of density ρ is streaming past a fixed spherical obstacles of radius a. The velocity is uniform and equal to U except in so far as it is disturbed by the sphere, and the pressure in the liquid at a great distance from the obstacle is Π. Show that the thrust on that half of the sphere on which the liquid impinges is $\Pi a^2\left[\Pi - \frac{\rho U^2}{16}\right]$.

Solution: If ϕ be the velocity potential of the liquid then

$$\phi = \frac{Ua^3}{2r^3}\cos\theta + Ur\cos\theta \qquad \text{... (1)}$$

Since the motion is steady and there are no extraneous forces, hence pressure equation is

$$\frac{p}{\rho} + \frac{1}{2}q^2 = C \qquad \text{... (2)}$$

Subject this to the condition $p = \Pi$, $q = U$, we get

$$C = \frac{\Pi}{\rho} + \frac{1}{2}U^2$$

Hence from (2), we have

$$\frac{p}{\rho} + \frac{1}{2}q^2 = \frac{\Pi}{\rho} + \frac{1}{2}U^2$$

$$\therefore \quad \frac{p}{\rho} - \Pi = \frac{1}{2}\left(U^2 - q^2\right)$$

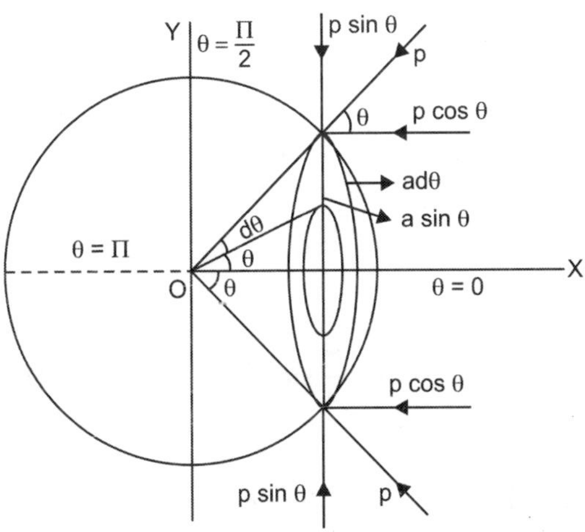

Fig. 5.10

$$\Rightarrow \quad p = \Pi + \frac{\rho}{2}\left(U^2 - q^2\right) \qquad \text{... (3)}$$

Let P be the required pressure on the half part of the sphere, then

$$P = \int_0^{\frac{\Pi}{2}} p \cos\theta \,(2\Pi\, a \sin\theta)\, a d\theta \qquad \text{... (4)}$$

From (1)

$$\left(\frac{\partial \phi}{\partial r}\right)_{r=a} = \left[U\left(1 - \frac{a^3}{r^3}\right)\cos\theta\right]_{r=a} = 0$$

and

$$\left(\frac{1}{r}\frac{\partial \phi}{\partial \theta}\right)_{r=a} = \left[\frac{U}{r}\left(r + \frac{a^3}{2r^2}\right)(-\sin\theta)\right]_{r=a}$$

$$= -\frac{3}{2} U \sin\theta$$

Let q be the velocity at any point of the boundary of the sphere $r = a$. Then we have

$$q^2 = \left[\left(-\frac{\partial \phi}{\partial r}\right)^2_{r=a} + \left(-\frac{1}{r}\frac{\partial \phi}{\partial \theta}\right)^2_{r=a}\right]$$

$$= \frac{9}{4} U^2 \sin^2\theta$$

Hence from (4), we have

$$P = \int_0^{\frac{\Pi}{2}} \left[\Pi + \frac{\rho}{2}\left(U^2 - q^2\right)\right] \cos\theta 2\Pi a^2 \sin\theta d\theta$$

$$= 2\Pi a^2 \int_0^{\frac{\Pi}{2}} \left[\Pi + \frac{\rho}{2}\left(1 - \frac{9}{4}\sin^2\theta\right)U^2\right] \cos\theta \times 2\Pi a^2 \sin\theta d\theta$$

$$= 2\Pi a^2 \int_0^{\frac{\Pi}{2}} \left[\left(\Pi + \frac{\rho}{2}U^2\right)\sin\theta d(\sin\theta) - \frac{9\rho}{8} U^2 \sin^3\theta d(\sin\theta)\right]$$

$$\text{or} \quad P = 2\Pi a^2\left[\left(\Pi + \frac{\rho U^2}{2}\right)\left\{\frac{1}{2}\sin^2\theta\right\}_0^{\frac{\Pi}{2}} - \frac{9}{8}\rho U^2\left\{\frac{1}{4}\sin^4\theta\right\}_0^{\frac{\Pi}{2}}\right]$$

$$= 2\Pi a^2\left[\left(\Pi + \frac{\rho U^2}{2}\right)\frac{1}{2} - \frac{9}{8}\rho U^2\frac{1}{4}\right]$$

$$= \Pi a^2\left[\Pi + \frac{\rho U^2}{2} - \frac{9U^2}{16}\right]$$

$$= \Pi a^2\left[\Pi - \frac{1}{16}\rho U^2\right]$$

2. A sphere of radius a is made to move in incompressible perfect fluid with non-uniform velocity u along x-axis. If the pressure at infinity is zero, prove that at a point x in advance of the centre.

$$p = \frac{1}{2}\rho a^3\left[\frac{U}{x^2} + U^2\left(\frac{2}{x^3} - \frac{a^3}{x^6}\right)\right]$$

Solution: Let (x_0, O, O) be the centre of the sphere.

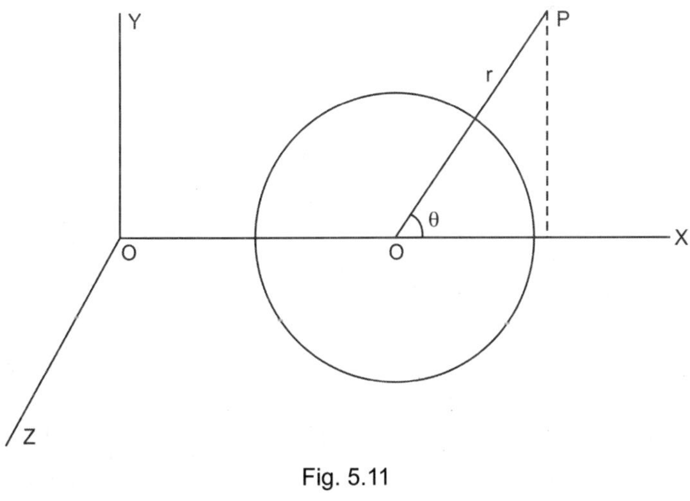

Fig. 5.11

When a sphere of radius a moves along x-axis with velocity u, then velocity potential at P is given by

$$\phi = \frac{1}{2} \frac{Ua^3}{r^2} \cos\theta \qquad \dots (1)$$

Where $O^1P = r$, $\angle PO^1X = \theta$.

$$r^2 = (x - x_0)^2 + y^2 + z^2 \qquad \dots (2)$$

and

$$\dot{x}_0 = U$$

Hence (1) becomes

$$\phi = \frac{1}{2} \frac{Ua^3(x - x_0)}{r^3}$$

Now

$$q^2 = \left(\frac{\partial\phi}{\partial r}\right)^2 + \left(\frac{1}{r}\frac{\partial\phi}{\partial\theta}\right)^2$$

$$= \left(\frac{1}{2} Ua^3\right)^2 \left[\left(\frac{2}{r^3}\cos\theta\right)^2 + \left(\frac{\sin\theta}{r^3}\right)^2\right]$$

$$= \frac{1}{4} U^2 \frac{a^6}{r^6} \left(4\cos^2\theta + \sin^2\theta\right)$$

From (2)

$$2r\dot{r} = 2(x - x_0)(-\dot{x}_0)$$

$$\Rightarrow \quad r = -\frac{U(x - x_0)}{r} = -U\cos\theta$$

Hence

$$\frac{\partial\phi}{\partial t} = \frac{1}{2}\frac{\dot{U}a^3}{r^3}(x - x_0) - \frac{3}{2} U \frac{a^3}{r^4} \dot{r}(x - x_0) + \frac{1}{2} U \frac{a^3}{r^3}(-\dot{x}_0)$$

$$= \frac{1}{2}\dot{U}\frac{a^3}{r^2}\cos\theta - \frac{3}{2}\frac{Ua^3}{r^3}(-U\cos\theta)\cos\theta - \frac{1}{2}\frac{U^2a^3}{r^3}$$

$$= \frac{1}{2}\dot{U}\frac{a^3}{r^2}\cos\theta + \frac{3}{2}\frac{U^2a^3}{r^3}\cos^2\theta - \frac{1}{2}\frac{U^2a^3}{r^3}$$

By pressure equation

$$\frac{p}{\rho}+\frac{1}{2}q^2-\frac{\partial\phi}{\partial t}=C$$

$$\Rightarrow \frac{p}{q}+\frac{1}{8}\frac{U^2a^6}{r^6}\left(4\cos^2\theta+\sin^2\theta\right)-\frac{1}{2}\frac{Ua^3}{r^2}\cos\theta$$

$$-\frac{3}{2}\frac{U^2a^3}{r^3}\cos^2\theta+\frac{1}{2}\frac{U^2a^3}{r^3}=C \qquad \dots (3)$$

Subjecting this to the condition $p = 0$ at $r = \infty$, hence from (3), we get $C = 0$.

Therefore,

$$\frac{p}{\rho}+\frac{1}{8}\frac{U^2a^6}{r^6}\left(4\cos^2\theta+\sin^2\theta\right)-\frac{1}{2}\frac{\dot{U}a^3}{r^2}\cos\theta$$

$$-\frac{3}{2}\frac{U^2a^3}{r^3}\cos^2\theta+\frac{1}{2}\frac{U^2a^3}{r^3}=0$$

Pressure at a distance x in advance to the centre is obtained by putting $\theta = 0$ and $r = x$ in the above equation so that

$$\frac{p}{\rho}+\frac{1}{2}\frac{U^2a^6}{x^6}-\frac{1}{2}\frac{Ua^3}{x^2}-\frac{3}{2}\frac{U^2a^3}{x^3}+\frac{1}{2}\frac{U^2a^3}{x^3}=0$$

$$\Rightarrow \frac{p}{\rho}=\frac{1}{2}a^3\left[\frac{\dot{U}}{x^2}+U^2\left(\frac{2}{x^3}-\frac{a^3}{x^6}\right)\right]$$

$$\Rightarrow p=\frac{\rho}{2}a^3\left[\frac{\dot{U}}{x^2}+U^2\left(\frac{2}{x^3}-\frac{a^3}{x^6}\right)\right]$$

3. Prove that at a point on a sphere moving through an infinite liquid the pressure is given by the formula

$$\frac{p-p_0}{\rho}=\frac{1}{2}af\cos\theta_1+\frac{1}{8}V^2\left(9\cos^2\theta-5\right)$$

Where V is the velocity, f the acceleration of the sphere and θ, θ_1 are the angles between the radius and directions of v, f respectively and p_0 is the hydrostatic pressure. (I.A.S. 87)

Solution: Let U, V, W be the components of the velocity v, (x_0, y_0, z_0) is the centre of the sphere and (x, y, z) be any point in the liquid such that

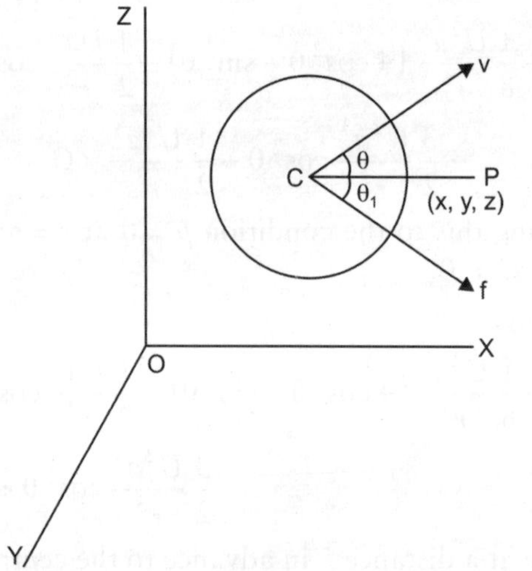

Fig. 5.12

$$r^2 = (x - x_0)^2 + (y - y_0)^2 + (z - z_0)^2$$
$$= CP^2 \quad \text{... (1)}$$

Let

$$v = U\vec{i} + V\vec{j} + W\vec{k}$$
$$= \dot{x}_0\vec{i} + \dot{y}_0\vec{j} + \dot{z}_0\vec{k}$$

and

$$v^2 = U^2 + V^2 + W^2 = \sum U^2$$

and

$$v \cos\theta = \frac{U(x - x_0)}{r} + \frac{V(y - y_0)}{r} + \frac{W(z - z_0)}{r} \quad \text{... (2)}$$

and

$$f \cos\theta_1 = \dot{U}\frac{(x - x_0)}{r} + \dot{V}\frac{(y - y_0)}{r} + \dot{W}\frac{(z - z_0)}{r} \quad \text{... (3)}$$

Since the direction cosine of CP are $\frac{x - x_0}{r}, \frac{y - y_0}{r}, \frac{z - z_0}{r}$

Also velocity potential ϕ at $P(x, y, z)$ is given by

$$\phi = \frac{a^3 v \cos\theta}{2r^2}$$

$$\Rightarrow \quad \phi = \frac{a^3}{2r^2}\left[\frac{U(x - x_0)}{r} + \frac{V(y - y_0)}{r} + \frac{W(z - z_0)}{r}\right]$$

$$\Rightarrow \quad \phi = \frac{a^3}{2r^3}\sum U(x - x_0) \qquad \dots (4)$$

$$\therefore \quad \frac{\partial\phi}{\partial t} = -\frac{3a^3}{2r^4}\dot{r}\sum U(x - x_0) + \frac{a^3}{2r^3}\sum \dot{U}(x - x_0) + \frac{a^3}{2r^3}\sum U(0 - \dot{x}_0) \qquad \dots (5)$$

From (1)

$$2r\dot{r} = \sum 2(x - x_0)(-\dot{x}_0)$$

$$\therefore \quad \dot{r} = -\frac{\sum U(x - x_0)}{r}$$

$$\Rightarrow \quad \dot{r} = -v\cos\theta$$

Hence from (5)

$$\frac{\partial\phi}{\partial t} = -\frac{3a^3}{2r^4}(-v\cos\theta)(rv\cos\theta) + \frac{a^3}{2r^3}fr\cos\theta_1 - \frac{a^3}{2r^3}\sum U^2$$

$$\therefore \quad \left(\frac{\partial\phi}{\partial t}\right)_{r=a} = \frac{3}{2}v^2\cos^2\theta + \frac{a}{2}f\cos\theta_1 - \frac{1}{2}v^2 \qquad \dots (6)$$

From (4)

$$\frac{\partial\phi}{\partial x} = \frac{-3a^3}{2r^4}\left(\frac{x - x_0}{r}\right)\sum U(x - x_0) + \frac{a^3}{2r^3}U$$

$$= -\frac{3a^3}{2r^4}(x - x_0)\,v\cos\theta + \frac{a^3 U}{2r^3}$$

$$\therefore \quad \left(\frac{\partial\phi}{\partial x}\right)_{r=a} = -\frac{3}{2a}(x - x_0)\,v\cos\theta + \frac{1}{2}U$$

Similarly

$$\left(\frac{\partial \phi}{\partial y}\right)_{r=a} = \frac{-3}{2a}(y - y_0)v \cos\theta + \frac{V}{2}$$

$$\left(\frac{\partial \phi}{\partial z}\right)_{r=a} = \frac{-3}{2a}(z - z_0)v \cos\theta + \frac{W}{2}$$

Now

$$\begin{aligned}(q^2)_{r=a} &= \sum\left(\frac{\partial \phi}{\partial x}\right)^2_{r=a} \\ &= \frac{U^2 + V^2 + W^2}{4} + \frac{9v^2 \cos^2\theta}{4a^2}\Big[(x - x_0)^2 \\ &\quad + (y - y_0)^2 + (z - z_0)^2\Big] - \frac{3v\cos\theta}{2a} \\ &\quad \big[U(x - x_0) + V(y - y_0) + W(z - z_0)\big] \\ &= \frac{U^2 + V^2 + W^2}{4} + \frac{9v^2\cos^2\theta}{4} - \frac{3v^2\cos^2\theta}{2} \\ &= \frac{U^2 + V^2 + W^2}{4} + \frac{3}{4}v^2\cos^2\theta \\ &= \frac{1}{4}v^2 + \frac{3}{4}v^2\cos^2\theta\end{aligned}$$

Also pressure equation is

$$\frac{p}{\rho} - \frac{\partial \phi}{\partial t} + \frac{1}{2}q^2 = C$$

At $r = \infty$, $p = p_0$, $q = 0$, $\frac{\partial \phi}{\partial t} = 0$

Hence $C = \frac{p_0}{\rho}$

Thus

$$\frac{p}{\rho} + \frac{1}{2}q^2 - \frac{\partial \phi}{\partial t} = \frac{p_0}{\rho}$$

$$\Rightarrow \quad \frac{p - p_0}{\rho} = \frac{\partial \phi}{\partial t} - \frac{1}{2}q^2$$

At $r = a$

$$\frac{p - p_0}{\rho} = \frac{3}{2}v^2 \cos^2 \theta + \frac{a}{2} f \cos \theta_1 - \frac{1}{2}v^2 - \frac{1}{2}\left(\frac{1}{4}v^2 + \frac{3}{4}v^2 \cos^2 \theta\right)$$

$$= \frac{1}{8}\left(9\cos^2 \theta - 5\right) v^2 + \frac{a}{2} f \cos \theta_1$$

4. Liquid of density ρ fills the space between solid sphere of radius a and density ρ^1 and fixed concentric spherical envelope of radius b. Prove that the work done by an impulse which starts the solid sphere with velocity V is

$$\frac{1}{3}\Pi a^3 V^2 \left[2\rho^1 + \left(\frac{2a^3 + b^3}{b^3 - a^3}\right)\rho\right] \qquad \text{(I.A.S. 83)}$$

Solution:

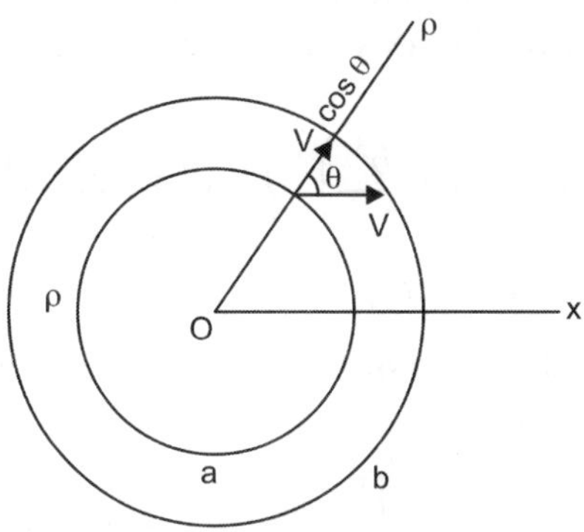

Fig. 5.13

We know that the impulse I necessary to start the solid sphere with velocity U is given by

$$I = MU + \frac{1}{2} M^1 U\left(\frac{2a^3 + b^3}{b^3 + a^3}\right) \qquad \dots (1)$$

Where M = mass of the inner sphere $= \frac{4}{3}\Pi a^3 \rho^1 U$

and M^1 = mass of the liquid displaced $= \frac{4}{3}\Pi a^3 \rho^1 V$

Hence from (1)

$$I = \frac{2\Pi a^3 U}{3}\left[2\rho^1 + \rho\left(\frac{2a^3 + b^3}{b^3 + a^3}\right)\right]$$

Therefore work done by the impulse

$$= \text{impulse} \times \text{mean of initial and final velocities}$$

$$= I \times \frac{O + U}{2}$$

$$= \frac{1}{2} IU$$

$$= \frac{1}{3}\Pi a^3 U^2\left[2\rho^1 + \frac{2a^3 + b^3}{b^3 - a^3}\rho\right]$$

5. Prove that for liquid contained between two instantaneously concentric spheres, when the outer (radius a) is moving parallel to the axis of x with velocity u and the inner (radius b) is moving parallel to the axis of y with velocity v, the velocity potential is

$$-\frac{1}{a^3 - b^3}\left[a^3 ux\left(1 + \frac{b^3}{2r^3}\right) - b^3 vy\left(1 + \frac{a^3}{2r^3}\right)\right]$$

and find the kinetic energy.

Solution: In the direction of OP, the resolved components of v, u are $v \sin\theta$, $u\cos\theta$.

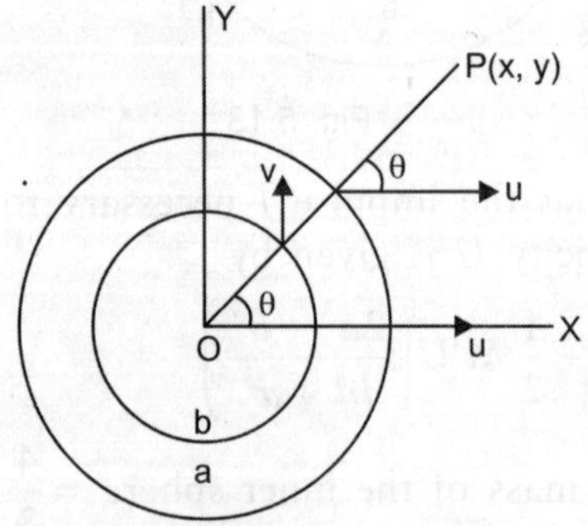

Fig. 5.14

Boundary conditions for ϕ are

(i) $\vec{\nabla}^2 \phi = 0$

(ii) $\left(\dfrac{-\partial\phi}{\partial r}\right)_{r=\infty} = 0 = \left(-\dfrac{1}{r}\dfrac{\partial\phi}{\partial\theta}\right)_{r=\infty}$

(iii) $\left(\dfrac{-\partial\phi}{\partial r}\right)_{r=a} = U\cos\theta$

(iv) $\left(\dfrac{-\partial\phi}{\partial r}\right)_{r=b} = v\sin\theta$

Let us assume

$$\phi = \left(Ar + \frac{B}{r^2}\right)\cos\theta + \left(Cr + \frac{D}{r^2}\right)\sin\theta$$

Hence

$$\frac{\partial\phi}{\partial r} = \left(A - \frac{2B}{r^3}\right)\cos\theta + \left(C - \frac{2D}{r^3}\right)\sin\theta$$

Subjecting this to (iii) and (iv), we get

$$\left(A - \frac{2B}{a^3}\right)\cos\theta + \left(C - \frac{2D}{a^3}\right)\sin\theta = -u\cos\theta$$

and

$$\left(A - \frac{2B}{b^3}\right)\cos\theta + \left(C - \frac{2D}{b^3}\right)\sin\theta = -v\sin\theta$$

Therefore

$$A - \frac{2B}{a^3} = -u \qquad \text{... (1)}$$

$$C - \frac{2D}{a^3} = 0 \qquad \text{... (2)}$$

and

$$A - \frac{2B}{b^3} = 0 \qquad \text{... (3)}$$

$$C - \frac{2D}{b^3} = 0 \qquad \text{... (4)}$$

Solving (1) and (3), we get

$$B = \frac{-Ua^3b^3}{2(a^3 - b^3)}, A = \frac{-Ua^3}{a^3 - b^3}$$

Also solving (2) and (4), we have

$$D = \frac{-va^3b^3}{2(b^3 - a^3)}, C = \frac{-vb^3}{b^3 - a^3}$$

Hence

$$\phi = -\frac{a^3u}{a^3 - b^3}\left(r + \frac{b^3}{2r^2}\right)\cos\theta - \frac{vb^3}{b^3 - a^3}\left(r + \frac{a^3}{2r^2}\right)\sin\theta$$

$$\Rightarrow \quad \phi = \frac{1}{a^3 - b^3}\left[-a^3u\left(r + \frac{b^3}{2r^2}\right)\cos\theta + vb^3\left(r + \frac{a^3}{2r^2}\right)\sin\theta\right] \quad \dots (5)$$

$$\Rightarrow \quad \phi = -\frac{1}{a^3 - b^3}\left[xa^3u\left(1 + \frac{b^3}{2r^3}\right) - vb^3y\left(1 + \frac{a^3}{2r^3}\right)\right]$$

Now we determine the kinetic energy. Let T be the kinetic energy of the liquid between the two spheres.

Hence

$$T = -\frac{1}{2}\rho \iint \phi \frac{\partial\phi}{\partial n}\, ds$$

Where the surface integral is carried over the whole boundary of the liquid and $-\frac{\partial\phi}{\partial n}$ denotes the outward normal velocity.

$$\Rightarrow \quad T = \frac{1}{2}\rho \iint \left(\phi \frac{\partial\phi}{\partial r}\right)_{r=a} ds - \frac{1}{2}\rho \iint \left(\phi \frac{\partial\phi}{\partial r}\right)_{r=b} ds$$

$$= \frac{1}{2}\rho \int_0^{\Pi} \left(\phi \frac{\partial\phi}{\partial r}\right)_{r=a} 2\Pi a \sin\theta a d\theta - \frac{1}{2}\rho \int_0^{\Pi} \left(\phi \frac{\partial\phi}{\partial r}\right)_{r=b} 2\Pi b \sin\theta b d\theta$$

$$= \Pi\rho a^2 \int_0^{\Pi} (\phi)_{r=a} (-u\cos\theta)\sin\theta\, d\theta$$

$$+ \Pi\rho b^2 \int_0^{\Pi} (\phi)_{r=b}\, v\sin^2\theta d\theta$$

[From (iii) and (iv)]

$$= -\Pi a^2 u\rho \int_0^{\Pi} (\phi)_{r=a} \sin\theta\cos\theta d\theta$$

$$+ \Pi b^2 v\rho \int_0^{\Pi} (\phi)_{r=b} \sin^2\theta d\theta$$

Using (5), we get

$$T = -\frac{\Pi a^2 u\rho}{a^3 - b^3} \int_0^{\Pi} \left[-a^3 u \left(a + \frac{b^3}{2a^3} \right) \sin\theta\cos^2\theta \right.$$

$$\left. + vb^3 \left(a + \frac{a}{2} \right) \sin^2\theta\cos\theta \right] d\theta$$

$$+ \frac{\Pi b^2 u\rho}{a^3 - b^3} \int_0^{\Pi} \left[-a^3 u \left(b + \frac{b}{2} \right) \sin^2\theta\cos\theta \right.$$

$$\left. + vb^3 \left(b + \frac{a^3}{2b^2} \right) \sin^3\theta \right] d\theta$$

Now

$$\int_0^{\Pi} \sin\theta\cos^2\theta d\theta = -\int_0^{\Pi} \cos^2\theta d(\cos\theta)$$

$$= -\left[\frac{\cos^3\theta}{3} \right]_0^{\Pi} = \frac{2}{3}$$

Also

$$\int_0^{\Pi} \sin^2\theta\cos\theta d\theta = \int_0^{\Pi} \sin^2\theta d(\sin\theta) = \left[\frac{\sin^3\theta}{3} \right]_0^{\Pi} = 0$$

and

$$\int_0^{\Pi} \sin^3\theta = 2\int_0^{\frac{\Pi}{2}} \sin^3\theta d\theta = 2 \cdot \frac{3-1}{3} = \frac{4}{3}$$

With these values, T becomes

$$T = -\frac{\Pi a^2 u\rho}{a^3 - b^3}\left[-a^3 u\left(\frac{b^3 + 2a^3}{2a^2}\right)\frac{2}{3} + 0\right]$$

$$+\frac{\Pi b^2 v\rho}{a^3 - b^3}\left[0 + vb^2\left(\frac{2b^3 + a^3}{2b^2}\right)\frac{4}{3}\right]$$

$$= \frac{\Pi\rho}{3(a^3 - b^3)}\left[u^2 a^3\left(b^3 + 2a^3\right) + 2v^2 b^3\left(2b^3 + a^3\right)\right]$$

$$= \frac{\Pi\rho}{3(a^3 - b^3)}\left[a^3 b^3\left(u^2 + 2v^2\right) + 2\left(u^2 a^6 + u^2 b^6\right)\right]$$

6. The space between two concentric spherical shells of radii a and b $(a > b)$ is filled with an incompressible fluid of density ρ and the shells suddenly begin to move with velocities U, V in the same direction. Prove that the resultant impulsive pressure on the inner shell is

$$\frac{2\Pi\rho b^3}{3(a^3 - b^3)}\left[3\dot{a}^2 u - \left(a^3 + 2b^3\right)V\right]$$

Further show that the kinetic energy of the liquid is

$$\frac{\Pi\rho}{3(a^3 - b^3)}\left[a^3 b^3 (V - U)^2 + 2\left(b^3 V - a^3 U\right)\right]$$

Solution: Let us assume that

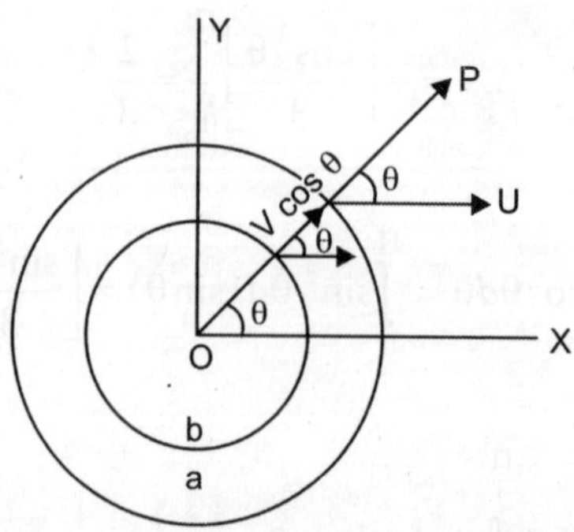

Fig. 5.15

$$\phi = \left(Ar + \frac{B}{r^2}\right)\cos\theta$$

Boundary conditions are

(i) $\vec{\nabla}^2\phi = 0$

(ii) $\left(\dfrac{-\partial\phi}{\partial r}\right)_{r=\infty} = 0 = \left(-\dfrac{1}{r}\dfrac{\partial\phi}{\partial\theta}\right)_{r=\infty}$

(iii) $\left(\dfrac{-\partial\phi}{\partial r}\right)_{r=a} = U\cos\theta$

(iv) $\left(\dfrac{-\partial\phi}{\partial r}\right)_{r=b} = V\cos\theta$

Since resolved components of U and V in the direction of OP are $U\cos\theta$ and $V\cos\theta$.

From (1)

$$\frac{\partial\phi}{\partial r} = \left(A - \frac{2B}{r^3}\right)\cos\theta$$

Hence from (iii) and (iv), we have

$$\left(A - \frac{2B}{a^3}\right)\cos\theta = -U\cos\theta$$

and $\left(A - \dfrac{2B}{b^3}\right)\cos\theta = -V\cos\theta$

$$\Rightarrow \quad A - \frac{2B}{a^3} = -U \qquad \text{... (1)}$$

and $A - \dfrac{2B}{b^3} = -V$ ... (2)

Solving these two, we get

$$A = \frac{Vb^3 - Ua^3}{a^3 - b^3},\ B = \frac{(V-U)a^3b^3}{2\left(a^3 - b^3\right)}$$

Hence

$$\phi = \frac{1}{\left(a^3 - b^3\right)}\left[\left(Vb^3 - Ua^3\right)r + (V-U)\frac{a^3b^3}{2r^2}\right]\cos\theta \qquad \text{... (3)}$$

If p be the impulsive pressure at any point (b, θ) of the inner cylinder, then

$p = (\rho\phi)_{r=b}$ so that

$$= \frac{\rho b}{a^3 - b^3}\left[Vb^3 - Ua^3 + (V - U)\frac{a^3}{2}\right]\cos\theta$$

$$= \frac{\rho b}{2(a^3 - b^3)}\left[V(2b^3 + a^3) - 3Ua^3\right]\cos\theta$$

The total impulsive pressure on the cylinder is

$$\int p\cos\theta ds = \int_0^{\Pi}\frac{\rho b}{2(a^3 - b^3)}\left[V(2b^3 + a^3) - 3Ua^3\right]\cos^2\theta \times 2\Pi b\sin\theta b d\theta$$

$$= \frac{\rho b 2\Pi b^2}{2(a^3 - b^3)}\left[V(2b^3 + a^3) - 3Ua^3\right]\left[-\frac{1}{3}\cos^3\theta\right]_0^{\Pi}$$

$$= -\frac{2}{3}\Pi\rho b^3\frac{1}{a^3 - b^3}\left[3Ua^3 - V(2b^3 + a^3)\right]$$

Neglecting negative sign, we get the required magnitude of impulsive pressure.

If T be the kinetic energy of the liquid, then

$$T = \frac{1}{2}\rho\int_{\theta=0}^{\Pi}\int_{r=b}^{a} q^2(2\Pi r\sin\theta)\, r d\theta dr$$

$$= \Pi\rho\int_0^{\Pi}\int_b^{a} r^2 q^2\sin\theta d\theta dr \qquad \ldots (4)$$

Also

$$q^2 = \left(\frac{-\partial\phi}{\partial r}\right)^2 + \left(-\frac{1}{r}\frac{\partial\phi}{\partial\theta}\right)^2$$

$$= \frac{1}{(a^3 - b^3)^2}\left[\left\{b^3V - a^3U - \frac{a^3b^3(V - U)}{r^3}\right\}^2\cos^2\theta + \left\{b^3V - a^3U + \frac{a^3b^3(V - U)}{2r^3}\right\}^2\sin^2\theta\right]$$

$$= \frac{1}{\left(a^3 - b^3\right)^2}\left[\left(Vb^3 - Ua^3\right)^2 + \frac{(V-U)^2 a^6 b^6}{4r^6}\left(3\cos^2\theta + 1\right)\right]$$

$$+ \frac{Vb^3 - Ua^3}{r^3\left(a^3 - b^3\right)^2}(V-U)a^3b^3\left(\sin^2\theta - 2\cos^2\theta\right)$$

Now

$$\int_0^{\Pi} \sin\theta d\theta = 2, \quad \int_0^{\Pi} \cos^2\theta \sin\theta d\theta = \frac{2}{3}$$

$$\int_0^{\Pi}\left(\sin^2\theta - 2\cos^2\theta\right)\sin\theta d\theta = \int_0^{\Pi}\left(1 - 3\cos^2\theta\right)\sin\theta d\theta$$

$$= 2 - 3\frac{2}{3} = 0$$

Hence

$$r^2\int_0^{\Pi} q^2 \sin\theta d\theta = \frac{r^2}{\left(a^3 - b^3\right)^2}\left[\left(Vb^3 - Ua^3\right)^2 2 + \frac{(V-U)^2 b^6 a^6}{4r^6}\left(2 + \frac{32}{3}\right)\right]$$

Therefore from (3), we have

$$T = \frac{\Pi\rho}{\left(a^3 - b^3\right)^2}\left[2\left(Vb^3 - Ua^3\right)^2 \frac{\left(a^3 - b^3\right)}{3} + \frac{(V-U)^2 a^6 b^6}{(-3)}\left(\frac{1}{a^3} - \frac{1}{b^3}\right)\right]$$

$$= \frac{\Pi\rho}{3\left(a^3 - b^3\right)}\left[2\left(Vb^3 - Ua^3\right) + (V-U)^2 a^3 b^3\right]$$

7. A stream of water of greater depth is flowing with a uniform velocity V over a plane level bottom. A hemisphere of weight w in water and of radius a rests with its base on the bottom. Prove that the average pressure between the base of the hemisphere and the bottom is less than the fluid pressure at any point of the bottom at a great distance from the hemisphere if

$$V^2 > \frac{32w}{11\Pi a^2 \rho}$$

Solution: Let water be flowing past a fixed hemisphere with velocity V along z-axis and (r, θ, t) be the spherical polar coordinates of a point referred to the centre of the hemisphere as the origin.

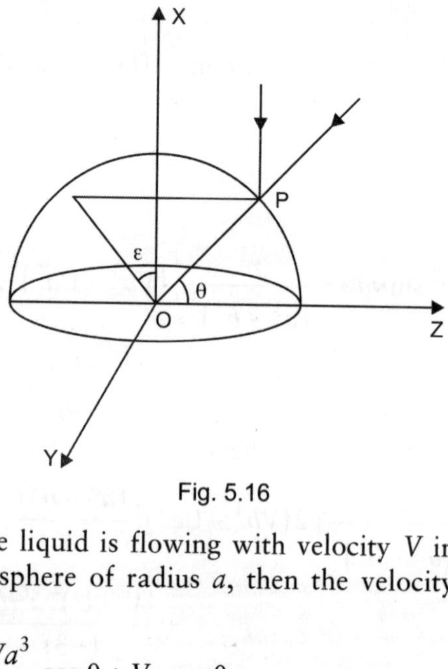

Fig. 5.16

When the liquid is flowing with velocity V in presence of a fixed hemisphere of radius a, then the velocity potential is given by

$$\phi = \frac{Va^3}{2r^2}\cos\theta + Vr\cos\theta$$

$$\therefore \quad \frac{\partial\phi}{\partial r} = -\frac{Va^3}{r^3}\cos\theta + V\cos\theta$$

and

$$\frac{1}{r}\frac{\partial\phi}{\partial\theta} = -\left(\frac{Va^3}{2r^3} + V\right)\sin\theta$$

$$\therefore \quad \left(q^2\right)_{r=a} = \left(\frac{-\partial\phi}{\partial r}\right)^2_{r=a} + \left(-\frac{1}{r}\frac{\partial\phi}{\partial\theta}\right)^2_{r=a}$$

$$= 0 + \frac{qV^2}{4}\sin^2\theta$$

Since the motion is steady, $\phi = 0$ and the pressure equation is

$$\frac{p}{\rho} + \frac{1}{2}q^2 = \text{const.} = C \qquad \text{... (1)}$$

When $q = V$, $p = p_0$ (say), $r = \infty$

Hence from (1)

$$\frac{p_0}{\rho} + \frac{1}{2}V^2 = C$$

Hence (1) becomes

$$\frac{p}{\rho} + \frac{1}{2}q^2 = \frac{p_0}{\rho} + \frac{1}{2}V^2$$

Hence

$$p = p_0 + \frac{\rho}{2}\left(V^2 - q^2\right)$$

$$= p_0 + \frac{1}{2}\rho V^2\left(1 - \frac{9}{4}\sin^2\theta\right) \qquad \text{... (2)}$$

Let l, m, n be the direction cosines of the normal to the surface at any point of the surface of the hemisphere.

Then

$$l = \frac{a\sin\theta\cos t}{a}, m = \frac{a\sin\theta\sin t}{a}, n = \frac{a\cos\theta}{a}$$

Let us consider an element $ad\theta a \sin\theta\, dt$ at P.

Hence downward pressure on the hemisphere i.e. along y-axis is

$$= \int_{\theta=0}^{\pi}\int_{t=0}^{\pi} pm\left(ad\theta a\sin\theta dt\right)$$

From (2), it becomes

$$= a^2 \int_0^{\pi}\int_0^{\pi}\left[p_0 + \frac{V^2}{2} + \left(1 - \frac{9}{4}\sin^2\theta\right)\right]\sin\theta\sin\theta\sin t d\theta dt$$

$$= 2a^2 2\int_0^{\frac{\pi}{2}}\left[p_0 + \frac{V^2}{2} + \left(1 - \frac{9}{4}\right)\sin^2\theta\right]\sin^2\theta d\theta$$

$$= 4a^2\left[p_0\frac{\pi}{4} + \frac{V^2\rho}{2}\frac{\pi}{4} - \frac{9}{4}\rho V^2\frac{3\pi}{16}\right]$$

$$= \pi a^2\left[p_0 - \frac{11}{32}\rho V^2\right]$$

Thus total force on the base is

$$= \pi a^2\left(p_0 - \frac{11}{32}\rho V^2\right) + w$$

Area of the base of the hemisphere $= \pi a^2$

Therefore, average pressure

$$= \left[\pi a^2\left(p_0 - \frac{11}{32}\rho V^2\right) + w\right]\frac{1}{\pi a^2}$$

This average pressure will be less than p_0 if

$$p_0 - \frac{11}{32}\rho V^2 + \frac{w}{\pi a^2} < p_0$$

or $$V^2 > \frac{32w}{11\pi a^2\rho}$$

8. A solid sphere is moving through frictionless liquid. Compare the velocities of slip of the liquid past it at different parts of its surface.

Prove that when the sphere is in motion with uniform velocity U, the pressure at the part of its surface where the radius makes angle θ with direction of motion is increased on account of the motion by the amount $\frac{\rho U^2}{16}(9\cos 2\theta - 1)$, where ρ is the density of the liquid.

Solution: Here the velocity potential

$$\phi = \frac{Ua^3}{2r^2}\cos\theta$$

$$\text{Velocity of slip } = \left(-\frac{1}{r}\frac{\partial\phi}{\partial\theta}\right)_{r=a} = \frac{1}{2}U\sin\theta$$

Giving different values to θ, we can compare the velocity of slip at different parts of its surface.

Let (x_0, y_0, z_0) be the coordinates of the centre of the sphere, so that

$$\dot{x}_0 = U;\ \cos\theta = \frac{x - x_0}{r}$$

Also $r^2 = (x - x_0)^2 + (y - y_0)^2 + (z - z_0)^2$

$$\therefore\quad 2r\dot{r} = 2(x - x_0)(-\dot{x}_0) = 2(x - x_0)(-U)$$

$$\Rightarrow\quad \dot{r} = -U\cos\theta$$

Also

$$q^2 = \left(\frac{-\partial\phi}{\partial r}\right)^2_{r=a} + \left(-\frac{1}{r}\frac{\partial\phi}{\partial\theta}\right)^2_{r=a}$$

$$= \left(\frac{Ua^3}{r^3}\cos\theta\right)^2_{r=a} + \left(\frac{Ua^3}{2r^3}\sin\theta\right)^2$$

$$= \frac{U^2}{4}\left(1 + 3\cos^2\theta\right)$$

Also

$$\phi = \frac{Ua^3\cos\theta}{2r^2} = \frac{Ua^3(x - x_0)}{2r^3}$$

$$\therefore\quad \frac{\partial\phi}{\partial t} = \frac{Ua^3}{2}\left[-\frac{3}{r^4}\dot{r}(x - x_0) - \frac{\dot{x}_0}{r^3}\right]$$

$$= \frac{Ua^3}{2r^3}\left[\frac{3U\cos\theta\cdot\cos\theta}{r} - U\right]$$

$$= \frac{U^2a^3}{2r^3}\left(3\cos^2\theta - 1\right)$$

$$\left(\frac{\partial \phi}{\partial t}\right)_{r=a} = \frac{U^2}{2}\left(3\cos^2\theta - 1\right)$$

Since the sphere is moving and so the motion is unsteady. By pressure equation

$$\frac{p}{\rho} + \frac{1}{2}q^2 - \frac{\partial \phi}{\partial t} = C \qquad \text{... (1)}$$

At $r = \infty$, $p = p_0$, $q = 0$, so that $\frac{\partial \phi}{\partial t} = 0$

$$C = \frac{p_0}{\rho}$$

Hence (1) becomes

$$\frac{p - p_0}{\rho} = \frac{\partial \phi}{\partial t} - \frac{1}{2}q^2$$

$$\frac{p - p_0}{\rho} = \frac{U^2}{2}\left(3\cos^2\theta - 1\right) - \frac{U^2}{8}\left(1 + 3\cos^2\theta\right)$$

$$= \frac{U^2}{8}\left[12\cos^2\theta - 4 - 1 - 3\cos^2\theta\right]$$

$$= \frac{U^2}{8}\left(9\cos^2\theta - 5\right)$$

$$= \frac{U^2}{16}\left[9\left(1 + \cos 2\theta\right) - 10\right]$$

or $$p - p_0 = \frac{\rho U^2}{16}\left(9\cos 2\theta - 1\right)$$

9. In compressible fluid, of density ρ, is contained between two rigid concentric spherical surfaces, the outer one of the mass M_1 and radius a, the inner one of mass M_2 and radius b. A normal blow P is given to the outer surface. Prove that the initial velocities of the two containing surfaces (U for the outer and V for the inner) are given by the equations

$$\left\{M_1 + \frac{2\Pi\rho a^3\left(b^3 + 2a^3\right)}{3\left(a^3 - b^3\right)}\right\}U - \frac{2\Pi\rho a^3 b^3}{a^3 - b^3}V = P$$

$$\left\{M_2 + \frac{2\pi\rho b^3\left(2b^3 + a^3\right)}{3\left(a^3 - b^3\right)}\right\} V = \frac{2\pi\rho a^3 b^3}{a^3 - b^3} U$$

Solution: We know that

$$\phi = \frac{1}{a^3 - b^3}\left[\left(Vb^3 - Ua^3\right) r + (V - U)\frac{a^3 b^3}{2r^2}\right]\cos\theta \quad \ldots (1)$$

Let P be the normal blow given to the outer sphere. Then

$$P = M_1 U - \iint_{r=a} \overline{w} \cos\theta \, ds \quad \ldots (2)$$

and

$$M_2 V = -\iint_{r=b} \overline{w} \cos\theta ds \quad \ldots (3)$$

where $\overline{w} = \rho\phi$ = impulsive pressure over an element δs of the surface.

From (1) and (2), we get

$$P = M_1 U - \int \frac{\rho}{a^3 - b^3}\left[\left(Vb^3 - Ua^3\right) a + (V - U)\frac{ab^3}{2}\right]$$

$$\times \cos^2\theta \left(2\pi\, a \sin\theta\right) a d\theta$$

$$= M_1 U - \frac{2\pi\rho a^3}{a^3 - b^3}\left[\left(Vb^3 - Ua^3\right) + \frac{(V - U)b^3}{2}\right]\frac{2}{3}$$

$$= \left[M_1 + \frac{2\pi\rho a^3}{3\left(a^3 - b^3\right)}\left(2a^3 + b^3\right)\right] U - \frac{2\pi\rho}{a^3 - b^3} \cdot a^3 b^3 V$$

Again from (1) and (3)

$$M_2 V = -\int_0^{\pi} \rho\phi \cos\theta ds$$

$$\Rightarrow \quad M_2 V = -\int_0^{\pi} \frac{1}{a^3 - b^3}\rho\left[\left(Vb^3 - Ua^3\right)b + (V - U)\frac{a^3 b}{2}\right]$$

$$\times \cos^2\theta \left(2\pi b \sin\theta\right) b d\theta$$

$$= \frac{-2\pi\rho b^3}{a^3 - b^3}\left[\left(Vb^3 - Ua^3\right) + (V - U)\frac{a^3}{2}\right]\int_0^\pi \cos^2\theta \sin\theta d\theta$$

$$= \frac{-2\pi b^3\rho}{a^3 - b^3}\left[\left(Vb^3 - Ua^3\right) + (V - U)\frac{a^3}{2}\right]\frac{2}{3}$$

$$= \frac{-2\pi b^3\rho}{a^3 - b^3}\left[\frac{V}{2}\left(2b^3 + a^3\right) - \frac{3Ua^3}{2}\right]\frac{2}{3}$$

Hence

$$\left[M_2 + \frac{2\pi b^3\rho\left(a^3 + 2b^3\right)}{3\left(a^3 - b^3\right)}\right]V = \frac{2\pi\rho}{a^3 - b^3}\, a^3b^3U$$

10. A spherical shell of internal radius a contains a concentric sphere of radius λa and density σ, the intervening space being filled with water and the whole system is at rest. If a velocity V is communicated to the shell, prove that the initial velocity U communicated to the sphere is given by

$$U = \frac{3V}{2\sigma\left(1 - \lambda^3\right) + 1 + 2\lambda^3}$$

Solution: The boundary conditions are

(i) $\vec{\nabla}^2\phi = 0$

(ii) $\left(\frac{-\partial\phi}{\partial r}\right)_{r=\infty} = \left(-\frac{1}{r}\frac{\partial\phi}{\partial\theta}\right)_{r=\infty} = 0$

(iii) $\left(\frac{-\partial\phi}{\partial r}\right)_{r=\lambda a} = U\cos\theta$

(iv) $\left(\frac{-\partial\phi}{\partial r}\right)_{r=a} = V\cos\theta$

Let us take

$$\phi = \left(Ar + \frac{B}{r^2}\right)\cos\theta \qquad \text{... (1)}$$

$$\therefore \quad \frac{\partial\phi}{\partial r} = \left(A - \frac{2B}{r^3}\right)\cos\theta$$

Hence

$$\left(A - \frac{2B}{\lambda^3 a^3}\right)\cos\theta = -U\cos\theta$$

and

$$\left(A - \frac{2B}{a^3}\right)\cos\theta = -V\cos\theta = 0$$

or $A - \dfrac{2B}{\lambda^3 a^3} = -U,\ A - \dfrac{2B}{a^3} = -V$

Solving these, we get

$$B = \frac{a^3\lambda^3(U-V)}{2(1-\lambda^3)},\ A = \frac{\lambda^3 U - V}{1-\lambda^3}$$

Hence from (1), we have

$$\phi = \frac{1}{1-\lambda^3}\left[(\lambda^3 U - V)r + \frac{a^3\lambda^3}{2r^2}(U-V)\right]\cos\theta$$

The impulsive pressure $\bar{w}$ at any point of the sphere $r = \lambda a$ is given by

$$\bar{w} = (\rho\phi)_{r=a\lambda}$$

$$= \frac{1}{1-\lambda^3}\left[(\lambda^3 U - V)\lambda a + \frac{a\lambda}{2}(U-V)\right]\cos\theta$$

Since density of water = 1

$$= \frac{\lambda a}{1-\lambda^3}\left[\lambda^3 U - V + \frac{U-V}{2}\right]\cos\theta$$

$$= \frac{\lambda a}{2(1-\lambda^3)}\left[(1+2\lambda^3)U - 3V\right]\cos\theta$$

Hence total impulsive pressure on the sphere

$$= \int_0^\pi \bar{w}\cos\theta\left[2\pi(\lambda a)\sin\theta\right](a\lambda)\,d\theta$$

$$= \frac{2\pi\lambda^2 a^2}{2(1-\lambda^3)}\left[(1+2\lambda^3)U - 3V\right]\int_0^\pi \sin\theta\cos^2\theta\,d\theta$$

$$= \frac{\pi\lambda^2 a^3}{1-\lambda^3}\left[(1+2\lambda^3)U - 3V\right]\frac{2}{3}$$

Since the sphere of density σ starts moving with velocity U, the equation of motion is

$$\frac{4}{3}\pi(\lambda a)^3\sigma U = \frac{-2\pi\lambda^3 a^3}{3(1-\lambda^3)}\left[(1+2\lambda^3)U - 3V\right]$$

$$\Rightarrow \quad U = \frac{3V}{(1+2\lambda^3)+2\sigma(1-\lambda^3)}$$

11. A rigid surface of radius a is moving in a straight line with velocity U and acceleration f through an infinite incompressible liquid, prove that the resultant fluid pressures over the two hemispheres, into which sphere is divided by a diametral plane perpendicular to the direction of motion are $\Pi\pi a^2 \pm \frac{1}{4}Mf - \frac{3}{64}\frac{MU^2}{a}$ where Π is the pressure at a great distance and M is the mass of the fluid displaced by the sphere.

Solution: The pressure p at any point $(a, 0)$ on the spherical boundary is given by

$$\frac{p}{\rho} = \frac{\Pi}{\rho} + \frac{1}{8}U^2\left(9\cos^2\theta - 5\right) + \frac{1}{2}af\cos\theta \qquad \ldots (1)$$

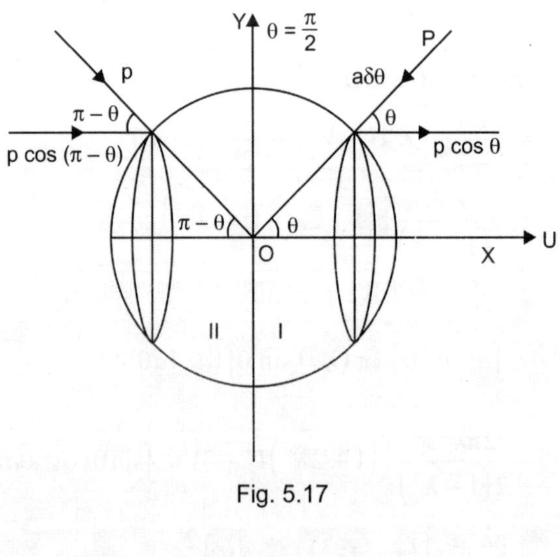

Fig. 5.17

Let YOY^1 be the diametral plane perpendicular to the direction of motion which divide the sphere into two hemispheres.

The pressure on the hemisphere at I part

$$= \int_0^{\frac{\pi}{2}} p \cos\theta 2\pi\, a\sin\theta a d\theta$$

$$= 2\pi a^2 p \int_0^{\frac{\pi}{2}} \left[\frac{\Pi}{\rho} + \frac{1}{2} af \cos\theta + \frac{U^2}{8}\left(9\cos^2\theta - 5\right)\right] \sin\theta\cos\theta d\theta$$

From (1)

$$= 2\pi a^2 \rho \left[\left(\frac{\Pi}{\rho} - \frac{5}{8}U^2\right)\left(\frac{\sin^2\theta}{2}\right) + \frac{1}{2}af\left(-\frac{1}{3}\cos^3\theta\right) + \frac{9}{8}U^2\left(-\frac{1}{4}\cos^2\theta\right)\right]_o^{\frac{\pi}{2}}$$

$$= 2\pi a^2 \rho \left[\left(\frac{\Pi}{\rho} - \frac{5}{8}U^2\right)\frac{1}{2} + \frac{af}{6} + \frac{9}{32}U^2\right]$$

$$= \Pi\pi a^2 + \frac{\pi a^3 \rho f}{3} - \frac{1}{16}\pi a^2 \rho U^2$$

$$= \Pi\pi a^2 + \frac{M}{4}f - \frac{3}{64a}MU^2 \qquad \dots (2)$$

Since

$$M = \frac{4}{3}\pi a^3 \rho$$

Again pressure on the hemisphere at the II part

$$= \int_{\frac{\pi}{2}}^{\pi} p \cos(\pi - \theta)\, 2\pi a \sin\theta a d\theta$$

$$= 2\pi a^2 \rho \int_{\frac{\pi}{2}}^{\pi} \left[\frac{\pi}{\rho} + \frac{a}{2} f \cos\theta + \frac{U^2}{8}\left(9\cos^2\theta - 5\right)\right] \times (-\cos\theta\sin\theta)\, d\theta \quad \text{From (1)}$$

$$= 2\pi a^2\rho\left[\left(\frac{\Pi}{\rho}-\frac{5}{8}U^2\right)\left(\frac{1}{2}\cos^2\theta\right)+\frac{1}{2}af\left(\frac{1}{3}\cos^3\theta\right)\right.$$

$$\left.+\frac{9}{8}U^2\left(\frac{1}{4}\cos^4\theta\right)\right]_{\frac{\pi}{2}}^{\pi}$$

$$= 2\pi a^2\rho\left[\left(\frac{\Pi}{\rho}-\frac{5}{8}U^2\right)\frac{1}{2}+\frac{1}{2}af\left(-\frac{1}{3}\right)+\frac{9}{8}U^2\frac{1}{4}\right]$$

$$= \Pi\pi a^2\rho - \frac{1}{16}\pi a^2\rho U^2 - \frac{1}{3}\pi a^3\rho f$$

$$= \Pi\pi a^2 - \frac{3}{64a}MU^2 - \frac{1}{4}Mf \quad \text{... (3)}$$

Hence the resultant pressure over the two hemispheres are

$$\Pi\pi a^2 \pm \frac{1}{4}Mf - \frac{3M}{64a}U^2$$

12. If AB be a uniform line source and A, B equal sinks of each strength that there is no total gain or loss of fluid, show that the stream function at any point is

$$\psi = c\left[(r_1 - r_2)^2 - c^2\right]\left(\frac{1}{r_1}-\frac{1}{r_2}\right)$$

Where c is the strength of AB and r_1, r_2 are the distances of the point considered from A and B.

Solution: Let m be the strength per unit length of line source AB. Let $-m^1$ be the strength of each sink placed at A and B. Since c is the length of line source, then the total strength of line source is $+mc$. Let $\angle PAB = \theta_1$ and $\angle PBX = \theta_2$.

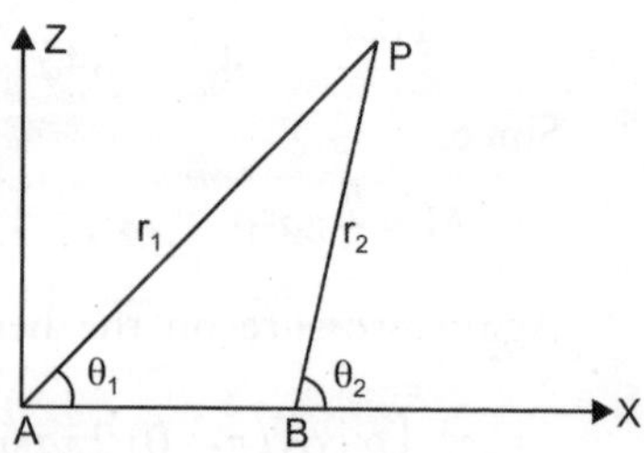

Fig. 5.18

Since the total flow is zero, then

$$-4\pi m^1 - 4\pi m^1 + 4\pi cm = 0$$

$$\Rightarrow \quad m^1 = \frac{cm}{2}$$

Now to neutralize the gain of fluid due to the line source, there must be two sinks of strength $-\frac{1}{2}mc$ each at A and B. If ψ is the Stoke's stream function at P, then

$$\psi = -m^1 \cos\theta_1 - m^1 \cos\theta_2 + m(AP - BP)$$

$$\Rightarrow \quad \psi = -\frac{1}{2}cm(\cos\theta_1 + \cos\theta_2) + m(r_1 - r_2) \qquad \dots (1)$$

Now

$$\cos\theta_1 + \cos\theta_2 = \cos\theta_1 - \cos(\pi - \theta_2)$$

$$= \frac{c^2 + r_1^2 - r_2^2}{2cr_1} - \frac{c^2 + r_2^2 - r_1^2}{2cr_2}$$

$$= \frac{c}{2}\left(\frac{1}{r_1} - \frac{1}{r_2}\right) + \frac{1}{2c}(r_1 - r_2) - \frac{1}{2c}\left(\frac{r_2^2}{r_1} - \frac{r_1^2}{r_2}\right)$$

Using this in (1), we have

$$\psi = -\frac{1}{4}mc^2\left(\frac{1}{r_1} - \frac{1}{r_2}\right) - \frac{1}{4}m(r_1 - r_2) + \frac{1}{4}m\frac{(r_2^3 - r_1^3)}{r_1 r_2} + m(r_1 - r_2)$$

$$= \frac{m}{4}\left[-c^2\left(\frac{1}{r_1} - \frac{1}{r_2}\right) + \frac{(r_2^3 - r_1^3) - 3r_1 r_2(r_2 - r_1)}{r_1 r_2}\right]$$

$$= \frac{m}{4}\left[-c^2\left(\frac{1}{r_1} - \frac{1}{r_2}\right) + \frac{(r_2 - r_1)^3}{r_1 r_2}\right]$$

$$= \frac{m}{4}\left(\frac{1}{r_1} - \frac{1}{r_2}\right)\left[-c^2 + (r_2 - r_1)^2\right]$$

$$= c\left[(r_1 - r_2)^2 - c^2\right]\left(\frac{1}{r_1} - \frac{1}{r_2}\right)$$

$$\text{Where } c = \frac{m}{4}$$

13. A solid of revolution is moving along its axis in an infinite liquid, show that the kinetic energy of the liquid is

$-\frac{1}{2}\pi\rho \int \frac{\psi}{w}\frac{\partial\psi}{\partial n}\, ds$, where ψ is the Stoke's stream function, w the distance of a point from the axis and the integral is taken round a meridian curve of the solid. Hence obtain the kinetic energy of infinite liquid due to the motion of a sphere through it with velocity V.

Solution: The kinetic energy T of liquid is given by

$$T = -\frac{1}{2}\rho \iint \phi \frac{\partial\phi}{\partial n}\, ds \quad \text{... (1)}$$

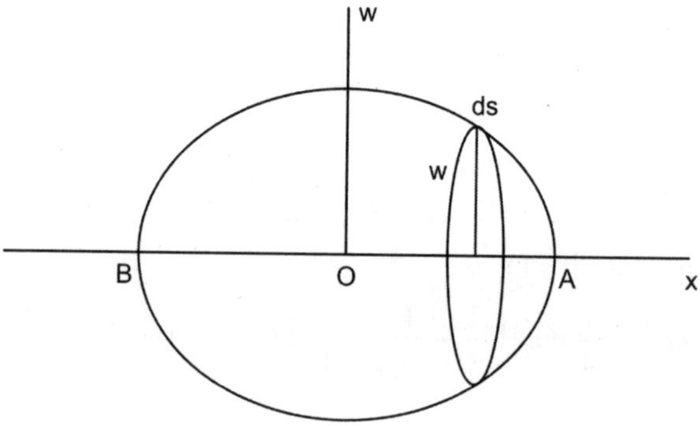

Fig. 5.19

We take an element ds of the meridian curve and revolving it about the x-axis, then

$$ds = 2\pi w ds$$

Also

$$-\frac{\partial\phi}{\partial n} = \text{outward normal velocity}$$

$$= \frac{1}{w}\frac{\partial\psi}{\partial s}$$

Hence (1) gives

$$T = -\frac{1}{2}\rho \int \phi \frac{\partial\phi}{\partial n} 2\pi w ds$$

$$= \frac{1}{2}\rho 2\pi \int w\phi \frac{1}{w}\frac{\partial\psi}{\partial s}\, ds$$

$$= \pi\rho \int \phi d\psi \qquad \text{... (2)}$$

$$= \pi\rho\phi\psi - \pi\rho \int \psi d\phi \qquad \text{(Integrating by parts)}$$

$$= -\pi\rho \int \psi d\phi$$

The first term is zero because ψ is constant along the curve in the meridian plane along which the integration is taken.

Also

$$-\frac{\partial\phi}{\partial s} = \frac{1}{w}\frac{\partial\psi}{\partial n}$$

Hence from (2), we get

$$T = -\pi\rho \int \frac{\psi}{w}\frac{\partial\psi}{\partial n} ds$$

Hence along the whole meridian

$$T = -\frac{\pi\rho}{2} \int \frac{\psi}{w}\frac{\partial\psi}{\partial n} ds$$

as half of the whole meridian curve is equal to meridian curve from A to B.

For the K.E. of a sphere moving with velocity V along the axis, we have

$$\phi = \frac{Va^3}{2r^2}\cos\theta \text{ and } \psi = -\frac{1}{2}\frac{Va^3}{r}\sin^2\theta$$

Hence

$$(\phi d\psi)_{r=a} = \left(\frac{Va^3}{2a}\cos\theta\right)\left(-\frac{1}{2r}Va^3 2\sin\theta\cos\theta d\theta\right)_{r=a}$$

$$= -\frac{1}{2}V^2a^2\sin\theta\cos^2\theta\, d\theta$$

Hence, by (2)

$$T = -\pi\rho \int_0^{\pi} -\frac{1}{2}V^2a^3\sin\theta\cos^2\theta d\theta$$

$$= \frac{\pi}{2}\rho V^2a^3\left(-\frac{1}{3}\cos^3\theta\right)_0^{\pi}$$

$$= \frac{\pi}{2}\rho V^2 a^3 \frac{2}{3}$$

$$= \frac{\pi}{3}\rho V^2 a^3$$

$$= \frac{1}{4} MV^2$$

where $M = \frac{4}{3}\pi a^3 \rho$ = mass of the liquid displaced by the sphere.

14. A thin stream of incompressible fluid is contained between two concentric spheres, show that the velocity at any point is equivalent to the components $-\frac{1}{\sin\theta}\frac{\partial\psi}{\partial\varpi}, \frac{\partial\psi}{\partial\theta}$ along the meridian and parallel to latitude respectively. Also if the fluid be homogeneous and the motion irrotational, prove that

$$\frac{\partial\phi}{\partial\theta} = \frac{1}{\sin\theta}\frac{\partial\psi}{\partial\varpi}, \frac{\partial\psi}{\partial\theta} = -\frac{1}{\sin\theta}\frac{\partial\phi}{\partial\varpi}$$

and deduce that $\phi + i\psi = F\left[e^{i\varpi}\tan\frac{\theta}{2}\right]$

Solution: Let *PQ* be the vertical axis of rotation and any great circle drawn with *PQ* as diameter is known as a meridian. Velocity parallel to *AB* is called velocity component parallel to latitude.

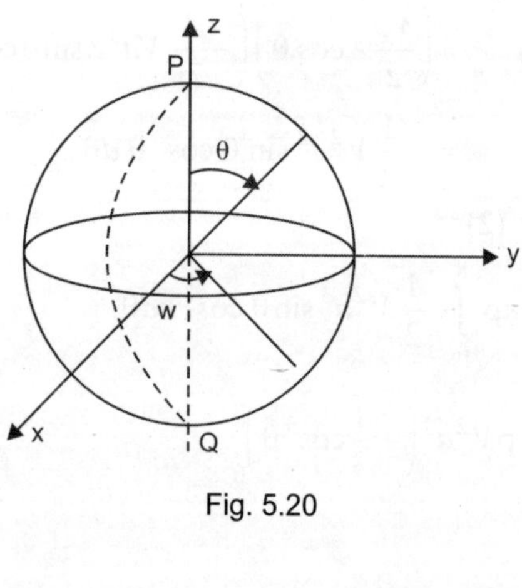

Fig. 5.20

Since the fluid stratum is thin, it follows that motion is on the surface of the sphere, so that there is no normal velocity i.e. $q_r = U = 0$. Also $q_\theta = V$, $q_w = w$ are velocity components parallel to meridian and latitude respectively. The equation of continuity becomes

$$\frac{\partial}{\partial \theta}(V \sin \theta) + \frac{\partial w}{\partial \bar{w}} = 0 \qquad \text{... (1)}$$

The streamlines are given by

$$\frac{a d\theta}{V} = \frac{a \sin \theta \, d\bar{w}}{w}$$

$$\Rightarrow \quad w d\theta - V \sin \theta \, d\bar{w} = 0 \qquad \text{... (2)}$$

The equation (2) is exact by virtue of (1). Hence we can take

$$w d\theta - V \sin \theta d\bar{w} = d\psi = \frac{\partial \psi}{\partial \theta} d\theta + \frac{\partial \psi}{\partial \bar{w}} d\bar{w}$$

Hence

$$w = \frac{\partial \psi}{\partial \theta}, -V \sin \theta = \frac{\partial \psi}{\partial \bar{w}} \text{ or } V = \frac{-1}{\sin \theta} \frac{\partial \psi}{\partial \bar{w}} \qquad \text{... (3)}$$

This proves the first part of the problem.

Again from (2)

$$d\psi = 0$$

$\Rightarrow \psi =$ const.

Hence ψ is a stream function

Let ϕ be the velocity potential, then

$$V = -\frac{1}{a} \frac{\partial \phi}{\partial \theta}, w = -\frac{1}{a \sin \theta} \frac{\partial \phi}{\partial \bar{w}}$$

Taking $a = 1$, so that

$$V = \frac{-\partial \phi}{\partial \theta}, w = \frac{-1}{\sin \theta} \frac{\partial \phi}{\partial \bar{w}} \qquad \text{... (4)}$$

From (3) and (4), we have

$$\frac{-\partial \phi}{\partial \theta} = -\frac{1}{\sin \theta} \frac{\partial \psi}{\partial \bar{w}} \text{ and } \frac{\partial \psi}{\partial \theta} = \frac{-1}{\sin \theta} \frac{\partial \phi}{\partial \bar{w}}$$

$$\Rightarrow \quad \frac{\partial \phi}{\partial \theta} = \frac{1}{\sin \theta} \frac{\partial \phi}{\partial \bar{w}} \text{ and } \frac{\partial \psi}{\partial \theta} = \frac{-1}{\sin \theta} \frac{\partial \phi}{\partial \bar{w}} \qquad \dots (5)$$

This completes the second part of the problem. Again from (5), we have

$$\frac{\partial}{\partial \theta}(\theta + i\psi) = \frac{i}{\sin \theta} \frac{\partial}{\partial \bar{w}}(\psi - i\phi)$$

$$= -\frac{i}{\sin \theta} \frac{\partial}{\partial \bar{w}}(\phi + i\psi)$$

Taking $\phi + i\psi = \zeta$, we have

$$\frac{\partial \rho}{\partial \theta} = -\frac{i}{\sin \theta} \frac{\partial \zeta}{\partial \bar{w}}$$

$$\Rightarrow \quad \frac{\partial \rho}{\partial \theta} + \frac{i}{\sin \theta} \frac{\partial \zeta}{\partial \bar{w}} = 0$$

Comparing this with Lagrange's partial differential equations $P\frac{\partial z}{\partial x} + Q\frac{\partial z}{\partial y} = R$ whose solution is $\frac{dx}{P} = \frac{dy}{Q} = \frac{dz}{R}$.

We obtain $\frac{d\theta}{\sin \theta} = \frac{d\bar{w}}{i} = \frac{d\zeta}{O}$

Taking first two ratios, we get

$$\operatorname{cosec} \theta \, d\theta = -i d\bar{w}$$

Integrating, we have

$$\log \tan \frac{\theta}{2} = -i\bar{w} + \log A$$

$$\Rightarrow \quad \tan \frac{\theta}{2} = Ae^{-i\bar{w}}$$

$$\Rightarrow \quad e^{i\bar{w}} \tan \frac{\theta}{2} = A$$

Taking last two ratios, we have $d\zeta = 0$

$$\Rightarrow \quad \zeta = \text{cons} = F(A) \text{ (say)}$$

$$\Rightarrow \quad \phi + i\psi = F\left(e^{i\bar{w}} \tan \frac{\theta}{2}\right).$$

15. A doublet of strength m is placed at the point $(0, a, 0)$ with its axis parallel to z-axis, prove that at points close to origin, the velocity potential of the doublet is approximately $\frac{mz}{a^3}+\frac{3myz}{a^4}$, neglecting terms of order $\frac{r^3}{a^5}$ and higher powers. Deduce that if a small sphere of radius c is placed with its centre at the origin, the velocity potential is increased by the terms

$$\frac{1}{2}\frac{mc^3z}{a^3r^3}+\frac{2mc^5yz}{a^4r^5}$$

Solution: Let a doublet of strength m be placed at A $(0, a, 0)$ with its axis parallel to the axis of z. We take a point $P(x, y, z)$ in the liquid and let AP make an angle θ with z-axis. Then

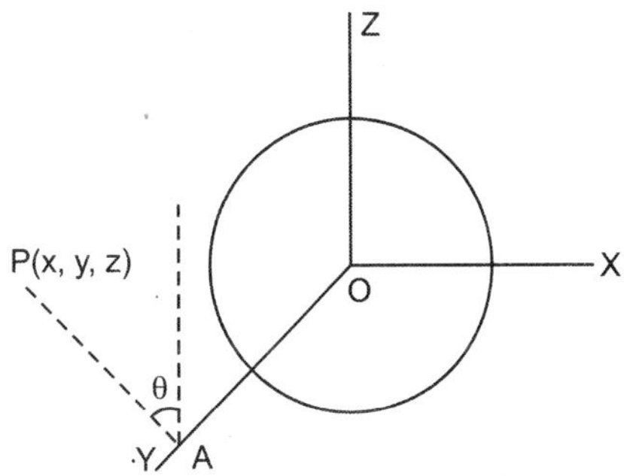

Fig. 5.21

$$\cos\theta=\frac{z}{AP}$$

Where

$$r^2 = x^2 + y^2 + z^2,\ AP^2 = x^2 + (y - a)^2 + z^2$$

The velocity potential ϕ due to this doublet at P is given by

$$\phi=\frac{m\cos\theta}{AP^2}=\frac{mz}{AP^3}$$

$$AP^2 = x^2+(y-a)^2+z^2 = r^2-2ay+a^2$$

$$AP^{-3} = \left(r^2-2ay+a^2\right)^{-3/2}$$

$$=a^{-3}\left[1+\left(\frac{r^2-2ay}{a^2}\right)\right]^{-3/2}$$

$$=a^{-3}\left[1-\frac{3}{2a^2}\left(r^2-2ay\right)\right] \text{ upto first approximation}$$

$$=a^{-3}\left(1+\frac{3y}{a}\right) \text{ neglecting } r^2 \text{ and higher powers of } r^2$$

Hence

$$\phi = \frac{mz}{a^3}\left(1 + \frac{3y}{a}\right) = \frac{mz}{a^3} + \frac{3myz}{a^4}$$

This proves the first part of the result.

Now when the sphere of radius c is inserted, let ϕ_1 be the velocity potential at any point. Then ϕ_1 is such that $\left(\frac{-\partial\phi_1}{\partial r}\right) = 0$ when $r = c$. Also liquid at infinity remains unaltered. Let the increase in velocity potential be $\frac{Az}{r^3} + \frac{Byz}{r^5}$. Then,

$$\phi_1 = \frac{mz}{a^3} + \frac{3myz}{a^4} + \frac{Az}{r^3} + \frac{Byz}{r^5}$$

$$\Rightarrow \quad \phi_1 = \frac{m}{a^3} r \cos\theta + \frac{3m}{a^4} r^2 \sin\theta \sin w \cos\theta + \frac{A\cos\theta}{r^2} + \frac{B\sin\theta}{r^3}\sin w \cos\theta$$

$$\therefore \quad \frac{\partial\phi_1}{\partial r} = \frac{m\cos\theta}{a^3} + \frac{6mr}{a^4}\sin\theta\sin w\cos\theta - \frac{2A}{r^3}\cos\theta - \frac{3B}{a^4}\sin\theta\sin w\cos\theta$$

$$\Rightarrow \quad \left(\frac{\partial\phi_1}{\partial r}\right)_{r=c} = \left(\frac{m}{a^3} - \frac{2A}{c^3}\right)\cos\theta + \left(\frac{6mc}{a^4} - \frac{3B}{r^4}\right)\sin\theta\sin w\cos\theta$$

$$\Rightarrow \quad 0 = \left(\frac{m}{a^3} - \frac{2A}{c^3}\right)\cos\theta + \sin\theta\sin w\cos\theta\left(\frac{6mc}{a^4} - \frac{3B}{c^4}\right)$$

$$\Rightarrow \quad \frac{m}{a^3} - \frac{2A}{c^3} = 0 \text{ and } \frac{6mc}{a^4} - \frac{3B}{c^4} = 0$$

$$\Rightarrow \quad A = \frac{ma^3}{a^3}, B = \frac{2mc^5}{a^4}$$

Therefore, increase in velocity potential

$$= \frac{Az}{r^3} + \frac{Byz}{r^5}$$

$$= \frac{mc^2 z}{2a^3 r^3} + \frac{2mc^5 yz}{r^5 a^4}$$

16. Doublets of strengths μ_1, μ_2 are situated at points A_1, $(0, 0, c_1)$ and A_2 $(0, 0, c_2)$, their axes being directed towards and away from the origin respectively. Find the condition that there is no transport of fluid over the surface of the sphere $x^2 + y^2 + z^2 = c_1 c_2$.

Solution: Let OA_1A_2 be the initial line. The axes of the doublets at A_1, A_2 make angles θ_1, θ_2 with A_1P, A_2P. Then the velocity potential at P (r, θ, ϕ) reduces to

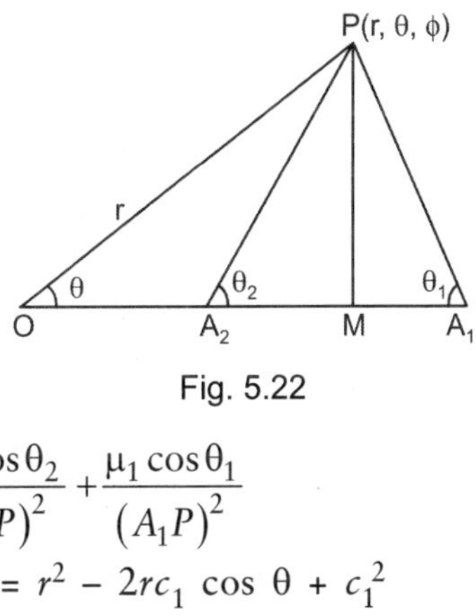

Fig. 5.22

$$\phi = \frac{\mu_2 \cos\theta_2}{(A_2P)^2} + \frac{\mu_1 \cos\theta_1}{(A_1P)^2} \quad \text{... (1)}$$

where $A_1P^2 = r^2 - 2rc_1 \cos\theta + c_1^2$

and $A_2P^2 = r^2 - 2rc_2 \cos\theta + c_2^2$

In ΔA_1PM

$$\cos\theta_1 = \frac{A_1M}{PP_1} = \frac{OA_1 - OM}{PA_1} = \frac{c_1 - r\cos\theta}{A_1P}$$

and

$$\cos\theta_2 = \frac{r\cos\theta - c_2}{A_2P}$$

Hence (1) becomes

$$\phi = \frac{\mu_2 (r \cos \theta - c_2)}{(A_2 P)^3} + \frac{\mu_1 (c_1 = r \cos \theta)}{(A_1 P)^3}$$

$$\Rightarrow \quad \phi = \frac{\mu_1 (r \cos \theta - c_2)}{(r^2 - 2rc_2 \cos \theta + c_2^2)^{3/2}} + \frac{\mu_1 (c_1 - r \cos \theta)}{(r^2 - 2rc_1 \cos \theta + c_1^2)^{3/2}} \qquad \dots (2)$$

We will show that $\frac{\partial \phi}{\partial r} = 0$ when

$$x^2 + y^2 + z^2 = c_1 c_2 \text{ or } r^2 = c_1 c_2 \text{ i.e. } r = \sqrt{c_1 c_2} \qquad \dots (3)$$

From (2)

$$\frac{\partial \phi}{\partial r} = \mu_2 \Big[\cos \theta (r^2 - 2rc_2 \cos \theta + c_2^2)^{-3/2}$$

$$- 3(r^2 - 2rc_2 \cos \theta + c_2^2)^{-5/2} (r - c_2 \cos \theta)$$

$$\times (r \cos \theta - c_2) \Big] + \mu_1 \Big[- \cos \theta (r^2 - 2rc_1 \cos\theta + c_1^2)^{-3/2}$$

$$- 3(c_1 - r \cos \theta)(r - c_1 \cos \theta)$$

$$(r^2 - 2rc_1 \cos \theta + c_1^2)^{-5/2} \Big] \qquad \dots (4)$$

From (3) and (4), we have

$$\mu_2 \Big[\cos \theta (c_1 c_2 - 2\sqrt{c_1 c_2}\, c_2 \cos \theta + c_2^2)^{-3/2}$$

$$- 3(c_1 c_2 - 2\sqrt{c_1 c_2}\, c_2 \cos \theta + c_2)^{-5/2}$$

$$(\sqrt{c_1 c_2} - c_2 \cos \theta) \times (\sqrt{c_1 c_2} \cos \theta - c_2) \Big]$$

$$= \mu_1 \Big[\cos \theta (c_1 c_2 - 2\sqrt{c_1 c_2}\, c_1 \cos \theta + c_1^2)^{-3/2}$$

$$-3(c_1 - \sqrt{c_1 c_2} \cos \theta)(\sqrt{c_1 c_2} - c_1 \cos\theta)$$

$$(c_1 c_2 - 2\sqrt{c_1 c_2}\, c_1 \cos \theta + c_1^2)^{-5/2} \Big]$$

$$\Rightarrow \quad \frac{\mu_2}{\mu_1} = (c_2 c_1)^{3/2}$$

17. A three-dimensional doublet of strength μ whose axis is in the direction OX is distant a from the rigid plane $x = 0$ which is the sole boundary of liquid of density ρ, infinite in extent. Find the pressure at a point on the boundary distant r from the doublet given that the pressure at infinity is P_∞. Show that the pressure on the plane is least at a distance $a\sqrt{5/2}$ from the doublet.

Solution: Let $\gamma\gamma^1$ be a rigid boundary. The image system of a doublet of strength μ placed at $A(a, 0)$ consists a doublet of same strength with regard to the rigid boundary at the corresponding point $A^1(-a, 0)$. Thus the velocity potential at preduces to

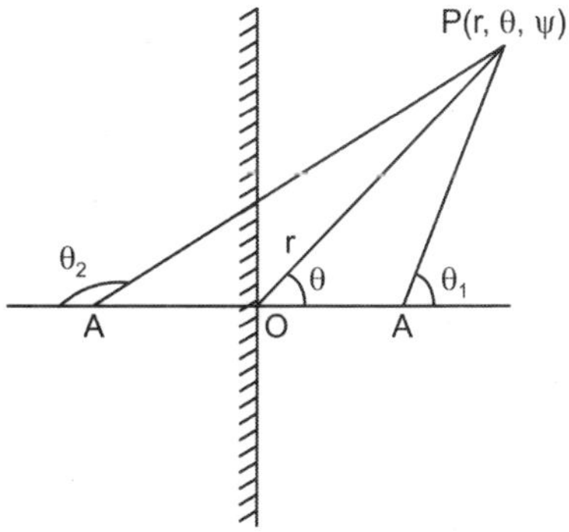

Fig. 5.23

$$\phi = \frac{\mu \cos \theta_1}{AP^2} + \frac{\mu \cos \theta_2}{A^1P^2}$$

$$\Rightarrow \quad \phi = \frac{\mu(r\cos\theta - a)}{(r^2 + a^2 - 2ra\cos\theta)^{3/2}} - \frac{\mu(r\cos\theta + a)}{(r^2 + a^2 + 2ra\cos\theta)^{1/2}}$$

Thus

$$q_r = \frac{-\partial\phi}{\partial r} = -\mu\Big[\cos\theta\,(r^2 + a^2 - 2ra\cos\theta)^{-3/2}$$

$$- 3(r - a\cos\theta)(r\cos\theta - a)(r^2 + a^2 - 2ra\cos\theta)^{-5/2}$$

$$- \cos\theta\,(r^2 + a^2 + 2ra\cos\theta)^{-1/2}$$

$$+ 3(r + a\cos\theta)(r\cos\theta + a)(r^2 + a^2 - 2ra\cos\theta)^{-5/2}\Big]$$

Also

$$q_s = -\frac{1}{r}\frac{\partial\phi}{\partial\theta}$$

$$= -\frac{\mu}{r}\Big[-r\sin\theta\left(r^2+a^2-2ra\cos\theta\right)^{-3/2}$$

$$-3ra\sin\theta\left(r\cos\theta-a\right)\times\left(r^2+a^2-2ra\cos\theta\right)^{-5/2}$$

$$+r\sin\theta\left(r^2+a^2+2ra\cos\theta\right)^{-3/2}$$

$$-3ra\sin\theta\left(r\cos\theta+a\right)\left(r^2+a^2+2ra\cos\theta\right)^{-5/2}\Big]$$

and

$$q_\psi = -\frac{1}{r\sin\theta}\frac{\partial\phi}{\partial\psi} = 0$$

At $\theta = \frac{\pi}{2}$, $q_r = -\sigma\,\mu ar\left(r^2+a^2\right)^{-5/2}$, $q_s = 0$, $q_\psi = 0$

From the Bernoulli's equation, we have

$$p = p_\infty - \frac{1}{2}\rho q^2 = p_\infty - 18\rho\mu^2 a^2 r^2\left(r^2+a^2\right)^{-5}$$

$$\frac{dp}{dr} = -18\rho\mu^2 a^2\left[2r\left(r^2+a^2\right)^{-5} - 5\left(r^2+a^2\right)^{-6}2r^3\right]$$

$$= -18\rho\mu^2 a^2\left(r^2+a^2\right)^{-6}2r\left[r^2+a^2-5r^2\right]$$

$$= -36\rho\mu^2 a^2\left(r^2+a^2\right)^{-6}r\left(a^2-4r^2\right)$$

Hence

$$\left(\frac{dp}{dr}\right)_{r=\frac{a}{2}} = 0$$

Also for $r = \frac{a}{2}+$, $\frac{dp}{dr} > 0$

and for $r = \frac{a}{2}-$, $\frac{dp}{dr} < 0$

Hence p is minimum at $r = \frac{a}{2}$ on the plane i.e. at distance $\frac{a}{2}\sqrt{5}$ from the doublet.

EXERCISE

1. Show that when a sphere of radius a moves with uniform velocity u through a perfect, incompressible, infinite fluid, the acceleration of a particle at (r, θ) is $3u^2\left(\frac{a^3}{r^4}-\frac{a^6}{r^7}\right)$.

2. A solid sphere moves through quiescent frictionless liquid whose boundaries are at a distance from it, great compared with its radius, prove that at each instant the motion in the liquid depends only on the position and velocity of the sphere at that instant. Prove that the liquid streams past the sides of the sphere with help of the velocity of the sphere.

3. When a sphere of radius a moves in an infinite liquid, show that the pressure at any point exceeds what would be the pressure if the sphere were at rest by

$$\frac{a^3}{2r^2}f-\frac{a^3}{8r^6}\left(4r^3+a^3\right)q^2+\frac{3a^3}{8r^6}\left(4r^3-a^3\right)q_1{}^2$$

where q is the velocity of the sphere and q_1 and f are the resolved parts of its velocity and acceleration in the direction of r and the density liquid is unity.

[**Hint.** In worked out 3, put $v = q$, $v\cos\theta = q^1$, $\rho = 1$]

4. Find the pressure at any point of a liquid of infinite extent and at rest at a great distance, through which a sphere is moving under no external forces with constant velocity u, and show that the mean pressure over the sphere is in defect of the pressure Π at a great distance by $\frac{1}{4}\rho u^2$, if being supposed that Π is sufficiently large for the pressure every where to the positive is that $\Pi > \frac{5}{8}\rho u^2$.

[**Hint.** See worked out 8]

5. A sphere of radius a is in motion in fluid, which is at rest at infinity, the pressure being Π, determine the pressure at any point of the fluid; and show that the pressure on the front hemisphere cut off by a plane perpendicular to the direction of motion is the resultant of the pressures.

$\Pi a^2\left[\Pi - \frac{1}{16}\rho V^2\right]$ and $\frac{1}{3}\Pi\rho a^3 f$ in the directions respectively opposite to those of the velocity V and the acceleration f of the centre to the sphere.

[**Hint.** See worked out 3]

6. Discuss the motion for which Stoke's stream function is given by

$$\psi = \frac{1}{2} V\left[a^4 r^{-2}\cos\theta - r^2\right]\sin^2\theta$$

where r is the distance from a fixed point and θ is the angle this distance makes with a fixed direction.

[**Ans.** The stream function gives the motion of a liquid flowing past a solid $r^4 = a^4 \cos\theta$, moving with velocity V along x-axis.]

7. Prove that the velocity potential at a point p due to a uniform finite line source AB of strength $+m$ per unit length is of the form $\phi = m \log f$, where $f = \frac{r_2 + x_2}{r_1 + x_1} = \frac{r_1 - x_1}{r_2 - x_2} = \frac{a + l}{a - l}$ in which $AB = 2l$, $PA = r_1$, $PB = r_2$, $NA = x_1$, $NB = x_2$ and $r_1 + r_2 = 2a$. N being the perpendicular drawn from p on the line AB.

8. An infinite homogeneous liquid is flowing steadily past a rigid boundary consisting partly of the horizontal plane $y = 0$, and partly of a hemispherical bass $x^2 + y^2 + z^2 = a^2$, with irrotational motion which tends, at a great distance from the origin, to uniform velocity V parallel to the axis of z. Find the velocity potential and the surface of equal pressure.

[**Ans.** $\phi = Vr\cos\theta + \frac{1}{2}\frac{Va^3}{r^2}\cos\theta$ and $\left(1 - \frac{a^3}{r^3}\right)^2\cos^2\theta + \left(1 + \frac{a^3}{2r^3}\right)^2\sin^2\theta = \text{const}$ determines the surface of equipressure.]

6 GENERAL THEORY OF IRROTATIONAL MOTION

6.1. Some Elementary Definitions

1. Connectivity. A region of space is said to be connected if a continuous curve joining any two points of the region lies entirely in the given region.

For example, the region interior and exterior to a sphere individually are connected.

2. Simply Connected Region. A region D is said to be simply connected if any closed curve in D can be shrunk to a point without crossing the boundary of D.

For example the region interior to a sphere, any region without holes or gaps, unbounded space are all simply connected regions.

A region of space which is not simply connected is said to be multiply connected.

3. Barriers. In a multiply connected region, it is generally possible to insert suitable barriers without breaking up the space into disconnected regions.

A region of space is said to be doubly connected if it can be made simply connected by the insertion of one barrier. Similarly a region of space is said to be connected if it can be made simply connected by the insertion of $(n - 1)$ barriers.

Examples of doubly connected regions are the inner portion of a ring, the space between two coaxial cylinders.

4. Flow. If A and B be any two points in a fluid, then the value of the integral

$$\int_A^B (udx + vdy + wdz) = \int_A^B \vec{q} \cdot d\vec{r}$$

taken along any path in the fluid is called a flow from A to B. If the motion is irrotational, then

$$\text{Flow } \int_A^B \vec{q}.d\vec{r} = \int_A^B -\vec{\nabla}\phi \cdot d\vec{r} = -\int_A^B d\phi = \phi_A - \phi_B$$

5. Circulation. The flow round a closed curve C is known as circulation. Thus

$$\text{Circulation} = \int_c \vec{q} \cdot d\vec{r} = \phi_A - \phi_B = \phi_A - \phi_A = 0$$

Since for a closed path in simply connected domain, A and B coincide.

6. Acyclic Motion. A fluid motion is said to be acyclic if there exists a single valued velocity potential ϕ so that the motion is irrotational. Hence in a simply connected region only acyclic irrotational motion is possible.

7. Cyclic Motion. A fluid motion in a domain D is said to be cyclic if the circulation along any closed path in D does not vanish.

6.2. Irrotational Motion

The necessary and sufficient condition for irrotational motion is that there exists a velocity potential ϕ such that $\vec{q} = -\vec{\nabla}\phi$, $\vec{q}$ being velocity vector.

Proof. Let the motion be irrotational in a simply connected region so that curl $\vec{q} = 0$.

To prove that there exists a scalar function ϕ such that $\vec{q} = -\vec{\nabla}\phi$.

Let A be a fixed point and P an arbitrary point in the region of space under consideration. Join A to P by two paths ABP and ACP. Let S be the surface area of the rim $ABPCA$.

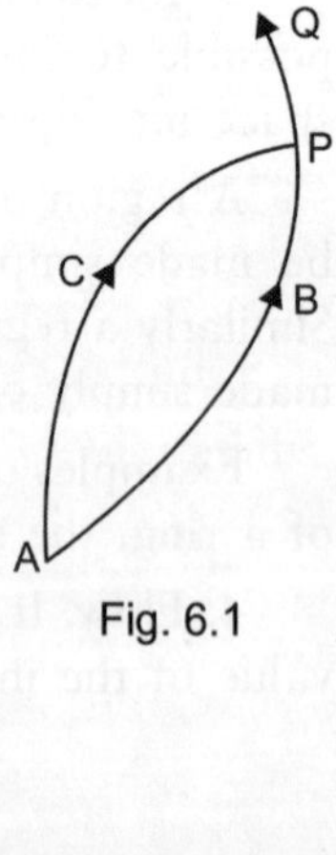

Fig. 6.1

By Stoke's theorem

$$\int_{ABPCA} \vec{q}\cdot d\vec{r} = \int_S \vec{n}\cdot \text{curl}\,\vec{q}ds = 0 \qquad \text{as curl } \vec{q} = 0$$

$$\Rightarrow \int_{ABP} \vec{q}\cdot d\vec{r} + \int_{PCA} \vec{q}\cdot d\vec{r} = 0$$

$$\Rightarrow \int_{ABP} \vec{q}\cdot d\vec{r} = \int_{ACP} \vec{q}\cdot d\vec{r} = -\phi_P\,(\text{say}) \qquad \text{... (1)}$$

Evidently ϕ_P is a scalar function which depends upon the position of P and not on A. We consider a point Q very near to P and assume that $\vec{q}$ is constant along the line PQ.

Let $\overrightarrow{PQ} = \eta$. Then from (1)

$$\int_{ABPQ} \vec{q}\cdot d\vec{r} = -\phi_Q \qquad \text{... (2)}$$

From (1) and (2)

$$-\phi_Q - (-\phi_P) = \int_{ABPQ} \vec{q}\cdot d\vec{r} - \int_{ABP} \vec{q}\cdot d\vec{r}$$

$$\Rightarrow -(\phi_Q - \phi_P) = \left[\int_{ABP} + \int_{PQ} - \int_{ABP}\right]\vec{q}\cdot d\vec{r}$$

$$\Rightarrow -(\phi_Q - \phi_P) = \int_{PQ} \vec{q}\cdot d\vec{r} = \vec{q}\cdot \int_{\overrightarrow{PQ}} d\vec{r}$$

as $\vec{q}$ is constant along PQ

$$= \vec{q}\cdot\vec{\eta} = \vec{q}\cdot\vec{a}PQ$$

Where $\vec{a}$ is the unit vector along PQ, so that

$$\vec{a}PQ = \overrightarrow{PQ} = \eta$$

$$\Rightarrow \frac{\phi_Q - \phi_P}{PQ} = -\vec{q}\cdot\vec{a}$$

But L.H.S. is a directional derivative of ϕ along the direction $\vec{a}$ which is $\vec{a}\cdot\vec{\nabla}\phi$

Hence

$$\vec{a}\cdot\vec{\nabla}\phi = -\vec{q}\cdot\vec{a}$$

$\Rightarrow \quad \vec{a}\cdot\left(\vec{\nabla}\phi + \vec{q}\right) = 0$

$\Rightarrow \quad \vec{\nabla}\phi + \vec{q} = 0$ as $\vec{a} \neq \vec{0}$

$\Rightarrow \quad \vec{q} = -\vec{\nabla}\phi$

Conversely let $\vec{q} = -\vec{\nabla}\phi$

To prove that motion is irrotational.

Then curl $\vec{q} = \text{curl}\left(-\vec{\nabla}\phi\right) = -$ curl grad ϕ

Hence the motion is irrotational $= 0$.

6.3. Kinetic Energy of a Finite Liquid

Let T be the K.E. of a liquid occupying a volume. Let the liquid be moving with velocity $\vec{q}$ and S be the surface enclosed by the volume V.

$$T = \int_V \frac{1}{2}(\rho dv)q^2 = \frac{1}{2}\rho \int_V q^2 dv \qquad \text{... (1)}$$

Let the motion be irrotational, then

$$\vec{q} = -\vec{\nabla}\phi$$

Equation of continuity is $\vec{\nabla}\cdot\vec{q} = 0$

$\Rightarrow \quad \vec{\nabla}\left(-\vec{\nabla}\phi\right) = 0$

$$\Rightarrow \quad \vec{\nabla}^2\phi = 0 \qquad \text{... (2)}$$

Hence from (1), we have

$$T = \frac{1}{2}\rho \int_V \vec{q}\cdot\vec{q}\,dv$$

$$= \frac{1}{2}\rho \int_V \left(-\vec{\nabla}\phi\right)\cdot\left(-\vec{\nabla}\phi\right)dv$$

$$= \frac{1}{2}\rho \int_V \left(\vec{\nabla}\phi\cdot\vec{\nabla}\phi\right)dv$$

$$= \frac{1}{2}\rho \int_S \left(\phi\,\frac{\partial\phi}{\partial n} - \phi\,\vec{\nabla}^2\phi\right) ds \qquad \text{(By Green's theorem)}$$

$$= \frac{1}{2} \phi \int_S \left(\phi \frac{\partial \phi}{\partial n} - 0 \right) \qquad \text{From (2)}$$

$$= \frac{1}{2} \rho \int_S \phi \frac{\partial \phi}{\partial n} ds$$

δn being an element of unit outward normal.

If $\hat{n}$ be the unit inward normal, then for simply connected domain

$$T = -\frac{1}{2} \rho \int_S \phi \frac{\partial \phi}{\partial n} ds \qquad \dots (3)$$

The equation (3) declares that the kinetic energy of a given mass of liquid moving irrotationally in a simply connected space depends only on the motion of its boundaries.

Deduction. To show that irrotational motion is not possible in a simply connected domain with fixed boundaries.

Proof. Since boundary is at rest.

So outward normal velocity $= -\dfrac{\partial \phi}{\partial n} = 0$

$$\Rightarrow \quad T = \frac{1}{2} \rho \int_V q^2 dV$$

$$= -\frac{1}{2} \rho \int \phi \frac{\partial \phi}{\partial n} ds = 0$$

So $q = 0$ everywhere in V. Hence liquid is at rest.

6.4. Kinetic Energy of Infinite Liquid

Let a liquid of density ρ be moving irrotationally with velocity $\vec{q}$ at rest at infinity. Let the liquid be bounded internally by a solid S and externally by a large surface S_1. Then kinetic energy T of finite liquid contained between S and S_1 is given by

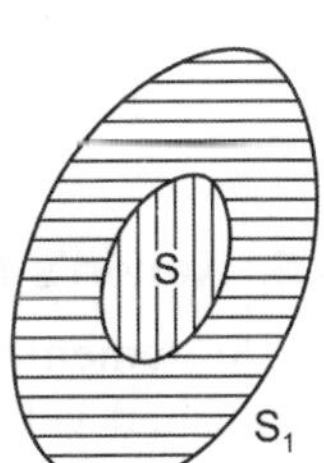

Fig. 6.2

$$T = -\frac{1}{2} \rho \int_S \phi \frac{\partial \phi}{\partial n} ds - \frac{1}{2} \rho \int_{S_1} \phi \frac{\partial \phi}{\partial n} ds \quad \dots (1)$$

Since the there is no flow into the region across S, so equation of continuity is

$$\int_S \rho(\vec{q}\cdot\hat{n})\,dS + \int_{S_1} \rho(\vec{q}\cdot\hat{n})\,dS_1 = 0 \qquad \text{... (2)}$$

But $\vec{q}\cdot\hat{n} = -\vec{\nabla}\phi\cdot\hat{n} = -\dfrac{\partial\phi}{\partial n}$

Hence (2) becomes

$$-\rho\int_S \frac{\partial\phi}{\partial n}\,dS - \rho\int_{S_1}\frac{\partial\phi}{\partial n}\,dS_1 = 0$$

$$\Rightarrow \quad \int_S \frac{\partial\phi}{\partial n}\,dS + \int_{S_1}\frac{\partial\phi}{\partial n}\,dS_1 = 0 \qquad \text{... (3)}$$

Since S_1 goes to infinity where liquid is at rest so that ϕ = constant = c, say on S_1. Then

$$\int_{S_1} \phi\,\frac{\partial\phi}{\partial n}\,dS_1 = c\int_{S_1}\frac{\partial\phi}{\partial n}\,dS_1 = -c\int_S \frac{\partial\phi}{\partial n}\,dS \qquad \text{From (3)}$$

Using this in (1), we get

$$T = -\frac{1}{2}\rho\int_S \phi\,\frac{\partial\phi}{\partial n}\,dS + \frac{1}{2}\rho c\int_S \frac{\partial\phi}{\partial n}\,dS$$

$$= -\frac{1}{2}\rho\int_S (\phi - c)\,\frac{\partial\phi}{\partial n}\,dS$$

If the flux of liquid across the inner surface is zero, i.e. if $-\rho\displaystyle\int_S \frac{\partial\phi}{\partial n}\,dS = 0$. Then

$$T = -\frac{1}{2}\rho\int_S \phi\,\frac{\partial\phi}{\partial n}\,dS$$

6.5. Kelvin's Minimum Energy Theorem

Statement. If a mass of liquid be set in motion by giving prescribed velocities to its boundaries, the kinetic energy in the actual motion is less than that in any other motion consistent with the same motion of boundaries.

Proof. Let T be the kinetic energy; $\vec{q}$ the fluid velocity, ϕ the velocity potential of actual irrotational motion, then $\vec{q} = -\vec{\nabla}\phi$.

Let T_1 be the kinetic energy and $\vec{q}$ the velocity in any other possible state of motion.

As the equations of continuity must be satisfied, we must have

$$\vec{\nabla}\cdot\vec{q} = 0, \vec{\nabla}\cdot\vec{q}_1 = 0 \qquad \text{... (1)}$$

Since the boundary has the same normal velocity in either case, hence

$$\vec{n}\cdot\vec{q} = \vec{n}\cdot\vec{q}_1 \qquad \text{... (2)}$$

We are to prove that $T_1 > T$.

Now

$$T_1 - T = \frac{1}{2}\rho\int_V q_1^2 dV - \frac{1}{2}\rho\int_V q^2 dV$$

$$= \frac{1}{2}\rho\int_V\left[2\vec{q}\cdot(\vec{q}_1 - \vec{q}) + (\vec{q}_1 - \vec{q})^2\right]dV$$

$$= \rho\int_V \vec{q}(\vec{q}_1 - \vec{q})\,dV + \frac{1}{2}\rho\int_V(\vec{q}_1 - \vec{q})^2\,dV$$

$$= -\rho\int_V(\vec{\nabla}\phi)\cdot(\vec{q}_1 - \vec{q})\,dV + \frac{1}{2}\rho\int_V(\vec{q}_1 - \vec{q})^2\,dV \qquad \text{... (3)}$$

Also

$$\vec{\nabla}\cdot(\phi\vec{u}) = \phi\vec{\nabla}\cdot\vec{u} + \vec{u}\,\vec{\nabla}\phi$$

$$\Rightarrow \quad \vec{\nabla}\cdot\left[\phi(\vec{q}_1 - \vec{q})\right] = \vec{\phi}\vec{\nabla}\cdot(\vec{q}_1 - \vec{q}) + (\vec{\nabla}\phi)(\vec{q}_1 - \vec{q})$$

$$\Rightarrow \quad -\int_V(\vec{\nabla}\phi)\cdot(\vec{q}_1 - \vec{q})\,dV = \int_V \phi\vec{\nabla}\cdot(\vec{q}_1 - \vec{q})\,dV - \int_V\vec{\nabla}\cdot\left[\phi(\vec{q}_1 - \vec{q})\right]dV$$

$$= 0 - \int_V\vec{\nabla}\cdot\left[\phi(\vec{q}_1 - \vec{q})\right]dV \text{ From (1)}$$

$$= -\int_S \phi(\vec{q}_1 - \vec{q}) \cdot \hat{n} dS$$

(by Gauss's Theorem)

$$= 0; \text{ by } (2)$$

Thus (3) becomes

$$T_1 - T = \frac{1}{2}\rho \int_V (\vec{q}_1 - \vec{q})^2 \, dV > 0$$

or $T_1 > T$

6.6. Mean Value of Velocity Potential over a Spherical Surface

If a region lying wholly in the liquid be bounded by a spherical surface throughout whose interior $\vec{\nabla}^2 \phi = 0$, the mean value of the velocity potential ϕ over the surface is equal to its value at the centre of the sphere.

Let S denote the surface of a sphere of radius r with its centre at P; the sphere lies wholly in a liquid which is at rest at infinity. We draw another concentric sphere of unit radius.

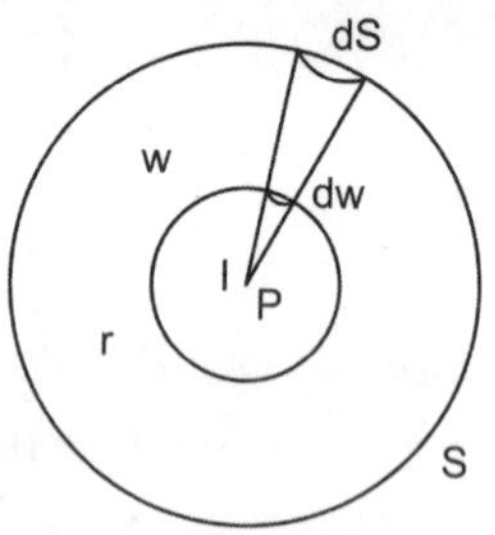

Fig. 6.3

Then a cone with vertex P which intercepts dS from the sphere S intercepts dW (the solid angle) from the sphere w, such that

$$\frac{dw}{dS} = \frac{r^2}{1^2}$$

$$\Rightarrow \; dS = r^2 dw$$

Let ϕ_C and ϕ_S denote the value of velocity potential at P and mean value of ϕ over S.

Then

$$\phi_S = \int_S \phi dS \Big/ \int_S dS$$

$$\Rightarrow \; \phi_S = \frac{1}{4\pi r^2} \int_S \phi dS = \frac{1}{4\pi} \int_S \phi dw$$

$$\Rightarrow \quad \frac{\partial \phi_S}{\partial r} = \frac{1}{4\pi} \int_S \frac{\partial \phi}{\partial r}\, dw$$

$$\Rightarrow \quad \frac{\partial \phi_S}{\partial r} = \frac{1}{4\pi r^2} \int_S \frac{\partial \phi}{\partial r}\, dS \qquad \ldots (1)$$

Now

$$\int_S \frac{\partial \phi}{\partial r}\, dS = -\int_S \frac{\partial \phi}{\partial n}\, dS = -\int_S \hat{n} \cdot \vec{\nabla}\phi\, dS$$

$$= \int_V \vec{\nabla}^2 \phi\, dV \qquad \text{By Gauss's Theorem}$$

$$= 0, \text{ by Laplace's equation } \vec{\nabla}^2 \phi = 0$$

$\therefore$ From (1)

$$\frac{\partial \phi_S}{\partial r} = 0$$

or ϕ_S is independent of r

Hence, when the sphere S shrinks to a point P, then

$$\phi_S = \phi_C$$

6.7. Uniqueness Theorem

1. *Acyclic irrotational is impossible in a liquid bounded wholly by fixed rigid walls.*

Since boundary is rigid, so normal velocity of boundary $= \dfrac{\partial \phi}{\partial n} = 0$.

$$\text{Hence K.E. } = \frac{1}{2}\rho \int_V q^2 dV$$

$$= -\frac{1}{2}\rho \int_S \phi \frac{\partial \phi}{\partial n}\, dS = 0$$

Thus $\vec{q} = \vec{0}$, everywhere on V.

Therefore liquid is at rest. From this the required result follows.

2. *There cannot be two different forms of acyclic irrotational motion for a given confined mass of liquid whose boundaries have prescribed velocities.*

Let ϕ_1 and ϕ_2 be velocity potential of two different motions.

Then $\dfrac{\partial \phi_1}{\partial n} = \dfrac{\partial \phi_2}{\partial n}$ at every point of the boundary, or $\dfrac{\partial \phi}{\partial n} = 0$

where $\phi = \phi_1 - \phi_2$

Then $\vec{\nabla}^2 \phi = \vec{\nabla}^2 (\phi_1 - \phi_2) = \vec{\nabla}^2 \phi_1 - \vec{\nabla}^2 \phi_2 = 0$

Thus ϕ is a solution of Laplace's equation and represents a possible form of irrotational in which $\dfrac{\partial \phi}{\partial n} = 0$.

Then K.E. $= \dfrac{1}{2}\rho \int\limits_V \vec{q}^2 dV = -\dfrac{1}{2}\rho \int\limits_S \phi \dfrac{\partial \phi}{\partial n} dS = 0$

Hence $\vec{q} = 0$

Therefore ϕ = constant

or $\phi_1 - \phi_2$ = constant

It means that ϕ_1 differs ϕ_2 by a constant. This proves the required result.

6.8. Green's Theorem

Statement. If ϕ_1, ϕ_2 are two continuously differentiable scalar point functions such that $\vec{\nabla}\phi_1$ and $\vec{\nabla}\phi_2$ are also continuously differentiable and S denotes a closed surface bounding any simply connected region of space, then

$$\int\limits_V (\vec{\nabla}\phi_1 \cdot \vec{\nabla}\phi_2) dV = -\int\limits_V \phi_1 \vec{\nabla}^2 \phi_2 dV - \int\limits_S \phi_1 \frac{\partial \phi_2}{\partial n} dS$$

$$= -\int\limits_V \phi_2 \vec{\nabla}^2 \phi_1 dV - \int\limits_S \phi_2 \frac{\partial \phi_1}{\partial n} dS$$

Proof. By vector calculus, we have

$$\vec{\nabla} \cdot (\phi \vec{a}) = \vec{a} \cdot (\vec{\nabla}\phi) + \phi (\vec{\nabla} \cdot \vec{a}) \qquad \text{... (1)}$$

Where ϕ is a scalar and $\vec{a}$ a vector point function.

Putting $\phi = \phi_1, \vec{a} = \vec{\nabla}\phi_2$ in (1), we obtain

$$\vec{\nabla} \cdot (\phi_1 \vec{\nabla}\phi_2) = \vec{\nabla}\phi_2 \cdot \vec{\nabla}\phi_1 + \phi_1 (\vec{\nabla} \cdot \vec{\nabla}\phi_2)$$

$$= \vec{\nabla}\phi_1 \cdot \vec{\nabla}\phi_2 + \phi_1 \vec{\nabla}^2 \phi_2$$

$$\Rightarrow \int_V \vec{\nabla}\cdot\left(\phi_1 \vec{\nabla}\phi_2\right)dV = \int_V \left(\vec{\nabla}\phi_1 \cdot \vec{\nabla}\phi_2\right)dV + \int_V \phi_1 \vec{\nabla}^2 \phi_2 dV$$

By Gauss's divergence theorem

$$-\int_S \hat{n}\cdot\left(\phi_1 \vec{\nabla}\phi_2\right)dS = \int_V \left(\vec{\nabla}\phi_1 \cdot \vec{\nabla}\phi_2\right)dV + \int_V \phi_1 \vec{\nabla}^2 \phi_2 dV \quad \dots (2)$$

Where $\hat{n}$ is the unit inward normal.

But $\hat{n}\cdot\left(\phi_1 \vec{\nabla}\phi_2\right) = \phi_1 \left(\hat{n}\cdot\vec{\nabla}\phi_2\right) = \phi_1 \dfrac{\partial \phi_2}{\partial n}$

Hence (2) becomes

$$-\int_S \phi_1 \frac{\partial \phi_2}{\partial n} dS = \int_V \left(\vec{\nabla}\phi_1 \cdot \vec{\nabla}\phi_2\right)dV + \int_V \phi_1 \vec{\nabla}^2 \phi_2 dV$$

$$\Rightarrow \int_V \left(\vec{\nabla}\phi_1 \cdot \vec{\nabla}\phi_2\right)dV = -\int_V \phi_1 \vec{\nabla}^2 \phi_2 dV - \int_S \phi_1 \frac{\partial \phi_2}{\partial n} dS \quad \dots (3)$$

Interchanging ϕ_1 and ϕ_2 and noting that

$$\vec{\nabla}\phi_1 \cdot \vec{\nabla}\phi_2 = \vec{\nabla}\phi_2 \cdot \vec{\nabla}\phi_1$$

We get

$$\int_V \left(\vec{\nabla}\phi_1 \cdot \vec{\nabla}\phi_2\right)dV = -\int_V \phi_2 \vec{\nabla}^2 \phi_1 dV - \int_S \phi_2 \frac{\partial \phi_1}{\partial n} dS \quad \dots (4)$$

From (3) and (4), the required result follow.

6.9. Kelvin's Circulation Theorem

Statement. The circulation in any closed path moving with the fluid is constant for all times, provided the external forces are conservative and derivable from a single valued potential Ω and the density ρ is a function of pressure only.

Proof. If C be a closed circuit consisting of the same fluid particles, then

$$\frac{d\Gamma}{dt} = \frac{d}{dt}(\text{circulation}) = \frac{d}{dt}\left[\int_C \vec{q}\cdot d\vec{r}\right]$$

$$= \int_C \left[\frac{d\vec{q}}{dt}\cdot d\vec{r} + \vec{q}\cdot d\left(\frac{d\vec{r}}{dt}\right)\right] \quad \dots (1)$$

Where $\Gamma = \int_C \vec{q}\cdot d\vec{r}$

Euler's equation of motion is

$$\frac{d\vec{q}}{dt} = \vec{F} - \frac{1}{\rho}\vec{\nabla}p = -\vec{\nabla}\Omega - \frac{1}{\rho}\vec{\nabla}p \qquad \text{... (2)}$$

From (1) and (2), we have

$$\frac{d\Gamma}{dt} = \int_C \left[\left(-\vec{\nabla}\Omega - \frac{1}{\rho}\vec{\nabla}p\right)\cdot d\vec{r} + \vec{q}\cdot d\vec{q}\right]$$

$$= \int_C \left[-d\Omega - \frac{dp}{\rho} + \vec{q}\cdot d\vec{q}\right]$$

$$\Rightarrow \quad \frac{d\Gamma}{dt} = -\int_C d\left[\Omega + \int\frac{dp}{\rho} - \frac{1}{2}q^2\right]$$

$$\text{or} \quad \frac{d\Gamma}{dt} = -\int\frac{dp}{\rho} \qquad \text{... (3)}$$

This gives the rate of change of flow along any line moving with the fluid. Since ρ is a single valued function of density, the integral in (3) vanishes.

Thus we have $\frac{d\Gamma}{dt} = 0.$

$\Rightarrow$ Γ = constant.

Therefore the circulation Γ is independent of time t.

Deduction. *If any portion of the moving fluid once irrotational under the above conditions, then it remains so for all subsequent times.*

If the motion is irrotational at any instant of time t, then for every closed curve circulation is zero i.e.

$$\Gamma = \int_C \vec{q}\cdot d\vec{r}$$

By Stoke's theorem

$$\Gamma = \iint_S \left(\vec{\nabla}\times\vec{q}\right)dS$$

Since the motion is irrotational at all points of S, therefore $\vec{\nabla}\times\vec{q} = 0$

$$\Rightarrow \quad \iint_S (\vec{\nabla} \times \vec{q})\, dS = 0$$

$$\Rightarrow \quad \Gamma = 0$$

It follows that the irrotational motion is permanent.

WORKED OUT EXAMPLES

1. Show that in the motion of a fluid in two dimensions of the coordinates (x, y) of an element at any time can be expressed in terms of initial coordinates (a, b) and the time, the motion is irrotational if

$$\frac{\partial(\dot{x}, x)}{\partial(a,b)} + \frac{\partial(\dot{y}, y)}{\partial(a,b)} = 0$$

Solution: Let u and v be the components of velocity parallel to coordinate axes, then

$$u = \dot{x} = \frac{dx}{dt} \text{ and } v = \dot{y} = \frac{dy}{dt}$$

Again

$$\frac{\partial u}{\partial a} = \frac{\partial u}{\partial x}\frac{\partial x}{\partial a} + \frac{\partial u}{\partial y}\frac{\partial y}{\partial a} = \frac{\partial u}{\partial x}\frac{\partial x}{\partial b} + \frac{\partial u}{\partial y}\frac{\partial y}{\partial b}$$

Similar expressions may be written for $\dfrac{\partial v}{\partial a}$ and $\dfrac{\partial v}{\partial b}$.

Now

$$\frac{\partial(\dot{x}, x)}{\partial(a,b)} = \frac{\partial(u, x)}{\partial(a,b)} = \frac{\partial u}{\partial a}\cdot\frac{\partial x}{\partial b} - \frac{\partial x}{\partial a}\frac{\partial u}{\partial b}$$

$$= \left(\frac{\partial u}{\partial x}\frac{\partial x}{\partial a} + \frac{\partial u}{\partial y}\frac{\partial y}{\partial a}\right)\frac{\partial x}{\partial b} - \left(\frac{\partial u}{\partial x}\frac{\partial x}{\partial b} + \frac{\partial u}{\partial y}\frac{\partial y}{\partial b}\right)\frac{\partial x}{\partial a}$$

$$= \frac{\partial u}{\partial y}\left(\frac{\partial x}{\partial b}\frac{\partial y}{\partial a} - \frac{\partial y}{\partial b}\frac{\partial y}{\partial a}\right) = \frac{\partial u}{\partial y}\frac{\partial(x, y)}{\partial(a,b)}$$

Hence

$$\frac{\partial(\dot{x}, x)}{\partial(a,b)} = -\frac{\partial u}{\partial y}\frac{\partial(x, y)}{\partial(a,b)} \qquad \dots (1)$$

Also

$$\frac{\partial(\dot{y},y)}{\partial(a,b)}=\frac{\partial(v,y)}{\partial(a,b)}=\frac{\partial v}{\partial a}\frac{\partial y}{\partial b}-\frac{\partial y}{\partial a}\frac{\partial v}{\partial b}$$

$$=\left(\frac{\partial v}{\partial x}\frac{\partial x}{\partial a}+\frac{\partial v}{\partial y}\frac{\partial y}{\partial a}\right)\frac{\partial y}{\partial b}-\left(\frac{\partial v}{\partial x}\frac{\partial x}{\partial b}+\frac{\partial v}{\partial y}\frac{\partial y}{\partial b}\right)\frac{\partial y}{\partial a}$$

$$=\frac{\partial v}{\partial x}\left(\frac{\partial x}{\partial a}\frac{\partial y}{\partial b}-\frac{\partial x}{\partial b}\frac{\partial y}{\partial a}\right)$$

$$=\frac{\partial v}{\partial x}\frac{\partial(x,y)}{\partial(a,b)}$$

Therefore

$$\frac{\partial(\dot{y},y)}{\partial(a,b)}=\frac{\partial v}{\partial x}\frac{\partial(x,y)}{\partial(a,b)} \quad \text{... (2)}$$

(1) + (2) gives

$$\frac{\partial(\dot{x},x)}{\partial(a,b)}+\frac{\partial(\dot{y},y)}{\partial(a,b)}=\frac{\partial(x,y)}{\partial(a,b)}\left[\frac{\partial v}{\partial x}-\frac{\partial u}{\partial y}\right]$$

$$=0 \qquad \text{iff } \frac{\partial v}{\partial x}-\frac{\partial u}{\partial y}=0$$

i.e. iff motion is irrotational.

2. If a, b, c, d, e, f are arbitrary constants, what type of fluid motion does the velocity $(a + by - cz, d - bx + ez, f + cx - ey)$ represent? (I.A.S. 95)

Solution: The velocity components are given as

$$u = a + by - cz,\ v = d - bx + ez,\ w = f + cx - ey$$

Now

$$\vec{q}=u\vec{i}+v\vec{j}+w\vec{k}$$

$$\vec{q}=(a+by-cz)\,\vec{i}+(d-bx+ez)\vec{j}+(f+cx-ey)\vec{k}$$

$$\frac{\partial\vec{q}}{\partial x}=-b\vec{j}+c\vec{k},\ \frac{\partial\vec{q}}{\partial y}=b\vec{i}-e\vec{k},\ \frac{\partial\vec{q}}{\partial z}=-c\vec{i}+e\vec{j}$$

Now

$$\vec{\nabla}\cdot\vec{q}=\vec{i}\cdot\frac{\partial\vec{q}}{\partial x}+\vec{j}\cdot\frac{\partial\vec{q}}{\partial y}+\vec{k}\cdot\frac{\partial\vec{q}}{\partial z}$$

$$= \vec{i} \cdot (c\vec{k} - b\vec{j}) + \vec{j} \cdot (b\vec{i} - c\vec{k}) + \vec{k} \cdot (e\vec{j} - c\vec{i})$$
$$= 0$$

Hence fluid motion is possible.

Rotational components are given as

$$w_x = \frac{1}{2}\left(\frac{\partial w}{\partial y} - \frac{\partial v}{\partial z}\right) = -e$$

$$w_y = \frac{1}{2}\left(\frac{\partial u}{\partial z} - \frac{\partial w}{\partial x}\right) = -c$$

and

$$w_z = \frac{1}{2}\left(\frac{\partial v}{\partial x} - \frac{\partial u}{\partial y}\right) = -b$$

Hence

$$\bar{w} = -(e\vec{i} + c\vec{j} + b\vec{k})$$

which shows that the velocity field represents a rigid body rotation of fluid at constant angular velocity.

Shear strain rate components are given as

$$\in_{xy} = \frac{1}{2}\left(\frac{\partial v}{\partial x} + \frac{\partial u}{\partial y}\right) = 0, \in_{yz} = \in_{zx} = 0$$

Also, the normal strain rate components are

$$\in_{xx} = 0 = \in_{yy} = \in_{zz}$$

It follows that the fluid motion is devoid of strain rates in any plane but rotation components are finite and constant. This is equivalent to pure rotation superimposed on pure translation.

3. Determine the different types of motion involved with velocity field.

$$\vec{q} = (Ax + 2By + U_0)\vec{i} + (Ay + V_0)\vec{j} + (-2Az + W_0)\vec{k}$$

where A, B, U_0, V_0 and W_0 are constants.

Solution: We know that the general motion of a fluid element is made up of three parts, viz., translation, rotation and deformation.

(1) Translation velocity

$$\vec{q} = (Ax + 2By + U_0)\vec{i} + (Ay + V_0)\vec{j} + (-2Az + W_0)\vec{k}$$

(2) Rotational velocity

$$= \vec{w} = \frac{1}{2}\text{curl}\,\vec{q}$$

$$= \frac{1}{2}\begin{vmatrix} \vec{i} & \vec{j} & \vec{k} \\ \frac{\partial}{\partial x} & \frac{\partial}{\partial y} & \frac{\partial}{\partial z} \\ Ax + 2By + U_0 & Ay + V_0 & -2Az + W_0 \end{vmatrix}$$

$$= \frac{1}{2}\left[\left\{\frac{\partial}{\partial y}(-2Az + W_0) - \frac{\partial}{\partial z}(Ay + V_0)\right\}\vec{i}\right.$$

$$-\left\{\frac{\partial}{\partial x}(-2Az + W_0) - \frac{\partial}{\partial z}(Ax + 2By + U_0)\right\}\vec{j}$$

$$\left.+\vec{k}\left\{\frac{\partial}{\partial x}(Ay + V_0) - \frac{\partial}{\partial y}(Ax + 2By + U_0)\right\}\right]$$

$$= \frac{1}{2}\left[0 - 2B\vec{k}\right]$$

$$= -B\vec{k}$$

which represents a constant angular velocity about the z-axis.

(3) Deformation. Rate of normal strain.

$$\in_{xx} = \frac{\partial u}{\partial x} = A, \in_{yy} = \frac{\partial v}{\partial y} = A, \in_{zz} = \frac{\partial w}{\partial z} = -2A$$

Rate of shear strain

$$\in_{xy} = \in_{yx} = \frac{1}{2}\left(\frac{\partial u}{\partial y} + \frac{\partial v}{\partial x}\right) = B$$

$$\in_{yz} = \in_{zy} = \frac{1}{2}\left(\frac{\partial v}{\partial z} + \frac{\partial w}{\partial y}\right) = 0$$

$$\in_{zx} = \in_{xz} = \frac{1}{2}\left(\frac{\partial w}{\partial x} + \frac{\partial u}{\partial z}\right) = 0$$

Thus

$$D = \vec{i}\,(Adx + Bdy) + \vec{j}\,(Bdx + Ady) + \vec{k}\,(-2Adz)$$

4. Find the vorticity of the fluid motion for the velocity components

$$u = \frac{-2xyz}{\left(x^2 + y^2\right)^2},\ v = \frac{\left(x^2 - y^2\right)z}{\left(x^2 + y^2\right)^2},\ w = \frac{y}{x^2 + y^2}$$

Solution: Let $\bar{\Omega}$ be the vorticity vector given as

$$\bar{\Omega} = wx\vec{i} + wy\vec{j} + wz\vec{k}$$

$$= \left(\frac{\partial w}{\partial y} - \frac{\partial v}{\partial z}\right)\vec{i} + \left(\frac{\partial u}{\partial z} - \frac{\partial w}{\partial x}\right)\vec{j} + \left(\frac{\partial v}{\partial x} - \frac{\partial u}{\partial y}\right)\vec{k}$$

$$= \left\{\frac{x^2 - y^2}{\left(x^2 + y^2\right)^2} - \frac{\left(x^2 - y^2\right)}{\left(x^2 + y^2\right)^2}\right\}\vec{i}$$

$$+ \left\{\frac{-2xy}{\left(x^2 + y^2\right)^2} + \frac{2xy}{\left(x^2 + y^2\right)^2}\right\}\vec{j}$$

$$+ \left\{\frac{2xz\left(3y^2 - x^2\right)}{\left(x^2 + y^2\right)^2} - \frac{2xz\left(3y^2 - x^2\right)}{\left(x^2 + y^2\right)^2}\right\}\vec{k}$$

$$\Rightarrow \quad \bar{\Omega} = 0$$

The vorticity vector of the fluid motion is zero, it follows that the flow is irrotational.

5. Find the vorticity of the fluid motion in spherical polar coordinates for the velocity components

$$q_r = \left(1 - \frac{A}{r^3}\right)\cos\theta,\ q_\theta = -\left(1 + \frac{A}{2r^3}\right)\sin\theta,\ q_\phi = 0$$

Solution: Let $\bar{\Omega}\,(\Omega r, \Omega\theta, \Omega\phi)$ be the vorticity vector in the spherical polar coordinates (r, θ, ϕ) then

$$\Omega_r = \frac{1}{r}\frac{\partial q_\phi}{\partial\theta} - \frac{1}{r\sin\theta}\frac{\partial q_\theta}{\partial\phi} + \frac{q_\phi}{r}\cot\theta$$

$$\Rightarrow \quad \Omega_r = 0$$

$$\Omega_\theta = \frac{1}{\partial \sin\theta}\frac{\partial q_r}{\partial \phi} - \frac{\partial q_\phi}{\partial r} - \frac{q_\phi}{r}$$

$$\Rightarrow \quad \Omega_\theta = 0$$

and

$$\Omega_\phi = \frac{\partial q_\theta}{\partial r} + \frac{q_\theta}{r} - \frac{1}{r}\frac{\partial q_r}{\partial \theta}$$

$$= \left(\frac{3A}{2r^4}\right)\sin\theta - \left(1 + \frac{A}{2r^3}\right)\frac{\sin\theta}{r} + \frac{1}{r}\left(1 - \frac{A}{r^3}\right)\sin\theta$$

$$= \sin\theta\left[\frac{3A}{2r^4} - \frac{1}{r} - \frac{A}{2r^4} + \frac{1}{r} - \frac{A}{r^4}\right]$$

$$= 0$$

Therefore $\Omega_r = \Omega_\theta = \Omega_\phi = 0$, it follows that the motion is irrotational.

6. A velocity field is given by

$$\vec{q} = \frac{\left(-\vec{i}y + \vec{j}x\right)}{x^2 + y^2}$$

Determine whether the flow is irrotational. Calculate the circulation round

(i) a square with its corners at (1,0), (2,0), (2,1), (1,1)

(ii) a unit circle with centre at the origin.

Solution: (i) We have

$$\text{Curl}\ \vec{q} = \begin{vmatrix} \vec{i} & \vec{j} & \vec{k} \\ \dfrac{\partial}{\partial x} & \dfrac{\partial}{\partial y} & \dfrac{\partial}{\partial z} \\ \dfrac{-y}{x^2+y^2} & \dfrac{x}{x^2+y^2} & 0 \end{vmatrix}$$

$$= \vec{k}\left\{\frac{\left(y^2 - x^2\right) + \left(x^2 - y^2\right)}{\left(x^2 + y^2\right)^2}\right\}$$

$$= 0$$

It follows that curl $\vec{q}$ is zero everywhere except at the origin. At origin it contains a singularity, therefore, the flow is irrotational.

We draw a square in the Cartesian plane as follows: $A(1,0)$, $B(2,0)$, $C(2,1)$, $D(1,1)$.

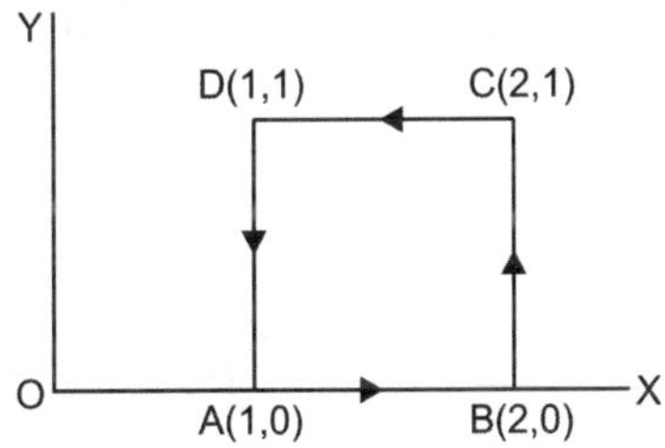

Fig. 6.4

The circulation Γ around the square $ABCD$ is given by

$$\Gamma = \int \vec{q}\cdot d\vec{r}$$

$$= \int_A^B \vec{q}\cdot d\vec{r} + \int_B^C \vec{q}\cdot d\vec{r} + \int_C^D \vec{q}\cdot d\vec{r} + \int_D^A \vec{q}\cdot d\vec{r} \qquad \dots (1)$$

Now

$$\int_A^B \vec{q}\cdot d\vec{r} = \int_{x=1}^{2} \left[\frac{(-\vec{i}y + \vec{j}x)}{x^2 + y^2}\right]_{y=0} (dx\vec{i})$$

Along AB $y = 0$, so $dy = 0$ and hence $d\vec{r} = dx\vec{i} + dy\vec{j} + dx\vec{i}$

$$= 0$$

Also

$$\int_C^D \vec{q}\cdot d\vec{r} = \int_{x=2}^{3} \left[\frac{-\vec{i}y + \vec{j}x}{x^2 + y^2}\right]_{y=1} (dx\vec{i})$$

$$= \int_1^2 \frac{dx}{x^2 + 1} = \tan^{-1} 2 - \tan^{-1} 1$$

Again

$$\int_B^C \vec{q}\cdot d\vec{r} = \int_0^1 \left[\frac{-\vec{i}y + \vec{j}x}{x^2 + y^2}\right]_{x=2} (\vec{j}dy)$$

$$= \int_0^1 \frac{2dy}{y^2 + 4} = \tan^{-1}\frac{1}{2}$$

Since along the line BC, which is parallel to y-axis, $x = 2$ and $dx = 0$.

Also $d\vec{r} = dx\vec{i} + dy\vec{j} = dy\vec{j}$

Now

$$\int_D^C \vec{q}\cdot d\vec{r} = \int_{y=1}^{y=0} \left\{\frac{-\vec{i}y + \vec{j}x}{x^2 + y^2}\right\}_{x=1} (\vec{j}dy)$$

$$= \int_1^0 \frac{dy}{y^2 + 1} = -\tan^{-1} 1$$

Putting these values (1) becomes

$$\Gamma = \tan^{-1}\frac{1}{2} + \tan^{-1} 2 - \tan^{-1} 1 - \tan^{-1} 1$$

$$= \tan^{-1}\frac{2 + \frac{1}{2}}{1 - \frac{1}{2}\cdot 2} - 2\tan^{-1} 1$$

$$= \frac{\pi}{2} - \frac{\pi}{2} = 0$$

It follows that curl $\vec{q}$ is zero everywhere inside the square.

(ii) Now, we shall determine the circulation around the unit circle with centre at the origin.

Let $x = r\cos\theta$, $y = r\sin\theta$, then

$$\vec{q} = \frac{-r\sin\theta\vec{i} + r\cos\theta\vec{j}}{r^2} = \frac{-\sin\theta}{r}\vec{i} + \frac{\cos\theta}{r}\vec{j}$$

So that $U = \dfrac{-\sin\theta}{r}$ and $V = \dfrac{\cos\theta}{r}$

Hence $q_r = U\cos\theta + V\sin\theta = 0$

and $q_\theta = -U\sin\theta + V\cos\theta = \dfrac{1}{r}$

Hence

$$\Gamma = \int \vec{q}\cdot d\vec{r} = \int_0^{2\pi} \frac{1}{r}\, r d\theta = 2\pi$$

7. Prove that if

$$\lambda = \frac{\partial u}{\partial t} - v\left(\frac{\partial v}{\partial x} - \frac{\partial u}{\partial y}\right) + w\left(\frac{\partial u}{\partial z} - \frac{\partial w}{\partial x}\right)$$

and μ, υ have two similar expressions, then $\lambda dx + \mu dy + \upsilon dz$ is a perfect differential, if the forces are conservative and density is constant.

Solution: Let the force F be conservative and ρ be constant. Then $\vec{F} = -\vec{\nabla}V$, V being force potential function.

Then by Euler's equation of motion

$$\frac{d\vec{q}}{dt} = \vec{F} - \frac{1}{\rho}\vec{\nabla}p = -\vec{\nabla}v - \vec{\nabla}\left(\frac{p}{\rho}\right)$$

Thus

$$\frac{du}{dt} = -\frac{\partial}{\partial x}\left(v + \frac{p}{\rho}\right) \text{ etc.} \qquad \dots (1)$$

Given that

$$\lambda = \frac{\partial u}{\partial t} - v\left(\frac{\partial v}{\partial x} - \frac{\partial u}{\partial y}\right) + w\left(\frac{\partial u}{\partial z} - \frac{\partial w}{\partial x}\right)$$

$$\Rightarrow \quad \lambda = \left(\frac{\partial}{\partial t} + u\frac{\partial}{\partial x} + v\frac{\partial}{\partial y} + w\frac{\partial}{\partial z}\right)u - \left(u\frac{\partial u}{\partial x} + v\frac{\partial v}{\partial x} + w\frac{\partial w}{\partial x}\right)$$

$$\Rightarrow \quad \lambda = \frac{du}{\partial t} - \frac{\partial}{\partial x}\left(\frac{u^2 + v^2 + w^2}{2}\right)$$

$$= \frac{-\partial}{\partial x}\left(v + \frac{p}{\rho}\right) - \frac{1}{2}\frac{\partial q^2}{\partial x} \qquad \text{From (1)}$$

$$= -\frac{\partial}{\partial x}\left(v + \frac{p}{\rho} + \frac{1}{2}q^2\right)$$

$$\Rightarrow \quad \lambda = \frac{-\partial Q}{\partial x} \qquad \text{where } Q = v + \frac{p}{\rho} + \frac{1}{2}q^2$$

$$\Rightarrow \quad \lambda dx = \frac{-\partial \phi}{\partial x}dx$$

In view of this, we have

$$\lambda dx + \upsilon dy + \mu dz = -\left(\frac{\partial Q}{\partial x}dx + \frac{\partial Q}{\partial y}dy + \frac{\partial Q}{\partial z}dz\right)$$

$$= -\,dQ$$

$$= \text{perfect differential.}$$

Hence $\lambda dx + \upsilon dy + \mu dz$ is a perfect differential.

8. Show that the theorem under certain conditions the motion of a frictionless fluid if once irrotational, will always be so, is true also when each particle is acted on by a resistance varying as the velocity.

Solution: We consider two adjacent points $P(x, y, z)$ and $Q\ (x + \delta x, y + \delta y, z + \delta z)$ such that $PQ = \delta s$. Then

$$\frac{d}{dt}(\text{Flow along } PQ) = \frac{d}{dt}(u\delta x + v\delta y + w\delta z)$$

$$\Rightarrow \quad \frac{d}{dt}(\text{Flow}) = \frac{d}{dt}\sum u\delta x \qquad \ldots (1)$$

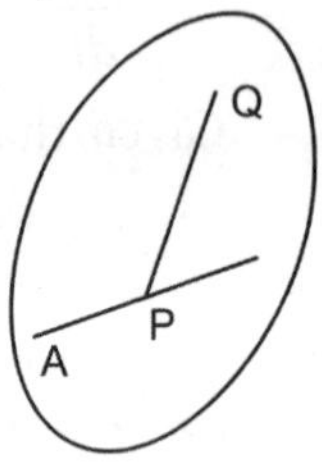

Fig. 6.5

But

$$\frac{d}{dt}(u\delta x) = \frac{du}{dt}\delta x + u\frac{d}{dt}\delta x$$

$$= \frac{du}{dt}\delta x + u\delta\left(\frac{dx}{dt}\right)$$

$$\Rightarrow \quad \frac{d}{dt}(u\delta x) = \frac{du}{dt}\delta x + u\delta x \qquad \ldots (2)$$

Given that there exists an extra force $k\vec{q}$.

By Euler's equation of motion

$$\frac{d\vec{q}}{dt} = -\vec{\nabla}v - \frac{1}{\rho}\vec{\nabla}p - k\vec{q}$$

Therefore

$$\frac{du}{dt} = \frac{-\partial u}{\partial x} - \frac{1}{\rho}\frac{\partial p}{\partial x} - ku$$

Hence (2) becomes

$$\frac{d}{dt}(u\delta x) = \left(-\frac{\partial v}{\partial x} - \frac{1}{\rho}\frac{\partial p}{\partial x} - ku\right)\delta x + u\delta u$$

$$\therefore \quad \sum \frac{d}{dt}(u\delta x) = -\sum \frac{\partial v}{\partial x}\delta x - \frac{1}{\rho}\sum \frac{\partial p}{\partial x}\delta x - k\sum u\delta x + \sum u\delta u$$

$$\Rightarrow \quad \frac{d}{dt}\sum u\delta x = -\delta v - \frac{1}{p}\delta p + \frac{1}{2}\delta q^2 - k\sum u\delta x$$

Hence (1) becomes

$$\frac{d}{dt}(\text{Flow}) = -\delta v - \frac{1}{\rho}\delta p + \frac{1}{2}\delta q^2 - k\sum u\delta x \qquad \dots (3)$$

If Γ be the circulation along APA, then

$$\Gamma = \int_A^A (u\delta x + v\delta y + w\delta z) = \int_A^A \left[\sum u\delta x\right] \qquad \dots (4)$$

$$\Rightarrow \quad \frac{d\Gamma}{dt} = \int_A^A \frac{d}{dt}\sum u\delta x = \int_A^A \frac{d}{dt}(\text{Flow})$$

$$= \int_A^A \left[-\delta v - \frac{1}{\rho}\delta p + \frac{1}{2}\delta q^2 - k\sum u\delta x\right] \qquad \text{by (3)}$$

$$= \left[-v + \frac{1}{2}q^2 - \int \frac{dp}{\rho}\right]_A^A - k\Gamma \qquad \text{by (4)}$$

$$= 0 - k\Gamma \text{ if } v \text{ is single valued and } f(\rho)$$

Thus

$$\frac{d\Gamma}{dt} = -k\Gamma$$

$$\Rightarrow \quad \frac{d\Gamma}{\Gamma} = -kdt$$

Integrating, we get $\Gamma = ce^{-kt}$

If $t = 0$, then $\Gamma = \Gamma_0$ (say) so that $c = \Gamma_0$.

Hence

$$\Gamma = \Gamma_0 e^{-kt} \qquad \dots (5)$$

Let the motion be irrotational at time $t = 0$, then $\Gamma_0 = 0$ (as circulation vanishes in case of irrotational motion).

Hence from (5) $\Gamma = 0 \;\forall t$.

It means that if the motion is once irrotational, then it is always irrotational.

9. In an irrotational motion in two dimensions, prove that

$$\left(\frac{\partial q}{\partial x}\right)^2+\left(\frac{\partial q}{\partial y}\right)^2=q\,\vec{\nabla}^2 q$$

Solution: In an irrotational motion, the velocity potential ϕ exists and satisfies the Laplace equation

$$\vec{\nabla}^2\phi=\frac{\partial^2\phi}{\partial x^2}+\frac{\partial^2\phi}{\partial y^2}=0 \qquad \dots (1)$$

Let $\vec{q}$ be the velocity at a point, then

$$q^2=\left(\frac{\partial\phi}{\partial x}\right)^2+\left(\frac{\partial\phi}{\partial y}\right)^2 \qquad \dots (2)$$

Differentiating (2) partially w.r.t. x and y, we get

$$q\frac{\partial q}{\partial x}=\frac{\partial\phi}{\partial x}\frac{\partial^2\phi}{\partial x^2}+\frac{\partial\phi}{\partial y}\frac{\partial^2\phi}{\partial y\partial x} \qquad \dots (3)$$

and

$$q\frac{\partial q}{\partial y}=\frac{\partial\phi}{\partial x}\frac{\partial^2\phi}{\partial x\partial y}+\frac{\partial\phi}{\partial y}\frac{\partial^2\phi}{\partial y^2} \qquad \dots (4)$$

Squaring and adding (3) and (4), we get

$$q^2\left[\left(\frac{\partial q}{\partial x}\right)^2+\left(\frac{\partial q}{\partial y}\right)^2\right]=\left(\frac{\partial\phi}{\partial x}\right)^2\left[\left(\frac{\partial^2\phi}{\partial x^2}\right)^2+\left(\frac{\partial^2\phi}{\partial x\partial y}\right)^2\right]$$
$$+\left(\frac{\partial\phi}{\partial y}\right)^2\left[\left(\frac{\partial^2\phi}{\partial x\partial y}\right)^2+\left(\frac{\partial^2\phi}{\partial y^2}\right)^2\right]$$
$$+2\frac{\partial\phi}{\partial x}\frac{\partial\phi}{\partial y}\frac{\partial^2\phi}{\partial x\partial y}\left(\frac{\partial^2\phi}{\partial x^2}+\frac{\partial^2\phi}{\partial y^2}\right)$$

Using (2), we get

$$q^2\left[\left(\frac{\partial q}{\partial x}\right)^2+\left(\frac{\partial q}{\partial y}\right)^2\right]=\left(\frac{\partial^2\phi}{\partial x\partial y}\right)^2\left[\left(\frac{\partial\phi}{\partial x}\right)^2+\left(\frac{\partial\phi}{\partial y}\right)^2\right]$$
$$+\left(\frac{\partial^2\phi}{\partial x^2}\right)^2\left[\left(\frac{\partial\phi}{\partial x}\right)^2+\left(\frac{\partial\phi}{\partial y}\right)^2\right]$$

$$= q^2\left[\left(\frac{\partial^2\phi}{\partial x\partial y}\right)^2 + \left(\frac{\partial^2\phi}{\partial x^2}\right)^2\right]$$

$$\Rightarrow \quad \left(\frac{\partial q}{\partial x}\right)^2 + \left(\frac{\partial q}{\partial y}\right)^2 = \left(\frac{\partial^2\phi}{\partial x\partial y}\right)^2 + \left(\frac{\partial^2\phi}{\partial x^2}\right)^2 \qquad \ldots (5)$$

Differentiating (3) and (4) partially with respect to x and y respectively, we get

$$q\frac{\partial^2 q}{\partial x^2} + \left(\frac{\partial q}{\partial x}\right)^2 = \left(\frac{\partial^2\phi}{\partial x^2}\right)^2 + \frac{\partial\phi}{\partial x}\frac{\partial^3\phi}{\partial x^3} + \left(\frac{\partial^2\phi}{\partial x\partial y}\right)^2 + \frac{\partial\phi}{\partial y}\frac{\partial^3\phi}{\partial x^2\partial y}$$

and

$$q\frac{\partial^2 q}{\partial y^2} + \left(\frac{\partial q}{\partial y}\right)^2 = \left(\frac{\partial^2\phi}{\partial x\partial y}\right)^2 + \frac{\partial\phi}{\partial x}\frac{\partial^3\phi}{\partial y^2\partial x} + \left(\frac{\partial^2\phi}{\partial y^2}\right)^2 + \frac{\partial\phi}{\partial y}\frac{\partial^3\phi}{\partial y^3}$$

Adding these two, we get

$$q\left(\frac{\partial^2 q}{\partial x^2} + \frac{\partial^2 q}{\partial y^2}\right) + \left(\frac{\partial q}{\partial x}\right)^2 + \left(\frac{\partial q}{\partial y}\right)^2$$

$$= \left[\left(\frac{\partial^2\phi}{\partial x^2}\right)^2 + \left(\frac{\partial^2\phi}{\partial y^2}\right)^2\right] + 2\left(\frac{\partial^2\phi}{\partial x\partial y}\right)^2$$

$$+ \frac{\partial\phi}{\partial x}\left(\frac{\partial^3\phi}{\partial x^3} + \frac{\partial^3\phi}{\partial y^2\partial x}\right) + \frac{\partial\phi}{\partial y}\left(\frac{\partial^3\phi}{\partial y^3} + \frac{\partial^3\phi}{\partial x^2\partial y}\right) \qquad \ldots (6)$$

But

$$\vec{\nabla}^2\phi = 0$$

$$\Rightarrow \quad \frac{\partial^2\phi}{\partial x^2} = -\frac{\partial^2\phi}{\partial y^2}$$

$$\Rightarrow \quad \frac{\partial^2\phi}{\partial x^3} = -\frac{\partial^2\phi}{\partial y^2\partial x}$$

and

$$\frac{\partial^3\phi}{\partial x^2 \partial y} = \frac{-\partial^3\phi}{\partial y^3}$$

Now (6) becomes

$$q\vec{\nabla}^2 q + \left[\left(\frac{\partial q}{\partial x}\right)^2 + \left(\frac{\partial q}{\partial y}\right)^2\right] = 2\left[\left(\frac{\partial^2\phi}{\partial x^2}\right)^2 + \left(\frac{\partial^2\phi}{\partial x \partial y}\right)^2\right]$$

$$= 2\left[\left(\frac{\partial q}{\partial x}\right)^2 + \left(\frac{\partial q}{\partial y}\right)^2\right] \text{ From (5)}$$

$$\Rightarrow \quad q\vec{\nabla}^2 q = \left(\frac{\partial q}{\partial x}\right)^2 + \left(\frac{\partial q}{\partial y}\right)^2.$$

10. If p denote the pressure, V the potential of the external forces and $\vec{q}$ the velocity of a homogeneous liquid moving irrotationally, show that $\vec{\nabla}^2 q^2$ is positive and $\vec{\nabla}^2 p$ is negative provided $\vec{\nabla}^2 V = 0$. Hence prove that the velocity potential cannot have a maximum value and the pressure cannot have a minimum value at a point in the interior of the liquid.

Solution: Since the motion is irrotational, the velocity potential ϕ exists, such that

$$\vec{q} = -\vec{\nabla}\phi$$

Also

$$q^2 = \left(\frac{\partial\phi}{\partial x}\right)^2 + \left(\frac{\partial\phi}{\partial y}\right)^2 + \left(\frac{\partial\phi}{\partial z}\right)^2 = \sum\left(\frac{\partial\phi}{\partial x}\right)^2$$

Since

$$\vec{\nabla} fg = f\vec{\nabla} g + g\vec{\nabla} f$$

$$\therefore \quad \vec{\nabla}\cdot\left(\vec{\nabla} fg\right) = \vec{\nabla}\cdot\left(f\vec{\nabla} g\right) + \vec{\nabla}\cdot\left(g\vec{\nabla} f\right)$$

$$\Rightarrow \quad \vec{\nabla}^2 fg = \left[\left(\vec{\nabla} f\cdot\vec{\nabla} g\right) + f\vec{\nabla}\cdot\vec{\nabla} g\right] + \left[\vec{\nabla} g\cdot\vec{\nabla} f + g\vec{\nabla}\cdot\vec{\nabla} f\right]$$

$$= 2\vec{\nabla} f\cdot\vec{\nabla} g + f\vec{\nabla}^2 g + g\vec{\nabla}^2 f$$

In view of this, we have

$$\vec{\nabla}^2\left(\frac{\partial\phi}{\partial x}\right)^2 = \vec{\nabla}^2\left(\frac{\partial\phi}{\partial x}\right)\cdot\left(\frac{\partial\phi}{\partial x}\right)$$

$$= 2\frac{\partial\phi}{\partial x}\vec{\nabla}^2\left(\frac{\partial\phi}{\partial x}\right) + 2\left(\vec{\nabla}\frac{\partial\phi}{\partial x}\right)\cdot\left(\vec{\nabla}\frac{\partial\phi}{\partial x}\right)$$

But

$$\vec{\nabla}^2\frac{\partial\phi}{\partial x} = \frac{\partial}{\partial x}\vec{\nabla}^2\phi = 0$$

Hence

$$\vec{\nabla}^2\left(\frac{\partial\phi}{\partial x}\right)^2 = 2\left[\left(\frac{\partial^2\phi}{\partial x^2}\right)^2 + \left(\frac{\partial^2\phi}{\partial y\partial x}\right)^2 + \left(\frac{\partial^2\phi}{\partial z\partial x}\right)^2\right] > 0$$

Hence

$$\vec{\nabla}^2\sum\left(\frac{\partial\phi}{\partial x}\right)^2 > 0$$

$$\Rightarrow \quad \vec{\nabla}^2 q^2 > 0 \qquad \ldots (1)$$

From Bernoulli's equation, we have

$$\frac{p}{\rho} - \frac{\partial\phi}{\partial f} + \frac{1}{2}q^2 + V = f(t)$$

Hence

$$\vec{\nabla}^2\left(\frac{p}{\rho}\right) - \frac{\partial}{\partial t}\left(\vec{\nabla}^2\phi\right) + \frac{1}{2}\vec{\nabla}^2 q^2 + \vec{\nabla}^2 V = 0$$

But

$$\vec{\nabla}^2\phi = 0 = \nabla^2 V$$

$$\Rightarrow \quad \frac{\vec{\nabla}^2 p}{\rho} = -\frac{1}{2}\vec{\nabla}^2 q^2 < 0 \qquad \text{as } \vec{\nabla}^2 q^2 > 0 \qquad \text{by (1)}$$

Thus

$$\vec{\nabla}^2 p < 0 \qquad \ldots (2)$$

To prove the second part let p be a point in the interior of the liquid. We enclose P by a surface S_1.

$$\vec{\nabla}^2 q^2 > 0$$

$$\Rightarrow \quad \vec{\nabla}\cdot\vec{\nabla}q^2 > 0$$

$$\Rightarrow \quad \int_{V_1} \vec{\nabla}\cdot\vec{\nabla}q^2 dV > 0$$

If $\hat{n}$ be the outward unit normal at P, then by Gauss divergence theorem

$$\int_{S_1} \hat{n}\cdot\vec{\nabla}q^2 ds > 0$$

$$\Rightarrow \quad \int_{S_1} \frac{\partial q^2}{\partial n} ds > 0$$

$$\Rightarrow \quad \frac{\partial q^2}{\partial n} > 0 \text{ for every outward normal}$$

Hence q^2 and so q cannot be maximum at an interior point.

Again

$$\vec{\nabla}^2 p < 0$$

$$\Rightarrow \quad \vec{\nabla}\cdot\left(\vec{\nabla}p\right) < 0$$

$$\Rightarrow \quad \int_{V_1} \vec{\nabla}\cdot\left(\vec{\nabla}p dV\right) < 0$$

$$\Rightarrow \quad \int_{S_1} \hat{n}\cdot\vec{\nabla}p ds < 0$$

$$\Rightarrow \quad \int_{S_1} \frac{\partial p}{\partial n} ds < 0$$

$$\Rightarrow \quad \frac{\partial p}{\partial n} < 0$$

Therefore p cannot be minimum at an interior point.

11. Deduce from the principle that kinetic energy set up is a minimum that, if a mass of incompressible liquid be given at rest, completely filling a closed vessel of any shape and if any motion of the liquid be produced suddenly by giving arbitrarily prescribed normal velocities at all points of its bounding surface subject to the condition of constant volume, the motion produced is irrotational.

Solution: Let $\vec{q} = u\vec{i} + v\vec{j} + w\vec{k}$

The K.E. of the liquid is

$$T = \frac{1}{2}\rho \int_V q^2 dV = \frac{1}{2}\rho \int (U^2 + V^2 + W^2)\, dV \qquad \text{... (1)}$$

Let $\hat{n}$ be unit outward normal at any point P on the liquid outward normal velocity = $\hat{n}\cdot\vec{q}$.

It is given that

(i) T is minimum so that $\delta T = 0$.

(ii) $\hat{n}\vec{q}$ is prescribed on the boundary so that $\hat{n}\cdot\vec{q}$ = const then $\delta(\hat{n}\cdot\vec{q}) - 0$.

Equation of continuity is $\vec{\nabla}\cdot\vec{q} = 0$

So $\vec{\nabla}\cdot\delta\vec{q} = 0$

$\Rightarrow \int_V \phi\vec{\nabla}\cdot(\delta\vec{q})\,dV = 0$, ϕ being any function

$$\Rightarrow \int_V \left[\vec{\nabla}\cdot(\phi\delta\vec{q}) - \vec{\nabla}\phi\cdot\delta\vec{q}\right] dV = 0$$

Using Gauss Divergence Theorem, we get

$$\int_S (\hat{n}\cdot\phi\delta\vec{q})\,ds - \int_V \vec{\nabla}\phi\cdot(\delta\vec{q})\,dV = 0 \qquad \text{... (2)}$$

But

$$\hat{n}(\phi\delta\vec{q}) = \phi\hat{n}\cdot\delta\vec{q}$$
$$= \phi\delta(\hat{n}\cdot\vec{q})$$
$$= \phi\cdot 0 = 0, \text{ by (ii)}$$

Hence (2) becomes

$$\int_V \vec{\nabla}\phi\cdot(\delta\vec{q})\,dV = 0 \qquad \text{... (3)}$$

Subjecting (1) to the condition (1), we get

$$\frac{1}{2}\rho \int_V \delta q^2 dV = 0$$

$$\Rightarrow \quad \rho \int_V \vec{q} \cdot \delta\vec{q}\, dV = 0$$

$$\Rightarrow \quad \int_V \vec{q} \cdot \delta\vec{q}\, dV = 0 \qquad \text{... (4)}$$

(3) + (4) gives

$$\int_V \left[\vec{\nabla}\phi + \vec{q}\right] \cdot \delta\vec{q}\, dV = 0$$

So $\vec{\nabla}\phi + \vec{q} = 0$

$$\Rightarrow \quad \vec{q} = -\vec{\nabla}\phi$$

So that ϕ is a velocity potential, showing there by motion is irrotational.

12. Show that if

$$\phi = -\frac{1}{2}\left(ax^2 + by^2 + cz^2\right)$$

$$V = \frac{1}{2}\left(lx^2 + my^2 + nz^2\right),$$

where a, b, c, l, m, n are functions of time and $a + b + c = 0$, irrotational motion is possible with a free surface of equi-pressue if

$$\left(l + a^2 + a\right) e^{2\int a dt}, \left(m + b^2 + b\right) e^{2\int b dt}, \left(n + c^2 + c\right) e^{2\int c dt}$$

are constants.

Solution: Since

$$\phi = -\frac{1}{2}\left(ax^2 + by^2 + cz^2\right)$$

$$\therefore \quad \frac{\partial\phi}{\partial x} = -ax, \frac{\partial\phi}{\partial y} = -by, \frac{\partial\phi}{\partial z} = -cz$$

and

$$\frac{\partial^2\phi}{\partial x^2} = -a, \frac{\partial^2\phi}{\partial y^2} = -b, \frac{\partial^2\phi}{\partial z^2} = -c$$

Therefore

$$\vec{\nabla}^2\phi = \frac{\partial^2\phi}{\partial x^2} + \frac{\partial^2\phi}{\partial y^2} + \frac{\partial^2\phi}{\partial z^2}$$

$$= -(a + b + c)$$
$$= 0 \text{ (Given)}$$

Thus the irrotational motion is possible.

Also

$$q^2 = \left(\frac{\partial \phi}{\partial x}\right)^2 + \left(\frac{\partial \phi}{\partial y}\right)^2 + \left(\frac{\partial \phi}{\partial z}\right)^2$$
$$= a^2x^2 + b^2y^2 + c^2z^2$$

and

$$\frac{\partial \phi}{\partial t} = -\frac{1}{2}\left(\dot{a}x^2 + \dot{b}y^2 + \dot{c}z^2\right)$$

Bernoulli's pressure equation is

$$\frac{p}{\rho} - \frac{\partial \phi}{\partial t} + \frac{1}{2}q^2 + V = f(t)$$

$$\Rightarrow \quad \frac{p}{\rho} = \frac{\partial \phi}{\partial t} - \frac{1}{2}q^2 - V + f(t)$$

$$= f(t) + \frac{\partial \phi}{\partial t} - \frac{1}{2}\left[\left(\frac{\partial \phi}{\partial x}\right)^2 + \left(\frac{\partial \phi}{\partial y}\right)^2 + \left(\frac{\partial \phi}{\partial z}\right)^2\right] - V$$

$$= f(t) - \frac{1}{2}\left(\dot{a}x^2 + \dot{b}y^2 + \dot{c}z^2\right)$$
$$- \frac{1}{2}\left(a^2x^2 + b^2y^2 + c^2z^2\right) - \frac{1}{2}\left(lx^2 + my^2 + nz^2\right)$$

$$\Rightarrow \quad \frac{p}{\rho} = f(t) - \frac{1}{2}\left[\left(l + a^2 + \dot{a}\right)x^2 + \left(m + b^2 + \dot{b}\right)y^2 + \left(n + c^2 + \dot{c}\right)z^2\right] \quad \text{... (1)}$$

For a free surface of equipressure p = const.

$$\Rightarrow \quad \frac{dp}{dt} = \frac{\partial p}{\partial t} + U\frac{\partial p}{\partial x} + V\frac{\partial p}{\partial y} + W\frac{\partial p}{\partial z} = 0 \quad \text{... (2)}$$

From (1)

$$\frac{\partial p}{\partial t} = -\frac{1}{2}\rho\left[x^2\left(\ddot{a} + 2a\dot{a} + \dot{l}\right) + y^2\left(\ddot{b} + 2b\dot{b} + \dot{m}\right) z^2\left(\ddot{c} + 2c\dot{c} + \dot{n}\right)\right] + \rho\dot{F}(t)$$

and

$$U\frac{\partial p}{\partial x}=\left(\frac{-\partial\phi}{\partial x}\right)\left(\frac{\partial p}{\partial x}\right)$$

$$=ax\rho(-x)\left(l+a^2+\dot{a}\right)$$

$$=-\rho ax^2\left(l+a^2+\dot{a}\right)$$

Similarly, we have

$$V\frac{\partial p}{\partial y}=-\rho by^2\left(m+b^2+\dot{b}\right)$$

and

$$W\frac{\partial p}{\partial z}=-\rho cz^2\left(n+c^2+\dot{c}\right)$$

Putting these values in (2), we have

$$\dot{F}(t)=\frac{1}{2}x^2\left[\left(\ddot{a}+2a\dot{a}+\dot{l}\right)+2a\left(l+a^2+\dot{a}\right)\right]$$
$$+\frac{1}{2}y^2\left[\left(\ddot{b}+2b\dot{b}+\dot{m}\right)+2b\left(m+b^2+\dot{b}\right)\right]$$
$$+\frac{1}{2}z^2\left[\left(\ddot{c}+2c\dot{c}+\dot{n}\right)+2c\left(n+c^2+\dot{c}\right)\right]$$

This equation is true for all values of t, so the coefficients of x^2, y^2, z^2 and the constant term must vanish separately, we have

$$\left(\ddot{a}+2a\dot{a}+\dot{l}\right)+2a\left(l+a^2+\dot{a}\right)=0 \qquad \dots (3)$$

$$\left(\ddot{b}+2b\dot{b}+\dot{m}\right)+2b\left(m+b^2+\dot{b}\right)=0 \qquad \dots (4)$$

and

$$\left(\ddot{c}+2c\dot{c}+\dot{n}\right)+2c\left(n+c^2+\dot{c}\right)=0 \qquad \dots (5)$$

From (3), we have

$$\frac{\ddot{a}+2a\dot{a}+\dot{l}}{\dot{a}+a^2+l}=-2a$$

Integrating, we get

$$\log\left(l+a^2+\dot{a}\right)+\int 2at\,dt=\text{constant}$$

$$\Rightarrow \left(l+a^2+\dot{a}\right)c^{\int 2at\,dt}=\text{constant}$$

Similarly, we get

$$\left(m+b^2+\dot{b}\right)e^{\int 2btdt}=\text{const}, \left(n+c^2+\dot{c}\right)e^{\int 2ctdt}=\text{const}.$$

13. Prove that if the velocity potential at any instant be λxyz, the velocity at any point $(x + \xi, y + \eta, z + \zeta)$ relative to the fluid at the point (x, y, z) where ξ, η, ζ are small, is normal to the quadric $x\eta\zeta + y\xi\zeta + z\xi\eta$ = constant, with centre at (x, y, z).

Solution: Let $\vec{q}\,(u, v, w)$ and $\vec{q}^{\,1}\left(u^1, v^1, w^1\right)$ be the velocities of the fluid particle at the points $P(x, y, z)$ and $Q\ (x + \xi, y + \eta, z + \zeta)$

Given $\phi = \lambda xyz$

Then

$$u=\frac{-\partial\phi}{\partial x}=-\lambda yz,\ v=-\lambda zx,\ w=-\lambda xy$$

The value of ϕ at P^1 is given by

$$\phi=\lambda(x+\xi)(y+\eta)(z+\zeta)$$

Also

$$\begin{aligned} u^1 &= u+\xi\frac{\partial u}{\partial x}+\eta\frac{\partial u}{\partial y}+\zeta\frac{\partial u}{\partial z} \\ &= u+\eta(-\lambda z)+\zeta(-\lambda y) \\ &= u-\lambda(\eta z+y\zeta) \end{aligned}$$

Similarly

$$v^1 = v-\lambda(\zeta x+\xi z)$$

$$w^1 = w-\lambda(\xi y+\eta x)$$

Velocity of P^1 relative to P is

$$(u^1 - u,\ v^1 - v,\ w^1 - w)$$

$$\Rightarrow \left\{-\lambda(\eta z+\zeta y), -\lambda(\zeta x+\xi z), -\lambda(\xi y+\eta x)\right\} \qquad \text{... (1)}$$

Let $F = x\eta\zeta + y\zeta\xi + z\xi\eta = \text{constant}$ ... (2)

Then direction ratios of the normal at P^1 are

$$\frac{\partial F}{\partial \xi}, \frac{\partial F}{\partial \eta}, \frac{\partial F}{\partial \zeta}$$

i.e. $y\zeta + z\eta,\ x\zeta + z\xi,\ x\eta + y\xi$ ... (3)

From (1) and (3), it follows that the velocity at P^1 relative to that at P is normal to the quadratic (2).

14. Liquid of density ρ is flowing in two dimensions between the oval curves $r_1r_2 = a^2$, $r_1r_2 = b^2$ where r_1, r_2 are the distances measured from two fixed points, if the motion is irrotational and quantity q per unit time cross any line joining the bounding curves, then the kinetic energy is $\pi\rho q^2 \Big/ \log\frac{b}{a}$.

Solution: Let us assume that

$$z - z_1 = r_1 e^{i\theta_1},\ z - z_2 = r_2 e^{i\theta_2}$$

Let the complex potential w is of the form

$$w = iA \log (z - z_1)(z - z_2)$$

$$= iA \log \left(r_1 e^{i\theta_1} r_2 e^{i\theta_2}\right)$$

$$= iA \log r_1 r_2 e^{i(\theta_1 + \theta_2)}$$

$$\Rightarrow \quad \phi + i\psi = iA\left[\log r_1r_2 + i(\theta_1 + \theta_2)\right]$$

Separating into real and imaginary parts, we have

$$\phi = -A(\theta_1 + \theta_2),\ \psi = A \log r_1 r_2$$

Now

$$q = \psi_b - \psi_a$$

$$= A \log b^2 - A \log a^2$$

$$= 2A \log \frac{b}{a} \qquad \text{... (1)}$$

$$\Rightarrow \quad A = \frac{q}{2 \log \frac{b}{a}}$$

Since the two-dimensional irrotational motion occurs in a doubly connected region. Hence the circulation k is given by

$$k = -A(2\pi + 2\pi) = -4\pi A \qquad \text{... (2)}$$

Hence the kinetic energy of cyclic irrotational motion becomes

$$T = -\frac{1}{2}\rho \int \phi \frac{\partial \phi}{\partial n} ds - \frac{1}{2}\rho k \int \frac{\partial \phi}{\partial n} ds$$

$$= 0 - \frac{1}{2}\rho k \int \frac{\partial \psi}{\partial n} ds$$

as $\int \phi \dfrac{\partial \phi}{\partial n} ds = 0$ on rigid boundary

$$= -4\pi A\rho \int_a^b d\psi \qquad \text{From (2)}$$

$$= -2\pi A\rho(\psi_b - \psi_a)$$

$$= 2\pi \rho q^2 \Big/ \log \frac{b}{a} \qquad \text{From (1)}$$

15. In compressible fluid of density p is contained between two coaxial circular cylinders of radii a and b $(a < b)$ and between two rigid planes perpendicular to the axis at a distance l apart. The cylinders are at rest and the fluid is circulating in irrotational motion, its velocity being V at the surface of the inner cylinder. Prove that the kinetic energy is $\pi\rho \big/ a^2V^2 \log\left(\dfrac{b}{a}\right)$.

Solution: For irrotational two-dimensional fluid motion, we have

$$\vec{\nabla}^2 \psi = \frac{\partial^2 \psi}{\partial r^2} + \frac{1}{r}\frac{\partial \psi}{\partial r} + \frac{1}{r^2}\frac{\partial^2 \psi}{\partial \theta^2} = 0 \qquad \text{... (1)}$$

Since ψ is a function of r only, so $\dfrac{\partial^2 \psi}{\partial \theta^2} = 0$

Hence equation (1) becomes

$$\frac{\partial^2 \psi}{\partial r^2} + \frac{1}{r}\frac{\partial \psi}{\partial r} = 0$$

$$\Rightarrow \quad \partial \frac{\partial^2 \psi}{\partial r^2} + \frac{\partial \psi}{\partial r} = 0$$

$$\Rightarrow \quad \frac{\partial}{\partial r}\left(\partial \frac{\partial \psi}{\partial r}\right) = 0$$

Integrating, we have

$$r\frac{\partial \psi}{\partial r} = c \qquad \text{... (2)}$$

When $r = a$, $\dfrac{\partial \psi}{\partial r} = V$

So from (2) $c = aV$

Now the radial velocity is zero as $\dfrac{\partial \psi}{\partial \theta} = 0$.

Thus $q = \dfrac{aV}{r}$

The kinetic enegy T of the fluid is given by

$$T = \int_a^b \frac{1}{2}(2\pi r dr l \rho)\left(\frac{aV}{r}\right)^2$$

$$= \pi \rho l a^2 V^2 \int_a^b \frac{1}{r} dr$$

$$= \pi \rho l a^2 V^2 \log \frac{b}{a}$$

16. Show that the curvature of a streamline in steady motion is $\dfrac{1}{q^2}\dfrac{\partial}{\partial V}\left(\dfrac{p}{\rho} + V\right)$, where p, ρ, q are the pressure, density and velocity of the liquid, V the potential of the external forces, and δV is an element of the principal normal to the streamline, and hence obtain the velocity potential of the two-dimensional irrotational motion for which the streamlines are confocal ellipse.

Solution: Let r be the radius of curvature of the streamline at the point P. The normal acceleration of the fluid element at P is $\dfrac{q^2}{r}$. Equating the external forces along the normal to the streamline, we have

$$\frac{q^2}{r} = -\frac{\partial V}{\partial v} - \frac{1}{\rho}\frac{\partial p}{\partial v}$$

$$\Rightarrow \quad -\frac{1}{r} = \frac{1}{q^2}\frac{\partial}{\partial \theta}\left(\frac{p}{\rho} + V\right)$$

which gives the curvature of a streamline in steady motion.

We know that $q = \dfrac{-\partial \phi}{\partial s}$ where δs is an element of length of the streamline passing through the point.

The velocity at the point P on the confocal ellipse is given by $= A\{(a - ex)(a + ex)\}^{\frac{-1}{2}}$

where A is an arbitrary constant and e is the eccentricity of the ellipse.

Now

$$\phi = -\int q\,ds$$

$$= -A\int \frac{ds}{\sqrt{a^2 - e^2x^2}}$$

$$\Rightarrow \quad \phi = -A\int \frac{1}{\sqrt{a^2 - e^2x^2}} \frac{ds}{d\theta}\, d\theta \qquad \text{... (1)}$$

In an ellipse $x = a \cos \theta$, $y = b \sin \theta$

Hence

$$\left(\frac{ds}{d\theta}\right)^2 = a^2 \sin^2 \theta + b^2 \cos^2 \theta$$

$$= a^2 \left[\sin^2 \theta + \left(1 - e^2\right)\cos^2 \theta\right]$$

$$= a^2 \left(1 - e^2 \cos^2 \theta\right)$$

Therefore from (1), we get

$$\phi = -A\int \frac{1}{\sqrt{a^2 - e^2a^2 \cos^2 \theta}}\, a\sqrt{1 - e^2 \cos^2 \theta}\, d\theta$$

$$= -A\theta = -A \tan^{-1}\left(\frac{ay}{bx}\right)$$

which gives the velocity potential of the two-dimensional irrotational motion for which the streamlinc are confocal ellipse.

17. If the fluid be in motion with a velocity potential $\phi = z \log r$ and if the density at a point fixed in space be independent of the time, show that the surfaces of equal density are of the forms $r^2\left(\log r - \frac{1}{2}\right) - z^2 = f(\theta, \rho)$ where ρ is the density at (z, r, θ).

Solution: For surfaces of equal density

$$\rho = \text{constant}, \quad \frac{d\rho}{dt} = 0$$

Thus $\dfrac{\partial \rho}{\partial t} = (\vec{q} \cdot \vec{\nabla})\rho = 0$

$$\Rightarrow \quad \frac{\partial \phi}{\partial r}\frac{\partial \rho}{\partial r} + \frac{1}{r^2}\frac{\partial \phi}{\partial \theta}\frac{\partial \rho}{\partial \theta} + \frac{\partial \phi}{\partial z}\frac{\partial p}{\partial z} = 0 \qquad \text{... (1)}$$

Also $\vec{q} = -\vec{\nabla}\phi$

The velocity potential ϕ is given by the relation

$$\phi = z \log r \qquad \text{... (2)}$$

From (1) and (2), we have

$$\frac{z}{r}\frac{\partial p}{\partial r} + \log r \frac{\partial \rho}{\partial z} = 0$$

$$\Rightarrow \quad z\frac{\partial \rho}{\partial r} + r \log r \frac{\partial \rho}{\partial z} = 0$$

Comparing with Lagrange's differential equation, we have

$$\frac{dr}{z} = \frac{dz}{r \log r} = \frac{d\rho}{0}$$

$\Rightarrow r \log r dr - z dz = 0, \; d\rho = 0$

Integrating, we get

$$r^2\left(\log r - \frac{1}{2}\right) - z^2 = f(\theta, \rho) \text{ and } \rho = \rho_\theta = \text{constant}$$

18. Obtain Cauchy's integral using circulation theorem.

Solution: Let c_0 be the initial position of the closed curve c in yz plane.

From circulation theorem

$$\int_{c_0} (v_0 db + w_0 dc) = \int_c (u dx + v dy + w dz)$$

By Stoke's theorem, we have

$$\iint_{s_0} \left(\frac{\partial w_0}{\partial b} - \frac{\partial v_0}{\partial c}\right) ds_0 = \iint_s \left\{ l\left(\frac{\partial w}{\partial y} - \frac{\partial v}{\partial z}\right) + m\left(\frac{\partial u}{\partial z} - \frac{\partial w}{\partial x}\right) + n\left(\frac{\partial v}{\partial x} - \frac{\partial u}{\partial y}\right) \right\} ds$$

$$\Rightarrow \quad \iint_{s_0} \xi_0 dbdc = \int_s (l\xi + m\eta + n\zeta) ds$$

$$\Rightarrow \quad \iint_{s_0} \xi_0 dbdc = \iint_s \left\{ \xi \frac{\partial(y,z)}{\partial(b,c)} + \eta \frac{\partial(z,x)}{\partial(b,c)} + \zeta \frac{\partial(x,y)}{\partial(c,a)} \right\} dbdc$$

$$= aslds = dydz = \frac{\partial(y,z)}{\partial(b,c)}$$

$$\Rightarrow \quad \xi_0 = \xi \frac{\partial(y,z)}{\partial(b,c)} + \eta \frac{\partial(z,x)}{\partial(b,c)} + \zeta \frac{\partial(x,y)}{\partial(b,c)} \qquad \dots (1)$$

Similarly, we have

$$\eta_0 = \xi \frac{\partial(y,z)}{\partial(c,a)} + \eta \frac{\partial(z,x)}{\partial(c,a)} + \zeta \frac{\partial(x,y)}{\partial(c,a)} \qquad \dots (2)$$

and

$$\zeta_0 = \xi \frac{\partial(y,z)}{\partial(a,b)} + \eta \frac{\partial(z,x)}{\partial(a,b)} + \zeta \frac{\partial(x,y)}{\partial(a,b)} \qquad \dots (3)$$

Multiplying (1), (2) and (3) by $\frac{\partial x}{\partial a}, \frac{\partial x}{\partial b}, \frac{\partial x}{\partial c}$ respectively and adding, we have

$$\xi_0 \frac{\partial x}{\partial a} + \eta_0 \frac{\partial x}{\partial b} + \zeta_0 \frac{\partial x}{\partial c} = \xi \left\{ \frac{\partial x}{\partial a} \frac{\partial(y,z)}{\partial(b,c)} + \frac{\partial x}{\partial b} \frac{\partial(y, z)}{\partial(c, a)} + \frac{\partial x}{\partial c} \frac{\partial(y, z)}{\partial(a, b)} \right\}$$

$$\Rightarrow \quad \xi_0 \frac{\partial x}{\partial a} + \eta_0 \frac{\partial x}{\partial b} + \zeta_0 \frac{\partial x}{\partial c} = \xi \frac{\partial(x,y,z)}{\partial(a,b,c)} \qquad \dots (4)$$

Also equation of continuity in Lagrangian form is given by

$$\rho \frac{\partial(x, y, z)}{\partial(a, b, c)} = \rho_0 \qquad \dots (5)$$

From (4) and (5), we get

$$\frac{\xi}{\rho} = \frac{\xi_0}{\rho_0} \frac{\partial x}{\partial a} + \frac{\eta_0}{\rho_0} \frac{\partial x}{\partial b} + \frac{\xi_0}{\rho_0} \frac{\partial x}{\partial c} \qquad \dots (6)$$

Similarly, we have

$$\frac{\eta}{\rho} = \frac{\xi_0}{\rho_0} \frac{\partial y}{\partial a} + \frac{\eta_0}{\rho_0} \frac{\partial y}{\partial b} + \frac{\xi_0}{\rho_0} \frac{\partial y}{\partial c} \qquad \dots (7)$$

and

$$\frac{\zeta}{\rho} = \frac{\xi_0}{\rho_0}\frac{\partial z}{\partial a} + \frac{\eta_0}{\rho_0}\frac{\partial z}{\partial b} + \frac{\xi_0}{\rho_0}\frac{\partial z}{\partial c} \quad \text{... (8)}$$

Equations (6), (7) and (8) are known as Cauchy's integral.

19. Prove that irrotational acyclic motion of a liquid contained in a boundary cannot be created or destroyed by application of impulses.

Solution: Equations of motion are

$$u^1 - u = X - \frac{1}{\rho}\frac{\partial \overline{w}}{\partial x},\ v^1 - v = Y - \frac{1}{\rho}\frac{\partial \overline{w}}{\partial y}$$

and

$$w^1 - w = Z - \frac{1}{\rho}\frac{\partial \overline{w}}{\partial z}$$

where X, Y, Z are the extraneous impulses per unit mass of a fluid at the point (x, y, z) (u, v, w) and (u^1, v^1, w^1) are the velocity components at this point just before and after the impulsive action and $\overline{w}$ is the impulsive pressure.

So $(u^1dx + v^1dy + w^1dz) - (udx + vdy + wdz)$

$$= (Xdx + Ydy + Zdz) - \frac{1}{\rho}\left(\frac{\partial \overline{w}}{\partial x}dx + \frac{\partial \overline{w}}{\partial y}dy + \frac{\partial \overline{w}}{\partial z}dz\right)$$

$$= -\,dV - \frac{1}{\rho}d\overline{w}$$

$$= -\,d\left(V + \frac{\overline{w}}{\rho}\right) \quad \text{where } \rho = \text{constant} \quad \text{... (1)}$$

The R.H.S. of (1) is an exact differential but if $udx + vdy + wdz$ is not an exact differential then $u^1dx + v^1dy + w^1dz$ will also not be an exact differential.

$$\left.\begin{aligned} &\text{Let } udx + vdy + wdz = -\,d\phi \\ &\text{and } u^1dx + v^1dy + w^1dz = -\,d\phi^1 \end{aligned}\right\} \quad \text{... (2)}$$

From (1) and (2), we have

$$-\,d\phi^1 + d\phi = -\,d\left(V + \frac{\overline{w}}{\rho}\right)$$

$$\Rightarrow \quad \phi^1 - \phi = v + \frac{\bar{w}}{\rho} + A \qquad \text{... (3)}$$

If the potential function v and the impulsive pressure $\bar{w}$ be single valued, $\phi^1 - \phi$ must also be single valued. Therefore, if ϕ be single valued, ϕ^1 must be and if ϕ be many valued, ϕ^1 must also be many valued. So that $\phi^1 - \phi$ is single valued.

20. Prove that the results of curvature R at any point of a streamline ψ = constant is given by

$$R = \left(u^2 + v^2\right)^{3/2} \Big/ \left| u^2 \frac{\partial v}{\partial x} - 2uv \frac{\partial u}{\partial x} - v^2 \frac{\partial u}{\partial y} \right|$$

Where (u, v) are respectively the velocity components of a fluid motion along OX, OY.

Solution: The radius of curvature R at a point (x, y) is given by

$$R = \left[1 + \left(\frac{\partial y}{\partial x}\right)^2\right]^{3/2} \Big/ \left|\frac{d^2 y}{dx^2}\right| \qquad \text{... (1)}$$

The equation of the streamline is

$$\psi = \text{const.} \qquad \text{... (2)}$$

Where $u = \dfrac{-\partial \psi}{\partial y}$ and $v = \dfrac{-\partial \psi}{\partial x}$ ... (3)

Differentiating (2) with respect to x, we have

$$\frac{\partial \psi}{\partial x} + \frac{\partial \psi}{\partial y} \frac{\partial y}{\partial x} = 0$$

$$\Rightarrow \quad \frac{dy}{dx} = \frac{v}{u}$$

$$\Rightarrow \quad \frac{d^2 y}{dx^2} = \frac{d}{dx}\left(\frac{v}{u}\right) = \frac{\partial}{\partial x}\left(\frac{v}{u}\right) + \frac{\partial}{\partial y}\left(\frac{v}{u}\right)\frac{dy}{dx}$$

$$\Rightarrow \quad \frac{d^2 y}{dx^2} = \frac{u \dfrac{\partial v}{\partial x} - v \dfrac{\partial u}{\partial x}}{u^2} + \frac{v}{u} \, \frac{u \dfrac{\partial v}{\partial y} - v \dfrac{\partial u}{\partial y}}{u^2}$$

$$\Rightarrow \quad \frac{d^2 y}{dx^2} = \frac{u^2 \dfrac{\partial v}{\partial x} - 2uv \dfrac{\partial u}{\partial x} - v^2 \dfrac{\partial u}{\partial y}}{u^3} \qquad \text{... (4)}$$

Since

$$\frac{\partial v}{\partial y}=\frac{\partial}{\partial y}\left(\frac{\partial \psi}{\partial x}\right)=\frac{\partial}{\partial x}\left(\frac{\partial \psi}{\partial y}\right)=-\frac{\partial u}{\partial x}$$

Putting the values of $\frac{dy}{dx}$ and $\frac{d^2y}{dx^2}$ into (1), we have

$$R=\left(1+\frac{v^2}{u^2}\right)^{3/2}\Bigg/\left|\frac{u^2\frac{\partial v}{\partial x}-2uv\frac{\partial u}{\partial x}-v^2\frac{\partial u}{\partial y}}{u^3}\right|$$

$$=\left(u^2+v^2\right)^{3/2}\Bigg/\left|u^2\frac{\partial v}{\partial x}-2uv\frac{\partial u}{\partial x}-v^2\frac{\partial u}{\partial y}\right|$$

EXERCISE

1. Find the vorticity of the fluid motion for the velocity components.

$$u=-\frac{2xyz}{\left(x^2+y^2\right)^2},\ v=\frac{\left(x^2-y^2\right)z}{\left(x^2+y^2\right)^2},\ w=\frac{y}{x^2+y^2}$$

[**Ans.** 0]

2. Find the vorticity of the fluid motion for the given velocity components

(i) $u=Ay^2+By+C, v=0$ $\left[\textbf{Ans. } -(2Ay+B)\vec{k}\right]$

(ii) $u=2Cxz, v=C\left(a^2+x^2-z^2\right)$

$\left[\textbf{Ans. } 2C\left(-z\vec{i}+x\vec{j}+y\vec{k}\right)\right]$

(iii) $u = C\ (x + y),\ v = -\ C\ (x + y)$ $\left[\textbf{Ans. } -2C\vec{k}\right]$

Where A, B, C are arbitrary constants.

3. Determine the vorticity of the fluid motion for the given velocity vector

$\vec{q}=\vec{i}\ (Az-By)+\vec{j}\ (Bx-Cz)+\vec{k}\ (Cy-Ax)$

where A, B and C are non-zero constants. Also determine the equation of the vortex lines.

[**Ans.** Vorticity vector is $2\left(C\vec{i}+A\vec{j}+B\vec{k}\right)$ and equation of the vortex lines are $Az - By = C_1$, $Bx - Cz = C_2$, $Cy - Ax = C_3$ where C_1, C_2 and C_3 are constants]

4. Find the shear strain rate for the fluid flows described by velocity field $\vec{q} = 5x^3\vec{i} - 15x^2y\vec{j}$.

[**Ans.** $-15xy$]

5. In the case of irrotational motion in two-dimensions on the surface of a sphere, show that the velocity potential is of the form

$$f\left(\frac{x+iy}{r+z}\right)+f\left(\frac{x-iy}{r+z}\right),$$

r being the radius of the sphere and x, y, z the coordinates of a point referred to rectangular axes through the centre of the sphere.

6. If λ denotes a variable parameter and f a given function, find the condition that $f(x, y, \lambda) = 0$ should be a possible system of streamlines for steady irrotational motion in two dimensions.

[**Ans.** y must be a function of λ]

7. A rigid envelope is filled with homogeneous frictionless liquid, show that it is not possible by any movements applied to the envelope, to set its contents into motion which will persist after the envelope has come to rest.

8. A space is bounded by an ideal fixed surfaces drawn in a homogeneous incompressible fluid, satisfying the condition for the continued existence of a velocity potential ϕ under conservative forces. Prove that the rate per unit time at which energy flows across s into the space bounded by c is $-\rho \iint \frac{\partial \phi}{\partial f}\frac{\partial \phi}{\partial n}\, ds$ where ρ is the density and δn an element of the normal to ds drawn into the space considered.

9. Prove that in a cyclic irrotational motion of a homogeneous fluid the total momentum of the fluid contained within the sphere of any radius is equivalent to a single vector through the centre of the sphere.

7 VORTEX MOTION

7.1. Introduction

The most general type of motion which a fluid is capable one which is a combination of irrotational and rotational motion. The component velocities may be regarded as consisting of two parts $\vec{q}$ and $\vec{q}_1$ the first is derivable from the velocity potential and the other depend upon the molecular rotation. Here we shall discuss the fluid motion in terms of vorticity rather in terms of velocity. This motion was first developed by Helmholtz and afterwards proof of some of his theorems were given by Kelvin, Krichhoff and other Mathematicians.

7.2. Some Definitions

1. Vorticity

If $\vec{q}\,(u, v, w)$ be the velocity of a fluid particle, then $\overline{W} = \frac{1}{2}\, curl\, \vec{q}$ is called vorticity vector or simply vorticity. As a matter of fact, vorticity is the angular velocity (velocity of rotation) of an infinitesimal fluid element.

If $\overrightarrow{W}(\xi, \eta, \zeta)$, then

$$\xi\vec{i} + \eta\vec{j} + \zeta\vec{k} = \frac{1}{2}\begin{vmatrix} \vec{i} & \vec{j} & \vec{k} \\ \frac{\partial}{\partial x} & \frac{\partial}{\partial y} & \frac{\partial}{\partial z} \\ u & v & w \end{vmatrix}$$

$$\xi = \frac{1}{2}\left(\frac{\partial w}{\partial y} - \frac{\partial v}{\partial z}\right)$$

$$\eta = \frac{1}{2}\left(\frac{\partial u}{\partial z} - \frac{\partial w}{\partial x}\right)$$

$$\zeta = \frac{1}{2}\left(\frac{\partial v}{\partial x} - \frac{\partial u}{\partial y}\right)$$

The components ξ, η, ζ are called components of spin.

2. Irrotational Motion

A fluid motion is said to be irrotational if $\overline{W} = 0$.

3. Vortex Line

A vortex line is a curve drawn in the fluid such that the tangent to it at each point is in the direction of the vorticity vector $\vec{W}$ at that point. The differential equation of vortex lines are

$$\frac{dx}{\xi} = \frac{dy}{\eta} = \frac{dz}{\zeta}$$

4. Vortex Tube, Vortex Filament

If vortex lines are drawn through every point of a small closed curve, then the space bounded by these lines is called vortex tube. The fluid within this tube is known as vortex filaments or simply vortex. Thus we can also say that boundary of a vortex filament is a vortex tube.

7.3. Properties of Vortex Filament

1. The product of the cross section and vorticity at any point on a vortex filament is constant along the filament.

We consider a vortex filament with vorticities $\overrightarrow{W_1}, \overrightarrow{W_2}$ at the ends of the filament where the cross-sectional areas are ds_1 and ds_2. Then by Gauss's divergence theorem

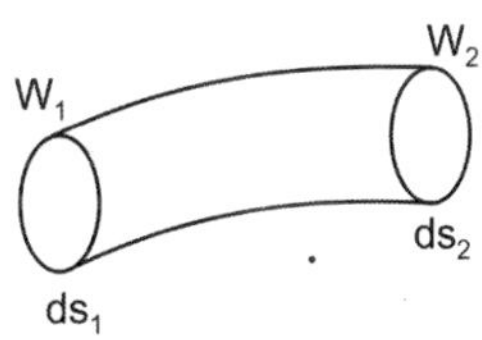

Fig. 7.1

$$\int_S (\vec{W} \cdot \hat{n})\, ds = \int_V (\vec{\nabla} \cdot \vec{W})\, dv$$

$$= \int_V \vec{\nabla} \cdot \left(\frac{1}{2} \vec{\nabla} \times \vec{q} \right) dv$$

$$= \frac{1}{2} \int_V (\text{div curl } \vec{q}) dv = 0 \qquad \text{as div curl} = 0$$

$$\therefore \quad \int_S \overrightarrow{W} \cdot \hat{n}\, ds = 0$$

where S is the closed surface enclosing volume V of the fluid in the vortex tube.

$$\Rightarrow \quad \int_{S_1} \overrightarrow{W} \cdot \hat{n} ds + \int_2 \overrightarrow{W} \cdot \hat{n} ds + \int_{\text{Walls}} \overrightarrow{W} \cdot \hat{n} ds = 0 \qquad \ldots (1)$$

But on the walls of the tube, $\overrightarrow{W}$ is along the tube and so

$$\int_{\text{Walls}} \overrightarrow{W} \cdot \hat{n} ds = 0 \text{ as } \overrightarrow{W} \cdot \hat{n} = 0 \text{ on the walls}$$

Hence, from (1), we have

$$\int_{S_1} \overrightarrow{W} \cdot \hat{n} ds + \int_2 \overrightarrow{W} \cdot \hat{n} ds = 0$$

$$\Rightarrow \quad \int_{S_1} \overrightarrow{W} \cdot \hat{n}_1 ds = \int_{S_2} \overrightarrow{W} \cdot \hat{n}_2 ds$$

where $\hat{n}_1$ and $\hat{n}_2$ are unit outward normals on the surfaces S_1 and S_2 drawn in the same sense.

$$\int_S \overrightarrow{W} \cdot \hat{n}\, ds = \text{constant.}$$

2. Vortex lines moves with the fluid i.e. vortex filaments are composed of the same elements of the fluid.

We know that circulation $= \int_C \vec{q} \cdot d\vec{r}$

$$= \int_S \hat{n} \cdot \text{curl } \vec{q} ds$$

$$= \int_S (l\xi + m\eta + n\zeta) ds$$

Now if circulation = 0, then

$$l\xi + m\eta + n\zeta = 0 \qquad \text{... (1)}$$

Let us assume that circulation = 0 along any closed path c over the surface S.

Then $l\xi + m\eta + n\zeta = 0$ at any point of S, by (1). It means that vortex lines all lie on S i.e. S is composed of the vortex lines. The particles which lie on S at any time t, lie on S^1 at time $t + \delta t$. Similarly C^1 is taken at time $t + \delta t$ in place of the path C at t. Hence circulation along C is equal to circulation along C^1 so that circulation along $C^1 = 0$. It means that S^1 is composed of vortex lines.

Hence any surface composed of vortex lines continues to be composed of vortex lines as it moves with the fluid but the intersection of two such surfaces must be vortex so that vortex lines move with the fluid.

7.4. Strength of Vortex

If K be the circulation round a closed curve C which encloses a vortex tube, then by Stoke's theorem

$$K = \int_C \vec{q} \cdot d\vec{r} = \int_S \text{curl } \vec{q} \cdot \hat{n} ds$$

$$\Rightarrow \quad K = \int_S 2\overrightarrow{W} \cdot \hat{n} ds$$

$$= 2\int_S (l\xi + m\eta + n\zeta)\, ds \qquad \text{as } \hat{n} = \hat{n}\,(\xi, \eta, \zeta)$$

If πa^2 be the cross section of the vortex tube (supposed small), then

$$K = 2(l\xi + m\eta + n\zeta) \int_S ds$$

$$= 2\pi a^2 (l\xi + m\eta + n\zeta) = \text{constant}$$

The quantity K is called the strength of the vortex. This strength is taken to be positive when the circulation round the vortex filament is anti-clockwise.

7.5. Rectilinear Vortices

The liquid within the cylinder of which the circle is a cross section is said to form a rectilinear or coloumnar vortex.

7.6. Complex Potential of a Two-Dimensional Vortex Motion

To determine the complex potential due to a rectilnear vortex of strength K.

We consider a cylindrical vortex tube in xy plane. Let the tube be surrounded by incompressible irrotational fluid. The motion is purely two-dimensional so that

$$w = 0,\ u = u\ (x, y),\ v = v\ (x, y)$$

Consequently $\xi = 0,\ \eta = 0,\ \zeta = \frac{1}{2}\left(\frac{\partial v}{\partial x} + \frac{\partial u}{\partial y}\right)$

But $u = -\frac{\partial \phi}{\partial x} = -\frac{\partial \psi}{\partial y},\ v = -\frac{\partial \phi}{\partial y} = \frac{\partial \psi}{\partial x}$

Hence

$$2\zeta = \frac{\partial^2 \psi}{\partial x^2} + \frac{\partial^2 \psi}{\partial y^2}$$

$$= \frac{\partial^2 \psi}{\partial r^2} + \frac{1}{r}\frac{\partial \psi}{\partial r} + \frac{1}{r^2}\frac{\partial^2 \psi}{\partial \theta^2} \qquad \text{(in polar coordinates)}$$

Let $\zeta = W$ = constant within the cylinder and $\zeta = 0$ outside the cylinder.

Symmetry exists about the origin and so ψ must be a functin of r only.

Then

$$\frac{d^2\psi}{dr^2} + \frac{1}{r}\frac{d\psi}{dr} = \frac{1}{r}\frac{d}{dr}\left(r\frac{d\psi}{dr}\right)$$

$$= \begin{cases} 2\zeta & \text{withing vortex} \\ O & \text{outside} \end{cases}$$

Hence

$$\frac{rd\psi}{dr} = \begin{cases} \zeta r^2 + A & \text{within vortex} \\ B & \text{outside} \end{cases}$$

Therefore $\psi = c \log r$ outside the vortex

Since the motion outside the vortex filament is irrotational, so there exists a velocity potential ϕ such that $\phi = -\ c\theta$.

Let w be the complex potential outside the filament, which is given by

$$w = \phi + i\psi = -c\theta + ic\log r$$
$$= ic\ (\log r + i\theta)$$
$$= ic\ \log\ (re^{i\theta})$$
$$= ic\ \log\ z$$

Let k be the strength of the vortex, then

$$K = \int_0^{2\pi}\left(-\frac{1}{r}\frac{\partial\phi}{\partial\theta}\right) r d\theta$$
$$= \int_0^{2\pi} c d\theta = 2\pi c$$

Hence complex potential becomes

$$w = \frac{ik}{2\pi}\log z$$

Also the velocity components at any point in the direction of r and θ are

$$q_r = -\frac{\partial\phi}{\partial r} = -\frac{1}{r}\frac{\partial\psi}{\partial\theta} = 0$$

and

$$q_\theta = -\frac{1}{r}\frac{\partial\phi}{\partial\theta} = \frac{\partial\psi}{\partial r} = \frac{k}{2\pi}$$

If the vortex be at $z = z_0$, then the complex potential at any point z outside the vortex is given by

$$w = \frac{ik}{2\pi}\log\left(z - z_0\right)$$

If there be a large number of vortices of strengths k_1, k_2, k_3, ... k_n at $z = z_1, z_2, \ldots\ z_n$ respectively, then the complex potential is given by

$$w = \frac{ik_1}{2\pi}\log\left(z - z_1\right) + \frac{ik_2}{2\pi}\log\left(z - z_2\right) + \ldots + \frac{ik_n}{2\pi}\log\left(z - z_n\right)$$

$$= \frac{i}{2\pi}\sum_{r=1}^{n} kr\log\left(z - z_r\right)$$

Deduction. If the vortex has a circular cross section, then also $w = \frac{ik}{2\pi} \log(z - z_0)$ for vortex at $z = z_0$. The same proof will be given for this also.

7.7. Motion due to *m* Vortices

To find velocity at a point $P(z)$ due to m vortices of strengths k_n at the points z_n $(n = 1, 2, \ldots.. m)$.

The complex potential at a point $P(z)$ due to m vortices of strengths k_n at the points z_n $(n = 1, 2, \ldots.. m)$.

The complex potential due to a vortex of strength k_1 at $z = z_1$ is

$$w = \frac{ik_1}{2\pi} \log(z - z_1)$$

$$\frac{dw}{dz} = \frac{ik_1}{2\pi} \frac{1}{z - z_1}$$

But

$$\frac{dw}{dz} = -u_1 + iv_1$$

$$\therefore \quad -u_1 + iv_1 = \frac{ik_1}{2\pi} \frac{1}{z - z_1} = \frac{ik_1}{2\pi} \frac{1}{(x - x_1) + i(y - y_1)}$$

Let $r_1^2 = (x - x_1)^2 + (y - y_1)^2$

Then

$$-u_1 + iv_1 = \frac{ik_1}{2\pi} \frac{(x - x_1) - i(y - y_1)}{r_1^2}$$

$$\Rightarrow \quad -u_1 + iv_1 = \frac{k_1}{2\pi} \frac{y - y_1}{r_1^2} + \frac{ik_1}{2\pi} \frac{x - x_1}{r_1^2}$$

Separating into real and imaginary parts, we get

$$u_1 = \frac{-k_1}{2\pi} \frac{y - y_1}{r_1^2}, \; v_1 = \frac{k_1}{2\pi} \frac{x - x_1}{r_1^2}$$

u_1, v_1 are velocity components at P due to the vortex of strength k_1. Similarly if u_n, v_n be velocity components due to vortex of strength k_n, then

$$u_n = \frac{-k_n}{2\pi}\left(\frac{y - y_n}{r_n^2}\right)$$

$$v_n = \frac{k_n}{2\pi}\left(\frac{x - x_n}{r_n^2}\right)$$

If u, v are velocity components along x and y axes at $P(z)$ due to vortices of strengths k_1, k_2 ..... k_m at the points z_1, z_2, ... z_n, then

$$u = \sum_{n=1}^{m} u_n = -\sum_{n=1}^{m} \frac{k_n}{2\pi}\left(\frac{y - y_n}{r_n^2}\right)$$

and

$$v = \sum_{n=1}^{m} v_n = \sum_{n=1}^{m} \frac{k_n}{2\pi}\left(\frac{x - x_n}{r_n^2}\right)$$

Also

$$-u + iv = \sum_{n=1}^{m} \frac{ik_n}{2\pi} \frac{1}{z - z_n}$$

and

$$w = \sum_{n=1}^{m} \frac{ik_n}{2\pi} \log(z - z_n)$$

7.8. Single Vortex in the Field of Several Vortices

The value of ψ at any point inside of a circular vortex tube is given by

$$\frac{1}{r}\frac{d}{dr}\left(r\frac{d\psi}{dr}\right) = 2\zeta$$

Its solution is $\psi = \frac{1}{2}\zeta r^2 + c \log r + a$ where c and a are constants of integration. Velocity at right angle to the radius vector.

$$= \frac{d\psi}{dr} = \zeta r + \frac{c}{r}$$

But velocity at the origin is finite so we take $c = 0$.

$$\therefore \quad \frac{d\psi}{dr} = \zeta r$$

So that $\left(\dfrac{d\psi}{dr}\right)_{r=0} = 0$

This shows that velocity at the origin due to a single vortex is zero. Thus the vortex filament (vortex) induces no velocity at its centre.

From this discussion we can say that if a vortex in the field of several vortices, then its motion in the liquid will be on account of other vertices.

The above article can also be expressed as: To show that a single rectilinear vortex is an unlimited mass of liquid remains stationary, and when such a vortex is in the presence of other vortices it has no tendency to move of itself but its motion through the liquid is entirely due to the velocities caused by the other vortices.

7.9. Centre of Vortices

We consider fluid motion due to n vortices of strengths k_n at points z_s (s = 1, 2, ... n). The complex potential at $P(z)$ outside the filament is

$$w = \sum_{s=1}^{n} \frac{ik_s}{2\pi} \log(z - z_s)$$

Hence

$$-u + iv = \frac{dw}{dz} = \sum_{s} \frac{iks}{2\pi} \frac{1}{z - z_s}$$

The velocity components u_r, v_r of the vortex of strength k_r at z_r produced by other vortices and so

$$-u_r + iv_r = \sum_{s=1}^{n} \frac{ik_s}{2\pi} \frac{1}{z_r - z_s} \qquad \dots (1)$$

Where s takes all values from 1 to n except r.

From (1)

$$\sum_{r=1}^{n} k_r(-u_r + iv_r) = \sum_{r=1}^{n} \sum_{\substack{s=1 \\ s \neq r}}^{n} \frac{ik_r k_s}{2\pi} \frac{1}{z_r - z_s}$$

$$\Rightarrow \sum_{r=1}^{n} k_r(-u_r + iv_r) = 0 \qquad \dots (2)$$

Since the terms $\frac{ik_r k_s}{2\pi(z_r - z_s)}$ and $\frac{ik_s k_r}{2\pi(z_s - z_r)}$ cancel each other.

Hence from (2)

$$\sum_{r=1}^{n} k_r u_r = 0, \sum_{r=1}^{n} k_r v_r = 0 \quad \text{... (3)}$$

The equation (3) shows that if we regard k_s as a mass, the centre of gravity G of the vortex filament remains stationary during their motion about one another. The point b is called the centre of the system of vortices.

7.10. System of Two Vortex Filaments

Let A_1, A_2 be two vortex filaments of strength K_1 and K_2 situated at the points z_1 and z_2, then complex potential w is given by

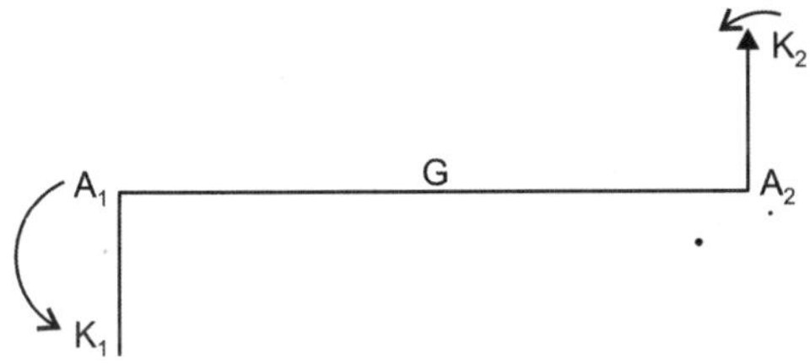

Fig. 7.2

$$w = \frac{iK_1}{2\pi} \log(z - z_1) + \frac{iK_2}{2\pi} \log(z - z_2) \quad \text{... (1)}$$

$$\Rightarrow \quad -u + iv = \frac{dw}{dz} = \frac{iK_1}{2\pi} \frac{1}{z - z_1} + \frac{iK_2}{2\pi} \frac{1}{z - z_2} \quad \text{... (2)}$$

The velocity at A_1 is due to the vortex A_2 and so

$$-u_1 + iv_1 = \left(\frac{iK_2}{2\pi} \frac{1}{z - z_2} \right)_{z = z_1} \quad \text{by (2)}$$

$$\Rightarrow \quad -u_1 + iv_1 = \frac{iK_2}{2\pi} \frac{1}{z_1 - z_2} \quad \text{... (3)}$$

$$\therefore \quad q_1 = |-u_1 + iv_1| = \frac{K_2}{2\pi A_1 A_2} \quad \text{... (4)}$$

The velocity of vortex at A_2 is due to vortex at A_1 and so

$$-u_2 + iv_2 = \frac{iK_1}{2\pi}\left(\frac{1}{z - z_1}\right)_{z = z_2} \quad \text{by (2)}$$

$$\Rightarrow \quad -u_2 + iv_2 = \frac{iK_1}{2\pi}\frac{1}{z_2 - z_1} \qquad \text{... (5)}$$

$$\Rightarrow \quad q_2 = \frac{K_1}{2\pi A_1 A_2} \qquad \text{... (6)}$$

From (3) and (5), we have

$$K_1(-u_1 + iv_1) + K_2(-u_2 + iv_2) = 0$$

$$\Rightarrow \quad -(K_1u_1 + K_2u_2) + i(K_1v_1 + K_2v_2) = 0$$

$$\Rightarrow \quad K_1u_1 + K_2u_2 = 0, K_1v_1 + K_2v_2 = 0$$

$$\Rightarrow \quad (K_1u_1 + K_2u_2) + i(K_1v_1 + K_2v_2) = 0$$

$$\Rightarrow \quad K_1(u_1 + iv_1) + K_2(u_2 + iv_2) = 0$$

$$\Rightarrow \quad K_1\dot{z}_1 + K_2\dot{z}_2 = 0$$

$$\Rightarrow \quad \frac{K_1\dot{z}_1 + K_2\dot{z}_2}{K_1 + K_2} = 0 \text{ if } K_1 + K_2 \neq 0$$

$$\Rightarrow \quad \frac{d}{dt}\left(\frac{K_1z_1 + K_2z_2}{K_1 + K_2}\right) = 0$$

$$\Rightarrow \quad \frac{K_1z_1 + K_2z_2}{K_1 + K_2} = \text{constant} \qquad \text{... (7)}$$

The point $\dfrac{K_1z_1 + K_2z_2}{K_1 + K_2}$ may be called the centre G of vortices by analogy with the centre of gravity. The point G is fixed by (7).

Since $K_1 \cdot A_1G = K_2 \cdot A_2G$

Hence

$$\frac{A_2G}{A_1G} = \frac{K_1}{K_2}$$

$$\Rightarrow \quad \frac{A_1G + A_2G}{A_1G} = \frac{K_1 + K_2}{K_2}$$

$$\Rightarrow \quad A_1G = \frac{K_2}{K_1 + K_2} A_1A_2$$

From (4)

$$q_1 = \frac{1}{2\pi} \frac{K_2 A_1 A_2}{K_1 + K_2} \frac{K_1 + K_2}{(A_1A_2)^2}$$

$$= \frac{1}{2\pi} A_1G \frac{K_1 + K_2}{(A_1A_2)^2}$$

$$= A_1G \cdot W$$

Since the angular velocity W of $A = \frac{K_1 + K_2}{2\pi(A_1A_2)^2}$ about G.

Similarly $q_2 = A_2G \cdot W$

Hence vortex A_1 moves with velocity $A_1G \cdot W$ perpendicular to A_1A_2 whereas vortex A_2 moves with velocity $A_2G \cdot W$ perpendicular to A_1A_2 in opposite direction. If $K_1 > K_2$, then G will be on A_1A_2 produced and if $K_2 > K_1$, it will be on the line A_2A_1 produced.

This shows that the line A_1A_2 rotates about G with angular velocity W. Also from (7), the system has no velocity along A_1A_2. Hence A_1A_2 remains constant in length.

7.11. Vortex Pair

A pair of vortices each of strength K but of opposite rotation is called a vortex pair.

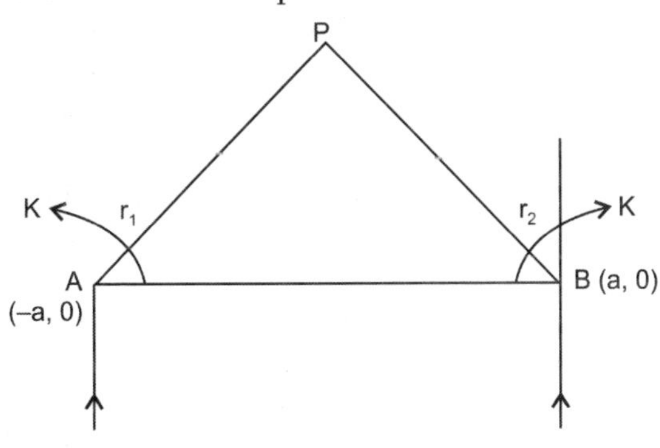

Fig. 7.3

Let A, B the vortex filaments of strengths K and $-K$ be situated at z_1 and z_2 then

$$W = \frac{iK}{2\pi} \log (z - z_1) - \frac{iK}{2\pi} \log (z - z_2) \quad \text{... (1)}$$

However, the vortices situated at A and B would start moving due to the presence of each other. Let u_1, v_1 be the components of the velocity q_1 of A, which is due to B alone.

Then we have

$$u_1 - iv_1 = \left[\frac{1}{2\pi} \frac{iK}{z - z_1} + \left(-\frac{dw}{dz}\right)\right]_{z = z_1}$$

$$= \left[\frac{iK}{2\pi (z - z_1)} - \left\{\frac{iK}{2\pi (z - z_1)} - \frac{iK}{2\pi (z - z_2)}\right\}\right]_{z = z_1}$$

$$= \frac{iK}{2\pi (z_1 - z_2)} \quad \text{... (2)}$$

$$\therefore \quad q_1 = |u_1 - iv_1| = \frac{K}{2\pi |z_1 - z_2|} = \frac{K}{2\pi AB} \quad \text{... (3)}$$

Similarly

$$q_2 = |u_2 - iv_2| = \frac{K}{2\pi AB} \quad \text{... (4)}$$

Let

$$q_1 = q_2 = q \text{ (say)} \quad \text{... (5)}$$

Thus the velocity q_1 of A_1 due to A_2 is q perpendicular to A_1A_2. Similarly, the velocity q_2 of A_2 due to A_1 is q perpendicular to A_1A_2 in the same sense as that of A_1. Hence the vortices situated at A_1 and A_2 move in the same direction perpendicular to A_1A_2 with uniform velocity q. However, the line may move forward or backward according to the direction of rotation.

Let $w = \phi + i\psi$

Again from (1)

$$w = \frac{iK}{2\pi} \log \left(\frac{z - z_1}{z - z_2}\right)$$

$$\Rightarrow \quad \phi + i\psi = \frac{iK}{2\pi} \log \left(\frac{r_1 e^{i\theta_1}}{r_2 e^{i\theta_2}} \right)$$

Equating imaginary parts of both sides, we get

$$\psi = \frac{K}{2\pi} \log \frac{r_1}{r_2} \qquad \text{Where } r_1 = A_1P \text{ and } r_2 = A_2P$$

Streamlines are given by ψ = constant

$$\text{i.e. } \frac{K}{2\pi} \log \left(\frac{r_1}{r_2} \right) = \text{constant}$$

$$\Rightarrow \quad \frac{r_1}{r_2} = \text{constant} = b$$

$$\Rightarrow \quad \frac{(x+a)^2 + y^2}{(x-a)^2 + y^2} = b^2$$

$$\Rightarrow \quad x^2 + y^2 + 2a \left(\frac{1+b^2}{1-b^2} \right) x + a^2 = 0$$

This shows that the streamlines are coaxial circles with A_1 and A_2 as limiting points.

7.12. Vortex Doublet or Dipole

A vortex doublet is a combination of a vortex of strength $+K$ and a vortex of strength $-K$ at a small distance δs apart such that

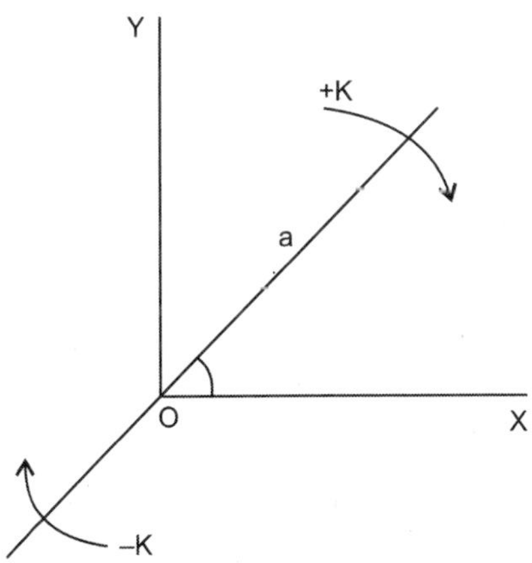

Fig. 7.4

$$\underset{\delta s \to 0}{Lt} \frac{K}{2\pi} \delta s = \text{a finite quantity}$$

$$= \mu \text{ (say)} \qquad \text{Where } K \to \infty$$

Then μ is called strength of the doublet and the axis of doublet is in the sense from $-K$ to K.

We consider a vortex pair of strength $+K$ at $z = ae^{i\alpha}$ and of strength $-K$ at $z = -ae^{i\alpha}$ respectively so that the axis of doublet is inclined at any angle α with x-axis.

The complex potential at any point $P(z)$ is given by

$$W = \frac{iK}{2\pi} \log\left(z - ae^{i\alpha}\right) - \frac{iK}{2\pi} \log\left(z + ae^{i\alpha}\right)$$

$$= \frac{-iK}{2\pi}\left[\log z\left(1 + \frac{ae^{i\alpha}}{z}\right) - \log z\left(1 - \frac{ae^{i\alpha}}{z}\right)\right]$$

$$= \frac{-iK}{2\pi}\left[\log\left(1 + \frac{ae^{i\alpha}}{z}\right) - \log\left(1 - \frac{ae^{i\alpha}}{z}\right)\right]$$

$$= \frac{-iK}{2\pi}\left[\left\{\frac{ae^{i\alpha}}{z} - \frac{1}{2}\left(\frac{ae^{i\alpha}}{z}\right)^2 + \frac{1}{3}\left(\frac{ae^{i\alpha}}{z}\right)^3\right\}\right.$$

$$\left. + \left\{\frac{ae^{i\alpha}}{z} + \frac{1}{2}\left(\frac{ae^{i\alpha}}{z}\right)^2 + \ldots\right\}\right]$$

$$= \frac{-2iK}{2\pi}\left[\frac{ae^{i\alpha}}{z} + \frac{1}{3}\left(\frac{ae^{i\alpha}}{z}\right)^3 + \ldots\right]$$

$$= \frac{-iKa}{\pi}\left[\frac{e^{i\alpha}}{z} + \frac{1}{3}\left(\frac{e^{i\alpha}}{z}\right)^3 a^2 + \ldots\right]$$

But $\frac{K}{2\pi} \cdot 2a = \mu$ as $K \to \infty, a \to 0$

Hence

$$W = -i\mu\left[\frac{e^{i\alpha}}{z} + 0 + 0 + \ldots\right]$$

or $$W = -i\mu \frac{e^{i\alpha}}{z} \qquad \ldots (1)$$

This is the required complex potential due to vortex doublet with its axis inclined at an angle α with x-axis.

Deductions. From (1)

$$w = \frac{-iK}{re^{i\theta}} e^{i\alpha} \qquad \text{Where } z = re^{i\theta}$$

$$\Rightarrow \quad w = \frac{-i\mu}{r} e^{i(\alpha-\theta)}$$

$$\therefore \quad \phi + i\psi = \frac{-i\mu}{r}\left[\cos(\alpha-\theta) + \sin(\alpha-\theta)\right]$$

Hence

$$\psi = \frac{-\mu}{r}\cos(\alpha-\theta)$$

Taking $\mu = Ub^2, \alpha = \frac{\pi}{2}$

Then

$$\psi = \frac{-Ub^2}{r}\sin\theta$$

Which is stream function for a circular cylinder of radius a moving with velocity U along the x-axis. This shows that the motion due to a vortex at the centre with its axis perpendicular to the axis of motion is the same as the motion due to a circular cylinder of radius b moving with velocity U along x-axis.

7.13. Spiral Vortex

The combination of a source and a vortex is known as spiral vortex.

Let there be a source of strength $+m$ and vortex $+K$ both at the origin. Then

$$w = -m\log z + \frac{ik}{2\pi}\log z$$

$$w = \left(\frac{ik}{2\pi} - m\right)\log z$$

To find pressure due to spiral vortex

$$\frac{dw}{dz} = -(m + iK)\frac{1}{2} \qquad \text{Where } K = \frac{k}{2\pi}$$

$$\Rightarrow \quad q^2 = \frac{dw}{dz} \cdot \frac{d\bar{w}}{d\bar{z}} = \left(\frac{-m+iK}{z}\right)\left(\frac{-m-iK}{\bar{z}}\right)$$

$$= \frac{m^2 + K^2}{|z|^2} = \frac{m^2 + K^2}{r^2}$$

For steady flow of pressure equation is given by

$$\frac{p}{\rho} + \frac{1}{2} q^2 + \Omega = C$$

$$\therefore \quad \frac{p}{\rho} = C - \Omega - \frac{1}{2}\frac{m_1^2}{r^2} \qquad \text{Where } m_1^2 = m^2 + K^2$$

This gives the pressure at any point z due to the given system.

7.14. Image of a Vortex Filament in a Plane

To show that the image of a vortex filament of strength K is a vortex of strength $-K$ placed at the optical image of the given vortex.

Proof. Let the vortex filaments of strength K and $-K$ be situated at $A_1(z = z_1)$ and $B(z = z_2)$ respectively.

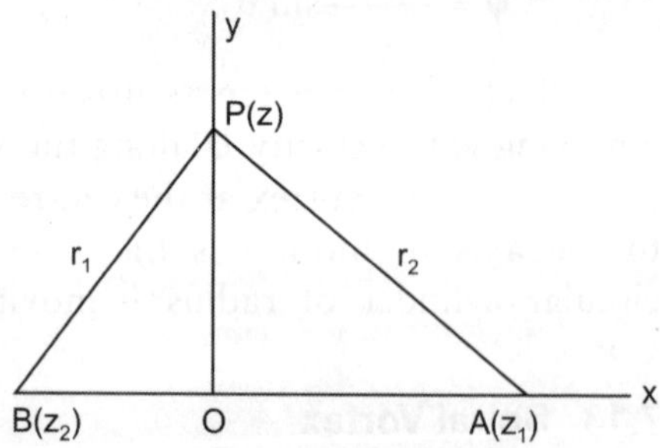

Fig. 7.5

The complex potential due to these vortices at any point $P(z)$ is given by

$$W = \frac{iK}{2\pi} \log(z - z_1) - \frac{iK}{2\pi} \log(z - z_2)$$

$$\Rightarrow \quad \phi + i\psi = \frac{iK}{2\pi} \log \frac{z - z_1}{z - z_2}$$

$$= \frac{iK}{2\pi} \log \frac{r_1 e^{i\theta_1}}{r_2 e^{i\theta_2}} \qquad \dots (1)$$

Where $z - z_1 = r_1 e^{i\theta_1}$ and $z - z_2 = r_2 e^{i\theta_2}$

Equating imaginary parts in (1), we get

$$\psi = \frac{K}{2\pi} \log \frac{r_1}{r_2}$$

If $r_1 = r_2$ then $OA = OB$ i.e. B is the optical image of A with respect to y-axis. Now $\psi = 0$. So that there is no flow across the plane. Hence the required result follows.

7.15. Image of a Vortex in a Quadrant

The image system of vortex of strength K, at the point $A(x, y)$ in xy plane with respect to quadrant XOY consists of

(i) a vortex of strength $-K$ at $B(-x, y)$

(ii) a vortex of strength $-K$ at $C(x, -y)$

(iii) a vortex of strength K at $D(-x, -y)$

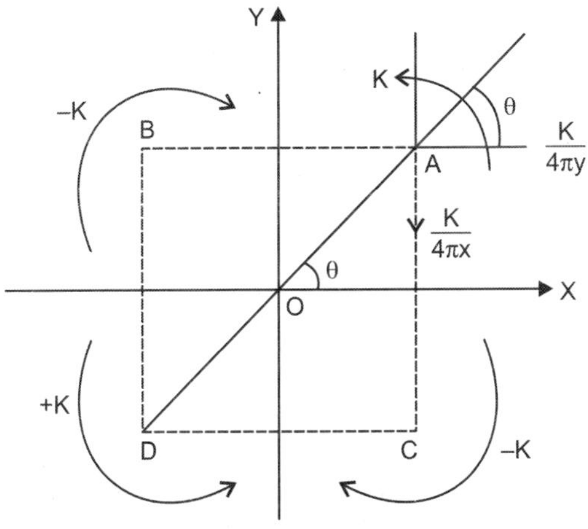

Fig. 7.6

The figure shows the arrangement of images. The velocity of the vortex at $A(x, y)$ is only on account of its images and its components are as indicated in the figure. Thus the radial and transverse components of velocity at A are given by

$$\frac{dr}{dt} = \frac{K}{4\pi y} \cos\theta - \frac{K}{4\pi x} \sin\theta = \frac{K\cos\theta}{4\pi r \sin\theta} - \frac{K \sin\theta}{4\pi r \cos\theta}$$

$$= \frac{K\left(\cos^2\theta - \sin^2\theta\right)}{2\pi r \sin 2\theta}$$

$$\Rightarrow \quad \frac{dr}{dt} = \frac{K \cos 2\theta}{2\pi r \sin 2\theta} \qquad \text{... (1)}$$

Also

$$r\frac{d\theta}{dt} = \frac{K}{4\pi r} - \frac{K \sin \theta}{4\pi y} - \frac{K \cos \theta}{4\pi x}$$

$$= \frac{K}{4\pi r} - \frac{K \sin \theta}{4\pi r \sin \theta} - \frac{K \cos \theta}{4\pi r \cos \theta}$$

$$\Rightarrow \quad r\frac{d\theta}{dt} = \frac{-K}{4\pi r} \qquad \text{... (2)}$$

On dividing (1) by (2), we get

$$\frac{1}{r}\frac{dr}{d\theta} = \frac{-2\cos 2\theta}{\sin 2\,\theta}$$

Integrating it, we get

$$\log r = -\log \sin 2\theta + \log c$$

$$\Rightarrow \quad r \sin 2\theta = c$$

which is Cote's spiral. Transforming into Cartesian, it becomes

$$2r \sin \theta \cos \theta = c$$

$$\Rightarrow \quad 4r^4 \cos^2\theta \sin^2\theta = c^2 r^2$$

$$\Rightarrow \quad 4x^2y^2 = c^2(x^2 + y^2)$$

$$\Rightarrow \quad \frac{1}{x^2} + \frac{1}{y^2} = \frac{4}{c^2}$$

7.16. Vortex Inside an Infinite Circular Cylinder

Let the vortex of strength $+K$ be situated at $A(OA = f)$ inside the circular cylinder of radius a with axis parallel to the axis of the cylinder.

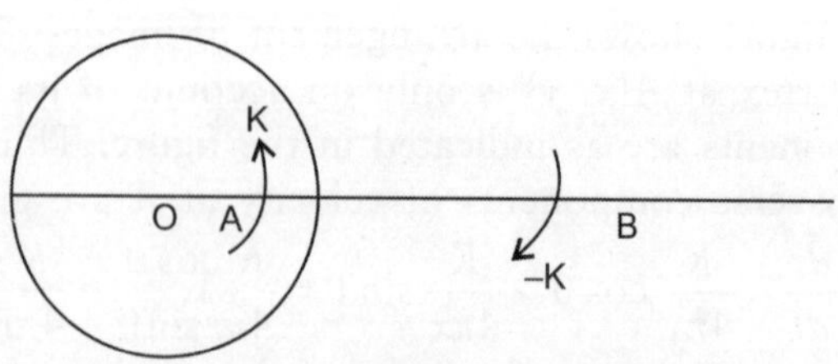

Fig. 7.7

An equal and opposite vortex is placed at an inverse point B such that $OA \cdot OB = a^2 \Rightarrow OB = \dfrac{a^2}{f}$.

The circle is one of the coaxial systems having A and B as limiting points and so it is a straight line.

The velocity of A is given by

$$= \frac{K}{2\pi \cdot AB} = \frac{K}{2\pi(OA - OB)}$$

$$= \frac{K}{2\pi\left(\dfrac{a^2}{f} - f\right)} = \frac{Kf}{2\pi\left(a^2 - f^2\right)}$$

which is perpendicular to OA. The velocity of B is the same as that of A, so OAB will not remain a straight line at the next instant. If A describes a circle about O with the above velocity, then at every instant the circle will be a streamline.

7.17. Vortex Outside a Circular Cylinder

Let the vortex of strength K be situated at $A(OA = f)$ outside the circular cylinder of radius a with axis parallel to the axis of the cylinder. Let a vortex of strength $-K$ be placed at B, where B is the inverse point of A with respect to the circular section of the cylinder so that $OB \cdot OA = a^2$ or $OB = \dfrac{a^2}{f}$.

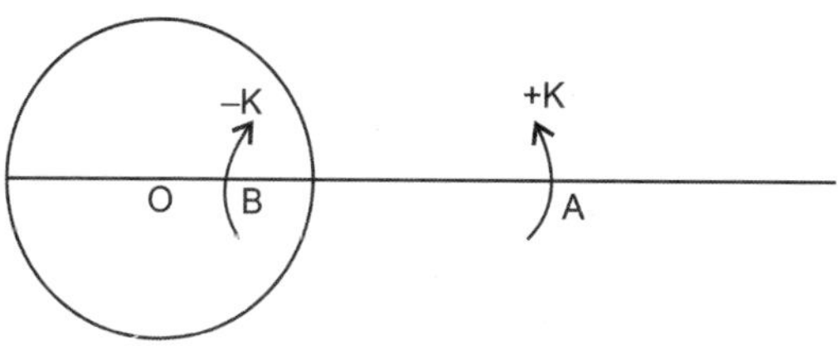

Fig. 7.8

The velocity of A is given by

$$= \frac{K}{2\pi\ AB} = \frac{K}{2\pi(OA - OB)}$$

$$= \frac{K}{2\pi\left(f - \frac{a^2}{f}\right)} = \frac{Kf}{2\pi(f^2 - a^2)}$$

Now the vortex of strength $-K$ at B gives a circulation $-K$ at about the cylinder. Let K_1 be the circulation about the cylinder. Thus we have the vortex of strength $+K$ at A, $-K$ at B and $K + K_1$ at the centre O. The velocity of A due to above system is given by

$$= \frac{K + K_1}{2\pi \cdot OA} - \frac{K}{2\pi \cdot AB}$$

$$= \frac{K + K_1}{2\pi f} - \frac{Kf}{2\pi\left(f^2 - a^2\right)}$$

Therefore A describes *a* circle with this velocity.

7.18. Image of Vortex Outside a Circular Cylinder

To show that the image system of a vortex K outside the circular cylinder consists of a vortex of strength $-K$ at the inverse point and a vortex of strength K at the centre.

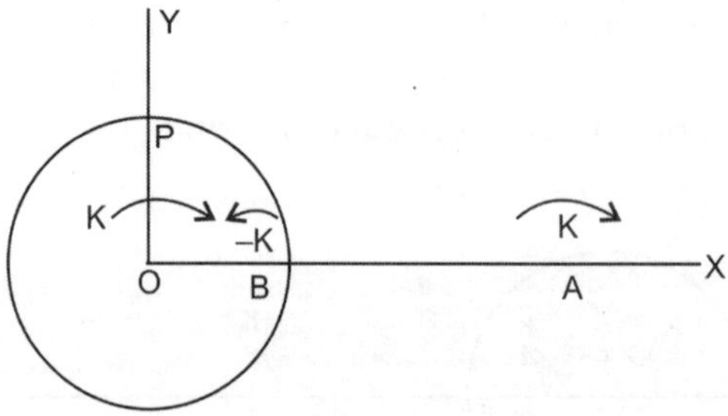

Fig. 7.9

Let us determine the image of a vortex filament of strength K placed at $(z = f > a)$ with respect to a circular cylinder $|z| = a$ with O as centre. Let B be the inverse point of A with respect to $|z| = a$ so that

$$OB = \frac{a^2}{OA} = \frac{a^2}{f}$$

The complex potential due to vortex $+K$ at A is

$$w = \frac{iK}{2\pi} \log (z - f)$$

When we insert a circular cylinder of radius $|z| = a$, the complex potential is

$$w = \frac{iK}{2\pi} \log (z - f) - \frac{iK}{2\pi} \log \left(\frac{a^2}{z} - f \right)$$

by circle theorem.

Addition of a constant term $iK \log (-f)$ does not change the nature of w. Then

$$w = \frac{iK}{2\pi} \log (z - f) - \frac{iK}{2\pi} \log \left(\frac{a^2}{z} - f \right) + iK \log (-f)$$

$$= \frac{iK}{2\pi} \log (z - f) - \frac{iK}{2\pi} \log \left(1 - \frac{a^2}{zf} \right)$$

But

$$1 - \frac{a^2}{zf} = \frac{1}{z} \left(z - \frac{a^2}{f} \right)$$

Hence

$$w = \frac{iK}{2\pi} \log (z - f) \frac{-iK}{2\pi} \log \left(z - \frac{a^2}{f} \right) + \frac{iK}{2\pi} \log (z - 0) \qquad \ldots (1)$$

This is the complex potential of

(i) vortex $+K$ at A $(z = f)$

(ii) vortex $-K$ at $B \left(z = \dfrac{a^2}{f} \right)$

(iii) vortex $+K$ at O $(z = 0)$

Since $OB = \dfrac{a^2}{f}$. Hence A and B are inverse points. If we put $z = ae^{i\theta}$ in (1), then $\psi = \text{const.}$

For

$$\psi = \frac{K}{2\pi}\log\frac{|z-f||z|}{\left|z-\frac{a^2}{f}\right|} = \frac{K}{2\pi}\log\frac{PA\cdot PO}{PB}$$

$$= \frac{K}{2\pi}\log a\,\frac{PA}{PB} = \frac{K}{2\pi}\log\frac{af}{a} = \text{constant}$$

Hence circle is a streamline. Thus the image of a vortex $+K$ at A outside the cylinder is a vortex $-K$ at B, the inverse point of A and $+K$ at O, the centre of the circle.

7.19. Image of a Vortex Inside a Circular Cylinder

To show that the image of a vortex inside a circular cylinder would be an equal and opposite vortex at the inverse point.

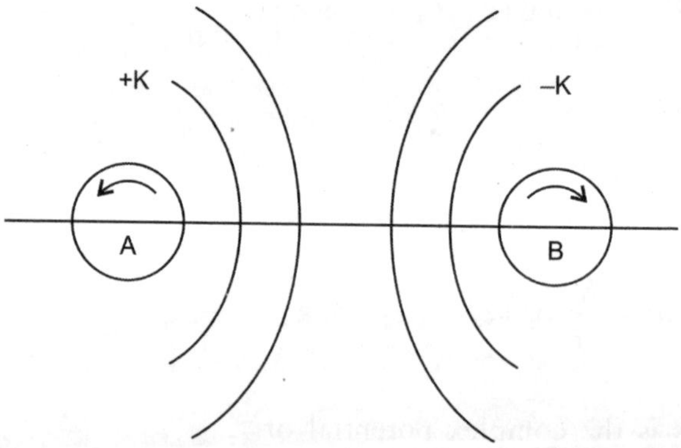

Fig. 7.10

Let there be a vortex pair consisting of two vortices of strength K at $A(z_1)$ and $-K$ at $B(z_2)$. Then the complex potential at any point is given by

$$w = \frac{iK}{2\pi}\log(z-z_1) - \frac{iK}{2\pi}\log(z-z_2)$$

$$\Rightarrow\quad \phi + i\psi = \frac{iK}{2\pi}\log\frac{z-z_1}{z-z_2} = \frac{iK}{2\pi}\log\frac{r_1 e^{i\theta_1}}{r_2 e^{i\theta_2}}$$

Equating imaginary parts on both the sides, we get

$$\psi = \frac{iK}{2\pi} \log \frac{r_1}{r_2}$$

Hence the streamlines are given by ψ = const. i.e. $\frac{r_1}{r_2} = c$ which represents a family of coaxial circles with A and B as limiting points. Moreover, the motion is unsteady and hence streamlines go on changing and following the vortices which move through the liquid. However, if a particular circle of the family of coaxial circle be replaced by a similar rigid boundary and held fixed, then it follows that the image of a vortex inside a circular cylinder would be an equal and opposite vortex at the inverse point.

Remark. Let O be the centre of the cylinder. Let $OA = c$. Then, if B is the inverse point of A, $OB = \frac{a^2}{c}$, where a is the radius of the circular cylinder. The vertex A will move round the circular cylinder with velocity q given by

$$q = \frac{K}{2\pi AB} = \frac{K}{2\pi(OB - OA)} = \frac{K}{2\pi\left(\frac{a^2}{c} - c\right)} = \frac{Kc}{2\pi\left(a^2 - c^2\right)}$$

Let w be the angular velocity of vortex at A, then

$$w = \frac{q}{OA} = \frac{q}{c} = \frac{K}{2\pi\left(a^2 - c^2\right)}$$

7.20. Vortex Rows

When a body moves slowly through a liquid rows of vortices are often generated in its walk. When these vortices are stable, then they can be photographed. In the next two articles we will consider infinite systems of parallel rectilinear vortices in two-dimensional flow.

7.21. Infinite Single Row of Vortices of Equal Strength

To show that the motion due to a set of line vortices of strength K at points $z = na$ (n = 0, 1, 2, ...) is given by the relation

$$w = \frac{iK}{2\pi} \log \sin \frac{\pi z}{a}$$

Also to get velocity components and streamlines

Let there be an infinite number of vortices of strength K each placed at the points $(0, 0)$, $(\pm a, 0)$, $(\pm 2a, 0)$ .....

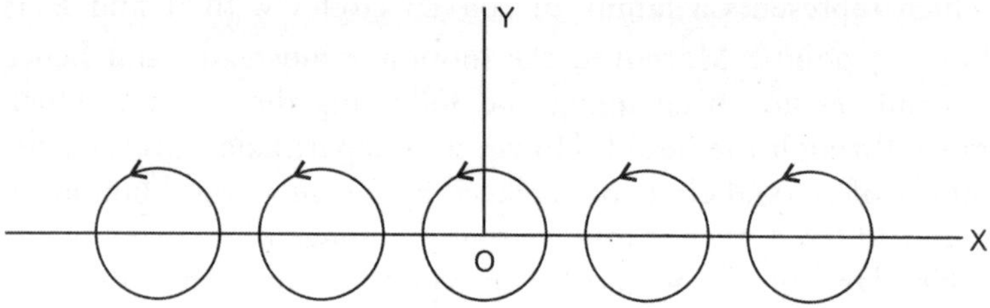

Fig. 7.11

The complex potential at any point z is therefore given by

$$w = \frac{iK}{2\pi} \Big[\log z + \log (z - a) + \log (z + a) + \log (z - 2a) + \log (z + 2a) + \ldots\ldots \Big]$$

$$= \frac{iK}{2\pi} \log \left[z \left(z^2 - a^2 \right) \left(z^2 - 2^2 a^2 \right) \ldots\ldots \right]$$

$$= \frac{iK}{2\pi} \log \left[\frac{\pi z}{a} \left(1 - \frac{z^2}{a^2} \right) \left(1 - \frac{z^2}{2^2 a^2} \right) \ldots\ldots + \frac{iK}{2\pi} \log \left\{ \frac{a}{\pi} a^2 2^2 a^2 \ldots\ldots \right\} \right]$$

$$= \frac{iK}{2\pi} \log \left[\frac{\pi z}{a} \left(1 - \frac{z^2}{a^2} \right) \left(1 - \frac{z^2}{2^2 a^2} \right) \ldots\ldots \right]$$

neglecting the constant

$$= \frac{iK}{2\pi} \log \sin \frac{\pi z}{a}$$

as $\sin \theta = \theta \left(1 - \frac{\theta^2}{\pi^2} \right) \left(1 - \frac{\theta^2}{2^2 \pi^2} \right) \ldots\ldots$

To determine the motion of vortex at $z = 0$

If w_0 be the complex potential at $z = 0$, then

$$w_0 = \frac{iK}{2\pi}\log \sin\frac{\pi z}{a} - \frac{iK}{2\pi}\log(z-0)$$

Since the motion of vortex at $z = 0$ is due to other vortices.

$$u_0 - iv_0 = -\left(\frac{dw_0}{dz}\right)_{z=0} = \frac{-iK}{2\pi}\left[\frac{\pi}{a}\cot\frac{\pi z}{a} - \frac{1}{2}\right]$$

$$= 0 \qquad \text{(evaluating the indeterminate form)}$$

Thus the vortex at the origin is at rest. Similarly all other vortices are at rest. Thus we may say that the vortex row induces no velocity in itself.

Velocity at any point of the fluid

If u and v are the velocity components at any point z, then

$$u - iv = -\frac{dw}{dz} = \frac{-iK}{2\pi}\frac{\pi}{a}\cot\frac{\pi z}{a}$$

or $$u - iv = \frac{-iK}{2a}\frac{\cos b(x+iy)\sin b(x-iy)}{\sin b(x+iy)\sin b(x-iy)}$$

Where $b = \dfrac{\pi}{a}$

$$= \frac{-iK}{2a}\frac{\sin 2bx - i\sinh 2by}{\cosh 2by - \cos 2bx}$$

Equating real and imaginary parts, we get

$$u = \frac{-K}{2a}\frac{\sinh\left(\dfrac{2\pi y}{a}\right)}{\cosh\left(\dfrac{2\pi y}{a}\right) - \cos\left(\dfrac{2\pi x}{a}\right)}$$

and

$$v = \frac{K}{2a}\frac{\sin\left(\dfrac{2\pi x}{a}\right)}{\cosh\left(\dfrac{2\pi y}{a}\right) - \cos\left(\dfrac{2\pi x}{a}\right)}$$

Also

$$\phi = i\psi - (\phi - i\psi) = \frac{iK}{2\pi}\log\sin\frac{\pi z}{a} - \left(-\frac{iK}{2\pi}\log\sin\frac{\pi z}{a}\right)$$

$$\Rightarrow \quad 2i\psi = \frac{iK}{2\pi}\log\sin\frac{\pi z}{a}\sin\frac{\pi\bar{z}}{a}$$

$$\Rightarrow \quad \psi = \frac{K}{4\pi}\log\left[\cosh\frac{2\pi y}{a} - \cos\frac{2\pi x}{a}\right] \quad \text{neglecting constant}$$

Streamlines are given by

$$\psi = \text{const.}$$

i.e. $\cosh\dfrac{2\pi y}{a} - \cos\dfrac{2\pi x}{a} = \text{const.}$

7.22. Two Infinite Rows of Vortices

We now consider two infinite rows of vortices one above the other at a distance b, the upper one having vortices each of strenth K and lower one each of strength $-K$, one vortex of the upper being above each of the lower.

Let the upper one be taken as x-axis and the lower one as $y = -b$.

Let the strengths of each of the vortices in the upper row be $+K$ and $-K$ that of the lower. Thus the system consists of

(i) vortices each of strength $+K$ at
$z = 0, \pm a, \pm 2a, \pm 3a, \ldots..$

(ii) vortices each of strength $-K$ at
$z = -b, \pm\, a - ib, \pm\, 2a - ib, \pm\, 3a - 2b, \ldots..$

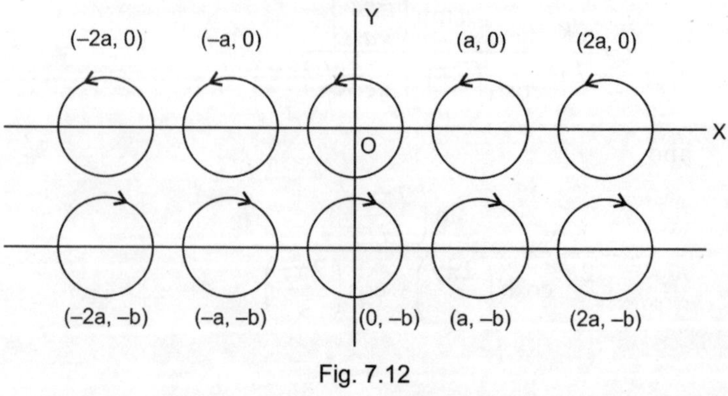

Fig. 7.12

The complex potential w at a point z is

$$w = w_1 + w_2 \qquad \text{... (1)}$$

Where

$$w_1 = \frac{iK}{2\pi}\left[\log z + \log(z-a) + \log(z+a) + \log(z-2a) + \log(z+2a) + ...\right]$$

$$= \frac{iK}{2\pi}\log\left[z(z^2-a^2)(z^2-2^2a^2).....\right]$$

$$= \frac{iK}{2\pi}\log\sin\left(\frac{\pi z}{a}\right)$$

Similarly

$$w_2 = \frac{-iK}{2\pi}\log\sin\frac{\pi}{a}(z+ib)$$

Hence from (1)

$$w = \frac{iK}{2\pi}\log\sin\frac{\pi z}{a} - \frac{iK}{2\pi}\log\sin\frac{\pi(z+ib)}{a}$$

The velocity components u, v at any point z is given by

$$u - iv = -\frac{dw}{dz} = -\frac{iK}{2a}\cot\frac{\pi z}{a} + \frac{iK}{2a}\log\cot\frac{\pi(z+ib)}{a}$$

If u_0 and v_0 are the velocity components of the vortex at origin $z = 0$, then

$$u_0 - iv_0 = \frac{-iK}{2\pi}\left[\frac{\pi}{a}\cot\frac{\pi z}{a} - \frac{\pi}{a}\cot(z+ib)\right]_{z=0}$$

$$= \frac{iK}{2\pi}\cot\frac{i\pi b}{a}$$

(Evaluating by L'Hospital's Rule)

$$= \frac{iK}{2a}(-i)\cot h\left(\frac{\pi b}{a}\right)$$

$$= \frac{K}{2a}\cot h\left(\frac{\pi b}{a}\right)$$

$$\therefore \quad u_0 = \frac{K}{2a}\cot h\left(\frac{\pi b}{a}\right),\ v_0 = 0$$

Thus the vortex system moves parallel to itself with velocity $\frac{K}{2a} \cot h \frac{\pi b}{a}$.

7.23. Karman's Vortex Street

An arrangement of two parallel rows of infinite vortices of the same spacing a such that each vortex K of one row is directly above the middle of the joining of two vortices of strengths $-K$ of another row, is said to be Karman's vortex street.

Let there are two parallel rows of vortices at a distance b apart such that each vortex in one row is opposite the centre of the interval between two consecutive vertices in the other row.

Thus vortices of strength K each are situated at (0, 0), ($\pm a$, 0), ($\pm 2a$, 0)..... and those of strength $-K$ each are situated at ($\pm a$, b), $\left(\pm \frac{3}{2}a, -b\right)$.....

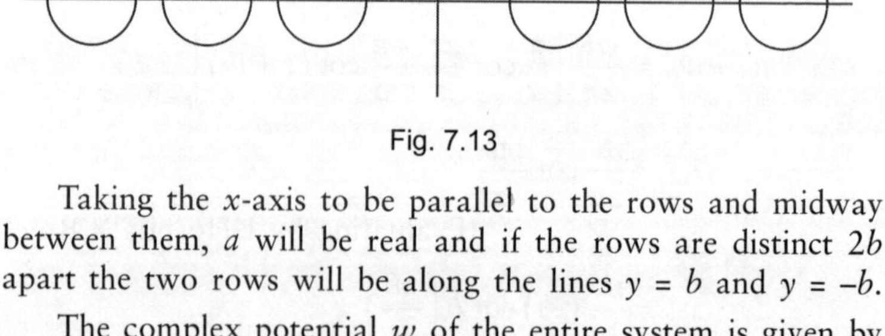

Fig. 7.13

Taking the x-axis to be parallel to the rows and midway between them, a will be real and if the rows are distinct $2b$ apart the two rows will be along the lines $y = b$ and $y = -b$.

The complex potential w of the entire system is given by

$$w = \frac{iK}{2\pi} \log \sin \frac{\pi z}{a} - \frac{iK}{2\pi} \log \sin \frac{\pi}{a}\left(z + \frac{a}{2} + ib\right)$$

Let u and v be the components of velocity at any point, then

$$u - iv = -\frac{dw}{dz} = -\frac{iK}{2a}\cot\frac{\pi z}{a} + \frac{iK}{2\pi}\cot\frac{\pi}{a}\left(z + \frac{a}{2} + ib\right)$$

Let u_0 and v_0 be velocity components of the vortex $+K$ at the origin, then

$$u_0 - iv_0 - -\frac{dw_0}{dz} = \frac{-iK}{2\pi}\left[\frac{\pi}{a}\cot\frac{\pi z}{a} - \frac{1}{z} - \frac{\pi}{a}\cot\frac{\pi}{2}\left(z + \frac{a}{2} + ib\right)\right]_{z=0}$$

$$= \frac{iK}{2a}\cot\left(\frac{\pi}{2} + \frac{i\pi b}{a}\right)$$

$$= -\frac{iK}{2a}\tan\frac{i\pi b}{a}$$

$$= \frac{K}{2a}\tan h\frac{\pi b}{a}$$

So that $u_0 = \dfrac{K}{2a}\tan h\dfrac{\pi b}{a}$ and $v_0 = 0$

It means that the vortex at the origin moves with the velocity $\dfrac{K}{2a}\tan h\dfrac{\pi b}{a}$.

7.24. Rectilinear Vortex with Circular Section

We consider a straight circular vortex tube of radius a in a liquid in a direction perpendicular to xy plane. The motion will be purely two-dimensional and there is only one component ζ of the vorticity in the direction of ζ-axis, its value is constant within the tube and zero outside.

Hence $u = u(x, y)$, $v = v(x, y)$, $w = 0$

Also $\xi = 0$, $\eta = 0$, $\zeta = \dfrac{1}{2}\left(\dfrac{\partial v}{\partial x} - \dfrac{\partial u}{\partial y}\right)$

But

$$u = \frac{-\partial\phi}{\partial x} = \frac{-\partial\psi}{\partial y}, \quad v = \frac{-\partial\phi}{\partial y} = \frac{\partial\psi}{\partial x}$$

$$\left.\begin{aligned}\therefore \quad \frac{\partial^2\psi}{\partial x^2}+\frac{\partial^2\psi}{\partial y^2}&=2\zeta, \quad \text{within the tube}\\ \text{and} \quad \frac{\partial^2\psi}{\partial x^2}+\frac{\partial^2\psi}{\partial y^2}&=0, \quad \text{outside the tube}\end{aligned}\right\} \qquad \dots (a)$$

In the spherical polar coordinates, the relation (1) reduces to

$$\frac{\partial^2\psi}{\partial r^2}+\frac{1}{r}\frac{\partial\psi}{\partial r}+\frac{1}{r^2}\frac{\partial^2\psi}{\partial\theta^2}=2\zeta \qquad \dots (1)$$

Symmetry exists about the origin and so ψ must be a function of r only.

Hence (1) reduces to

$$\frac{d^2\psi}{dr^2}+\frac{1}{r}\frac{d\psi}{dr}=\frac{1}{r}\frac{d}{dr}\left(r\frac{d\psi}{dr}\right)=\begin{cases}2\zeta \text{ when } r<a\\ 0 \text{ when } r>a\end{cases}$$

Thus

$$r\frac{d\psi}{dr}=\zeta r^2+A \quad \text{when } r<a$$

$$=B \qquad \text{when } r>a$$

Again integrating, we get

$$\psi=\zeta\frac{r^2}{2}+A\log r+C \text{ for } r<a \qquad \dots (2)$$

and

$$\psi_2=B\log r+D \text{ for } r>a \qquad \dots (3)$$

We now require that

(i) ψ is finite at the origin so that $A = 0$

(ii) solutions for $r < a$ and $r > a$ be coincident at the boundary of vortex, which requires that ψ and $\dfrac{d\psi}{dr}$ to be continuous at $r = a$.

Hence from (2) and (3), we have

$$\zeta\frac{a^2}{2}+0+C=B\log a+D \text{ as } A=0 \qquad \dots (4)$$

and

$$\zeta a = \frac{B}{a} \qquad \text{... (5)}$$

For a particular solution of (2), we can take $D = 0$.

From (4) and (5), we have

$$B = \zeta a^2$$

and

$$C = \zeta a^2 \log a - \frac{\zeta a^2}{2}$$

Now (2) and (3) becomes

$$\psi_1 = \frac{\zeta r^2}{2} + \zeta a^2 \log a - \frac{\zeta a^2}{2}$$

$$\Rightarrow \quad \psi_1 = \frac{\zeta}{a}\left(r^2 - a^2\right) + \zeta a^2 \log a, \; r < a \qquad \text{... (6)}$$

and

$$\psi_2 = \zeta a^2 \log r, \; r > a \qquad \text{... (7)}$$

The motion outside the cylindrical vortex is irrotational and ϕ_2 exists such that

$$V = -\frac{1}{r}\frac{\partial \phi_2}{\partial \theta} = \frac{-\partial \phi_2}{\partial y} = \frac{\partial \psi_2}{\partial x} = \frac{\partial \psi_2}{\partial r}$$

$$\Rightarrow \quad -\frac{1}{r}\frac{\partial \phi_2}{\partial \theta} = \frac{\zeta a^2}{r}$$

Thus

$$\phi_2 = \zeta a^2 \theta$$

Let K be the strength of the vortex so that K is a circulation round the tube. Then

$$K = 2\pi a^2 \zeta \text{ and so } \psi_1 = \frac{K}{4\pi a^2}\left(r^2 - a^2\right) + \frac{K}{2\pi}\log a$$

$$\psi_2 = \frac{K}{2\pi}\log r, \quad \phi_2 = \frac{-K\theta}{2\pi}$$

$$w_2 = \phi_2 + i\psi_2 = \frac{iK}{2\pi}\left[i\theta + \log r\right] = \frac{iK}{2\pi}\log\left(re^{i\theta}\right)$$

$$\Rightarrow \quad w_2 = \frac{iK}{2\pi} \log z$$

Finally

$$\psi = \frac{K}{4\pi a^2}\left(r^2 - a^2\right) + \frac{K}{2\pi} \log a, r < a \qquad \text{... (8)}$$

$$w = \frac{iK}{2\pi} \log z, r > a \qquad \text{... (9)}$$

Also

$$q = \left|\frac{dw}{dz}\right| = \left|\frac{iK}{2\pi z}\right| = \frac{K}{2\pi r}, r > a$$

$$q = \frac{d\psi}{dr} = \frac{Kr}{2\pi a^2}, r < a$$

7.25. Rankine's Combined Vortex

Rankine's combined vortex consists of a circular cylindrical vortex with its axis vertical in a mass of liquid which is moving irrotationally under the action of gravity, only the upper surface being exposed to atmospheric pressure π. The external forces are derivable from the potential Ω, the potential energy function per unit mass i.e.

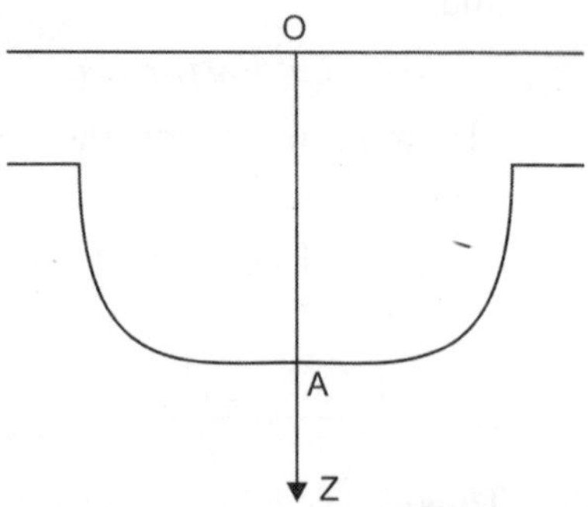

Fig. 7.14

$$\Omega = 1gh = -gz, \text{ where } OA = h = -z$$

We take the origin in the axis of the vortex and in the level of the liquid at infinity. We measure z in downward direction.

We know that $q = \dfrac{K}{2\pi r}, r > a$

and $q = \dfrac{Kr}{2\pi a^2}, r < a$

Also pressure equation for steady motion is

$$\frac{p}{\rho}+\frac{1}{2}q^2+\Omega=C$$

$$\therefore \quad \frac{p}{\rho}=A-\frac{K^2}{8\pi^2 r^2}+gz \text{ for } r>a \qquad \dots (1)$$

and

$$\frac{p}{\rho}=B+\frac{K^2 r^2}{8\pi^2 a^4}+gz \text{ for } r<a \qquad \dots (2)$$

In order to preserve the continuity, p is the same on $r = a$ so that

$$A-\frac{K^2}{8\pi^2 a^2}+gz=B+\frac{K^2 a^2}{8\pi^2 a^4}+gz$$

$$\Rightarrow \quad A=B+\frac{K^2}{4\pi^2 a^2}$$

Hence

$$\frac{p}{\rho}=B+\frac{K^2}{4\pi^2 a^2}-\frac{K^2}{8\pi^2 r^2}+gz,\ r>a \qquad \dots (3)$$

and

$$\frac{p}{\rho}=B+\frac{K^2 r^2}{8\pi^2 a^4}+gz\zeta,\ r<a \qquad \dots (4)$$

Considering the origin in the general level of the free surface when $r > a$ such that $z = 0$, $p = \pi$, when $r = \infty$.

Then from (3),

$$\frac{\pi}{\rho}=B+\frac{K^2}{4\pi^2 a^2}$$

Hence (3) and (4) becomes

$$\frac{p-\pi}{\rho}=\frac{-K^2}{8\pi^2 r^2}+gz,\ r>a$$

and

$$\frac{p-\pi}{\rho}=-\frac{K^2}{4\pi^2 a^2}+\frac{K^2 r^2}{8\pi^2 a^2}+gz,\ r<a$$

On the free surface $p = \pi$, hence free surface is given by

$$gz = \frac{K^2}{8\pi^2 r^2}, r > a$$

and

$$gz = \frac{K^2}{4\pi^2 a^2} - \frac{K^2 r^2}{8\pi^2 a^2}, r < a$$

To obtain the depth of depression at A, we put $r = 0$ in the second result so that

$$gz = \frac{K^2}{4\pi^2 a^2} \quad \text{or} \quad z = \frac{K^2}{4\pi^2 a^2 g}$$

7.26. Rectilinear Vortex with Elliptic Section

To show that a rectilinear vortex whose cross section is an ellipse and whose spin is constant can maintain its form rotating as if it were a solid cylinder in an infinite mass of liquid.

Let the elliptic cross section $\frac{x^2}{a^2} + \frac{y^2}{b^2} = 1$ ... (1)

be given by $\xi = \alpha$ where $x + iy = c \cos h\,(\xi + i\eta)$

So that $a = c \cos h\,\alpha$, $b = c \sin h\,\alpha$

$$a - b = ce^{-\alpha},\ c^2 = a^2 - b^2 \quad \text{... (2)}$$

Let Ω be the constant spin. Then the stream function satisfies the equations

$$\frac{\partial^2 \psi}{\partial x^2} + \frac{\partial^2 \psi}{\partial y^2} = 2\Omega, \text{ inside the boundary } \xi = \alpha \quad \text{... (3)}$$

and

$$\frac{\partial^2 \psi}{\partial x^2} + \frac{\partial^2 \psi}{\partial y^2} = 0, \text{ outside the boundary } \xi = \alpha \quad \text{... (4)}$$

We suppose that the motion is due to rotation of an elliptic cylinder about the axis with uniform angular velocity $\bar{w}$. The complex potential for irrotational motion with circulation K at an outside point is given by

$$w = \frac{i\bar{w}}{4}(a+b)^2 e^{-2(\xi + i\eta)} + \frac{iK}{2\pi}(\xi + i\eta)$$

Hence

$$\psi = \frac{\overline{w}}{4}(a+b)^2 e^{-2\xi}\cos 2\eta + \frac{K\xi}{2\pi} \qquad \text{... (5)}$$

Outside vortex

Circulation K is the strength of vortex so that

K = 2 angular velocity of vortex area of cross section of vortex

$$= 2\Omega\pi ab \qquad \text{... (6)}$$

Let us suppose that inside the boundary $\xi = \alpha$

$$\psi = \Omega\,(Ax^2 + By^2) \qquad \text{.. (7)}$$

From (3) and (7), we have

$$A + B = 1 \qquad \text{... (8)}$$

If l, m be the direction cosine of the normal to the section (1), then

$$l = \frac{x}{pa^2}, m = \frac{y}{pb^2} \quad \text{where } b = \sqrt{\frac{x^2}{a^2} + \frac{y^2}{b^2}} \qquad \text{... (9)}$$

The boundary condition is that the velocity of the liquid normal to boundary must be equal to that of the boundary.

Now the velocity components of any point of the elliptic section are $-\overline{w}y$ and $\overline{w}x$. Again from (7), the velocity components of any point inside the boundary are

$$u = \frac{-\partial\psi}{\partial y} = -2By, v = \frac{\partial\psi}{\partial x} = 2Ax$$

So the boundary condition $lu + mv = -(-\overline{w}y)l + (\overline{w}x)n$ reduces to

$$\Rightarrow \quad \frac{x}{pa^2}(-2By) + \frac{y}{pb^2}(2Ax) = (-\overline{w}y)\frac{x}{pa^2} + (\overline{w}x)\frac{y}{pb^2}$$

$$\Rightarrow \quad Aa^2 - Bb^2 = \frac{\overline{w}}{2}\left(a^2 - b^2\right) \qquad \text{... (10)}$$

Now the condition of continuity of the tangential velocity at the boundary makes the value of $\frac{\partial\psi}{\partial\xi}$ from (4) and (7) the same.

Putting $x = c \cosh \xi \cos \eta$ and $y = \sinh \xi \sin \eta$ in (7), we have

$$-\frac{1}{2}\bar{w}(a+b)^2 e^{-2\alpha} \cos 2\eta + \Omega ab$$
$$= \Omega c^2 \cosh\alpha \sinh\alpha \{A + B + (A - B)\cos 2\eta\}$$

for all values of η from 0 to 2π. Equating coefficients of $\cos 2\eta$, we get

$$-\frac{1}{2}\bar{w}(a+b)^2 e^{-2\alpha} = \Omega c^2 (A - B)\cosh\alpha \sinh\alpha$$

$$\Rightarrow \quad -\frac{1}{2}\bar{w}(a+b)^2 \frac{(a-b)^2}{c^2} = \Omega c^2 (A-B)\frac{ab}{c^2} \qquad \text{[by (2)]}$$

$$\Rightarrow \quad A - B = -\frac{\bar{w}}{2\Omega}\frac{a^2 - b^2}{ab} \qquad \dots (11)$$

From (8), (10) and (11), we have

$$A = \frac{b}{a+b}, \; B = \frac{a}{a+b} \qquad \dots (12)$$

and

$$w = \frac{2ab\Omega}{(a+b)^2} = \frac{2\sqrt{1-e^2}\,\Omega}{\left(1+\sqrt{1-e^2}\right)^2} \qquad \dots (13)$$

Equation (13) gives the velocity of rotation of the cylinder as a whole in terms of the spin Ω and eccentricity e of the elliptic section.

To find the path of the particle

Let (x, y) be the coordinates of any fluid particle in the vortex referred to the axes of the section. Then

$$\dot{x} - \bar{w}y = u = -\frac{\partial\psi}{\partial y} = -2\Omega By = -\frac{\bar{w}y(a+b)}{b}$$

and

$$\dot{y} + \bar{w}x = v = \frac{\partial\psi}{\partial x} = 2\Omega Ax = \frac{\bar{w}x(a+b)}{a}$$

Hence

$$\dot{x} = y\bar{w}\left(1 - \frac{a+b}{b}\right)$$

$$\dot{x} = -\frac{\bar{w}ya}{b} \text{ and } \dot{y} = \frac{\bar{w}xb}{a} \quad \dots (14)$$

$$\left.\begin{aligned} &\text{From (14)} \quad \ddot{x} = \frac{-\bar{w}\dot{y}a}{b} = -\frac{\bar{w}^2 b}{a}x \\ &\text{Also} \quad \ddot{y} = \frac{\bar{w}\dot{x}b}{a} = -\frac{\bar{w}^2 a}{b}y \end{aligned}\right\} \quad \dots (15)$$

which represents the S.H.M. whose time period is same i.e. the period of the relative motion is the same as that of the rotation of the cylinder.

By integrating (15), we have

$$x = La\cos(\bar{w}t + A) \text{ and } y = Lb\sin(\bar{w}t + A)$$

which shows that the paths of the particles of the vortex parallel to the boundary are similar ellipse.

7.27. Vortex between Parallel Walls

We consider a vortex K in the liquid midway between two parallel walls at a distance apart. Let the vortex $+K$ be at the origin and the line through it but parallel to the walls as x-axis.

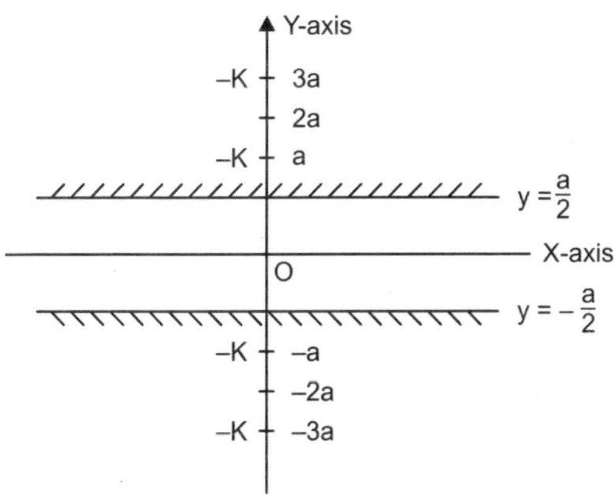

Fig. 7.15

Then, the images of the vortex in the walls $y = \frac{a}{2}, y = -\frac{a}{2}$ are vortices $-K$ at ia, $-ia$ respectively.

These vortices will have vortex images of strengths $+K$ at $z = 2ia$, $-2ia$ w.r.t. the two walls. It means that the image system is a row of vortices along y-axis each at a distance apart, alternating in signs. Thus

(i) $+K$ at $z = 0, \pm 2ia, \pm 4ia, \ldots\ldots, \pm 2nia, \ldots\ldots$

(ii) $-K$ at $z = \pm ia, \pm 3ia, \pm 5ia, \ldots\ldots, \pm(2n-1)ia, \ldots\ldots$

The complex potential at any point z due to vortices of strength K is given by

$$= \frac{iK}{2\pi} \log (z - 2nia)(z + 2nia)$$

$$= \frac{iK}{2\pi} \log \left(1 + \frac{z^2}{4n^2a^2}\right), \text{ ignoring the constants}$$

Hence the potential for the whole row can be written as

$$w = \frac{iK}{2\pi} \log \left[\frac{\overset{\infty}{\underset{1}{\pi}}\left(1 + \frac{z^2}{4n^2a^2}\right)}{\overset{\infty}{\underset{1}{\pi}}\left(1 + \frac{z^2}{(2n-1)^2 a^2}\right)}\right]$$

$$= \frac{iK}{2\pi} \log \frac{\sin \dfrac{h\pi z}{2a}}{\cos h \dfrac{\pi z}{2a}}$$

$$= \frac{iK}{2\pi} \log \tan h \left(\frac{\pi z}{2a}\right)$$

Also

$$2i\psi = w - \bar{w}$$

$$= \frac{iK}{2\pi} \log \left[\tan h\left(\frac{\pi z}{2a}\right) \tan h\left(\frac{\pi \bar{z}}{2a}\right)\right]$$

$$\Rightarrow \quad \psi = \frac{K}{4\pi} \log \left[\frac{(-i)^2 \sin \frac{i\pi}{2a}(x+iy) \sin \frac{i\pi}{2a}(x+iy)}{\cos \frac{i\pi}{2a}(x+iy) \cos \frac{i\pi}{2a}(x-iy)}\right]$$

$$\Rightarrow \quad \psi = \frac{K}{4\pi} \log \left[\frac{\sin \frac{\pi}{2a}(ix - y) \sin \frac{\pi}{2a}(ix + y)}{\cos \frac{\pi}{2a}(ix - y) \cos \frac{\pi}{2a}(ix + y)} \right]$$

$$\Rightarrow \quad \psi = \frac{K}{4\pi} \log \left[\frac{\cos h \frac{\pi x}{a} - \cos \frac{\pi y}{a}}{\cos h \left(\frac{\pi x}{a} \right) + \cos \frac{\pi y}{a}} \right]$$

Streamlines are given by

$$\psi = \text{constant}$$

$$\Rightarrow \quad \cos \frac{\pi y}{a} = \cos b \frac{\pi x}{a}$$

where b is a constant.

The motion of the vortex $+K$ at O is given by $\left(\frac{dw_0}{dz} \right)_{z=0}$

Such that

$$u_0 - iv_0 = -\left(\frac{dw_0}{dz} \right)_{z=0}$$

$$= \frac{-d}{dz} \left[w - \frac{iK}{2\pi} \log z \right]_{z=0}$$

Since the motion is due to other vortices.

$$\Rightarrow \quad u_0 - iv_0 = -\frac{iK}{2\pi} \left[\frac{2\pi}{2a} \operatorname{cosec} h \left(\frac{2\pi z}{2a} \right) - \frac{1}{z} \right]_{z=0}$$

$$= 0$$

Hence the vortex at O is at rest.

7.28. Conformal Transformation, Routh's Theorem

Let K be the strength of a vortex at Q in the ζ-plane and let P be the point corresponding to Q in the z-plane by the conformal mapping $\zeta = f(z)$. Since the complex potential w, stream function ψ and velocity potential ϕ passes equal values at the corresponding points in the two planes. It follows that $-\int d\phi$ taken round corresponding curves in the two planes must be the same, and hence the strengths of vortices at the

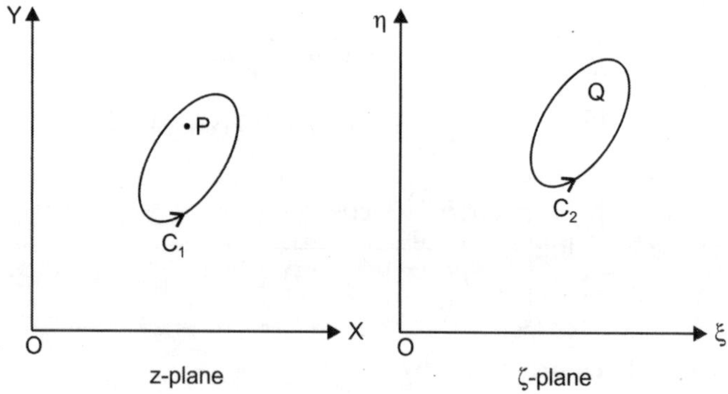

Fig. 7.16

corresponding points must be equal. Let P be z_1 and Q be ζ_1. Then the complex potential w in ζ-plane at any point ζ is given by

$$w = w\zeta + \frac{iK}{2\pi}\log(\zeta - \zeta_1) \quad \ldots (1)$$

where $w\zeta$ is the complex potential omitting the term due to vortex ζ_1.

Similarly, for the z-plane, we have

$$w = w_z + \frac{iK}{2\pi}\log(z - z_1) \quad \ldots (2)$$

Equating the values of w given by (1) and (2), we get

$$w_z = w\zeta + \frac{iK}{2\pi}\log\frac{\zeta - \zeta_1}{z - z_1}$$

The velocity of vortex ζ_1 is obtained from

$$u - iv = -\left(\frac{dw\zeta}{d\zeta}\right)_{\zeta = \zeta_1}$$

Similarly the velocity of vortex z_1 can be obtained.

Then for the motion of the vortex z_1, we get

$$w_{z_1} = w_{\zeta_1} + \left[\frac{iK}{2\pi}\log\frac{\zeta - \zeta_1}{z - z_1}\right]_{\substack{z = z_1 \\ \zeta = \zeta_1}}$$

If ψ_1 (ζ_1, η_1) and ψ_2 be the stream functions giving the motion of the vortices, then we have

$$\psi_2 = \psi_1 + \psi$$

Where

$$\frac{\partial \psi}{\partial y_1} = \text{real part of} \left[\frac{\partial}{\partial y} \cdot \frac{K}{2\pi} \log \frac{\zeta - \zeta_1}{z - z_1}\right]_{z = z_1}$$

$$= \text{real part of} \left[\frac{iK}{2\pi} \frac{d}{dz} \log \frac{\zeta - \zeta_1}{z - z_1}\right]_{z = z_1}$$

$$\left[\because \frac{\partial}{\partial y} = i \frac{d}{dz}\right]$$

Now expanding $\zeta - \zeta_1$ in terms of $z - z_1$, we get

$$\zeta - \zeta_1 = (z - z_1)\left(\frac{d\zeta}{dz}\right)_1 + \frac{1}{2}(z - z_1)^2\left(\frac{d^2\zeta}{dz^2}\right)_1 + ...$$

$$\Rightarrow \quad \frac{\zeta - \zeta_1}{z - z_1} = \left(\frac{d\zeta}{dz}\right)_1 + \frac{1}{2}(z - z_1)^2\left(\frac{d^2\zeta}{dz^2}\right)_1 + ...$$

$$\therefore \quad \text{real part of } \frac{iK}{2\pi} \frac{d}{dz} \log \frac{\zeta - \zeta_1}{z - z_1}$$

$$= \text{real part of } \frac{iK}{2\pi} \frac{d}{dz} \log \left[\left(\frac{d\zeta}{dz}\right)_1 + \frac{1}{2}(z - z_1)\left(\frac{d^2\zeta}{dz^2}\right) + ...\right]$$

$$= \text{real part of } \frac{iK}{4\pi} \frac{\frac{1}{2}\left(\frac{d^2\zeta}{dz^2}\right)_1}{\left(\frac{d\zeta}{dz}\right)_1 + \frac{1}{2}(z - z_1)\left(\frac{d^2\zeta}{dz^2}\right)_1 + ...}$$

$$= \text{real part of } \frac{iK}{4\pi} \frac{\frac{d^2\zeta}{dz^2}}{\left(\frac{d\zeta}{dz}\right)_1 + \frac{1}{2}(z - z_1)\left(\frac{d^2\zeta}{dz^2}\right)_1 + ...}$$

$$= \text{real part of } \frac{iK}{4\pi} \frac{\left(\frac{d^2\zeta}{dz^2}\right)_1}{\left(\frac{d\zeta}{dz}\right)_1}$$

approximately as $z \to z_1$, $\zeta \to \zeta_1$

$$= \frac{K}{4\pi} \frac{\partial}{\partial y} \log \left| \frac{d\zeta}{dz} \right|_1$$

Since

$$i\frac{d}{dz} = \frac{\partial}{\partial y}$$

Hence

$$\psi = \frac{K}{4\pi} \log \left| \frac{d\zeta}{dz} \right|_1$$

Thus (3) reduces to

$$\psi_2(x_1, y_1) = \psi_1(\xi_1, \eta_1) + \frac{K}{4\pi} \log \left| \frac{d\zeta}{dz} \right|_1$$

Which is the well-known Routh's theorem.

7.29. Vortex Sheet

Let a fluid be moving irrotationally everywhere except in the domain which lies between the surfaces S_1 and S_3. We

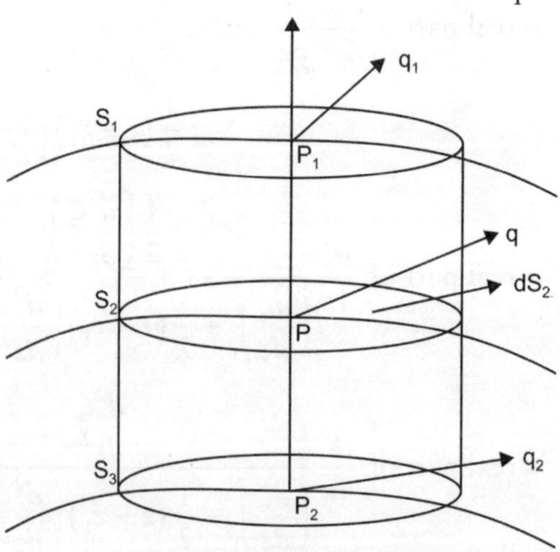

Fig. 7.17

consider an element dS_2 of the surface S_2 such that P is the centre of gravity of dS_2 and the points P_1 and P_2 are taken on normal at P such that $PP_1 = \frac{t\hat{n}}{2}, PP_2 = \frac{-t\hat{n}}{2}$, where $\hat{n}$ is a vector normal to S_2 at P and $t > 0$ be very small number. As P describes the surface S_2 the points P_1 and P_2 describe S_1 and S_3 respectively parallel to S_2 which is midway between them. The normals at the boundaries of dS_1, dS_2, dS_3 enclose a cylindrical element of volume $dv = \in dS_2$. Let $\vec{w}$ be the vorticity vector at P, then $\vec{w}dv = \vec{w}tdS_2 = w^1 dS_2$, where $\vec{w}^1 = \in \vec{w}$. Now if $\in \to 0$ and $\vec{w} \to \infty$ in such a way that $\vec{w}^1$ remains unaltered. Now we define the surface S_2 as the vortex sheet of vorticity vector w^1 per unit area. It can be proved that the normal components of velocity are continuous across the vortex sheet.

Working Rules

1. If there exists two vortices of strengths K_1, K_2 at A_1, A_2 respectively, then the line A_1A_2 rotates with angular velocity $w = \frac{K_1 + K_2}{2\pi(A_1A_2)^2}$ about O. Whereas the points A_1, A_2 move with velocity $\frac{K_2}{2\pi A_1A_2}, \frac{K_1}{2\pi A_1A_2}$ along a line (in opposite directions) respectively to the line A_1A_2.

WORKED OUT EXAMPLES

1. When an infinite liquid contains two parallel and equal vortices of the same strength at a disance $2b$ apart and the spin is in the same sense in both, show that the relative streamlines are given by

$$\log\left(r^4 + b^4 - 2b^2r^2\cos 2\theta\right) - \frac{r^2}{2b^2} = \text{const.}$$

O being measured from the join of vortices, the origin being its middle point.

Solution: Let there be two vortices each of strength K at A_1 $(a, 0)$ and A_2 $(-a, 0)$ and let O, the middle point of A_1A_2 be taken as the origin as shown in figure 7.18.

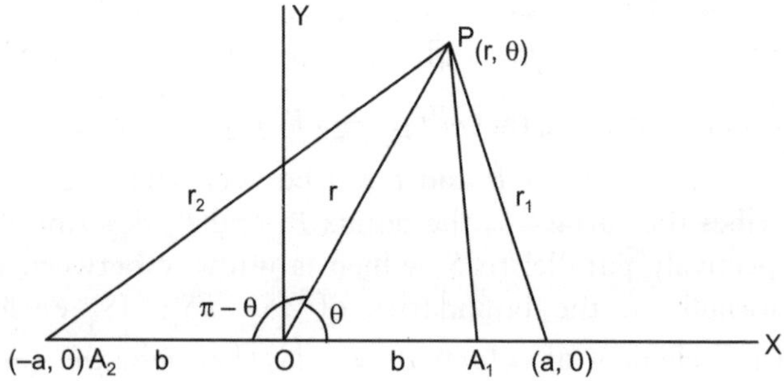

Fig. 7.18

Let $P(r, \theta)$ be any point in the fluid. Then we have

$$r_1^2 = A_1P^2 = r^2 + b^2 - 2rb\cos\theta$$

and

$$r_2^2 = A_2P^2 = r^2 + b^2 + 2rb\cos\theta$$

Hence

$$\begin{aligned} r_1^2 r_2^2 &= \left(r^2 + b^2\right)^2 - 4r^2b^2\cos^2\theta \\ &= r^4 + b^4 + 2r^2b^2 - 4r^2b^2\cos^2\theta \\ &= r^4 + b^4 - 2r^2b^2\cos 2\theta \end{aligned}$$

The stream function ψ at $P(r, \theta)$ is given by

$$\begin{aligned} \psi &= \frac{K}{2\pi}\log r_1 + \frac{K}{2\pi}\log r_2 \\ &= \frac{K}{2\pi}\log r_1 r_2 \\ &= \frac{K}{2\pi}\log r_1^2 r_2^2 \\ &= \frac{K}{4\pi}\log\left(r^4 + b^4 - 2r^2b^2\cos 2\theta\right) \qquad \text{... (1)} \end{aligned}$$

The line A_1A_2 revolves about O with angular velocity

$$W = \frac{K_1 + K_2}{2\pi(A_1A_2)^2} = \frac{2K}{2\pi(2b)^2} = \frac{K}{4\pi b^2}$$

So that the velocity at any point P due to this motion is

$$Wr = \frac{Kr}{4\pi b^2} \text{ as } V = rW$$

To reduce the vortex system to rest we impose a velocity $\frac{-Kr}{4\pi b^2}$ and if ψ^1 be the stream function due to this addition, then

$$\frac{\partial \psi^1}{\partial r} = -\frac{1}{r}\frac{\partial \phi^1}{\partial \theta} = -\frac{Kr}{4\pi b^2}$$

$$\therefore \quad \psi^1 = \frac{-Kr^2}{8\pi b^2}$$

Hence the streamlines relative to the vortices are given by

$$\psi = \frac{K}{4\pi}\log\left(r^4 + b^4 - 2r^2b^2\cos 2\theta\right) - \frac{Kr^2}{8\pi b^2}$$

$$= \text{const.}$$

$$\text{or} \quad \log\left(r^4 + b^4 - 2r^2b^2\cos 2\theta\right) - \frac{r^2}{2b^2} = \text{const.}$$

2. When an infinite liquid contain two parallel and opposite vertices at a distance $2b$, prove that the streamlines relative to the vortices are given by the equation

$$\log\left[\frac{x^2 + (y-b)^2}{x^2 + (y+b)^2}\right] + \frac{y}{b} = c$$

the origin being the middle point of the join which is taken for the axis of y.

Solution: Let us assume that there are two vortices of strengths K, $-K$ at A, B respectively such that origin O is middle point of $AB = 2b$ and AB lies along y-axis. Both vortices will move along a line parallel to x-axis with velocity

$$q = \frac{K}{2\pi AB} = \frac{K}{4\pi b}$$

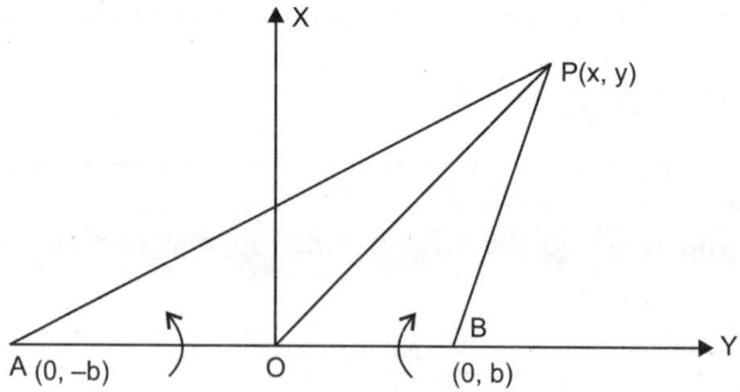

Fig. 7.19

The complex potential w at P due to these two vortices is given by

$$w = \frac{Ki}{2\pi}\log(z - ib) - \frac{K}{2\pi}\log(z + ib)$$

$$\Rightarrow \quad \phi + i\psi = \frac{Ki}{2\pi}\log\left[x + i(y-b)\right] - \frac{iK}{2\pi}\log\left[x + i(y+b)\right]$$

Equating imaginary parts from both the sides, we get

$$\psi = \frac{K}{4\pi}\log\left[x^2 + (y-b)^2\right] - \frac{K}{4\pi}\log\left[x^2 + (y+b)^2\right]$$

$$= \frac{K}{4\pi}\log\left[\frac{x^2 + (y-b)^2}{x^2 + (y+b)^2}\right]$$

To reduce the vortex system to rest, we superimpose a velocity $\dfrac{-K}{4\pi b}$ along x-axis to the system and ψ^1 be the stream function due to this addition, then

$$-\frac{\partial\psi^1}{\partial y} = \frac{-\partial\phi^1}{\partial x} = -\frac{K}{4\pi b}$$

$$\Rightarrow \quad \psi^1 = \frac{Ky}{4\pi b}$$

Hence the streamlines relative to vortices are given by

$$\psi = \frac{K}{4\pi}\log\left[\frac{x^2 + (y-b)^2}{x^2 + (y+b)^2}\right] + \frac{Ky}{4\pi b} = \text{const.}$$

$$\Rightarrow \log\left[\frac{x^2+(y-b)^2}{x^2+(y+b)^2}\right]+\frac{y}{b}=c$$

3. If n rectilinear vortices of the same strength K are symmetrically arranged as generators of a circular cylinder of radius a in an infinite liquid, prove that the vortices will move round the cylinder uniformly in time $\dfrac{8\pi^2a^2}{(n-1)K}$ and find the velocity at any point of the liquid.

Solution: Let us take the line through centre of the vortices as the x-axis. Since n vortices are symmetrically distributed; these will be spread over in such a way that the angular distance between any two consecutive vortices is $\dfrac{2\pi}{n}$.

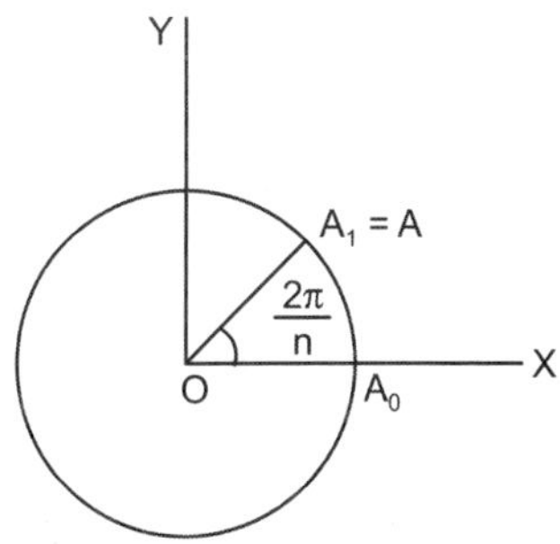

Fig. 7.20

The coordinates of the points A_r are given by

$$z = z_r = ae^{\frac{2\pi ir}{n}}$$

where $r = 0, 1, 2, \ldots, n-1$

These are n roots of the equation $z^n - a^n = 0$.

Hence

$$z^n - a^n = (z-z_0)(z-z_1)\ldots(z-z_{n-1}) \qquad \ldots (1)$$

The complex potential due to n vortices at P is given by

$$w = \frac{iK}{2\pi}\left[\log(z-z_0)+\log(z-z_1) + \ldots + \log(z-z_{n-1})\right]$$

$$= \frac{iK}{2\pi}\log\left[(z-z_0)(z-z_1)\ldots(z-z_{n-1})\right]$$

$$= \frac{iK}{2\pi}\log(z^n-a^n), \text{ From (1)} \qquad \ldots (2)$$

$$\Rightarrow \quad \phi + i\psi = \frac{iK}{2\pi} \log\left[r^n (\cos n\theta + i\sin n\theta) - a^n\right]$$

Equating imaginary parts, we get

$$\psi = \frac{K}{4\pi} \log\left[\left(r^n \cos n\theta - a^n\right)^2 + \left(r^n \sin n\theta\right)^2\right]$$

$$= \frac{K}{4\pi} \log\left[r^{2n} - 2r^n a^n \cos n\theta + a^{2n}\right]$$

This gives the current function at any point (r, θ)

If q is the velocity at any point, then

$$q^2 = \left|\frac{dw}{dz}\right|^2 = \frac{K^2}{4\pi^2} \frac{\left|nz^{n-1}\right|^2}{\left|z^n - a^n\right|^2}$$

$$= \frac{n^2K^2}{4\pi^2} \frac{\left|r^{n-1}e^{i(n-1)\theta}\right|^2}{\left|r^n e^{in\theta} - a^n\right|^2}$$

$$= \frac{n^2K^2}{4\pi^2} \frac{r^{2n-2}}{r^{2n} - 2a^n r^n \cos n\theta + a^{2n}}$$

We shall now find the velocity of any one of the vortices say at $z = a$.

For the motion of this vortex

$$w^1 = \frac{iK}{2\pi} \log\left(z^n - a^n\right) - \frac{iK}{2\pi} \log(z - a)$$

$$\Rightarrow \quad \psi^1 = \frac{K}{4\pi} \log\left(r^{2n} - 2a^n r^n \cos n\theta + a^{2n}\right)$$

$$- \frac{K}{4\pi} \log\left(r^2 - 2ar \cos\theta + a^2\right)$$

Now if ψ_0 be the stream function of the vortex $(a, 0)$, then

$$\frac{\partial \psi_0}{\partial a} = \left\{\frac{\partial \psi^1}{\partial r}\right\}_{r=a,\, \theta=0}$$

and

$$\frac{\partial \psi_0}{a\partial \theta} = \left\{\frac{1}{r}\frac{\partial \psi^1}{\partial \theta}\right\}_{r=a,\, \theta=0}$$

Now

$$\frac{\partial \psi^1}{\partial r} = \frac{K}{4\pi}\left[\frac{2nr^{2n-1} - 2nr^{n-1}a^n \cos n\theta}{r^{2n} + a^{2n} - 2r^n a^n \cos n\theta} - \frac{2r - 2a\cos\theta}{r^2 + a^2 - 2ra\cos\theta}\right]$$

Hence

$$\left(\frac{\partial \psi^1}{\partial r}\right)_{r=a} = \frac{K}{4\pi}\left[\frac{2na^{2n-1}(1-\cos n\theta)}{2a^{2n}(1-\cos n\theta)} - \frac{2a(1-\cos\theta)}{2a^2(1-\cos\theta)}\right]$$

$$= \frac{K(n-1)}{4\pi a} \qquad \text{as } \theta \to 0$$

Also

$$\frac{\partial \psi^1}{\partial \theta} = \frac{K}{4\pi}\left[\frac{2nr^n a^n \sin n\theta}{r^{2n} - 2r^n a^n \cos n\theta + a^{2n}} + \frac{2ra\sin\theta}{r^2 + a^2 - 2ar\cos\theta}\right]$$

$$\left(\frac{\partial \psi^1}{\partial \theta}\right)_{r=a} = \frac{K}{4\pi}\left[\frac{2na^{2n}\sin n\theta}{2a^{2n}(1-\cos n\theta)} + \frac{2a^2\sin\theta}{2a^2(1-\cos\theta)}\right]$$

$$\left[\text{form } \frac{0}{0} \text{ as } \theta \to 0\right]$$

$$= \frac{K}{4\pi}\left[\frac{n^2\cos n\theta}{n\sin\theta} + \frac{\cos\theta}{\sin\theta}\right] \qquad \text{as } \theta \to 0$$

$$= \frac{K}{4\pi}\left[\frac{-n\sin n\theta}{\cos 0} + \frac{(-\sin\theta)}{\cos 0}\right] \qquad \text{as } \theta \to 0$$

$$= 0$$

Finally

$$\frac{\partial \psi^1}{\partial \theta} = \frac{K}{4\pi a}(n-1),\ \frac{\partial \psi^1}{\partial \theta} = 0 \qquad \text{as } r \to a,\ \theta \to 0$$

Consequently the velocity q_0 of the vortex A_0 is given by

$$q_0 = \left[\left(\frac{\partial \psi^1}{\partial r}\right)^2 + \left(\frac{1}{r}\frac{\partial \psi^1}{\partial \theta}\right)^2\right]^{\frac{1}{2}} = \frac{K(n-1)}{4\pi a}$$

Thus the velocity is only along the tangent, there being no velocity along the normal to the circle. Therefore, the vortices will move round the cylinder uniformly with velocity $\dfrac{K(n-1)}{4\pi a}$.

The time of one complete revolution

$$= \frac{\text{distance}}{\text{velocity}} = \frac{2\pi a}{\dfrac{K(n-1)}{4\pi a}} = \frac{8\pi^2 a^2}{K(n-1)}$$

4. If (r_1, θ_1), (r_2, θ_2)..... be polar coordinates at time t of a system of rectilinear vortices of strengths K_1, K_2,....., prove that $\Sigma K r^2$ = constant and $\Sigma K r^2 \theta = \dfrac{1}{2\pi}\sum K_1 K_2$.

Solution: The complex potential due to all the vortices K_1, K_2 ..... at z_1, z_2 ..... is given by

$$w = \sum \frac{iKr}{2\pi} \log(z - z_r)$$

If u_i, v_i are the velocity components of the vortex K_i, then since the motion of K_i is due to other vortices only, the complex potential for the motion of this vortex is

$$w_i = \sum_{r \neq i} \frac{iKr}{2\pi} \log(z - z_r)$$

and

$$u_i - iv_i = -\left(\frac{dw_i}{dz}\right)_{z = z_i}$$

$$= -\sum_{r \neq i}^{\infty} \frac{iKr}{2\pi} \frac{1}{z_i - z_r}$$

Now we consider

$$\sum K_i z_i \left(\frac{dw_i}{dz}\right)_{z = z_i} = \frac{i}{2\pi} \sum_i \left\{K_i z_i \sum_{r \neq i}^{\infty} \frac{Kr}{z_i - z_r}\right\}$$

On the right-hand side every product $K_i K_r$ occurs twice and the two terms containing these are

$$\frac{K_i z_i K_r}{z_i - z_r} + \frac{K_i z_r K_r}{z_r - z_i} = K_i K_r$$

Also putting $-u_i + iv_i$ for $\left\{\frac{dw_i}{dz}\right\}_{z=z_i}$ on the left, the above equation gives

$$\sum K_i z_i \left(-u_i + iv_i\right) = \frac{i}{2\pi} \sum_{r \neq i} K_i K_r$$

$$\Rightarrow \quad \sum K_i \left(x_i + iy_i\right)\left(-u_i + iv_i\right) = \frac{i}{2\pi} \sum_{r \neq i} K_i K_r$$

Equating real and imaginary parts, we get

$$\sum K_i \left(x_i u_i + y_i v_i\right) = 0 \qquad \text{... (1)}$$

and

$$\sum K_i \left(x_i v_i - y_i u_i\right) = \frac{1}{2\pi} \sum_{r \neq i} K_i K_r \qquad \text{... (2)}$$

Now (1) gives

$$\sum K_i \left\{x_i \frac{dx_i}{dt} + y_i \frac{dy_i}{dt}\right\} = 0 \quad \text{as } u_i = \frac{dr_i}{dt}, v_i = \frac{dy_i}{dt}$$

$$\Rightarrow \quad \sum K_i \left(x_i dx_i + y_i dy_i\right) = 0$$

Integrating, we have

$$\sum K_i \left(x_i^2 + y_i^2\right) = \text{const.}$$

$$\Rightarrow \quad \sum K_i r_i^2 = \text{const.}$$

This proves the first part.

Again (2) can be written as

$$\sum K_i \left(x_i \frac{dy_i}{dt} - y_i \frac{dx_i}{dt}\right) = \frac{1}{2\pi} \sum_{r \neq i} K_i K_r \qquad \text{... (3)}$$

Also in polar coordinates, $\tan \theta_i = \frac{y_i}{x_i}$

i.e. $\sec^2 \theta_i \, \dot{\theta}_i = \frac{\dot{y}_i x_i - \dot{x}_i y_i}{x_i^2}$

$$\Rightarrow \quad \dot{x}_i y_i - \dot{x}_i y_i = x_i^2 \sec^2 \theta_i \dot{\theta}_i$$
$$= r_i^2 \cos^2 \theta_i \sec^2 \theta_i \dot{\theta}_i$$
$$= r_i^2 \dot{\theta}_i$$

Hence (3) gives

$$\sum K_i r_i^2 \dot{\theta}_i = \frac{1}{2\pi} \sum K_1 K_2$$

$$\Rightarrow \quad \sum \dot{K} r^2 \dot{\theta} = \frac{1}{2\pi} \sum K_1 K_2$$

This proves the second result.

5. An infinitely long line vortex of strength m, parallel to the axis of z, is situated in infinite liquid bounded by a rigid wall in the plane $y = 0$. Prove that, if there be no field of force, the surfaces of equal pressure are given by

$$\left\{(x-a)^2 + (y-b)^2\right\}\left\{(x-a)^2 + (y+b)^2\right\} = c\left\{y^2 + b^2 - (x-a)^2\right\}$$

where (a, b) are the coordinates of the vortex and c is a parametric constant.

Solution: The image of the vortex of strength m at $A_1(a, b)$ relative to the wall $y = 0$ i.e. x-axis is a vortex of strength $-m$ at $A_2(a, -b)$. These two vortices form a vortex pair.

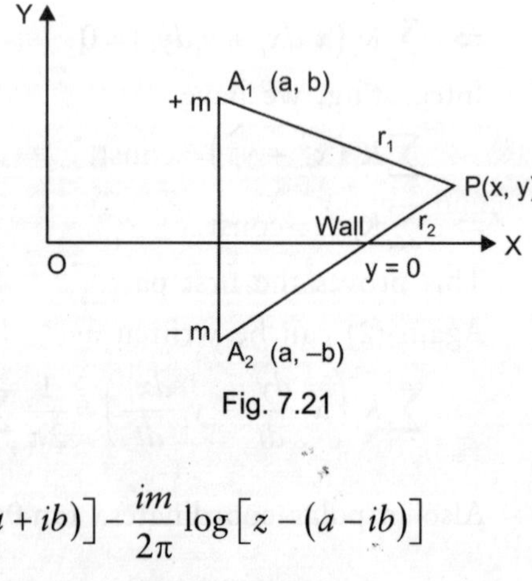

Fig. 7.21

The complex potential due to this system is given by

$$w = \frac{im}{2\pi} \log\left[z - (a+ib)\right] - \frac{im}{2\pi} \log\left[z - (a-ib)\right]$$

$$w = \frac{im}{2\pi}\log\left[(x-a)+i(y-b)\right]$$
$$-\frac{im}{2\pi}\log\left[(x-a)+i(y+b)\right]$$

$$\therefore \quad -\frac{dw}{dz} = \frac{-im}{2\pi}\frac{1}{\left[z-(a+ib)\right]} + \frac{im}{2\pi}\frac{1}{\left[z-(a-ib)\right]}$$

$$\Rightarrow \quad u - iv = \frac{-im}{2\pi}\frac{1}{(x-a)+i(y-b)}$$
$$+\frac{im}{2\pi}\frac{1}{(x-a)+i(y+b)} \qquad \text{... (1)}$$

Let $A_1P = r_1$, $A_2P = r_2$

Then

$$r_2^2 - r_1^2 = (x-a)^2 + (y+b)^2 - \left[(x-a)^2 + (y-b)^2\right]$$
$$= 4yb \qquad \text{... (2)}$$

$$u - iv = \frac{-im}{2\pi r_1^2}\left[(x-a) - i\,(y-b)\right]$$
$$+\frac{im}{2\pi r_2^2}\left[(x-a) - i\,(y+b)\right]$$

Equating real and imaginary parts, we get

$$u = \frac{m}{2\pi}\left[-\frac{(y-b)}{r_1^2} + \frac{y+b}{r_2^2}\right]$$

and

$$-v = \frac{m}{2\pi}\left[\frac{-(x-a)}{r_1^2} + \frac{x-a}{r_2^2}\right]$$

Each of these vortices will move with velocity $q = \frac{m}{2\pi(A_1A_2)} = \frac{m}{2\pi\cdot 2b} = \frac{m}{4\pi b}$ along a line perpendicular to A_1A_2 i.e. along x-axis.

To reduce the vortex system to rest, we superimpose a velocity $-\frac{m}{4\pi b}$ along x-axis to the system. In this case

$$\left.\begin{aligned} u &= \frac{-m}{2\pi}\left[\frac{y-b}{r_1^2} - \frac{y+b}{r_2^2}\right] - \frac{m}{4\pi b} \\ v &= \frac{m}{2\pi}(x-a)\left(\frac{1}{r_1^2} - \frac{1}{r_2^2}\right) \end{aligned}\right\} \quad \ldots (3)$$

Also Bernoulli's equation for steady motion is

$$\frac{p}{\rho} + \frac{1}{2}q^2 = \text{constant} \quad \ldots (4)$$

Surface of equal pressure are given by

$$p = \text{const.} \quad \ldots (5)$$

From (4) and (5), we have

$$q^2 = \text{constant}$$

$$\Rightarrow \quad u^2 + v^2 = \text{constant} = c$$

$$\Rightarrow \quad \frac{m^2}{4\pi^2}\left[\frac{y-b}{r_1^2} - \frac{y+b}{r_2^2}\right]^2 + \frac{m^2}{16\pi^2 b^2} + \frac{2m^2}{8\pi^2 b}\left[\frac{y-b}{r_1^2} - \frac{y+b}{r_2^2}\right] + \frac{m^2}{4\pi^2}(x-a)^2\left(\frac{1}{r_1^2} - \frac{1}{r_2^2}\right)^2 = c$$

$$\Rightarrow \quad \frac{(y-b)^2}{r_1^4} + \frac{(y+b)^2}{r_2^4} - \frac{2(y^2-b^2)}{r_1^2 r_2^2} + \frac{1}{b}\left[\frac{y-b}{r_1^2} - \frac{y+b}{r_2^2}\right] + (x-a)^2\left[\frac{1}{r_1^4} + \frac{1}{r_2^4} - \frac{2}{r_1^2 r_2^2}\right] = \frac{4\pi^2 c}{m^2} - \frac{1}{4b^2} = c_1$$

$$\Rightarrow \quad \frac{(x-a)^2 + (y-b)^2}{r_1^4} + \frac{(x-a)^2 + (y+b)^2}{r_2^4} - \frac{2}{r_1^2 r_2^2}\left[(x-a)^2 + y^2 - b^2\right] + \frac{1}{b}\left[\frac{y-b}{r_1^2} - \frac{y+b}{r_2^2}\right] = c_1$$

$$\Rightarrow \quad \frac{1}{r_1^2} + \frac{1}{r_2^2} - \frac{2\left[(x-a)^2 + y^2 - b^2\right]}{r_1^2 r_2^2} + \frac{\left[y\left(r_2^2 - r_1^2\right) - b\left(r_2^2 + r_1^2\right)\right]}{b r_1^2 r_2^2} = c_1$$

$$\Rightarrow \quad r_2^2 + r_1^2 - 2\left[(x-a)^2 + y^2 - b^2\right] + \frac{1}{b}(4yb)b$$
$$- \left(r_2^2 + r_1^2\right) - c_1 r_1^2 r_2^2$$

$$\Rightarrow \quad r_1^2 r_2^2 = -\frac{2}{c_1}\left[(x-a)^2 + y^2 - b^2 - 2y^2\right]$$

$$\Rightarrow \quad r_1^2 r_2^2 = c_3\left[y^2 + b^2 - (x-a)^2\right]$$

Where $\dfrac{-2}{c_1} = c_3$

6. A long fixed cylinder of radius a is surrounded by infinite incompressible liquid and there is in the liquid a vortex filament of strength k which is parallel to the axis of the cylinder at a distance c ($c > a$) from this axis. Given that there is no circulation round any circuit enclosing the cylinder but not the filament, show that the speed q of the fluid at the surface of the cylinder is $\dfrac{k}{2\pi a}\left(1 - \dfrac{c^2 - a^2}{r^2}\right)$, r being the distance of the point considered from the filament.

Solution: Let B be the inverse point of PA with regard to the cylinder such that

$$OB \cdot OA = a^2$$

$$\therefore \quad f = OB = \frac{a^2}{c} \text{ so that } cf = a^2.$$

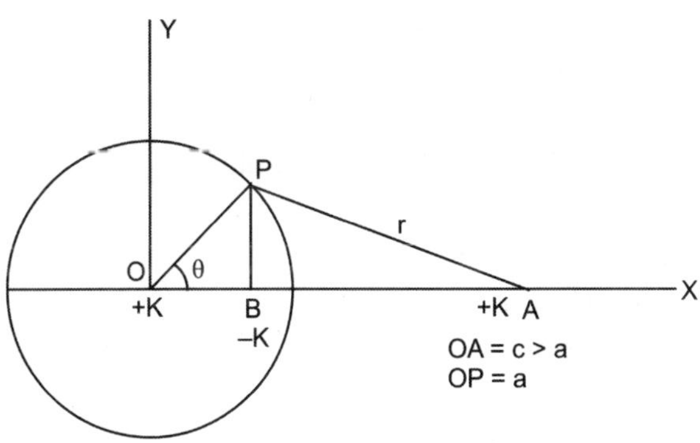

Fig. 7.22

Thus vortex of strength $+K$ at A gives rise to a vortex of strength $-K$ at the inverse point B and a vortex of strength $+K$ at O, the centre.

In ΔOAP

$$\cos\theta = \frac{a^2 + c^2 - r^2}{2ac}$$

$$\Rightarrow \quad a^2 + c^2 - r^2 = 2ac\cos\theta \qquad \text{... (1)}$$

The complex potential of the whole system is given by

$$w = \frac{iK}{2\pi}\log(z-c) - \frac{iK}{2\pi}\log(z-f) + \frac{iK}{2\pi}\log(z-0)$$

$$\therefore \quad \frac{dw}{dz} = \frac{iK}{2\pi}\left[\frac{1}{z-c} - \frac{1}{z-f} + \frac{1}{z}\right]$$

$$= \frac{iK}{2\pi}\frac{z^2 - 2fz + a^2}{z(z-c)(z-f)} \text{ as } a^2 = cf$$

$$q = \left|\frac{dw}{dz}\right| = \frac{K}{2\pi}\frac{|z^2 - 2fz + a^2|}{|z||z-c||z-f|}$$

$$= \frac{K}{2\pi}\frac{|z^2 - 2fz + a^2|}{OP \cdot BP\, AP}$$

$$\Rightarrow \quad q = \frac{K}{2\pi}\frac{|z^2 - 2fz + a^2|}{arBP} \qquad \text{... (2)}$$

Now

$$OB \cdot OA = a^2 = OP^2$$

$$\Rightarrow \quad \frac{OB}{OP} = \frac{OP}{OA} \text{ Also } \angle BOP = \angle POA$$

Hence ΔOBP and ΔOPA are similar.

$$\therefore \quad \frac{OB}{OP} = \frac{OP}{OA} = \frac{BP}{AP}$$

$$\Rightarrow \quad BP = \frac{AP \cdot OP}{OA} = \frac{ar}{c}$$

Hence

$$arBP = \frac{a^2r^2}{c}$$

Now

$$z^2 - 2fz + a^2 = (z - f)^2 + a^2 - f^2$$
$$= (ae^{i\theta} - f)^2 + a^2 - f^2$$
$$= (a \cos \theta - f + ia \sin \theta)^2 + a^2 - f^2$$
$$= (a \cos \theta - f)^2 - a^2 \sin^2 \theta + 2ia \sin \theta\,(a \cos \theta - f) + a^2 - f^2$$
$$= 2a^2 \cos^2\theta - 2af \cos \theta + 2ia \sin \theta\,(a \cos \theta - f)$$
$$= 2a\,[\cos \theta\,(a \cos \theta - f) + i \sin \theta\,(a \cos \theta - f)]$$
$$= 2a\,(a \cos \theta - f)\,(\cos \theta + i \sin \theta)$$

$$\therefore \quad |z^2 - 2fz + a^2| = 2a(f - a\cos\theta)$$
$$= 2a\left(\frac{a^2}{c} - a \cos \theta\right)$$
$$= 2\frac{a^2}{c}(a - c \cos \theta)$$
$$= \frac{2a^2}{c}\left[a - \frac{(a^2 + c^2 - r^2)}{2a}\right]$$
$$= \frac{2a^2}{2ca}\left[a^2 - c^2 + r^2\right]$$
$$= \frac{a}{c}r^2\left[1 - \frac{c^2 - a^2}{r^2}\right]$$

Hence (2) becomes

$$q = \frac{K}{2\pi}\frac{c}{a^2r^2}\frac{ar^2}{c}\left[1 - \frac{c^2 - a^2}{r^2}\right]$$
$$= \frac{K}{2\pi a}\left[1 - \frac{(c^2 - a^2)}{r^2}\right]$$

7. Find the motion of a straight vortex filament in an infinite region plane wall to which the filament is parallel, and prove that the pressure defect at any point of the wall due to the filament is proportional to $\cos^2\theta \cos 2\theta$, where θ is the inclination of the plane through the filament and the point to the plane through the filament perpendicular to the wall.

Solution: Let $A_1(x_1, y_1)$ be the position of the vortex of strength $+K$ at any instant with respect to the section of the plane XOX^1. The image system consists of an equal and opposite vortex of strength $-K$ at the point $A_2(x_1, -y_1)$.

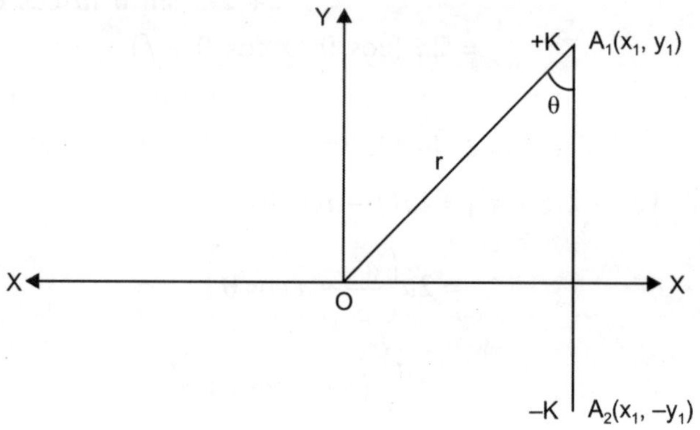

Fig. 7.23

These two vortices together form a vortex pair which will move parallel to x-axis with velocity $\dfrac{K}{2\pi A_1 A_2} = \dfrac{K}{4\pi y_1}$.

Each vortex moves with this velocity parallel to x-axis so that x_1 changes with regard to time but y_1 remains constant.

Hence $\dot{x}_1 = \dfrac{K}{4\pi y_1}$

The complex potential due to the vortex pair is given by

$$w = \frac{iK}{2\pi} \log(z - z_1) - \frac{iK}{2\pi} \log(z - \bar{z}_1) \quad \text{... (1)}$$

$$\therefore \quad \frac{dw}{dz} = \frac{iK}{2\pi}\left[\frac{1}{z - z_1} - \frac{1}{z - \bar{z}_1}\right]$$

Now

$$u_0 - iv_0 = -\left(\frac{dw}{dz}\right)_{z=0} = \frac{iK}{2\pi}\left(\frac{1}{z_1} - \frac{1}{\bar{z}_1}\right)$$

where $u_0 - iv_0$ is complex velocity at O. We shall study the pressure defect at some point say O on x-axis.

Let $OA_1 = r$, $\angle OA_1A_2 = \theta$, then

$$x_1 = r \sin \theta,\ y_1 = r \cos \theta,\ x_1^2 + y_1^2 = r^2$$

$$u_0 - iv_0 = \frac{iK}{2\pi}\left[\frac{x_1 - iy_1}{x_1^2 + y_1^2} - \frac{x_1 + iy_1}{x_1^2 + y_1^2}\right]$$

$$= \frac{Ky_1}{\pi r^2} = \frac{K \cos \theta}{\pi r}$$

Hence $u_0 = \dfrac{K}{\pi r}\cos\theta,\ v_0 = 0$

Now $q^2 = u_0^2 + v_0^2$

$$\therefore \quad q = u_0 = \frac{K \cos \theta}{\pi r}$$

Where q is the fluid velocity at O.

From (1)

$$(w)_{z=0} = \frac{iK}{2\pi}\left[\log(-z_1) - \log(-\bar{z}_1)\right]$$

$$= \frac{iK}{2\pi}\left[\log(-x_1 - iy_1) - \log(-x_1 + iy_1)\right]$$

Equating real parts, we have

$$(\phi)_{z=0} = \frac{-K}{2\pi}\left[\tan^{-1}\left(\frac{y_1}{x_1}\right) - \tan^{-1}\left(\frac{y_1}{-x_1}\right)\right]$$

$$= \frac{-K}{2\pi}\left[\tan^{-1}\left(\frac{y_1}{x_1}\right) + \tan^{-1}\left(\frac{y_1}{x_1}\right)\right]$$

$$= \frac{-K}{\pi}\tan^{-1}\left(\frac{y_1}{x_1}\right)$$

$$\left(\frac{\partial\phi}{\partial t}\right)_{z=0} = \frac{-K}{\pi}\frac{1}{1+\frac{y_1^2}{x_1^2}}\frac{-y_1\dot{x}_1}{x_1^2}$$

$$= \frac{K}{\pi}\frac{x_1^2}{x_1^2+y_1^2}\frac{y_1}{x_1^2}\frac{K}{4\pi y_1}$$

$$= \frac{K}{\pi}\frac{y_1}{r^2}\frac{K}{4\pi y_1}$$

$$= \frac{K^2}{4\pi^2 r^2}$$

By pressure equation,

$$\frac{p}{\rho} + \frac{1}{2}q^2 - \frac{\partial\phi}{\partial t} = c \qquad \text{... (2)}$$

Let $p = p_0$, when the vortex is not present i.e. when $q = 0$, $\frac{\partial\phi}{\partial t} = 0$.

From (2) $\frac{p_0}{\rho} = c$

Hence

$$\frac{p - p_0}{\rho} = \frac{\partial\phi}{\partial t} - \frac{1}{2}q^2$$

$$= \frac{K^2}{4\pi^2 r^2} - \frac{1}{2}\frac{K^2\cos^2\theta}{\pi^2 r^2}$$

$$= \frac{K^2}{4\pi^2 r^2}\left(1 - 2\cos^2\theta\right)$$

$$= \frac{K^2}{4\pi^2 r^2}(-\cos 2\theta)$$

$$\Rightarrow \quad p - p_0 = \frac{-K^2\rho}{4\pi^2 y_1^2}\cos^2\theta\cos 2\theta$$

$$\Rightarrow \quad p_0 - p = \frac{K^2\rho}{4\pi^2 y_1^2}\cos^2\theta\cos 2\theta$$

Hence $p_0 - p$ is proportional to $\cos^2\theta \cos 2\theta$.

8. When a pair of equal and opposite rectilinear vortices are situated in a long circular cylinder at equal distances from its axis, show that path of each vortex is given by the equation $(r^2 \sin^2\theta - b^2)(r^2 - a^2)^2 = 4a^2b^2r^2 \sin^2\theta$. θ being measured from the line through the centre perpendicular to the join of the vortices.

Solution: Let a pair of equal and opposite vortices of strength $+K$ and $-K$ are placed at the points $A\ (r, \theta)$ and $B\ (r, -\theta)$ inside the cylinder such that distances of A and B from the axis are equal. Clearly AB is perpendicular to x-axis.

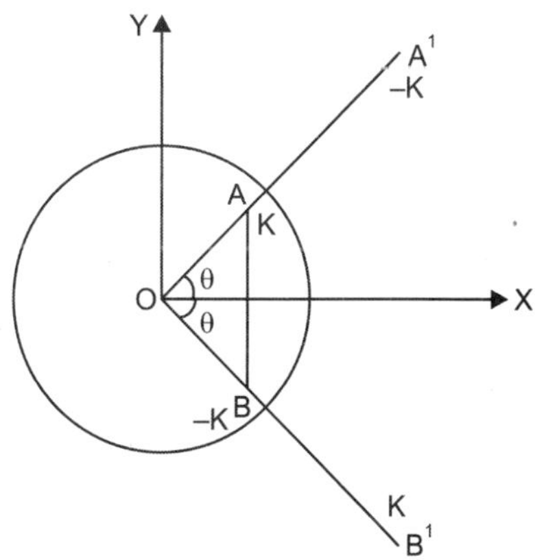

Fig. 7.24

The image system consists of

(i) a vortex of strength $-K$ at A^1, the inverse point of A

(ii) a vortex of strength $+K$ at B^1 where $OB \cdot OB^1 = a^2 = OA \cdot OA^1$

a is the radius of the cylinder.

The complex potential due to the system of any point z is given by

$$w = \frac{iK}{2\pi}\left[\log\left(z - re^{i\theta}\right) - \log\left(z - \frac{a^2}{r}e^{i\theta}\right) - \log\left(z - re^{-i\theta}\right) + \log\left(z - \frac{a^2}{r}e^{-i\theta}\right)\right]$$

The motion of the vortex at A is due to other vortices.

If w^1 be the complex potential for the motion of A, then

$$w^1 = w - \frac{iK}{2\pi}\log\left(z - re^{i\theta}\right) \text{ at } z = re^{i\theta}$$

$$= \frac{iK}{2\pi}\left[-\log\left(z - \frac{a^2 e^{i\theta}}{r}\right) - \log\left(z - re^{-i\theta}\right)\right.$$

$$\left. - \log\left(z - \frac{a^2}{r} e^{-i\theta}\right)\right] \text{ at } z = re^{i\theta}$$

$$w^1 = \frac{-iK}{2\pi}\left[\log\left(re^{i\theta} - \frac{a^2}{r} e^{i\theta}\right) + \log\left(re^{i\theta} - re^{i\theta}\right)\right.$$

$$\left. - \log\left(e^{i\theta} - \frac{a^2}{r} e^{-i\theta}\right)\right]$$

$$= \frac{-iK}{2\pi}\left[\log\left(r^2 - a^2\right)e^{i\theta} - \log r + \log\left(2ir\sin\theta\right)\right.$$

$$\left. - \log\left\{\left(r^2 - a^2\right)\cos\theta + i\sin\theta\left(r^2 + a^2\right)\right\} + \log r\right]$$

$$= \frac{-iK}{2\pi}\left[\log\left|\left(r^2 - a^2\right)e^{i\theta}\right| + \log\left|2ir\sin\theta\right|\right.$$

$$\left. - \log\left|\left\{\left(r^2 - a^2\right)\cos\theta + i\sin\theta\left(r^2 + a^2\right)\right\}\right|\right]$$

$$\Rightarrow \quad \psi = \frac{-K}{2\pi}\left[\log\left(r^2 - a^2\right) + \log 2r\sin\theta\right.$$

$$\left. - \frac{1}{2}\log\left\{\left(r^2 - a^2\right)\cos^2\theta + \sin^2\theta\left(r^2 + a^2\right)^2\right\}\right]$$

Streamlines are given by ψ = constant

$$\text{i.e. } \log\left\{\frac{\left(r^2 - a^2\right)^2 (2r\sin\theta)^2}{\left(r^2 - a^2\right)^2 \cos^2\theta + \sin^2\theta\left(r^2 + a^2\right)^2}\right\} = \text{const.}$$

$$= \log 4b^2$$

$$\Rightarrow \quad \left(r^2 - a^2\right)^2 r^2 \sin^2\theta = b^2\left[\left(r^2 - a^2\right)^2 + 4r^2a^2\sin^2\theta\right]$$

$$\Rightarrow \quad \left(r^2 - a^2\right)^2\left(r^2\sin^2\theta - b^2\right) = 4r^2a^2b^2\sin^2\theta$$

9. If a vortex pair is situated within a cylinder, show that it will remain at rest if the distance of either from the centre is given by $a\left(\sqrt{5} - 2\right)^{\frac{1}{2}}$, where a is the radius of the cylinder.

Solution: Let P, Q be the vortex pair situated with the cylinder at an equidistant r ($OP = OQ = r$).

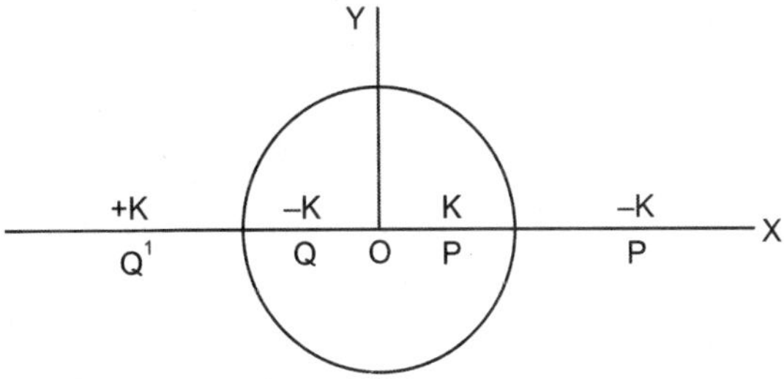

Fig. 7.25

The image system consists of a vortex of strength $-K$ at the inverse point P^1 and a vortex of strength $+K$ at an inverse point Q^1 such that

$$OP \cdot OP^1 = a^2 = OQ \cdot OQ^1$$

$$\Rightarrow \quad OP^1 = \frac{a^2}{r} = OQ^1$$

The vortex will remain at rest if its velocity due to other three vortices be zero.

$$\text{i.e.} \quad \frac{K}{2\pi}\left[\frac{1}{PP^1} - \frac{1}{PQ} + \frac{1}{Q^1P}\right] = 0$$

$$\Rightarrow \quad \frac{K}{2\pi}\left[\frac{1}{\frac{a^2}{r} - r} - \frac{1}{2r} + \frac{1}{\frac{a^2}{r} + r}\right] = 0$$

$$\Rightarrow \quad 2r\left(\frac{a^2}{r} + r\right) - \left(\frac{a^2}{r} + r\right)\left(\frac{a^2}{r} - r\right) + 2r\left(\frac{a^2}{r} - r\right) = 0$$

$$\Rightarrow \quad 2a^2 + 2r^2 - \left(\frac{a^4}{r^2} - r^2\right) + 2a^2 - 2r^2 = 0$$

$$\Rightarrow \quad 4a^2 - \frac{a^4}{r^2} + r^2 = 0$$

$$\Rightarrow \quad 4a^2r^2 - a^4 + r^4 = 0$$

$$\Rightarrow \quad \left(\frac{r^2}{a^2}\right)^2 + 4\left(\frac{r^2}{a^2}\right) - 1 = 0$$

or $\frac{r^2}{a^2} = \sqrt{5} - 2$

or $r = a\left(\sqrt{5} - 2\right)^{\frac{1}{2}}$

10. Two point vortices each of strength K are situated at $(\pm a, 0)$ and a point vortex of strength $\frac{-K}{2}$ is situated at the origin. Show that the fluid motion is stationary and find the equations of streamlines. Show that the streamline which passes through the stagnation points meet the x-axis at $(\pm b, 0)$ where $3\sqrt{3}\left(b^2 - a^2\right)^2 = 16a^3b$.

Solution: The figure is self-explanatory.

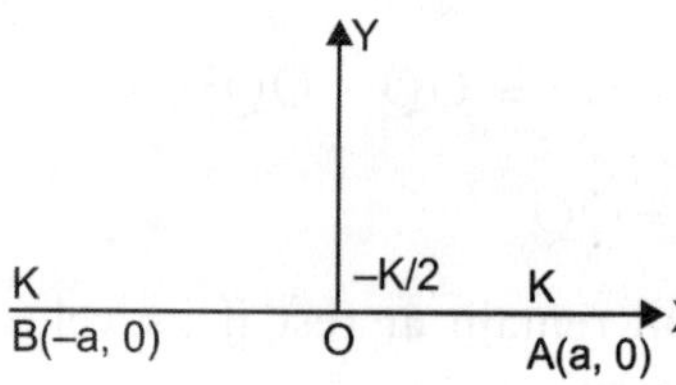

Fig. 7.26

The complex potential of the fluid motion is given by

$$w = \frac{iK}{2\pi}\log(z-a) + \frac{iK}{2\pi}\log(z+a) - \frac{iK}{4\pi}\log z$$

$$\Rightarrow \quad w = \frac{iK}{2\pi}\left[\log\left(z^2 - a^2\right) - \frac{1}{2}\log z\right] \qquad \text{... (1)}$$

For the motion of vortex at A the complex potential is

$$w^1 = w - \frac{iK}{2\pi}\log(z-a)$$

$$= \frac{iK}{2\pi}\left[\log(z+a) - \frac{1}{2}\log z\right]$$

Now

$$\left(\frac{-dw^1}{dz}\right) = \frac{-iK}{2\pi}\left[\frac{1}{z+a} - \frac{1}{2z}\right]_{z=a} = 0$$

Hence

$$U_A - iV_A = 0$$

Thus vortex at A is at rest.

The same fact is true for O, B also. Hence the fluid motion is stationary.

Now we shall determine streamlines.

From (1)

$$\phi + i\psi = \frac{iK}{2\pi}\left[\log\left(x^2 - y^2 - a^2 + 2ixy\right) - \frac{1}{2}\log\left(x + iy\right)\right]$$

Equating imaginary parts, we get

$$\psi = \frac{K}{4\pi}\left[\log\left\{\left(x^2 - y^2 - a^2\right)^2 + 4x^2y^2\right\} - \frac{1}{2}\log\left(x^2 + y^2\right)\right]$$

Streamlines are given by ψ = const.

$$\Rightarrow \quad \log\frac{\left(x^2 - y^2 - a^2\right)^2 + 4x^2y^2}{\sqrt{x^2 + y^2}} = \log c$$

$$\Rightarrow \quad \left(x^2 - y^2 - a^2\right)^2 + 4x^2y^2 = c\sqrt{x^2 + y^2}$$

$$\Rightarrow \quad \left(x^2 - y^2\right)^2 + a^4 - 2a^2\left(x^2 - y^2\right) + 4x^2y^2 = c\sqrt{x^2 + y^2}$$

$$\Rightarrow \quad \left(x^2 + y^2\right)^2 - 2a^2\left(x^2 - y^2\right) + a^4 = c\sqrt{x^2 + y^2} \qquad \dots (2)$$

These are required streamlines.

Finally we shall find the streamline which passes through stagnation points.

The stagnation points are given by

$$\frac{dw}{dz} = 0 \text{ i.e. } \frac{2z}{z^2 - a^2} - \frac{1}{2z} = 0, \text{ From (1)}$$

$$\Rightarrow \quad 3z^2 + a^2 = 0$$

$$\Rightarrow \quad z = \frac{\pm ai}{\sqrt{3}}$$

Stagnation points are at $\left(0, \frac{a}{\sqrt{3}}\right)$ and $\left(0, \frac{-a}{\sqrt{3}}\right)$.

The streamlines, given by (2), will pass through stagnation points only if

$$\left(\frac{a^2}{3}\right)^2 - 2a^2\left(0 - \frac{a}{3}\right) + a^4 = c\left(0 + \frac{a^2}{3}\right)^{\frac{1}{2}}$$

$$\Rightarrow \quad c = \frac{16a^3}{3\sqrt{3}}$$

The streamlines (2) will pass through $(\pm b, 0)$ if

$$b^4 - 2a^2 (b^2 - 0) + a^4 = c (b^2 + 0)$$

$$\Rightarrow \quad b^4 - 2a^2b^2 + a^4 = \frac{16a^3b^2}{3\sqrt{3}}$$

$$\Rightarrow \quad 3\sqrt{3}\left(b^2 - a^2\right)^2 = 16a^3b^2$$

11. A fixed cylinder of radius a is surrounded by incompressible homogeneous fluid extending to infinity symmetrically arranged round it as generators on a cylinder of radius $c(c > a)$ coaxial with the given one are n straight parallel vortex filaments each of strength K. Show that the filaments will remain on this cylinder throughout the motion and revolve round its axis with angular velocity

$$\frac{K}{4\pi c^2}\frac{(n+1)c^{2n} + (n-1)a^{2n}}{c^{2n} - a^{2n}}$$

where $a^2 = bc$.

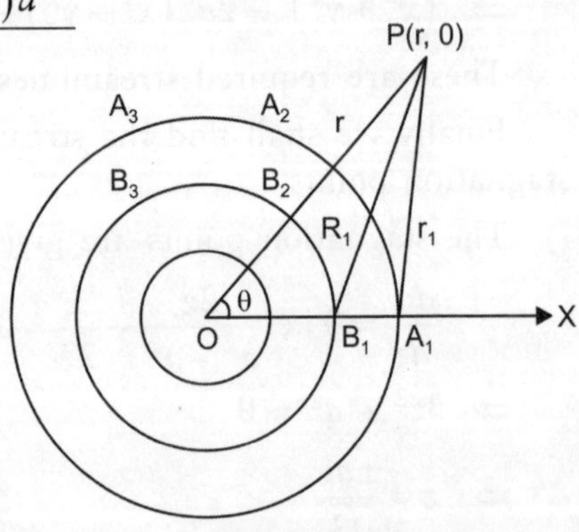

Fig. 7.27

Solution: Let the given vortices of the same strength K be situated at the points A_1, A_2.....A_n and arranged symmetrically round a circle of radius c $(c > a)$. Then

$$\angle A_1OA_2 = \frac{2\pi}{n}$$

Let $A_iP = r_i$ for $i = 1, 2, \ldots.. n$, $OP = r$

and $\angle POX = 0$

In ΔOPA_1

$r_1^2 = r^2 + c^2 - 2rc \cos \theta$

The image of vortex $+K$ at A_1 is a vortex $-K$ at B_i, the inverse point of A_i so that $OA_i \; OB_i = a^2$. Let $OB_i = b$ for $i = 1, 2, \ldots.., n$.

Then $a^2 = bc$

Thus there are vortices of the same strength $-K$ are placed at $B_1, B_2, \ldots.. B_n$ on a circle of radius b.

Let $B_iP = h_i$ for $i = 1, 2, \ldots.. n$

Then $h_i^2 = r^2 + b^2 - 2rb \cos \theta$

The stream function ψ at P due to this system is given by

$$\psi = \frac{K}{2\pi}\left[\log r_1 + \log r_2 + \ldots.. + \log r_n\right]$$

$$- \frac{K}{2\pi}\left[\log h_1 + \log h_2 + \ldots.. + \log h_n\right]$$

$$= \frac{K}{4\pi}\left[\log r_1^2 r_2^2 - r_n^2 - \log h_1^2 h_2^2 - h_n^2\right]$$

$$= \frac{K}{4\pi}\left[\log\left(r^2 + c^2 - 2rc\cos\theta\right)\left\{r^2 + c^2 - 2rc\cos\left(\theta + \frac{2\pi}{n}\right)\right\} - \log\left(r^2 + b^2 - 2rb\cos\theta\right)\left\{r^2 + b^2 - 2rb\cos\left(\theta + \frac{2\pi}{n}\right)\right\}\right]$$

$$= \frac{K}{4\pi}\log\frac{r^{2n} + c^{2n} - 2r^n c^n \cos n\theta}{r^{2n} + b^{2n} - 2r^n b^n \cos n\theta}$$

The motion of A_1 is due to the other vortices. For the motion of A_1,

$$\psi_1 = \psi - \frac{K}{4\pi}\log\left(r^2 + c^2 - 2rc\cos\theta\right)$$

$$\Rightarrow \quad \psi_1 = \frac{K}{4\pi}\log\left(r^{2n} + c^{2n} - 2r^n c^n \cos n\theta\right)$$

$$\frac{-K}{4\pi}\log\left(r^{2n}+b^{2n}-2r^n b^n\cos n\theta\right)$$

$$-\frac{K}{4\pi}\log\left(r^2+c^2-2rc\cos\theta\right) \qquad \text{... (1)}$$

Now

$$\frac{\partial\psi_1}{\partial r}=\frac{K}{4\pi}\left[\frac{2nr^{2n-1}-2nr^{n-1}c^n\cos n\theta}{r^{2n}+c^{2n}-2r^n c^n\cos n\theta}-\frac{2nr^{2n-1}-2nr^{n-1}b^n\cos n\theta}{r^{2n}+b^{2n}-2r^n b^n\cos n\theta}\right]-\frac{K}{4\pi}\frac{2r-2c\cos\theta}{r^2+c^2-2rc\cos\theta}$$

$$\Rightarrow \left(\frac{\partial\psi_1}{\partial r}\right)_{\partial=c}=\frac{K}{4\pi}\left[\frac{n}{c}-\frac{2nc^{2n-1}2nc^{n-1}b^n\cos n\theta}{c^{2n}+b^{2n}-2c^n b^n\cos n\theta}\right]-\frac{K}{4\pi c}$$

Also $\left(\frac{\partial\psi_1}{\partial r}\right)_{r=c,\,\theta=0}=\frac{K}{4\pi}\left[\frac{n-1}{c}-\frac{2nc^{n-1}}{c^n-b^n}\right]$

Also

$$\frac{1}{r}\frac{\partial\psi_1}{\partial\theta}=\frac{1}{r}\frac{K}{4\pi}\left[\frac{2nr^n c^n\sin n\theta}{r^{2n}+c^{2n}-2r^n c^n\cos n\theta}-\frac{2nr^n b^n\sin n\theta}{r^{2n}+b^{2n}-2r^n b^n\cos n\theta}\right]-\frac{K}{4\pi r}\frac{2rc\sin\theta}{r^2+c^2-2rc\cos\theta}$$

$$\left(\frac{1}{r}\frac{\partial\psi_1}{\partial\theta}\right)_{r=c}=\frac{K}{4\pi}\left[\frac{r\sin n\theta}{1-\cos n\theta}-\frac{r\sin n\theta}{1-\cos n\theta}-\frac{\sin\theta}{1-\cos\theta}\right]$$

$$\therefore \left(\frac{1}{r}\frac{\partial\psi_1}{\partial\theta}\right)_{\substack{r=c\\ \theta=0}}=\frac{-K}{4\pi c}\frac{\cos\theta}{\sin\theta}\left(\text{form }\frac{0}{0}\right)$$

$$=\frac{-K}{4\pi c}\left[\frac{-\sin\theta}{\cos\theta}\right]\text{ as }\theta\to 0$$

$$=0$$

Now

$$q_{A_1}^2 = \left(\frac{\partial \psi_1}{\partial c}\right)^2 + \left(\frac{1}{c}\frac{\partial \psi_1}{\partial \theta}\right)^2$$

$$= \left(\frac{\partial \psi_1}{\partial c}\right)^2$$

Hence angular velocity of A_1

$$= q_{A_1} \cdot \frac{1}{c}$$

$$= \frac{K}{4\pi c^2}\left[\frac{(n-1)(c^n - b^n) - 2nc^n}{c^n - b^n}\right]\frac{c^n}{c^n}$$

$$= \frac{K}{4\pi c^2}\left[\frac{(n+1)c^{2n} + (n-1)a^{2n}}{c^{2n} - a^{2n}}\right]$$

as $a^2 = bc$ and so $a^{2n} = b^n c^n$

12. A rectilinear vortex filament of strength K is in an infinite liquid bounded by two perpendicular infinite plane walls whose line of intersection is parallel to the filament. Show that the filament will retrace out a curve (in a plane at right angles to the wall) $r \sin 2\theta =$ const, where r is the distance of the vortex from the line of intersection of the walls and θ the angle between one of the walls and plane containing the filament and line of intersection.

Solution: Let there be a vortex $+K$ at z_1. Here we have two rigid boundaries at right angle to each other, say x-axis and y-axis. The image of $+K$ at z_1 w.r.t. y-axis is $-K$ at $-\bar{z}_1$.

Now the images of $+K$ at z_1 and $-K$ at $-\bar{z}_1$ have their images at $-K$ at $\bar{z}_1$ and $+K$ at $-z_1$. The complex potential due to this system at any point z is given by

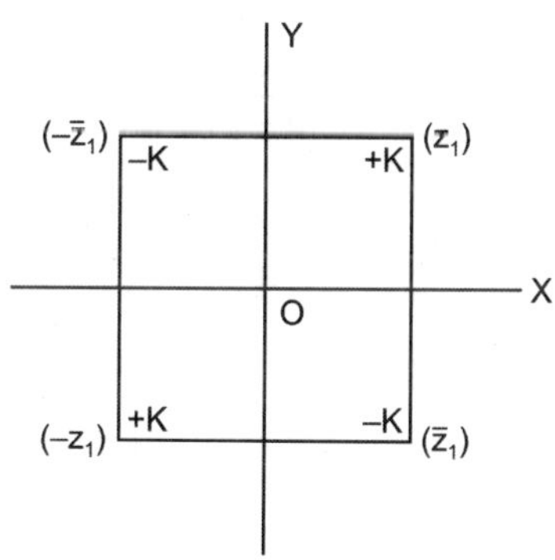

Fig. 7.28

$$w = \frac{iK}{2\pi}\log(z - z_1) + \frac{iK}{2\pi}\log(z + z_1) - \frac{iK}{2\pi}\log(z + \bar{z}_1) - \frac{iK}{2\pi}\log(z - \bar{z}_1)$$

$$\Rightarrow \quad w = \frac{iK}{2\pi}\left[\log(z - z_1)(z + z_1) - \log(z + \bar{z}_1)(z - \bar{z}_1)\right]$$

For the motion of vortex $+K$ at z_1

$$w^1 = w - \frac{iK}{2\pi}\log(z - z_1)$$

$$= \frac{iK}{2\pi}\left[\log\frac{(z - z_1)(z + z_1)}{(z + \bar{z}_1)(z - \bar{z}_1)(z - z_1)}\right]$$

$$= \frac{iK}{2\pi}\left[\log(z + z_1) - \log(z^2 - \bar{z}_1^2)\right]$$

Now

$$u_1 - iv_1 = \left(-\frac{dw^1}{dz}\right)_{z = z_1}$$

$$= \frac{-iK}{2\pi}\left[\frac{1}{z + z_1} - \frac{2z}{z^2 - \bar{z}_1^2}\right]_{z = z_1}$$

$$= \frac{-iK}{4\pi}\left[\frac{x_1 - iy_1}{x_1^2 + y_1^2} - \frac{x_1 + iy_1}{ix_1 y_1}\right]$$

$$u_1 - iv_1 = \frac{-K}{4\pi}\left[\frac{ix_1 + y_1}{x_1^2 + y_1^2} - \frac{x_1 + iy_1}{x_1 y_1}\right]$$

So $$u_1 = \frac{-K}{4\pi}\left[\frac{y_1}{x_1^2 + y_1^2} - \frac{1}{y_1}\right]$$

$$= -\frac{K}{4\pi}\frac{(-x_1^2)}{(x_1^2 + y_1^2)y_1} = \frac{dx_1}{df} \quad \text{... (1)}$$

and

$$v_1 = \frac{dy_1}{df} = \frac{K}{4\pi}\left[\frac{-x_1}{x_1^2 + y_1^2} + \frac{1}{x_1}\right]$$

$$= \frac{-Ky_1^2}{4\pi(x_1^2 + y_1^2)x_1} \quad \text{... (2)}$$

Dividing (1) by (2), we get

$$\frac{dx_1}{dy_1} = -\frac{x_1^3}{y_1^3}$$

$$\Rightarrow \quad \frac{dx}{x_1^3} + \frac{dy_1}{y_1^3} = 0$$

Integrating, we get

$$-\frac{1}{2}\left(\frac{1}{x_1^2} + \frac{1}{y_1^2}\right) = -a.$$

$$\Rightarrow \quad x_1^2 + y_1^2 = 2ax_1^2 y_1^2$$

$$\Rightarrow \quad r_1^2 = 2ar_1^4 \sin^2\theta_1 \cos^2\theta_1$$

$$\Rightarrow \quad (r_1 \sin 2\theta_1)^2 = \frac{2r_1^2}{a}$$

$$\Rightarrow \quad r_1 \sin 2\theta_1 = b = a \text{ constant.}$$

13. To find the paths of a particle due to a vortex in a liquid filling the space between two parallel planes, the vortex midway between them.

Solution: Let there be a vortex $+K$ at the origin O which fills the space between the parallel planes $y = \frac{a}{2}, y = -\frac{a}{2}$.

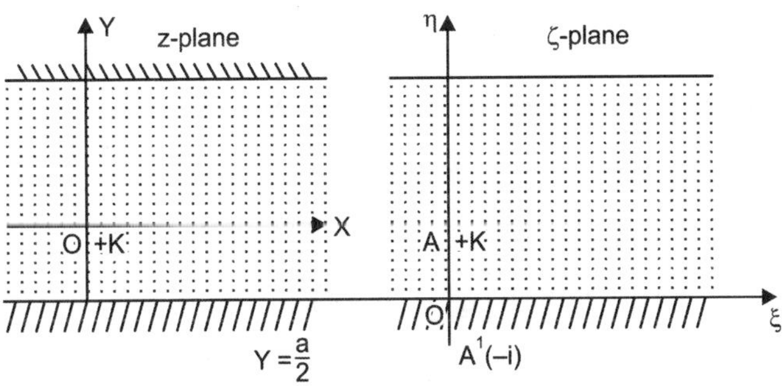

Fig. 7.29

We consider the transformation $\zeta = ie^{\pi z/a}$

Then $\xi + i\eta = ie^{\frac{\pi}{a}(x+iy)} = ie^{\pi x/a} e^{i\pi y/a}$

Then $\xi = -e^{\frac{\pi x}{a}} \sin\frac{\pi y}{a}, \eta = e^{\pi x/a} \cos\frac{\pi y}{a}$

Planes $y = \pm a/_2 \Rightarrow \xi = \pm e^{\pi x/a}, \eta = 0$

The points $\left(x, \pm\frac{a}{2}\right)$ in z-plane become $(\xi, 0)$ in ζ-plane where $\xi = \pm e^{\pi x/a}$.

Also $z = 0 \Rightarrow \xi = 0, \eta = 1 \Rightarrow \zeta = i$

Thus the space between the planes $y = \pm\frac{a}{2}$ in z-plane correspond to entire ξ-axis in ζ-plane. That is to say the space in z-plane corresponds to upper half of ζ-plane. The point O in z-plane corresponds to the point A in ζ-plane. The image of $+K$ at A relative to ξ-axis is a vortex $-K$ at A^1. The complex potential in ζ-plane is

$$w = \frac{iK}{2\pi}\log(\zeta - i) - \frac{iK}{2\pi}\log(\zeta + i)$$

$$\Rightarrow \quad \phi + i\psi = \frac{iK}{2\pi}\log\left[\frac{\xi + i(\eta - 1)}{\xi + i(\eta + 1)}\right]$$

Hence equating imaginary parts, we get

$$\psi = \frac{K}{4\pi}\log\left[\frac{\xi^2 + (n-1)^2}{\xi^2 + (n+1)^2}\right]$$

Paths of fluid particles are given by

$$\psi = \text{constant} = \frac{K}{4\pi}\log b \text{ (say)}$$

Then

$$\xi^2 + (\eta - 1)^2 = b\left[\xi^2 + (\eta + 1)^2\right]$$

$$\Rightarrow \quad \left(\xi^2 + \eta^2\right)(b - 1) + 2\eta(b + 1) + (b - 1) = 0$$

Dividing by $b - 1$ and writing $c = \frac{2(b+1)}{b-1}$, we have

$$\xi^2 + \eta^2 + \eta c + 1 = 0$$

Putting the values of ξ, η, we obtain

$$e^{2\pi x/a} + ce^{\pi x/a}\cos\frac{\pi y}{a} + 1 = 0$$

$$\Rightarrow \quad e^{\pi x/a} + c\cos\frac{\pi y}{a} + e^{-\frac{\pi x}{a}} = 0$$

$$\Rightarrow \quad 2\cosh\frac{\pi x}{a} + c\cos\frac{\pi y}{a} = 0$$

$$\Rightarrow \quad \cosh\frac{\pi x}{a} = \lambda\cos\frac{\pi y}{a} \quad \text{where } \lambda \text{ is a constant.}$$

This is the required path.

14. The space enclosed between the planes $x = 0$, $x = a$, $y = 0$ on the positive side of $y = 0$ is filled with uniform incompressible fluid. A rectilinear vortex parallel to the axis of z has coordinates (x_1, y_1). Determine the velocity at any point of the liquid and show that the path of the vortex is given by

$$\cot^2\frac{\pi x}{a} + \coth^2\frac{\pi y}{a} = \text{const.}$$

Solution: We consider the mapping

$$\zeta = -\cos\frac{\pi z}{a} \qquad \dots (1)$$

from z-plane into ξ-plane, we have

$$\xi + i\eta = -\cos\frac{\pi}{a}(x + iy)$$

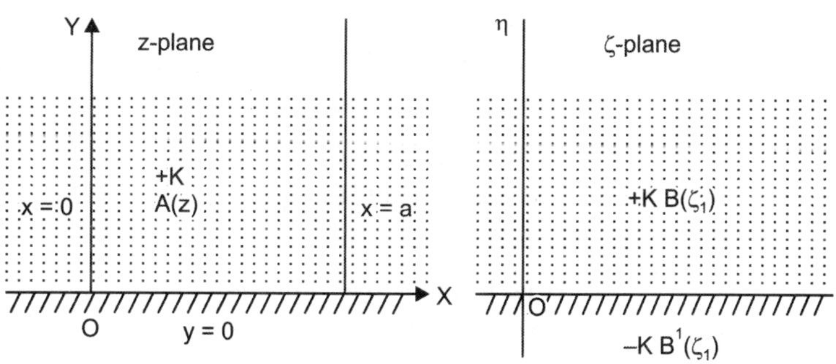

Fig. 7.30

Hence

$$\xi = -\cos\frac{\pi x}{a}\cosh\frac{\pi y}{a}$$

$$\eta = -\sin\frac{\pi x}{a}\sinh\frac{\pi y}{a}$$

The image of vortex of strength K at B relative to the boundary ξ-axis is a vortex of strength $-K$ at $B^1(\zeta_1)$. The complex potential at any point $P(\zeta)$ (not occupied by the vortex) is given by

$$w = \frac{iK}{2\pi}\log(\zeta-\zeta_1) - \frac{iK}{2\pi}\log(\zeta-\bar{\zeta}_1)$$

$$w = \frac{iK}{2\pi}\left[\log\left(-\cos\frac{\pi z}{a}+\cos\frac{\pi z_1}{a}\right) - \log\left(-\cos\frac{\pi z}{a}+\cos\frac{\pi \bar{z}_1}{a}\right)\right]$$

$$\frac{dw}{dz} = \frac{iK}{2\pi}\left[\frac{\sin\frac{\pi z}{a}}{-\cos\frac{\pi z}{a}+\cos\frac{\pi z_1}{a}} - \frac{\sin\frac{\pi z}{a}}{-\cos\frac{\pi z}{a}+\cos\frac{\pi \bar{z}_1}{a}}\right]\frac{\pi}{a}$$

$$\frac{dw}{dz} = \frac{iK}{2a}\,\frac{\sin\frac{\pi z}{a}\left(\cos\frac{\pi \bar{z}_1}{a}-\cos\frac{\pi z_1}{a}\right)}{\left(-\cos\frac{\pi z}{a}+\cos\frac{\pi z_1}{a}\right)\left(-\cos\frac{\pi z}{a}+\cos\frac{\pi \bar{z}_1}{a}\right)}$$

Taking $\lambda = \dfrac{\pi}{2a}$, we have

$$\frac{dw}{dz} = \frac{iK}{2a}\,\frac{\sin 2\lambda z\,\sin\lambda(z_1+\bar{z}_1)\sin\lambda(z_1-\bar{z}_1)}{\sin\lambda(z+z_1)\sin\lambda(z-z_1)\,\sin\lambda(z+\bar{z}_1)\sin\lambda(z-\bar{z}_1)}$$

$$\frac{dw}{dz} = \frac{iK}{2a}\,\frac{\sin 2\lambda z\sin 2\lambda x_1\sin\lambda(2iy_1)}{\sin\lambda(z+z_1)\sin\lambda(z-z_1)\,\sin\lambda(z+\bar{z}_1)\sin\lambda(z-\bar{z}_1)}$$

$$\Rightarrow \left|\frac{dw}{dz}\right| = \frac{iK}{2a}\,\frac{\sin 2\lambda x_1\sinh(2\lambda y_1)|\sin 2\lambda z|}{|\sin\lambda(z+z_1)\sin\lambda(z-z_1)\,\sin\lambda(z+\bar{z}_1)\sin\lambda(z-\bar{z}_1)|}$$

This gives velocity at any point.

Second Part. The stream function $\psi_1(\xi_1, \eta_1)$ due to vortex at (ξ_1, η_1) is ζ-plane is given by

$$\psi_1(\xi_1, \eta_1) = -\frac{K}{4\pi} \log \eta_1 \quad \text{... (2)}$$

From (1)

$$\xi_1 + i\eta_1 = -\cos\frac{\pi}{a}(x_1 + iy_1)$$

$$= -\left[\cos\frac{\pi x_1}{a}\cos\frac{i\pi y_1}{a} - \sin\frac{\pi x_1}{a}\sin\frac{i\pi y_1}{a}\right]$$

$$= -\left[\cos\frac{\pi x_1}{a}\cosh\frac{\pi y_1}{a} - i\sin\frac{\pi x_1}{a}\sinh\frac{\pi y_1}{a}\right]$$

$$\therefore \quad \eta_1 = \sin\frac{\pi x_1}{a}\sinh\frac{\pi y_1}{a}$$

Hence from (2)

$$\psi_1(\xi_1, \eta_1) = \frac{-K}{4\pi}\log\left(\sin\frac{\pi x_1}{a}\sinh\frac{\pi y_1}{a}\right) \quad \text{... (3)}$$

From (1)

$$\frac{d\zeta}{dz} = \sin\frac{\pi z}{a}\frac{\pi}{a}$$

$$\left|\frac{d\zeta}{dz}\right|_1 = \left|\frac{\pi}{a}\sin\frac{\pi z_1}{a}\right| = \left|\frac{\pi}{a}\sin\frac{\pi}{a}(x_1 + iy_1)\right|$$

$$= \frac{\pi}{a}\left|\sin\frac{\pi x_1}{a}\cos\frac{i\pi y_1}{a} + \cos\frac{\pi x_1}{a}\sin\frac{i\pi y_1}{a}\right|$$

$$= \frac{\pi}{a}\left|\sin\frac{\pi x_1}{a}\cosh\frac{\pi y_1}{a} + i\cos\frac{\pi x_1}{a}\sinh\frac{\pi y_1}{a}\right|$$

$$= \frac{\pi}{a}\sqrt{\sin^2\frac{\pi x_1}{a}\cosh^2\frac{\pi y_1}{a} + \cos^2\frac{\pi x_1}{a}\sinh^2\frac{\pi y_1}{a}}$$

By Routh's theorem, we have

$$\psi_2(x_1, y_1) = \psi_1(\xi_1, \eta_1) + \frac{K}{4\pi}\log\left|\frac{d\zeta}{dz}\right|_1$$

$$= \frac{-K}{4\pi}\log\left(\sin\frac{\pi x_1}{a}\sinh\frac{\pi y_1}{a}\right)$$

$$+\frac{K}{4\pi}\log\left[\frac{\pi}{a}\left(\sin^2\frac{\pi x_1}{a}\cosh^2\frac{\pi y_1}{a}+\cos^2\frac{\pi x_1}{a}\sinh^2\frac{\pi y_1}{a}\right)\right]^{\frac{1}{2}}$$

$$\Rightarrow \quad \psi_2(x_1,y_1)=\frac{K}{4\pi}\log\left[\frac{\sin^2\frac{\pi x_1}{a}\cosh^2\frac{\pi y_1}{a}+\cos^2\frac{\pi x_1}{a}\sinh^2\frac{\pi y_1}{a}}{\sin^2\frac{\pi x_1}{a}\sinh^2\frac{\pi y_1}{a}}\right]^{\frac{1}{2}}+\frac{K}{4\pi}\log\frac{\pi}{a}$$

$$\Rightarrow \quad \psi_2(x_1,y_1)=\frac{K}{4\pi}\left[\log\left(\coth^2\frac{\pi y_1}{a}+\cot^2\frac{\pi x_1}{a}\right)^{\frac{1}{2}}+\log\frac{\pi}{a}\right]$$

The path of the vortex (x_1, y_1) is

$$\psi_2(x_1,y_1)=\text{constant}$$

$$\Rightarrow \quad \frac{K}{4\pi}\left[\frac{1}{2}\log\left(\coth^2\frac{\pi y_1}{a}+\cot^2\frac{\pi x_1}{a}\right)+\log\frac{\pi}{a}\right]=\text{const.}$$

$$\Rightarrow \quad \coth^2\frac{\pi y_1}{a}+\cot^2\frac{\pi x_1}{a}=\text{constant}$$

Hence the path of the vortex at any point is

$$\coth^2\frac{\pi y}{a}+\cot^2\frac{\pi x}{a}=\text{constant.}$$

15. A vortex is an infinite liquid occupying the upper half of the z-plane bounded by a circle of radius a, centre O and parts of the x-axis outside the circle. Find the path of the vortex.

Solution:

$$\zeta=z+\frac{a^2}{z}$$

We consider the transformation

$$\zeta=z+\frac{a^2}{z} \qquad \dots (1)$$

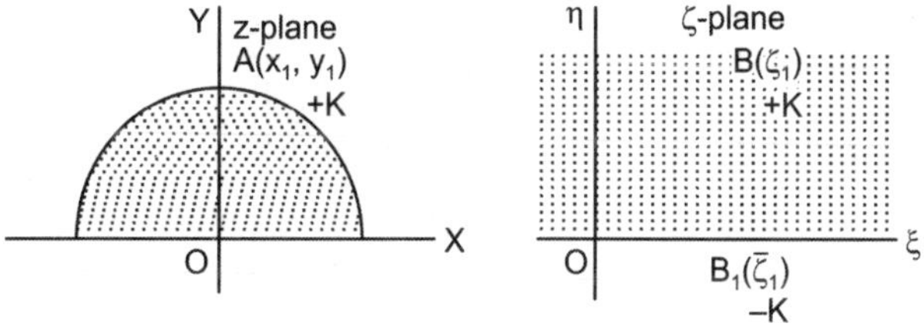

Fig. 7.31

which transforms the portion in the z-plane occupied by the liquid into the upper half of the ζ-plane. Let $A(x_1, y_1)$ in z-plane be mapped on $B(\xi_1, \eta_1)$ in ζ-plane.

The image of $B(\zeta_1)$ with respect to ξ-axis is a vortex $-K$ at an equal distance on either side of ξ-axis i.e. at $B_1(\bar{\zeta}_1)$.

The stream function due to the vortex $+K$ at B is

$$\psi_1(\xi_1, \eta_1) = -\frac{K}{4\pi} \log \eta_1 \quad \text{... (2)}$$

From (1)

$$\xi_1 + i\eta_1 = x_1 + iy_1 + \frac{a^2}{x_1 + iy_1}$$

$$= x_1 + iy_1 + \frac{a^2(x_1 - iy_1)}{x_1^2 + y_1^2}$$

$$\Rightarrow \quad \eta_1 = y_1 - \frac{a^2 y_1}{x_1^2 + y_1^2} = y_1 \frac{(x_1^2 + y_1^2 - a^2)}{x_1^2 + y_1^2}$$

Hence from (2)

$$\psi_1(\xi_1, \eta_1) = \frac{-K}{4\pi} \log \frac{y_1(x_1^2 + y_1^2 - a^2)}{x_1^2 + y_1^2} \quad \text{... (3)}$$

From (1)

$$\frac{d\zeta}{dz} = 1 - \frac{a^2}{z^2} = \frac{z^2 - a^2}{z^2}$$

$$\Rightarrow \quad \frac{d\zeta}{dz} = \frac{x^2 - y^2 - a^2 + 2ixy}{x^2 - y^2 + 2ixy}$$

$$\therefore \quad \left|\frac{d\zeta}{dz}\right|_1 = \left[\frac{\left(x_1^2 - y_1^2 - a^2\right)^2 + 4x_1^2 y_1^2}{\left(x_1^2 - y_1^2\right)^2 + 4x_1^2 y_1^2}\right]^{\frac{1}{2}}$$

$$= \left[\frac{\left(x_1^2 + y_1^2 - a^2\right)^2 + 4a^2 y_1^2}{\left(x_1^2 + y_1^2\right)^2}\right]^{\frac{1}{2}}$$

By Routh's theorem, we have

$$\psi_2(x_1, y_1) = \psi_1(\xi_1, \eta_1) + \log\left|\frac{d\zeta}{dz}\right|_1$$

$$\therefore \quad \psi_2(x_1, y_1) = \frac{-K}{4\pi}\log\frac{y_1\left(x_1^2 + y_1^2 - a^2\right)}{x_1^2 + y_1^2}$$

$$+ \frac{K}{4\pi}\log\frac{\left[\left(x_1^2 + y_1^2 - a^2\right) + 4a^2 y_1^2\right]^{\frac{1}{2}}}{x_1^2 + y_1^2}$$

$$= \frac{K}{4\pi}\log\frac{\left[\left(x_1^2 + y_1^2 - a^2\right)^2 + 4a^2 y_1^2\right]^{\frac{1}{2}}}{y_1\left(x_1^2 + y_1^2 - a^2\right)}$$

Hence the path of the vortex at (x_1, y_1) is given by

$$\psi_2(x_1, y_1) = \text{constant}$$

$$\Rightarrow \quad \frac{K}{4\pi}\log\frac{\left[\left(x_1^2 + y_1^2 - a^2\right)^2 + 4a^2 y_1^2\right]^{\frac{1}{2}}}{y_1\left(x_1^2 + y_1^2 - a^2\right)} = \frac{K}{4\pi}\log c^{\frac{1}{2}}$$

$$\Rightarrow \quad \left(x_1^2 + y_1^2 - a^2\right)^2 + 4a^2 y_1^2 = c y_1^2\left(x_1^2 + y_1^2 - a^2\right)^2$$

Hence the path of the vortex at any point is given by

$$\left(x^2 + y^2 - a^2\right)^2 + 4a^2 y^2 = c y^2\left(x^2 + y^2 - a^2\right)^2$$

16. To find the path of a vortex in the angle between two planes to which it is parallel.

Solution: Let the two planes in z-plane be inclined at an angle $\frac{\pi}{n}$. By the transformation

$$\zeta = cz^n \text{ i.e. } Re^{i\delta} = cr^n e^{in\theta} \quad \dots (1)$$

these two boundaries of z-plane are transformed onto ξ-axis of ζ-plane.

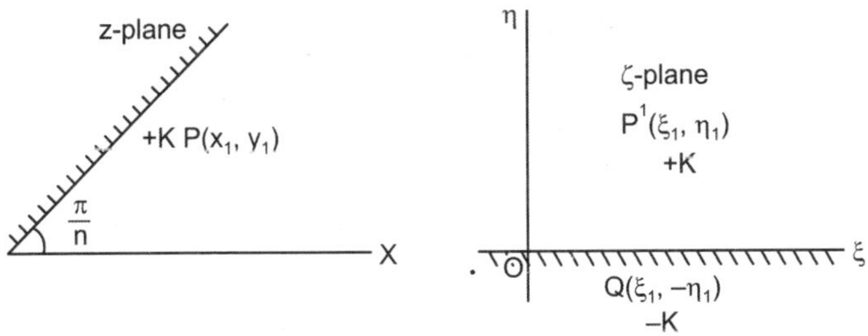

Fig. 7.32

For $\delta = n\theta$, $\theta = 0, \dfrac{\pi}{n}$

Then $\delta = 0, \pi$ in ξ-axis.

By this transformation the vortex $+K$ at $P(x_1, y_1)$ is transformed onto vortex $+K$ at $P^1(\zeta_1)$. The image of $+K$ at P^1 with respect to ξ-axis a vortex $-K$ at $Q(\overline{\zeta}_1)$. The complex potential w at any point ζ in ζ-plane is given by

$$w = \frac{iK}{2\pi}\log(\zeta - \zeta_1) - \frac{iK}{2\pi}\log(\zeta - \overline{\zeta}_1)$$

$$\Rightarrow \quad \phi + i\psi = \frac{iK}{2\pi}\log\left[\frac{(\xi - \xi_1) + i(\eta - \eta_1)}{(\xi - \xi_1) + i(\eta + \eta_1)}\right]$$

$$\therefore \quad \psi = \frac{K}{4\pi}\log\left[\frac{(\xi - \xi_1)^2 + (\eta - \eta_1)^2}{(\xi - \xi_1)^2 + (\eta + \eta_1)^2}\right]$$

The stream function $y_1(x_1, \eta_1)$ at P^1 is given by

$$\psi_1(\xi_1, \eta_1) = \frac{-K}{4\pi}\log\eta_1$$

By Routh's theorem

$$\psi_2(x_1, y_1) = \psi_1(\xi_1, \eta_1) + \log\left|\frac{d\zeta}{dz}\right|_1$$

$$= \frac{-K}{4\pi}\log\eta_1 + \log\left|ncz^{n-1}\right|$$

$$= \frac{K}{4\pi} \log \frac{cnr^{n-1}}{\eta_1}$$

Path of vortex at $P(x_1, y_1)$ is given by

$$\psi_2(x_1, y_1) = \text{constant}$$

$$\Rightarrow \quad \frac{cnr_1^{n-1}}{\eta_1} = \text{constant}$$

$$\Rightarrow \quad r_1^{n-1} = c\eta_1 = ca\, r_1^n \sin n\theta_1 \qquad (\because \zeta = cz^n)$$

$$\Rightarrow \quad r_1 \sin n\theta_1 = b \qquad \text{where } b = \frac{1}{ac}$$

Hence the required path is $r \sin n\,\theta = b$.

17. An infinite row of equidistant rectilinear vortices are at a distance a apart. The vortices are of the same numerical strength K but they are alternatively of opposite signs. Find the complex function that determine the velocity potential and the stream function. Show that the vortices remain at rest and draw streamlines. Show also that, if α be the radius of a vortex, the amount of flow between any vortex and the next is $\dfrac{K}{\pi} \log \cot \dfrac{\pi x}{2a}$.

Solution: Let the vortices each of strength K are placed at (0, 0), ($\pm 2a$, 0), ($\pm 4a$, 0)..... and vortices each of strength $-K$ are placed at ($\pm a$, 0), ($\pm 3a$, 0) ($\pm 5a$, 0).....

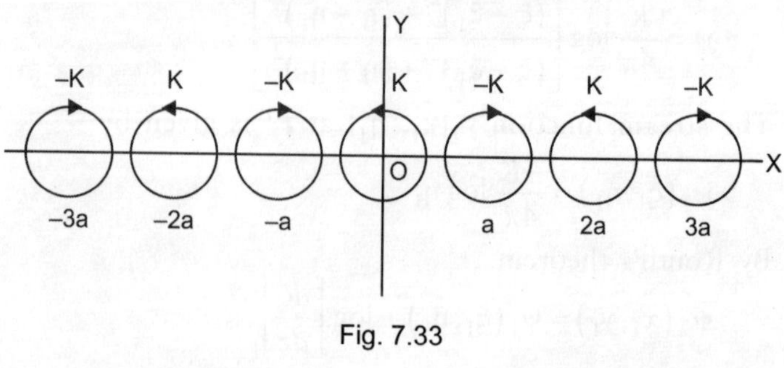

Fig. 7.33

The complex potential of the system at any point z is given by

$$w = \frac{iK}{2\pi}\log z + \frac{iK}{2\pi}\left[\log(z-2a)+\log(z+2a)+\ldots\right]$$
$$-\frac{iK}{2\pi}\left[\log(z-a)+\log(z+a)\ldots..\right]$$

$$= \frac{iK}{2\pi}\log\frac{z(z^2-2^2a^2)(z^2-4^2a^2)\ldots..}{(z^2-a^2)(z^2-3^2a^2)\ldots..}$$

$$= \frac{iK}{2\pi}\log\left[\frac{\frac{z}{2a}\left\{1-\left(\frac{z}{2a}\right)^2\right\}\left\{1-\left(\frac{z}{4a}\right)^2\right\}}{\left\{1-\left(\frac{z}{a}\right)^2\right\}\left\{1-\left(\frac{z}{3a}\right)^2\right\}}\right] + \text{constant}$$

$$\Rightarrow \quad w = \frac{iK}{2\pi}\log\frac{\sin\left(\frac{\pi z}{2a}\right)}{\cos\left(\frac{\pi z}{2a}\right)}$$

$$\Rightarrow \quad w = \frac{iK}{2\pi}\log\tan\frac{\pi z}{2a} \qquad \ldots (1)$$

$$\Rightarrow \quad \phi + i\psi = \frac{iK}{2\pi}\log\tan\frac{\pi z}{2a} \qquad \ldots (2)$$

Also

$$\phi - i\psi = \frac{iK}{2\pi}\log\tan\frac{\pi\bar{z}}{2a} \qquad \ldots (3)$$

(2) – (3) gives

$$2i\psi = \frac{iK}{2\pi}\log\left[\left(\tan\frac{\pi z}{2a}\right)\left(\tan\frac{\pi\bar{z}}{2a}\right)\right]$$

$$\Rightarrow \quad \psi = \frac{K}{4\pi}\log\left[\frac{\sin\frac{\pi z}{2a}\sin\frac{\pi\bar{z}}{2a}}{\cos\frac{\pi z}{2a}\cos\frac{\pi\bar{z}}{2a}}\right]$$

$$\Rightarrow \quad \psi = \frac{K}{4\pi}\log\left[\frac{\sin\left\{\frac{\pi(x+iy)}{2a}\right\}\sin\left\{\frac{\pi(x-iy)}{2a}\right\}}{\cos\left\{\frac{\pi(x+iy)}{2a}\right\}\cos\left\{\frac{\pi(x-iy)}{2a}\right\}}\right]$$

$$\Rightarrow \quad \psi = \frac{K}{4\pi} \log \frac{\cosh\left(\frac{\pi y}{a}\right) - \cos\left(\frac{\pi x}{a}\right)}{\cosh\left(\frac{\pi y}{a}\right) + \cos\left(\frac{\pi x}{a}\right)} \qquad \ldots (4)$$

Streamlines are given by ψ = constant, i.e.

$$\cosh \frac{\pi y}{a} = b \cos \frac{\pi x}{a}$$

Again (2) + (3) gives

$$2\phi = \frac{iK}{2\pi} \log \left[\tan \frac{\pi z}{2a} \cot \frac{\pi \bar{z}}{2a} \right]$$

$$\Rightarrow \quad \phi = \frac{iK}{4\pi} \log \left[\frac{\sin \frac{\pi z}{2a} \cos \frac{\pi \bar{z}}{2a}}{\sin \frac{\pi \bar{z}}{2a} \cos \frac{\pi z}{2a}} \right]$$

$$= \frac{iK}{4\pi} \log \frac{\sin \frac{\pi x}{a} + i \sinh\left(\frac{\pi y}{a}\right)}{\sin \frac{\pi x}{a} - i \sinh\left(\frac{\pi y}{a}\right)}$$

$$\Rightarrow \quad \phi = \frac{-K}{4\pi} \left[\tan^{-1} \frac{\sinh\left(\frac{\pi y}{a}\right)}{\sin\left(\frac{\pi x}{a}\right)} + \tan^{-1} \frac{\sinh\left(\frac{\pi y}{a}\right)}{\sin\left(\frac{\pi x}{a}\right)} \right]$$

$$\Rightarrow \quad \phi = \frac{-K}{2\pi} \tan^{-1} \frac{\sinh\left(\frac{\pi y}{a}\right)}{\sin\left(\frac{\pi x}{a}\right)} \qquad \ldots (5)$$

Required velocity potential and stream function are given by (4) and (5).

Now we consider the motion of the vortex $+K$ at the origin.

Then $w_1 = w - \frac{iK}{2\pi} \log z$

Let u_0, v_0 be the velocity components of the vortex at (0, 0)

$$u_0 - iv_0 = -\left(\frac{dw_1}{dz}\right)_{z=0}$$

$$= \frac{iK}{2\pi}\left[\frac{\sec\frac{\pi z}{2a}}{\tan\frac{\pi z}{2a}}\frac{\pi}{2a} - \frac{1}{z}\right]_{z=0} = 0$$

This shows that the vortex at (0, 0) is at rest. Similarly we can prove that every vortex is at rest.

Now we shall determine flow. ψ at any point of x-axis i.e. at $(x, 0)$ is

$$\psi = \frac{K}{4\pi}\log\left[\frac{1-\cos\frac{\pi x}{a}}{1+\cos\frac{\pi x}{a}}\right] \quad \text{From (4)}$$

$$\Rightarrow \quad \psi(x, 0) = \frac{K}{4\pi}2\log\tan\frac{\pi x}{2a}$$

Hence flow between two consecutive vortices

$$= 2 \text{ flow across } (a-\alpha, 0) \text{ to } (\alpha, 0)$$

$$= \psi(a-\alpha, 0) - \psi(\alpha, 0)$$

$$= \frac{K}{2\pi}\log\left[\frac{\tan\frac{\pi}{2a}(a-\alpha)}{\tan\frac{\pi\alpha}{2a}}\right]$$

$$= \frac{K}{2\pi}\log\left(\cot\frac{\pi\alpha}{2a}\right)^2$$

$$= \frac{K}{\pi}\log\cot\left(\frac{\pi\alpha}{2a}\right)$$

18. Prove that a thin cylindrical vortex of strength σ, running parallel to a plane boundary at a distance a will travel with velocity $\dfrac{\sigma}{4\pi a}$ and show that a stream of fluid will flow past between the travelling vortex and the boundary of total amount

$\frac{\sigma}{2\pi}\left[\log\left(\frac{2a}{c}\right)-\frac{1}{2}\right]$ per unit length along the vortex, where c is the small radius of the cross section of the vortex.

Solution: Let A be the vortex of strength σ and BOB^1 be the plane boundary. The line AOA^1 is the perpendicular drawn from A on OX intersecting at O, A^1 is the image of the vortex at A.

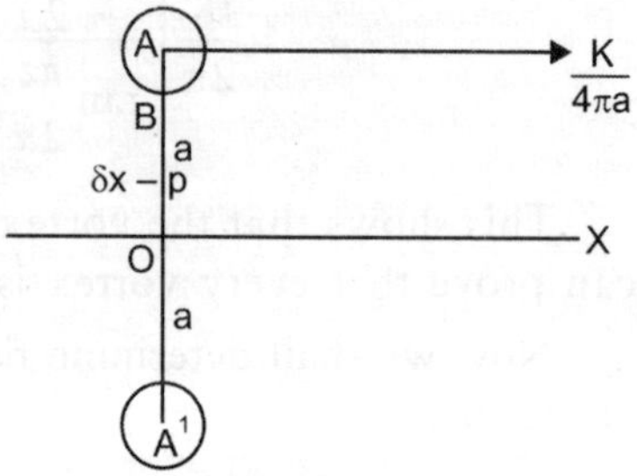

Fig. 7.34

The velocity of the vortex at A is $\frac{K}{2\pi AA^1}$.

We consider an element δx at P at a distance x from A. The velocity at the point P becomes

$$=\frac{K}{2\pi}\frac{1}{x}+\frac{K}{2\pi}\frac{1}{2a-x}$$

Then the fluid flows across OA is

$$=\frac{K}{2\pi}\int_c^a\left(\frac{1}{x}+\frac{1}{2a-x}\right)dx$$

$$=\frac{K}{2\pi}\left[\log x-\log(2a-x)\right]_c^a$$

$$=\frac{K}{2\pi}\log\frac{2a-c}{c}$$

$$=\frac{K}{2\pi}\log\frac{2a}{c}\qquad c \text{ being small}$$

But AA^1 moves with velocity $\frac{K}{4\pi a}$, so the vortex system can be reduced to rest by applying a velocity $\frac{-K}{4\pi}$. Hence the flow past between the travelling vortex and the boundary of total amount is

$$=\frac{K}{2\pi}\log\frac{2a}{c}-\frac{K}{4\pi}=\frac{K}{2\pi}\left[\log\left(\frac{2a}{c}\right)-\frac{1}{2}\right]$$

19. If $udx + vdy + wdz = d\theta + \lambda d\mu$ where θ, λ, μ are functions of x, y, z, t prove that the vortex lines at any time are the lines of intersection of the surfaces λ = constant and μ = constant.

Solution: Given $udx + vdy + wdz = d\theta + \lambda d\mu$

$$\Rightarrow \quad udx + vdy + wdz = \frac{\partial\theta}{\partial x}dx + \frac{\partial\theta}{\partial y}dy + \frac{\partial\theta}{\partial z}dz + \frac{\partial\theta}{\partial t}dt + \lambda\frac{\partial\mu}{\partial x}dx + \lambda\frac{\partial\mu}{\partial y}dy + \lambda\frac{\partial\mu}{\partial z}dz + \lambda\frac{\partial\mu}{\partial t}dt$$

$$\Rightarrow \quad u = \frac{\partial\theta}{\partial x} + \lambda\frac{\partial\mu}{\partial x}, \quad v = \frac{\partial\theta}{\partial y} + \lambda\frac{\partial\mu}{\partial y}$$

$$w = \frac{\partial\theta}{\partial z} + \lambda\frac{\partial\mu}{\partial z}, \quad O = \frac{\partial\theta}{\partial t} + \lambda\frac{\partial\mu}{\partial t}$$

The components of spin are

$$2\xi = \frac{\partial w}{\partial y} - \frac{\partial v}{\partial z}$$

$$2\xi = \frac{\partial^2\theta}{\partial y\partial z} + \lambda\frac{\partial^2\mu}{\partial y\partial z} + \frac{\partial\lambda}{\partial y}\frac{\partial\mu}{\partial z} - \frac{\partial^2\theta}{\partial y\partial z} - \lambda\frac{\partial^2\mu}{\partial y\partial z} - \frac{\partial\lambda\partial\mu}{\partial z\partial y}$$

$$= \frac{\partial\lambda}{\partial y}\frac{\partial\mu}{\partial z} - \frac{\partial\lambda}{\partial z}\frac{\partial\mu}{\partial y}$$

$$= \begin{vmatrix} \frac{\partial\lambda}{\partial y} & \frac{\partial\lambda}{\partial z} \\ \frac{\partial\mu}{\partial y} & \frac{\partial\mu}{\partial z} \end{vmatrix}$$

Similarly

$$2\eta = \begin{vmatrix} \frac{\partial\lambda}{\partial z} & \frac{\partial\lambda}{\partial x} \\ \frac{\partial\mu}{\partial z} & \frac{\partial\mu}{\partial x} \end{vmatrix}$$

and

$$2\zeta = \begin{vmatrix} \dfrac{\partial \lambda}{\partial x} & \dfrac{\partial \lambda}{\partial y} \\ \dfrac{\partial \mu}{\partial x} & \dfrac{\partial \mu}{\partial y} \end{vmatrix}$$

Therefore

$$2\left(\xi \frac{\partial \lambda}{\partial x} + \eta \frac{\partial \lambda}{\partial y} + \zeta \frac{\partial \lambda}{\partial z}\right) = \begin{vmatrix} \lambda x & \lambda y & \lambda z \\ \lambda x & \lambda y & \lambda z \\ \mu x & \mu y & \mu z \end{vmatrix} = 0$$

$$\Rightarrow \quad \xi \lambda x + \eta \lambda y + \zeta \lambda z = 0$$

Similarly

$$\xi \mu_x + \eta \mu_y + \zeta \mu_z = 0$$

The last two equations prove that the vortex lines lie on the surfaces θ = const and μ = const.

20. Prove that in a steady motion of a liquid

$$H = \frac{p}{\rho} + \frac{1}{2} q^2 + V = \text{constant along a streamline}$$

If this constant has the same value everywhere in the liquid, then prove that the motion must be either irrotational or the vortex lines must coincide with the streamlines. In two-dimensional motion of a liquid with constant vorticity ζ, prove that $\Delta(H - 2\zeta\psi) = 0$. Show also that if the motion be steady the pressure is given by

$$\frac{p}{\rho} + \frac{1}{2} q^2 + V - 2\zeta\psi = \text{const}$$

where Δ is Laplace's operator.

Solution: Euler's equation of motion is

$$\frac{d\vec{q}}{dt} = \vec{F} - \frac{1}{\rho} \vec{\nabla} p$$

$$\Rightarrow \quad \frac{\partial \vec{q}}{dt} + (\vec{q} \cdot \vec{\nabla})\vec{q} = -\vec{\nabla} V - \frac{1}{\rho} \vec{\nabla} p$$

Since the motion is steady, so that $\dfrac{\partial \vec{q}}{\partial t} = 0.$

Hence $(\vec{q}\cdot\vec{\nabla})\vec{q} = -\vec{\nabla}V - \frac{1}{\rho}\vec{\nabla}p$

$$\Rightarrow \left(u\frac{\partial}{\partial x} + v\frac{\partial}{\partial y} + w\frac{\partial}{\partial z}\right)u = -\frac{\partial v}{\partial x} - \frac{1}{\rho}\frac{\partial p}{\partial x}$$

and two similar expressions for v and w

$$\Rightarrow u\frac{\partial u}{\partial x} + v\frac{\partial v}{\partial x} + w\frac{\partial w}{\partial x} + v\left(\frac{\partial u}{\partial y} - \frac{\partial v}{\partial x}\right) + w\left(\frac{\partial u}{\partial z} - \frac{\partial w}{\partial x}\right) + \frac{1}{\rho}\frac{\partial p}{\partial x} + \frac{\partial v}{\partial x} = 0$$

$$\Rightarrow \frac{\partial}{\partial x}(u^2 + v^2 + w^2) + v(-2\zeta) + w(2\eta) + \frac{1}{\rho}\frac{\partial p}{\partial x} + \frac{\partial v}{\partial x} = 0 \quad \dots (1)$$

as $$\vec{w}(\xi, \eta, \zeta) = \frac{1}{2}\begin{vmatrix} \vec{i} & \vec{j} & \vec{k} \\ \frac{\partial}{\partial x} & \frac{\partial}{\partial y} & \frac{\partial}{\partial z} \\ u & v & w \end{vmatrix}$$

Taking $H = \int\frac{dp}{\rho} + v + \frac{1}{2}q^2$, we get from (1)

$$\frac{\partial H}{\partial x} = 2(v\zeta - w\eta) \quad \dots (2)$$

Similarly

$$\frac{\partial H}{\partial y} = 2(w\xi - u\zeta) \quad \dots (3)$$

$$\frac{\partial H}{\partial z} = 2(u\eta - v\xi) \quad \dots (4)$$

Multiplying (2), (3), (4) by u, v, w and then adding, we get

$$u\frac{\partial H}{\partial x} + v\frac{\partial H}{\partial y} + w\frac{\partial H}{\partial z} = 0 \quad \dots (5)$$

Also multiplying (2), (3) and (4) by ξ, η, ζ and then adding, we get

$$\xi\frac{\partial H}{\partial x} + \eta\frac{\partial H}{\partial y} + \zeta\frac{\partial H}{\partial z} = 0 \quad \dots (6)$$

(5) and (6) shows that the surface H = constant contains the streamline and vortex lines.

If H has the same value everywhere, then

$$\frac{\partial H}{\partial x} = 0, \frac{\partial H}{\partial y} = 0, \frac{\partial H}{\partial z} = 0$$

So that $v\zeta - w\eta = 0, w\xi - u\zeta = 0, u\eta - v\xi = 0$

Hence

$$\frac{u}{\xi} = \frac{v}{\eta} = \frac{w}{\zeta} \text{ or } \xi = \eta = \zeta = 0$$

which shows that the motion is irrotational. Also streamlines are given by

$$\frac{dx}{u} = \frac{dy}{v} = \frac{dz}{w}$$

coincide with vortex lines given by

$$\frac{dx}{\xi} = \frac{dy}{\eta} = \frac{dz}{\zeta}$$

Now we shall prove that $\Delta(H - 2\zeta\psi) = 0$

Now

$$2\zeta = \left(\frac{\partial v}{\partial x} - \frac{\partial u}{\partial y}\right), u = \frac{-\partial\phi}{\partial x} = \frac{-\partial\psi}{\partial y}, v = \frac{-\partial\phi}{\partial y} = \frac{\partial\psi}{\partial x}$$

$$\therefore \quad 2\zeta = \frac{\partial^2\psi}{\partial x^2} + \frac{\partial^2\psi}{\partial y^2} = \vec{\nabla}^2\psi$$

Equation of motion is $\dfrac{d\vec{q}}{dt} = -\vec{\nabla}v - \dfrac{1}{\rho}\vec{\nabla}p$

$$\Rightarrow \quad \left(\frac{\partial}{\partial t} = u\frac{\partial}{\partial x} + v\frac{\partial}{\partial y}\right)u = -\frac{\partial v}{\partial x} - \frac{1}{\rho}\frac{\partial p}{\partial x} \qquad \text{as } w = 0$$

$$\Rightarrow \quad \frac{\partial u}{\partial t} + u\frac{\partial u}{\partial x} + v\frac{\partial v}{\partial x} + v\left(\frac{\partial u}{\partial y} - \frac{\partial v}{\partial x}\right) = \frac{-\partial v}{\partial x} - \frac{1}{\rho}\frac{\partial p}{\partial x}$$

Also $q^2 = u^2 + v^2$

So that $\dfrac{\partial q^2}{\partial x} = 2\left(u\dfrac{\partial u}{\partial x} + v\dfrac{\partial v}{\partial x}\right)$

Hence

$$\frac{\partial u}{\partial t}+\frac{1}{2}\frac{\partial q^2}{\partial x}+v(-2\zeta)=\frac{-\partial v}{\partial x}-\frac{1}{\rho}\frac{\partial p}{\partial x}$$

$$\Rightarrow \quad \frac{\partial u}{\partial t}-2v\zeta=\frac{-\partial H}{\partial x} \qquad \text{... (7)}$$

Similarly

$$\frac{\partial v}{\partial t}+2u\zeta=\frac{-\partial H}{\partial y} \qquad \text{... (8)}$$

Differentiating (7) and (8) with respect to x and y respectively and then adding, we have

$$-\Delta H=\frac{\partial}{\partial t}\left(\frac{\partial u}{\partial x}+\frac{\partial v}{\partial y}\right)+2\zeta\left(\frac{-\partial v}{\partial x}+\frac{\partial u}{\partial y}\right)$$

But $\frac{\partial u}{\partial x}+\frac{\partial v}{\partial y}=0$ by equation of continuity.

Hence

$$-\Delta H=-2\zeta\left(\frac{\partial^2\psi}{\partial x^2}-\frac{\partial^2\psi}{\partial y^2}\right)$$

$$=-2\zeta\Delta\psi$$

$$\Rightarrow \quad \Delta(H-2\zeta\psi)=0 \qquad \text{... (9) as } \zeta = \text{constant}$$

Since the motion is steady, so that $\frac{\partial u}{\partial t}=0=\frac{\partial v}{\partial t}$

Integrating (9), we get $H - 2\zeta\psi$ = const = c (say)

Putting the value of H, we get

$$\frac{p}{\rho}+\frac{1}{2}q^2+v-2\zeta\psi=c.$$

21. A mass of liquid whose outer boundary is an infinitely long cylinder of radius b is in a state of cyclic irrotational motion is under the action of a uniform pressure P over the external surface. Prove that there must be a concentric cylindrical hollow whose radius a is given by $8\pi^3a^2b^2\,P = MK^2$ where M is the mass of unit length of the liquid and K the circulation.

Solution: The complex potential is given by

$$w = \frac{iK}{2\pi}\log(z - z_0) = \frac{iK}{2\pi}\log re^{i\theta}$$

or $\phi + i\psi = \frac{iK}{2\pi}[\log r + i\theta]$

Hence

$$\Rightarrow \quad \psi = \frac{K}{2\pi}\log r, \phi = \frac{-\theta K}{2\pi}$$

$$u = \frac{-\partial\phi}{\partial r} = 0, v = -\frac{1}{r}\frac{\partial\phi}{\partial\theta} = -\frac{1}{r}\left(\frac{-K}{2\pi}\right) = \frac{K}{2\pi r}$$

$$q^2 = u^2 + v^2 = 0 + v^2$$

$$q = v = \frac{K}{2\pi r}$$

By Bernoulli's equation for steady motion

$$\frac{p}{\rho} + \frac{1}{2}q^2 = c \qquad \text{... (1)}$$

Subjecting this to the boundary $p = P$, $r = b$, $q = \frac{K}{2\pi b}$

Hence from (1)

$$\frac{P}{\rho} + \frac{K^2}{8\pi^2 b^2} = c$$

Hence (1) becomes

$$\frac{p}{\rho} + \frac{1}{2}q^2 = \frac{P}{\rho} + \frac{K^2}{8\pi^2 b^2}$$

$$\Rightarrow \quad p = P + \frac{K^2\rho}{8\pi^2}\left(\frac{1}{b^2} - \frac{1}{\partial^2}\right) \qquad \text{... (2)}$$

If r is very small then $p < 0$. Hence there must be a cavity, say $r = a$. Then when $r = a$, $p = a$ since pressure vanishes on the surface of cavity. Thus (2) becomes

$$P + \frac{K^2\rho}{8\pi^2}\left(\frac{1}{b^2} - \frac{1}{a^2}\right) = 0$$

$$\Rightarrow \quad 8\pi^3 a^2 b^2 P = MK^2 \qquad \text{where } M = \pi\rho\,(b^2 - a^2)$$

22. An elliptic cylinder is filled with liquid which has molecular rotation w at every point and whose particles move in a plane perpendicular to the axis, prove that the streamlines are similar ellipses described in period distance.

$$\frac{\pi}{w}\frac{a^2+b^2}{ab}$$

Solution: There exists no liquid outside the cylinder. The stream function ψ satisfies the condition

$$\frac{\partial^2\psi}{\partial x^2}+\frac{\partial^2\psi}{\partial y^2}=2w, \text{ inside the boundary} \qquad \dots (1)$$

The cross section is given by

$$\frac{x^2}{a^2}+\frac{y^2}{b^2}-1=F(x,y)=0$$

Assume $\psi = Ax^2 + By^2$ ... (2)

Subjecting ψ to (1) we get $A + B = w$

The boundary condition is

normal component of velocity of liquid at boundary

= normal component of velocity of the surface

Then

$$u\frac{px}{a^2}+\frac{vpy}{b^2}=0 \text{ as boundary is fixed} \qquad \dots (3)$$

Also

$$u=\frac{-\partial\psi}{\partial y}=-2By$$

$$v=\frac{\partial\psi}{\partial x}=2Ax$$

Hence from (3)

$$-2By\frac{x}{a^2}+2Ax\frac{y}{b^2}=0$$

$\Rightarrow$ $a^2A - b^2B = 0$

$$\Rightarrow \frac{A}{b^2}=\frac{B}{a^2}=\frac{A+B}{a^2+b^2}=\frac{w}{a^2+b^2}$$

Hence

$$\psi = \frac{wb^2x^2}{a^2+b^2} + \frac{a^2wy^2}{a^2+b^2}$$

$$= \frac{wa^2b^2}{a^2+b^2}\left(\frac{x^2}{a^2} + \frac{y^2}{b^2}\right)$$

The streamlines are given by ψ = constant

i.e. $$\frac{wa^2 \cdot b^2}{a^2+b^2}\left(\frac{x^2}{a^2} + \frac{y^2}{b^2}\right) = \text{constant}$$

$$\Rightarrow \quad \frac{x^2}{a^2} + \frac{y^2}{b^2} = \text{constant}$$

which are similar ellipses.

Since the cylinder is fixed so that

$$\dot{x} = u = -\frac{\partial\psi}{\partial y} = -\frac{wa^2b^2}{a^2+b^2}\left(\frac{2y}{b^2}\right)$$

$$\dot{y} = v = \frac{\partial\psi}{\partial x} = \frac{wa^2b^2}{a^2+b^2}\left(\frac{2x}{a^2}\right)$$

Hence

$$\ddot{x} = \frac{-2wa^2b^2}{a^2+b^2}\frac{2}{b^2}\dot{y}$$

$$= -\left(\frac{wa^2b^2}{a^2+b^2}\right)^2 \frac{2}{b^2}\frac{2x}{a^2}$$

$$\Rightarrow \quad \ddot{x} = -\mu^2 x$$

Similarly $\dot{y} = -\mu^2 y$ where $\mu = \dfrac{2wab}{a^2+b^2}$

By integrating, we have

$$x = aL\cos(\mu t + t),\ y = bL\sin(\mu t + t)$$

Hence

$$\frac{x^2}{a^2} + \frac{y^2}{b^2} = L^2$$

which is similar ellipse.

Also periodic time $T = \frac{2\pi}{\sqrt{\mu^2}} = \frac{\pi(a^2+b^2)}{wab}$

23. Investigate the nature of the liquid motion given by

$$u = \frac{ax - by}{x^2 + y^2},\ v = \frac{ay + bx}{x^2 + y^2},\ w = 0$$

Also determine the velocity potential and the pressure p at any point (x, y).

Solution: Here

$$u = \frac{ax - by}{x^2 + y^2},\ v - \frac{ay + bx}{x^2 + y^2},\ w - 0 \qquad \text{... (1)}$$

$$\therefore \quad \frac{\partial u}{\partial x} = \frac{a(x^2+y^2) - 2x(ax-by)}{(x^2+y^2)^2}$$

$$= \frac{a(y^2 - x^2) + 2bxy}{(x^2+y^2)^2} \qquad \text{... (2)}$$

Also

$$\frac{\partial v}{\partial y} = \frac{a(x^2 - y^2) - 2bxy}{(x^2+y^2)^2} \qquad \text{... (3)}$$

(2) + (3) gives $\frac{\partial u}{\partial x} + \frac{\partial v}{\partial y} = 0$

Hence the liquid motion represented by (1) is a possible motion.

Let Ω be the velocity then

$$\Omega = \Omega x\vec{i} + \Omega y\vec{j} + \Omega z\vec{k} \qquad \text{... (4)}$$

where $\Omega x = \frac{\partial w}{\partial y} - \frac{\partial v}{\partial z},\ \Omega y = \frac{\partial u}{\partial z} - \frac{\partial w}{\partial x}$

$$\Omega z = \frac{\partial v}{\partial x} - \frac{\partial u}{\partial y}$$

From (1), we have

$$\Omega x = \Omega y = \Omega z = 0$$

$$\Rightarrow \quad \bar{\Omega} = 0$$

It follows that the nature of the liquid motion is irrotational.

Let ϕ be the velocity potential, then

$$d\phi = \frac{\partial \phi}{\partial x} dx + \frac{\partial \phi}{\partial y} dy = -udx - vdy$$

$$\Rightarrow \quad d\phi = -\left\{\frac{ax - by}{x^2 + y^2} dx + \frac{ay + bx}{x^2 + y^2} dy\right\}$$

$$\Rightarrow \quad d\phi = -\left\{\frac{a(xdx + ydy)}{x^2 + y^2} + \frac{b(xdy - ydx)}{x^2 + y^2}\right\}$$

Integrating, we get

$$\phi = -\frac{a}{2}\log\left(x^2 + y^2\right) + b\tan^{-1}\left(\frac{y}{x}\right)$$

Again let q be the fluid velocity and p be the pressure at any point (x, y) then by Bernoulli's equation, we have

$$p = \text{constant} - \frac{1}{2}\rho q^2$$

$$\Rightarrow \quad p = \text{constant} - \frac{1}{2}\rho\left(u^2 + v^2\right)$$

$$\Rightarrow \quad p = \text{constant} - \frac{1}{2}\rho\left[\frac{(ax - by)^2 + (ay + bx)^2}{\left(x^2 + y^2\right)^2}\right]$$

$$\Rightarrow \quad p = \text{constant} - \frac{1}{2}\rho\left(\frac{a^2 + b^2}{x^2 + y^2}\right).$$

24. Two parallel rectilinear vortices of strengths K_1 and K_2 $(K_1 > K_2)$ are at a distance $2a$ apart in an infinite mass of liquid. If the vortices intersect a plane perpendicular to their lengths at points A and B. Show that the point on AB at a distance b from the mid-point on the same side of mid-point as the vortex of strength K_1, is always occupied by the same fluid element if

$$\frac{K_1 - K_2}{K_1 + K_2} = \frac{b^3 - 5a^2 b}{ab^2 + 3a^3}$$

Solution: If G be the centre of vortex, then

$$AB = \frac{K_1}{K_1 + K_2},\ AG = \frac{2aK_2}{K_1 + K_2}$$

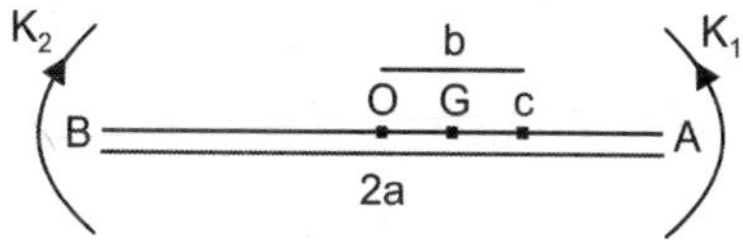

Fig. 7.35

Similarly $BG = \dfrac{2aK_1}{K_1 + K_2}$

Let O be the middle point of AB. Let $OC = b$.

$$\therefore \quad OG = OA - AG = a - \frac{2aK_2}{K_1 + K_2} = \frac{K_1 - K_2}{K_1 + K_2}a$$

and $GC = OC - OG = b - \dfrac{K_1 - K_2}{K_1 + K_2}a$

$$= \frac{(K_1 + K_2)b - (K_1 - K_2)a}{K_1 + K_2}$$

Now the line BA rotates about G with angular velocity

$$= \frac{K_1 + K_2}{2\pi AB^2} = \frac{K_1 + K_2}{8\pi a^2} \qquad \text{as } AB = 2a$$

Hence the velocity of the point C of AB

$$= GC\,\frac{K_1 + K_2}{8\pi a^2} = \frac{(K_1 + K_2)b - (K_1 - K_2)a}{8\pi a^2} \qquad \text{... (1)}$$

Taking O as origin and OA as x-axis. Then, we have two rectilinear vortices of strength K_1 and K_2 at A ($z = z_1 = a$) and $B(z = z_2 = -a)$. Hence the complex potential at any point $P(z = x + iy)$ is given by

$$w = \frac{i}{2\pi}\left[K_1 \log(z - z_1) + K_2 \log(z - z_2)\right]$$

$$= \frac{i}{2\pi}\left[K_1 \log(z - a) + K_2 \log(z + a)\right] \qquad \text{... (2)}$$

$$\therefore \quad u - iv = \frac{-dw}{dz} = \frac{-i}{2\pi}\left[\frac{K_1}{z-a} + \frac{K_2}{z+a}\right]$$

$$\Rightarrow \quad u - iv = -\frac{i}{2\pi}\left[\frac{(K_1+K_2)z + K_1 a - K_2 a}{(z-a)(z+a)}\right]$$

Using (2) for the point C for which $z = b$, we have

$$u - iv = \frac{-i}{2\pi}\left[\frac{(K_1+K_2)b + (K_1-K_2)a}{(b-a)(b+a)}\right]$$

giving $u = 0$,

$$v = \frac{1}{2\pi}\frac{(K_1+K_2)b + (K_1-K_2)a}{b^2 - a^2}$$

Hence velocity at C

$$= \sqrt{u^2+v^2} = \frac{1}{2\pi}\frac{(K_1+K_2)b + (K_1-K_2)a}{b^2-a^2} \qquad \text{... (3)}$$

By the question, relative velocity of two C's must be zero and so the two different expressions (1) and (3) must be the same, hence

$$\frac{(K_1+K_2)b - (K_1-K_2)a}{8\pi a^2} = \frac{1}{2\pi}\frac{(K_1+K_2)b + (K_1-K_2)a}{b^2-a^2}$$

$$\Rightarrow \quad (K_1-K_2)\left[4a^3 + a(b^2-a^2)\right] = (K_1+K_2)\left[b(b^2-a^2) - 4a^2b^2\right]$$

$$\Rightarrow \quad \frac{K_1-K_2}{K_1+K_2} = \frac{b^3 - 5a^2 b}{ab^2 + 3a^3}$$

EXERCISE

1. In an incompressible fluid the vorticity at every point is constant in magnitude and direction. Show that the components of velocity u, v, w are solutions of Laplace's equation.
2. Three parallel rectilinear vortices of the same strength K and in the same sense meet any plane perpendicular to them in an equilateral triangle of side a. Show that the

vortices all move round the same cylinder with uniform speed $2\pi a^2/3K$.

3. Prove that the necessary and sufficient condition that the vortex line may be at right angles to the streamlines are

$$u, v, w = \mu\left(\frac{\partial \phi}{\partial x}, \frac{\partial \phi}{\partial y}, \frac{\partial \phi}{\partial z}\right)$$

4. An infinite liquid contains two parallel, cqual and opposite rectilinear vortex filaments at a distance $2b$. Show that the paths of the fluid particles relative to the vortices can be represented by the equation

$$\log\left(\frac{r^2 + b^2 - 2rb\cos\theta}{r^2 + b^2 + 2rb\cos\theta}\right) + \frac{r\cos\theta}{b} = \text{constant}$$

5. Three vortex filaments, each of strength K, are symmetrically placed inside a circular cylinder of radius a and pass through the corners of an equilateral triangle of side $b\sqrt{3}$. If there is no circulation in the fluid other than that due to the vortices, show that they will revolve about the axis of the cylinder with angular velocity $\dfrac{K(a^6 + 2b^6)}{b^2(a^6 - b^6)}$.

6. A rectilinear vortex moves parallel to two rigid planes which intersect at right angles, prove that on the line of intersection of the planes the excess of pressure due to the vortex varies inversly as the square of the distance of the vortex from the line of intersection.

8 WAVES

8.1. Introduction

When a stone is thrown into a pound, some disturbances is seen to travel over the surface of water. This disturbance is called waterwaves. It is also a well known fact that sound waves spread through the room when a radio is played upon. The energy received from the sun is transmitted by waves in the ether. Other examples of wave motion in liquids are high and low tides, standing waves in lakes, etc. All types of waves have got the following two points in common:

(i) energy is propagated to distant points

(ii) disturbance travels through the medium without any transference of the medium itself.

8.2. Wave Motion

All types of waves satisfy the equation $\vec{\nabla} f = \frac{1}{c^2}\frac{\partial^2 f}{\partial t^2}$.

This equation is called wave equation. One dimensional form of wave equation is

$$\frac{\partial^2 f}{\partial x^2} = \frac{1}{c^2}\frac{\partial^2 f}{\partial t^2} \qquad \text{... (1)}$$

where $y = f(x, t)$ represents displacement at a distance x at time t.

To solve (1), we change the independent variables x, t to x_1, x_2 by the transformation

$$x_1 = x - ct, \; x_2 = x + ct$$

Now (1) becomes

$$\frac{\partial^2 f}{\partial x_1 \partial x_2} = 0$$

$$\Rightarrow \quad \frac{\partial}{\partial x_1}\left(\frac{\partial f}{\partial x_2}\right) = 0$$

Integrating w.r.t. x_1, we get

$$\frac{\partial f}{\partial x_2} = \text{const} = f_1(x_2)$$

Again integrating w.r.t. x_2

$$f = \int f_1(x_2)\,dx_2 + \text{const.}$$

$$= f_2(x_2) + f_3(x_1)$$

where $f_3(x_1)$ is constant of integration

Finally, we have

$$f = f_2(x + ct) + f_3(x - ct)$$

Taking $y = f$, then

$$y = F(x + ct) + \varsigma(x - ct) \qquad \text{... (2)}$$

where $F = f_2$, $\varsigma = f_3$

We consider the second term in (2).

G $(x - ct)$ remains unchanged if we replace x, t by $x + ct$, $t + T$ respectively.

For $y = G(x - ct) = G[x + ct - c(t + T)]$

It means that the disturbance occurring at the point x at time t is transferred to the point $x + cT$ at time $t + T$, while the ordinate remains unchanged. It also declares that the wave profile $y = f(x)$ moves with velocity in the positive direction of x-axis. Notice that the curve $y = G(x - ct) = G(x)$ at $t = 0$ is generally called wave profile. Similarly the first term in (2) namely $F(x + ct)$ represents a wave motion in the negative direction of x-axis with velocity c.

8.3. Some Definitions

A simple harmonic progression wave is represented by

$$y = a \sin (mx - nt) \quad \text{... (1)}$$

where a, m, n are constants.

(1) can also be written as

$$y = a \sin m\left(x - \frac{n}{mt}\right)$$

Comparing this with $y = f(x - ct)$, we find that the profile $y = a \sin mx$ at time $t = 0$ moves with velocity $c = \frac{n}{m}$ along the positive direction of x-axis. The velocity $c = \frac{n}{m}$ is called *velocity of propagation.*

A simple harmonic progression wave is represented by a curve in which the equation of vertical section of the free surface at time t is given by

$$y = a \sin (mx - nt) \quad \text{... (2)}$$

where a, m, n are constants.

The maximum value of the disturbance y namely a is called the amplitude.

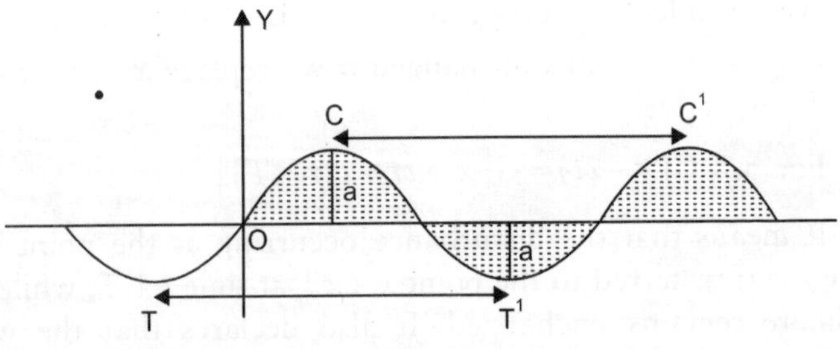

Fig. 8.1

The points C and C^1 of maximum elevation are called crests, and the points T and T^1 of maximum depression are called troughs of the wave.

If we replace x by $x + \frac{2\pi}{m}$, then (2) remains unchanged so that the wave profile repeats itself at regular distances $\frac{2\pi}{m}$.

The number $\frac{2\pi}{m}$ is called wavelength and is denoted by λ. Thus

$$\lambda = \frac{2\pi}{m} \qquad \dots (3)$$

That is to say, wavelength is a disturbance between two successive crests. Obviously the wave profile at C, C^1 or T, T^1 is the same, so that

$$CC^1 = TT^1 = \frac{2\pi}{m} = \lambda$$

Again, the nature of the free surface is unaltered if we replace t by $t + \frac{2\pi}{n}$. The time $T = \frac{2\pi}{n}$ is called the period of the wave.

Its reciprocal $\frac{1}{T} = \frac{n}{2\pi}$ is called the frequency, it denotes the number of oscillation per second.

The angle $(mx - nt)$ is called the phase angle and n is called the phase rate. If the equation of a progressive wave can be written in the form $y = a \sin (mx - nt + t)$, then t is called phase.

The number of waves per unit distance is called wave number and is denoted by K.

Also $cT = \frac{n}{m} \frac{2\pi}{n} = \frac{2\pi}{m} = \lambda$

Hence $cT = \lambda$

Remark. Even if a wave is not a harmonic wave, but the nature of wave profile is a regularly repeating pattern, the definitions of wavelength, period, frequency, etc. are all applicable.

8.4. Standing or Stationary Waves

Two simple harmonic progressive waves of equal amplitude, wavelength and period travelling in opposite directions, are given by the equations.

$$y_1 = a \sin (mx - nt), \; y_2 = a \sin (mx + nt)$$

Taking $y = y_1 + y_2$

Then $y = (2\,a \cos nt) \sin mx$ ... (1)

A wave motion of the type (1) is called stationary wave, for the resultant wave profile does not move forward. It is also called standing wave for the surface of the liquid moves up and down at a point x at any time t. The equation (1) represents sine curve of amplitude $2a \cos nt$ which varies from 0 to $2a \cos nt$.

Notice that each particle in the progressive wave vibrates with the same amplitude whereas the amplitude is not the same at all points in standing wave. Amplitude has maximum value $2a$ at points where

$$\sin mx = \sin(2p+1)\frac{\pi}{2}$$

$$\Rightarrow \quad mx = (2p+1)\frac{\pi}{2}$$

$$\Rightarrow x = (2p+1)\frac{\pi}{2m} = (2p+1)\frac{\lambda}{4}$$

where $p = 0, 1, 2, 3.....$

$$\Rightarrow \quad x = \frac{\lambda}{4}, \frac{3\lambda}{4}, \frac{5\lambda}{4}$$

These points are called antinodes. Also amplitude has a minimum value 0 at points x where $mx = p\pi$.

$$\Rightarrow \quad x = \frac{p\pi}{m} = p\pi\frac{\lambda}{2\pi} = \frac{p\lambda}{2}$$

for $p = 1, 2, 3.....$

$$\Rightarrow x = \frac{\lambda}{2}, \lambda, \frac{3\lambda}{2}.....$$

These points are called nodes.

8.5. Kinds of Wave Motion

For the sake of convenience wave motion in liquid is roughly divided into two groups: (1) Tidal Waves and (2) Surface Waves.

1. Tidal Waves. Tidal waves arise when the wavelength of oscillation is much greater than the depth of the liquid and the disturbance affects the motion of the whole of the liquid. Tidal waves are also known as long waves in shallow water. In tidal waves the vertical acceleration of the liquid is negligible compared with the horizontal acceleration and the plane of the liquid moves as a whole.

2. Surface Waves. Surface waves arise when the wavelength of the oscillation is much lesser than the depth of the liquid and the disturbance does not extend far below the surface. The vertical acceleration of the liquid is no longer negligible. Examples of surface waves are wind waves and surface tension waves.

8.6. Kinematical Surface Condition for Wave Profiles of Small Elevation

Let us consider a stream of constant depth h in which waves of height $\eta = \eta\,(x, t) = PN$ above the equilibrium level are propagated. We take x-axis on the undisturbed surface and y-axis vertically upwards.

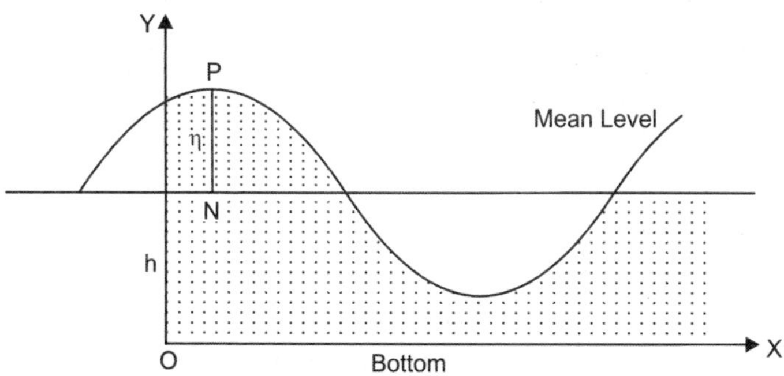

Fig. 8.2

The equation of the free surface is

$$y = h + \eta$$

Since the surface never moves with the fluid, so that

$$\frac{d}{dt}(y - h - \eta) = 0$$

$$\Rightarrow \quad \frac{dy}{dt} = \frac{dh}{dt} + \frac{d\eta}{dt} = 0 + \frac{d\eta}{dt} = \frac{\partial \eta}{\partial t} + u\,\frac{\partial \eta}{\partial x}$$

$$\Rightarrow \quad v = \frac{\partial \eta}{\partial t} + u \frac{\partial \eta}{\partial x} \qquad \text{... (1)}$$

We consider the disturbances to be very small so that the squares and products of small quantities can be neglected so that (1) becomes $v = \frac{\partial \eta}{\partial t}$. But $v = -\frac{\partial \phi}{\partial y}$. Hence

$$\frac{\partial \eta}{\partial t} = v = \frac{-\partial \phi}{\partial y} \qquad \text{... (2)}$$

$$\Rightarrow \quad \frac{\partial \eta}{\partial t} = \frac{\partial \psi}{\partial x}$$

The equation $\frac{\partial \eta}{\partial t} = \frac{\partial \psi}{\partial x}$ is the Kinematical Surface Condition for wave profiles of small height η and the slope $\frac{\partial \eta}{\partial x}$.

Pressure Condition at the Free Surface. If Ω be the force potential for waves under gravity, then $-g = F = -\frac{\partial \Omega}{\partial \eta}$ as g acts in downward direction.

Hence $\Omega = g\eta$.

By Bernoulli's pressure equation,

$$\frac{p}{\rho} + \frac{1}{2} q^2 - \frac{\partial \phi}{\partial t} + g\eta = f(t)$$

Neglecting q^2, we get

$$\frac{-\partial \phi}{\partial t} + g\eta = f(t) - \frac{p}{\rho} \qquad \text{... (1)}$$

Free surface (i.e. surface of equipressure) is given by p = constant.

Hence from (1)

$$\frac{-\partial \phi}{\partial t} + g\eta = f_1$$

$$\Rightarrow \quad \frac{-\partial \phi}{\partial t} + g\eta = 0 \qquad \text{on including the constant in } \phi.$$

Hence

$$\eta = \frac{1}{g} \frac{\partial \phi}{\partial t}$$

$$\therefore \quad \frac{\partial \eta}{\partial t} = \frac{1}{g} \frac{\partial^2 \phi}{\partial t^2}$$

Also

$$\frac{\partial \eta}{\partial t} + \frac{\partial \phi}{\partial y} = 0$$

$$\Rightarrow \quad \frac{1}{g} \frac{\partial^2 \phi}{\partial t^2} + \frac{\partial \phi}{\partial y} = 0$$

$$\Rightarrow \quad \frac{\partial^2 \phi}{\partial t^2} + g \frac{\partial \phi}{\partial y} = 0 \text{ at } y = \eta$$

Since the free surface is given by $y = \eta$. But η is very small. Hence we taken $y = 0$ for the free surface. Thus

$$\frac{\partial^2 \phi}{\partial t^2} + g \frac{\partial \phi}{\partial y} = 0 \text{ at } y = 0$$

This is the required condition.

8.7. Velocity of Propagation of Liquid Waves

We shall now prove that the velocity of propagation c of wave $\eta = a \sin (mx - nt)$ at the surface of water of uniform depth is given by $c^2 = \frac{g}{m} \tan hmh$.

See figure 8.2 of article 8.6.

Since for a sine wave of elevation

$$\eta = a \sin (mx - nt) \quad \text{... (1)}$$

at the surface of a canal of uniform depth h with parallel vertical walls where η is the elevation of free surface from the undisturbed level. The waves are under gravity control. We take the equilibrium as x-axis and y-axis vertically upwards as shown in the figure. The velocity potential ϕ satisfies the following conditions:

$$\frac{\partial \eta}{\partial t} = v = \frac{-\partial \phi}{\partial y} \text{ at } y = 0 \quad \text{... (2)}$$

$$\vec{\nabla}^2 \phi = \frac{\partial^2 \phi}{\partial x^2} + \frac{\partial^2 \phi}{\partial y^2} = 0 \quad \text{... (3)}$$

$$\frac{\partial \phi}{\partial y} = 0 \text{ at } y = -h \qquad \ldots (4)$$

$$\frac{\partial^2 \phi}{\partial t^2} + g\frac{\partial \phi}{\partial y} = 0 \text{ at } y = 0 \qquad \ldots (5)$$

From (1) and (2)

$$\frac{-\partial \phi}{\partial y} = \frac{\partial \eta}{\partial t} = -an\cos(mx - nt) \text{ at } y = 0$$

This suggests that ϕ must be of the form

$$\phi = f(y)\cos(mx - nt) \qquad \ldots (6)$$

Putting this in (3), we get

$$\frac{d^2 f}{dy^2} - m^2 f = 0$$

So that $f(y) = Ae^{my} + Be^{-my}$

Now (6) becomes

$$\phi = \left(Ae^{my} + Be^{-my}\right)\cos(mx - nt) \qquad \ldots (7)$$

Using (4), (7) becomes

$$0 = m\left(Ae^{-mh} - Be^{mh}\right)\cos(mx - nt)$$

So that $Ae^{-mh} - Be^{mh} = 0$

$$\Rightarrow \quad Ae^{-mh} = Be^{mh} = \frac{c}{2}(\text{say})$$

Hence

$$\phi = \frac{c}{2}\left[e^{m(y+h)} + e^{-m(y+h)}\right]\cos(mx - nt)$$

$$\Rightarrow \phi = c\cos hm(y + h)\cos(mx - nt) \qquad \ldots (8)$$

From (1) and (8), we get

$$\frac{\partial \eta}{\partial t} = -an\cos(mx - nt)$$

and

$$\frac{\partial \phi}{\partial y} = mc\sin hm(y + h)\cos(mx - nt)$$

Putting this in (2), we have

$$-an\cos(mx - nt) = -mc\sin hmh\cos(mx - nt)$$

$$\Rightarrow \quad c = \frac{an}{m \sin hmh} = \frac{ac}{\sin hmx} \qquad \text{where } c = \frac{n}{m}$$

Now (8) becomes

$$\phi = \frac{ac}{\sin hmh} \cos hm(y+h) \cos(mx - nt) \qquad \text{... (9)}$$

This is the required expression for ϕ.

Now we shall deter nine an expression for velocity of propagation.

Using (9), (5) becomes

$$-n^2 \cos hm(y+h) \cos(mx - nt)$$
$$+ gm \sin hm(y+h) \cos(mx - nt) = 0 \text{ at } y = 0$$

$$\Rightarrow \quad -n^2 \cos hmh + gm \sin hmh = 0$$

$$\Rightarrow \quad c^2 = \frac{n^2}{m^2} = \frac{g}{m} \tan hmh \qquad \text{... (10)}$$

where c is the velocity of propagation.

Using (10) in (9), we get

$$\phi = \frac{ag}{n \cos hmh} \cos hm(y+h) \cos(mx - nt) \qquad \text{... (11)}$$

This an alternate form of (9).

Corollary 1. To show that the complex potential for the simple harmonic progresive wave of small elevation η is $w = \dfrac{ac \cos(mz - nt)}{\sin hmh}$.

From (9)

$$\phi = c \cos hm(y+h) \cos(mx - nt) \qquad \text{where } c = \frac{ac}{\sin hmh}$$

Since

$$\frac{\partial \phi}{\partial x} = \frac{\partial \psi}{\partial y}$$

$$\therefore \quad \frac{\partial \psi}{\partial y} = -m\, c \cos hm(y+h) \sin(mx - nt)$$

Integrating, we get

$$\psi = -c \sin hm(y+h) \sin(mx - nt)$$

Now $w = \phi + i\psi$

$$= c\left[\cos(mx - nt)\cos hm(y + h) - i\sin hm(y + h)\sin(mx - nt)\right]$$

$$= c\left[\cos\{(mx - nt) + im(y + h)\}\right]$$

$$= c\cos\left[m(x + iy) + imh - nt\right]$$

or $$w = \frac{ac}{\sin hmh}\cos m(z + ch - ct) \qquad \text{... (12)}$$

relative to the origin O

Remark. On the bottom coordinates of O^1 is $z = 0 - ih = -ih$. Hence if we take the origin at O^1 in the bottom, then z will be replaced by $z - ih$ so that (12) becomes

$$w = \frac{ac}{\sin hmh}\cos m(z - ct) \quad \text{relative to the origin } O^1.$$

If we take the equation of simple harmonic wave in the form $\eta = a\cos(mx - nt)$, then

$$w = \frac{-ac}{\sin hmh}\cos m(z - ct)$$

Corollary 2. For deep sea $h \to \infty$, so that $mh \to \infty$ consequently $\tan hmh \to 1$.

Now $c^2 = \frac{g}{m}\tan hmh$ becomes

$$c^2 = \frac{g}{m}$$

or $$(\text{velocity of propagation})^2 = \frac{g}{m} = \frac{gh}{2\pi} = c^2$$

Corollary 3. For tidal wave, λ is larger in comparison to h. Hence $\frac{h}{\lambda} \to 0$ i.e. $mh \to 0$.

Then $\tan hmh \to mh$

$$\therefore \quad c^2 = \frac{g}{m}\tan hmh = \frac{g}{m}mh = gh$$

$$\therefore \quad \text{Velocity of propagation} = c = \sqrt{gh}$$

Remark. If h, b, A are respectively depth, breadth and cross section area, then $A = bh$ or $h = \frac{A}{b}$.

Hence $c = \sqrt{gh} = \sqrt{\frac{gA}{b}}$

Corollary 4. Variation in velocity with wavelength

$$c^2 = \frac{g}{m} \tan hmh = gh \frac{\tan hmh}{mh}$$

$$\Rightarrow \quad \frac{c^2}{gw} = \frac{\tan hmh}{mh}$$

Putting $\frac{c^2}{gh} = y, mh = \frac{2\pi h}{\lambda} = x$

We get $y = \frac{\tan hx}{x}$

$$\frac{dy}{dx} = \frac{x \sec h^2 x - \tan hx}{x^2}$$

$$= \frac{x - \tan hx \cos h^2 x}{x^2 \cos h^2 x}$$

$$= \frac{2x - \sin \text{h}\, 2x}{2x^2 \cos h^2 x}$$

Since $\sin h2x > 2x$ when $x > 0$ so that $\frac{dy}{dx} < 0$ when $x > 0$.

It means that y decreases as x increases. We can also say that y increases i.e. velocity increases when x decreases, i.e. λ increases as $x = \frac{2\pi h}{\lambda}$.

Thus larger is the wavelength the faster it moves.

8.8. Paths of Particles in a Canal of Finite Depth

To show that each fluid particle describes an ellipse about its equilibrium position in case of wave propagation in a canal of uniform depth.

We have $\phi = c \cos hm\ (y + h) \cos\ (mx - nt)$ ... (1)

where $c = \dfrac{ac}{\sin hmh}$

Let (X, Y) be the coordinates of a fluid particle relative to the mean position (x, y)

$$\dot{X} = -\frac{\partial \phi}{\partial x} = mc \cos hm(y+h)\sin(mx-nt)$$

$$\dot{Y} = -\frac{\partial \phi}{\partial y} = -mc \sin hm(y+h)\sin(mx-nt)$$

$\Rightarrow\ dX = mc \cos hm(y+h)\sin(mx-nt)\,dt$

and $dY = -mc \sin hm(y+h)\cos(mx-nt)\,dt$

Integrating, we get

$$X = \frac{mc}{n}\cos hm(y+h)\cos(mx-nt)$$

and $Y = \dfrac{mc}{n}\sin hm(y+h)\sin(mx-nt)$

We take $A = \dfrac{mc}{n}\cos hm(y+h),\ B = \dfrac{mc}{n}\sin hm(y+h)$

Then

$$\frac{X^2}{A^2} + \frac{Y^2}{B^2} = \cos^2(mx-nt) + \sin^2(mx-nt) = 1$$

This shows that the particles describe ellipse about its mean position. If E is the eccentricity of the ellipse, then

$$B^2 = A^2\left(1-E^2\right) \Rightarrow AE = \sqrt{A^2 - B^2} = \frac{mc}{n}$$

$$\Rightarrow\ AE = \frac{m}{n}\,\frac{ac}{\sin hmh} = \frac{a}{\sin hmh} \text{ as } c = \frac{n}{m}$$

Distance between the focii $= 2AE = \dfrac{2a}{\sin hmh} =$ constant

Also length of minor axis $= B = O$ when $y = -h$ as $\sin h\ O = 0$.

Thus fluid particles describe ellipse about its mean position, the distance between the focii is $2a$ cosec hmh which is the same for all such ellipses. The major and minor both axes decreases as depth increases. At the bottom, the ellipse degenerates into a straight line, here the particles have to and fro motion.

8.9. Progressive Wave on Deep Water

To find velocity of propagation c and velocity potential in a canal of uniform breadth and depth h, h being very large in comparison to wave length λ.

First we derive equation (7) of 8.7 namely

$$\phi = \left(Ae^{my} + Be^{-my}\right)\cos(mx - nt) \qquad \dots (7)$$

Putting this in (4), we get

$$Ae^{-mh} - Be^{mh} = 0$$

$$\Rightarrow \quad B = Ae^{-2mh}$$

But $h \to \infty$

Hence $B = A \cdot O = O$

Now (7) becomes

$$\phi = Ae^{my}\cos(mx - nt) \qquad \dots (8)$$

From (1) and (8), we get

$$\frac{\partial \eta}{\partial t} = -\, an\cos(mx - nt)$$

and

$$\frac{\partial \phi}{\partial y} = mAe^{my}\cos(mx - nt)$$

Putting these values in (2), we get

$$-an\cos(mx - nt) = -\, mA\cos(mx - nt)$$

$$\Rightarrow \quad A = \frac{an}{m} = ac$$

Hence (8) becomes

$$\phi = ace^{my}\cos(mx - nt) \qquad \dots (9)$$

This is the required expression for ϕ.

Putting (9) in (5), we have

$$\left[-n^2\cos(mx - nt) + gm\cos(mx - nt)\right]ac = 0$$

$$\Rightarrow \quad n^2 = gm$$

Hence

$$c^2 = \frac{n^2}{m^2} = \frac{gm}{m^2} = \frac{-g}{m} \qquad \text{... (10)}$$

Using (10) in (9), we have

$$\phi = \frac{ag}{n} e^{my} \cos(mx - nt)$$

Corollary. To show that the potential for a progressive wave in a deep canal of uniform depth is

$$w = ace^{-i(mz - nt)}$$

Proof. We have $\phi = ace^{my} \cos (mx - nt)$

$$\therefore \quad \frac{\partial \phi}{\partial x} = \frac{\partial \psi}{\partial y} \text{ gives}$$

$$\frac{\partial \psi}{\partial y} = -\, acme^{my} \sin(mx - nt)$$

Integrating w.r.t. y, we get

$$\psi = -\, ace^{my} \sin(mx - nt)$$

Hence

$$\begin{aligned} w &= \phi + i\psi \\ &= ace^{my}\left[\cos(mx - nt) - i \sin(mx - nt)\right] \\ &= ace^{my} e^{-i(mx - nt)} \\ &= ace^{my - i\,(mx - nt)} \\ &= ace^{-im(x + iy) + int} \\ w &= ace^{-imz + int} \end{aligned}$$

or $\quad w = ace^{-i(mz - nt)}$

8.10. Path of Particles in a Deep Canal

To show that the path of particles relative to mean position in a deep canal is a circle.

Proof. The velocity potential is given by

$$\phi = ace^{my} \cos(mx - nt)$$

Let (X, Y) be the coordinates of a fluid particle relative to mean position (x, y).

Then

$$\dot{X} = \frac{-\partial\phi}{\partial x} = ace^{my} m \sin(mx - nt)$$

$$\dot{Y} = \frac{-\partial\phi}{\partial y} = -acme^{my}\cos(mx - nt)$$

$$\Rightarrow \quad dX = acme^{my}\sin(mx - nt)dt$$

and $dY = -acme^{my}\cos(mx - nt)\,dt$

Integrating, we get

$$X = \frac{mac}{n} e^{my}\cos(mx - nt)$$

$$= ae^{my}\cos(mx - nt)$$

and

$$Y = \frac{acm}{n} e^{my}\sin(mx - nt)$$

$$= ae^{my}\sin(mx - nt)$$

or $\quad X^2 + Y^2 = (ae^{my})^2$

This prove that the path of a fluid particle is a circle of radius ae^{my}.

8.11. Standing or Stationary Waves on the Surface of a Canal of Finite Depth

Now we consider the waves which remain stationary, the surface moves vertically only. The equation for a stationary wave is given by

$$\eta = a \sin mx \cos nt \quad \text{... (1)}$$

Being an irrotational motion, the velocity potential ϕ satisfies the following conditions:

$$\vec{\nabla}^2\phi = \frac{\partial^2\phi}{\partial x^2} + \frac{\partial^2\phi}{\partial y^2} = 0 \quad \text{... (2)}$$

$$\frac{\partial\eta}{\partial t} = V = \frac{-\partial\phi}{\partial y} \text{ at } y = 0 \quad \text{... (3)}$$

$$\frac{\partial\phi}{\partial y}=0 \text{ at } y=-h \qquad \text{... (4)}$$

$$\frac{\partial^2\phi}{\partial t^2}+g\,\frac{\partial\phi}{\partial y}=0 \text{ at } y=0 \qquad \text{... (5)}$$

From (1)

$$\frac{\partial\eta}{\partial t}=-\,an\sin mx\sin nt$$

Using (3), we get

$$\frac{\partial\phi}{\partial y}=an\sin mx\sin nt \qquad \text{at } y=0$$

This suggests that ϕ must be of the form

$$\phi=f(y)\sin mx\sin nt \qquad \text{... (6)}$$

Using this in (2) and simplifying, we get

$$-m^2f+\frac{d^2f}{dy^2}=0$$

$$\Rightarrow\quad f=Ae^{my}+Be^{-my}$$

Now (6) becomes

$$\phi=\left(Ae^{my}+Be^{-my}\right)\sin mx\sin nt$$

$$\frac{\partial\phi}{\partial y}=m\left(Ae^{my}-Be^{-my}\right)\sin mx\sin nt \qquad \text{... (7)}$$

Using (4) in (7), we have

$$0=m\left(Ae^{-mh}-Be^{mh}\right)\sin mx\sin nt$$

$$\Rightarrow\quad Ae^{-mh}-Be^{mh}=0$$

$$\Rightarrow\quad Ae^{-mh}\;-Be^{mh}\;=\frac{1}{2}c\text{ (say)}$$

$$\therefore\quad \phi=\frac{c}{2}\left(e^{my}e^{mh}\;+e^{-mx}e^{-mh}\right)$$

$$=\frac{c}{2}\cosh m(y+h)$$

$$\therefore\quad \phi=\frac{c}{2}\left(e^{my}e^{mh}\;+e^{-my}e^{-mh}\right)\sin mx\sin nt$$

$$\Rightarrow \quad \phi = \frac{c}{2}\cos hm(y+h)\sin mx \sin nt \qquad \text{... (8)}$$

$$\Rightarrow \quad \frac{\partial \phi}{\partial y} = mc \sin hmh \sin mx \sin nt \text{ at } y = 0$$

From (1)

$$\Rightarrow \quad \frac{\partial \eta}{\partial t} = -\, an \sin mx \sin nt$$

Using (3)

$$-an \sin mx \sin nt = -m(\sin hmh \sin mx \sin nt)$$

$$\Rightarrow \quad C = \frac{an}{m \sin hmh} = \frac{ac}{\sin hmh} \qquad \text{where } C = \frac{n}{m}$$

Hence (8) becomes

$$\phi = \frac{aC}{\sin hmh}\cos hm(y+h)\sin mx \sin nt \qquad \text{... (9)}$$

This is the required expression for ϕ.

Using (9) in (5), we get

$$\frac{ac}{\sin hmh}\sin mx \sin nt\left[-n^2 \cos hm(y+h) + gm \sin hm(y+h)\right] = 0 \text{ at } y = 0$$

$$\Rightarrow \quad -n^2 \cos hmh + gm \sin hmh = 0$$

$$\Rightarrow \quad n^2 = gm \tan hmh \qquad \text{... (10)}$$

$$\Rightarrow \quad c^2 = \frac{n^2}{m^2} = \frac{g}{m}\tan hmh \qquad \text{... (11)}$$

This is the required expression for velocity of propagation.

From (9) and (11), we have

$$\phi = \frac{ag}{n}\frac{\cos hm(y+h)}{\cos hmh}\sin mx \sin nt \qquad \text{... (12)}$$

(12) is an alternate form of (9)

Corollary. To determine complex potential.

From (9)

$$\phi = c \cos hm(y+h)\sin mx \sin nt \qquad \text{Where } C = \frac{ac}{\sin hmh}$$

Now

$$\frac{\partial \psi}{\partial y} = \frac{\partial \phi}{\partial x} = C \cos hm(y+h)\, m \cos mx \sin nt$$

Integrating, we have

$$\psi = C \sin hm(y+h) \cos mx \sin nt$$

Hence

$$w = \phi = i\psi$$

$$= C \sin nt \left[\sin mx \cos hm(y+h) + i \sin hm\,(y+h) \cos mx\right]$$

$$= C \sin nt\left[\sin mx \cos im\,(y+h) + \sin im\,(y+h) \cos mx\right]$$

$$= C \sin nt \sin\left[m(x+iy) + imh\right]$$

$$= C \sin nt \sin m\,(z+ih)$$

$$\Rightarrow \quad w = \frac{ac}{\sin hmh} \sin nt \sin m(z+ih) \qquad \ldots (13)$$

Remark. Coordinates of O^1 is $z = 0 - ih$

On shifting the origin to O^1, (13) becomes

$$w = \frac{ac}{\sin hmh} \sin nt \sin mz$$

Corollary 2. Path of fluid particles.

Let (X, Y) be the coordinates of a fluid particle relative to the mean position (x, y). Then, we have

$$\phi = C \cos hm(y+h) \sin mx \sin nt \quad \text{Where } C = \frac{ac}{\sin hmh}$$

$$\therefore \quad \frac{dX}{dt} = \frac{-\partial \phi}{\partial x} = -\, mc \cos hm(y+h) \cos mx \sin nt$$

and

$$\frac{dY}{dt} = \frac{-\partial \phi}{\partial t} = -\, mc \sin hm(y+h) \sin mx \sin nt$$

$$\Rightarrow \quad dX = -\, mc \cos hm(y+h) \cos mx \sin nt\, dt$$

and

$$dY = -\, mc \sin hm\,(y+h) \sin mx \sin nt\, dt$$

Integrating, we get

$$X = \frac{m}{n} C \cos hm(y+h) \cos mx \cos nt$$

and

$$Y = \frac{m}{n} C \sin hm(y+h) \sin mx \cos nt$$

Dividing these two, we have

$$\frac{Y}{X} = \tan hm(y+h) \tan mx$$

which is independent of the time t, thus the motion of each particle is rectilinear. The direction of the particle varies from vertical below the crests and troughs; to the horizontal below the nodes.

We can also say that particle describe straight line with simple harmonic motion of period $\frac{2\pi}{n}$.

8.12. Standing or Stationary Wave on the Surface of a Deep Canal

First we derive equation (7) of Article 8.11 namely

$$\phi = \left(Ae^{my} + Be^{-my}\right) \sin mx \sin nt \qquad \ldots (7)$$

Here if h be the depth of the canal then $h \to \infty$.

Subjecting (7) to (4), we get

$$Ae^{-mh} - Be^{mh} = 0$$

$$\Rightarrow \quad B = Ae^{-2mh}$$

If $h \to \infty$ then $B = 0$

Hence (7) becomes

$$\phi = Ae^{my} \sin mx \sin nt \qquad \ldots (8)$$

At $y = 0$, $\frac{\partial \phi}{\partial y} = mA \sin mx \sin nt$

From (1)

$$\frac{\partial \eta}{\partial t} = -na \sin mx \sin nt$$

From (3)

$$\frac{\partial \eta}{\partial t} = -\frac{\partial \phi}{\partial y}$$

$\Rightarrow \quad -na \sin mx \sin nt = mA \sin mx \sin nt$

$\Rightarrow \quad A = \dfrac{an}{m} = ac$

Hence from (8)

Hence $\phi = ace^{my} \sin mx \sin nt$ ... (9)

This is the required expression for ϕ.

Using (9) in (5), we get

$$\left(-n^2 + gm\right) ace^{my} \sin mx \sin nt = 0 \text{ at } y = 0$$

$\Rightarrow \quad n^2 = gm \Rightarrow c^2 = \dfrac{n^2}{m^2} = \dfrac{g}{m}$ ... (10)

This is the required expression for velocity of propagation.

From (9) and (10)

$$\phi = \frac{ag}{n} e^{my} \sin mx \sin nt \quad \text{... (11)}$$

This is an alternate form of (9)

Corollary 1. To determine complex potential.

We have $\phi = ace^{my} \sin mx \sin nt$

$$\frac{\partial \psi}{\partial y} = \frac{\partial \phi}{\partial x} = acme^{my} \cos mx \sin nt$$

Integrating, we get

$$\psi = ace^{my} \cos mx \sin nt$$

Now

$$\begin{aligned} w &= \phi + i\psi \\ &= ace^{my} \sin nt (\sin mx + i \cos mx) \\ &= iace^{my} \sin nt (\cos mx - i \sin mx) \\ &= iace^{my} e^{-imx} \sin nt \\ &= iace^{-im(x+iy)} \sin nt \end{aligned}$$

$\Rightarrow \quad w = iac \sin nte^{-imz}$ ... (12)

Corollary 2. Path of Fluid Particles.

Let (X, Y) be the coordinates of a fluid particle relative to the mean position (x, y). We have $\phi = ace^{my} \sin mx \sin nt$.

$$\frac{dX}{dt} = \frac{-\partial\phi}{\partial x} = -\,acme^{my}\cos mx \sin nt$$

and

$$\frac{dY}{dt} = \frac{-\partial\phi}{\partial y} = -\,acme^{my}\sin mx \sin nt$$

$$\Rightarrow \quad dX = -acme^{my}\cos mx \sin nt\, dt$$

and

$$dY = -acme^{my}\sin mx \sin nt\, dt$$

Integrating, we get

$$X = ae^{my}\cos mx \cos nt$$

and $Y = ae^{my}\sin mx \cos nt$

Dividing these two, we have $\dfrac{Y}{X} = \tan mx$

which is independent of the time t. Thus the trajectory of the particle is a straight line with slope tan mx. The particle oscillates on the vertical direction at the antinodes and in the horizontal direction at the nodes. The amplitude of the oscillation ae^{my} decreases as depth increases.

8.13. Kinetic Energy and Potential Energy of a Wave per Unit Thickness

1. Potential Energy. The potential energy per unit thickness of a progressive wave means the P.E. due to the elevated liquid in a wavelength. We measure potential energy from the undisturbed level.

Let η be the elevation of the liquid. Then mass of the liquid above a base dx in xy-plane is $\rho\eta\, dx$ and its centre of mass is at a height $\dfrac{\eta}{2}$ so that the total potential energy is

$$V = Mgh = \int_0^{\lambda}(\rho\eta dx)g\left(\frac{1}{2}\eta\right)$$

$$= \frac{1}{2}\rho g \int_0^{\lambda}\eta^2 dx$$

2. Kinetic Energy. The kinetic energy possessed by the liquid (per unit thickness) stretching between two vertical planes placed at a distance of one wavelength and perpendicular to the direction of flow is called the kinetic energy of a progressive wave.

We know that kinetic energy

$$= T = -\frac{1}{2}\rho \int \phi \frac{\partial \phi}{\partial \eta} ds$$

$$= \frac{1}{2}\rho \int \phi \left(-\frac{\partial \phi}{\partial \eta}\right) ds$$

Integrated over a wave profile, $\partial\eta$ being the normal drawn into the liquid. Upto first order of small quantities we may write

$$T = \frac{1}{2}\rho \int_0^{\lambda} \left[\phi \frac{\partial \phi}{\partial y}\right]_{y=0} dx \quad \text{as } ds = 1dx$$

8.14. Energy of a Progressive Wave

To show that the total energy of a progressive wave is half K.E. and half P.E.

Proof. Equation to a simple harmonic progressive wave is $\eta = a \sin(mx - nt)$ and the corresponding velocity potential is given by

$$\phi = \frac{ac}{\sin hmh} \cos hm(y + h) \cos(mx - nt)$$

$$\text{Where } c^2 = \frac{g}{m} \tan hmh$$

Hence Potential Energy

$$= \frac{1}{2}\rho g \int_0^{\lambda} \eta^2 dx$$

$$= \frac{1}{2}\rho g \int_0^{\lambda} a^2 \sin^2(mx - nt)\, dx$$

$$= \frac{a^2}{4}\rho g \int_0^{\lambda} [1 - \cos 2(mx - nt)]\, dx$$

$$= \frac{a^2}{4}\rho g\left[x - \frac{\sin 2(mx - nt)}{2m}\right]_0^\lambda$$

$$= \frac{a^2}{4}\rho g\left[\lambda - \frac{1}{2m}\{\sin 2(m\lambda - nt) + \sin 2nt\}\right]$$

$$= \frac{a^2}{4}\rho g\left[\lambda - \frac{1}{2m}\{\sin(4\pi - 2nt) + \sin 2nt\}\right]$$

Since $m = \frac{2\pi}{\lambda}$

$$= \frac{a^2}{4}\rho g\left[\lambda - \frac{1}{2m}(-\sin 2nt + \sin 2nt)\right]$$

Hence

$$P.E. = \frac{1}{4}\rho g a^2 \lambda$$

At $y = 0$, $\phi\frac{\partial\phi}{\partial y} = (ac)^2\{\cos(mx - nt)\}^2 m \cot hmh$

But $c^2 = \frac{g}{m}\tan hmh$

so that $mc^2 \cot hmh = g$

At $y = 0$, $\phi\frac{\partial\phi}{\partial y} = a^2 g \cos^2(mx - nt)$

Hence Kinetic Energy

$$= \frac{1}{2}\rho\int_0^\lambda\left(\phi\frac{\partial\phi}{\partial y}\right)_{y=0} dx$$

$$= \frac{1}{2}\rho a^2 g\int_0^\lambda \cos^2(mx - nt)\, dx$$

$$= \frac{1}{4}\rho a^2 g\int_0^\lambda [1 + \cos 2(mx - nt)]\, dx$$

$$= \frac{1}{4}\rho a^2 g\lambda$$ as in case of P.E.

Finally, total energy

$= P.E. + K.E.$

$$= \frac{1}{4}\rho g a^2 \lambda + \frac{1}{4}\rho g a^2 \lambda$$

$$= \frac{1}{2}\rho a^2 g \lambda$$

This proves that energy is half the K.E. and half the P.E.

8.15. Energy of Stationary Waves

Equation to a stationary wave is

$$\eta = a \sin mx \cos nt \qquad \dots (1)$$

In this case the velocity potential is given by

$$\phi = \frac{ac}{\sin hmh} \cos hm(y+h) \sin mx \sin nt$$

where $c^2 = \frac{g}{m} \tan hmh$

Hence

$$\phi \frac{\partial \phi}{\partial y} = \left(\frac{ac}{\sin hmh}\right)^2 (\sin mx \sin nt)^2 m \cos hm(y+h) \times \sin hm\,(y+h)$$

At $y = 0$,

$$\phi \frac{\partial \phi}{\partial y} = ma^2 c^2 (\sin mx \sin nt)^2 \cot hmh$$

But $c^2 = \frac{g}{m} \tan hmh$

$\Rightarrow \quad mc^2 \cot hmh = g$

Hence at $y = 0$,

$$\phi \frac{\partial \phi}{\partial y} = a^2 g (\sin mx \sin nt)^2$$

Now Kinetic Energy

$$= T = \frac{1}{2}\rho \int_0^\lambda \left(\phi \frac{\partial \phi}{\partial y}\right)_{y=0} dx$$

$$= \frac{1}{2}\rho a^2 g \sin^2 nt \int_0^\lambda \sin^2 mx dx$$

$$= \frac{1}{4}\rho a^2 g \sin^2 nt \int_0^{\lambda} (1 - \cos 2mx)\, dx$$

$$= \frac{1}{4}\rho a^2 g \sin^2 nt \left[\frac{x - \sin 2mx}{2m}\right]_0^{\lambda}$$

$$= \frac{1}{4}\rho a^2 g \sin^2 nt \left[\lambda - \frac{1}{2m}\sin 2m\lambda\right]$$

$$= \frac{1}{4}\rho a^2 g \sin^2 nt \left[\lambda - \frac{1}{2m}\sin 4\pi\right]$$

Since $m = \dfrac{2\pi}{\lambda}$

$$= \frac{1}{4}\rho a^2 g (\sin^2 nt)\lambda$$

Also potential energy

$$= V = \frac{1}{2}\rho g \int_0^{\lambda} \eta^2 dx$$

$$= \frac{1}{2}\rho g a^2 \int_0^{\lambda} \sin^2 mx \cos^2 nt dx$$

$$= \frac{1}{4}\rho g a^2 \cos^2 nt \int_0^{\lambda} (1 - \cos 2\, mx)\, dx$$

$$= \frac{1}{4}\rho g a^2 \cos^2 nt$$

Hence total energy

$$= T + V$$

$$= \frac{1}{4}\rho g a^2 \lambda (\sin^2 nt + \cos^2 nt)$$

$$= \frac{1}{4}\rho g a^2 \lambda$$

It means that kinetic energy and potential energy change with time.

8.16. Steady Motion of Progressive Waves

The complex potential of a simple harmonic progressive wave is given by

$$W_0 = \frac{ac}{\sin hmh} \cos m\,(z + ih - ct)$$

relative to the origin O lying in the equilibrium level which is taken as x-axis, c being velocity of propagation parallel to x-axis. We take

$$A = \frac{ac}{\sin hmh} \qquad \text{(see figure of 8.3)}$$

If the frame of reference moves, then the complex potential w_0 will change to w_1 which will be obtained by replacing z in w_0 by $z + ct$, so that

$$w_1 = A \cos m(z + ct + ih - ct) = A \cos m\,(z + ih)$$

Let us superimpose on the whole system (i.e. liquid and wave both) a velocity equal and opposite to the velocity of propagation. Then the progressive wave is

$$\frac{-dw}{dz} = -c$$

or $\quad w = cz$ on integration

reduced to rest and the liquid begins to move with velocity c parallel to the negative direction of x-axis. This contributes a term cz in w_1. Hence the final form of complex potential w is given by

$$w = cz + A \cos m\,(z + ih) \qquad \text{... (1)}$$

$$\Rightarrow \quad \phi + i\psi = c(x + iy) + A \cos m\left[x + i(y + h)\right]$$

$$= c(x + iy) + A\left[\cos mx \cos mi\,(y + h) - \sin mx \sin mi(y + h)\right]$$

$$\therefore \quad \phi + i\psi = c(x + iy) + A\left[\cos mx \cos hm(y + h) - \sin mx\, i \sin hm(y + h)\right]$$

Hence

$$\phi = cx + A \cos mx \cos hm\,(y + h) \qquad \text{... (2)}$$

$$\text{and } \psi = cy - A \sin mx \sin hm(y + h) \qquad \text{... (3)}$$

At the bottom, $y = -h$, so that (3) gives

$$\psi = -ch - A \sin mx \sin h(-h + h) = -ch = \text{const.}$$

This proves that bottom is a streamline.

For the free surface $y = \eta = a \sin mx$ ... (4)

Putting (4) in (3), we get

$$\psi = c\eta - A\frac{\eta}{a} \sin hm(\eta + h)$$

Also $\sin hm\eta = m\eta$, $\cos hm\eta = 1$ and η^2 is negligible, we get

$$\psi - c\eta - \frac{ac}{\sin hmh}\frac{\eta}{a}(m\eta \cos hmh + \sin hmh)$$

$$\Rightarrow \quad \psi = c\eta - \frac{c}{\sin hmh}(0 + \eta \sin hmh)$$

$$= 0 = \text{const.}$$

It means that free surface is a streamline.

Pressure Distribution. Since the motion is steady and so by Bernoulli's equation

$$\frac{p}{\rho} + \frac{1}{2}q^2 + gy = \text{constant} \qquad \text{... (5)}$$

From (1) $\bar{w} = c\bar{z} + A \cos m(\bar{z} - ih)$

Also

$$q^2 = \frac{dw}{dz}\frac{d\bar{w}}{d\bar{z}}$$

$$= [c - mA \sin(z + ih)][c - mA \sin m(\bar{z} - ih)]$$

Neglecting a^2, we get

$$q^2 = c^2 - mAc[\sin m(z + ih) + \sin m(\bar{z} - ih)]$$

$$= c^2\left[1 - \frac{2ma}{\sin hmh} \cosh m(y + h) \sin mx\right]$$

At the free surface $y = a \sin mx = \eta$, $\psi = 0$

So that $cy - \dfrac{ac}{\sin hmh} \sin mx \sin hm(y + h) = 0$

$\Rightarrow \quad \sin hmh = \sin hm(y + h)$

So that $\cos hmh = \cos hm(y + h)$

$\therefore \quad q^2 = c^2[1 - 2my \cot hmh]$

Hence (5) becomes

$$\frac{p}{\rho}+\frac{c^2}{2}[1-2\,my\cot hmh]+gy=\text{constant}$$

$$\Rightarrow\quad \frac{p}{\rho}+y\left(g-mc^2\cot hmh\right)=\text{const}-\frac{c^2}{2}=\text{constant}$$

On free surface p = constant so that the last gives

$$g-mc^2\cot hmh=0$$

$$\Rightarrow\quad c^2=\frac{g}{m}\tan hmh \qquad \text{... (6)}$$

Which is true.

Hence pressure is constant along each streamline.

8.17. Steady Motion of Progressive Waves on Deep Water

The complex potential for steady motion is

$$w=cz+\frac{ac}{\sin hmh}\cos m\,(z+ih) \qquad \text{... (1)}$$

(Prove this as in Article 8.16)

$$=cz+\frac{ac}{\sin hmh}\left[\cos\left\{mx+i\left(my+hm\right)\right\}\right]$$

$$=cz+\frac{ac}{\sin hmh}\left[\cos mx\cos h\,(my+hm)\right.$$

$$\left.-\sin mx\; i\sin h\left(my+hm\right)\right]$$

$$\Rightarrow\quad w=cz+ac\left[\frac{\cos mx\cos h\left(my+h\right)}{\sin hmh}\right.$$

$$\left.-\,i\sin mx\,\frac{\sin h\,(my+hm)}{\sin hmh}\right] \qquad \text{... (2)}$$

Now

$$\frac{\cos hm\,(y+h)}{\sin hmh}=\frac{e^{m(y+h)}+e^{-m(y+h)}}{e^{mh}-e^{-mh}}$$

$$=e^{my}\,\frac{\left[1+e^{-m(y+h)}\right]}{1-e^{-2mh}}$$

$$=e^{my}\quad \text{as}\rightarrow\infty$$

Similarly

$$\frac{\sin hm(y+h)}{\sin hmh} = e^{my} \qquad \text{as } h \to \infty \qquad \text{... (3)}$$

Hence (2) becomes

$$w = cz + ace^{my}(\cos mx - i \sin mx)$$

$$= cz + ace^{my}e^{-imx}$$

$$\Rightarrow \quad w = cz + ace^{-imz} \qquad \text{... (4)}$$

$$\Rightarrow \quad \phi + i\psi = c(x+iy) + ace^{my}(\cos mx - i \sin mx)$$

Hence

$$\left.\begin{aligned} \phi &= cx + ace^{my}\cos mx \\ \text{and} \quad \psi &= cy - ace^{my}\sin mx \end{aligned}\right\} \qquad \text{... (5)}$$

Pressure Distribution. Since the motion is steady and so by pressure equation

$$\frac{p}{\rho} + \frac{1}{2}q^2 + gy = \text{constant} \qquad \text{... (6)}$$

From (4)

$$\frac{dw}{dz} = c - imace^{-imz}$$

$$= c(1 - 1\,mae^{-imz})$$

and $\bar{w} = c\bar{z} + ace^{im\bar{z}}$

so that

$$\frac{d\bar{w}}{dz} = c + ac\,ime^{im\bar{z}}$$

$$= c\left[1 + imae^{im\bar{z}}\right]$$

$$q^2 = \frac{dw}{dz}\frac{d\bar{w}}{dz}$$

$$= c^2\left[1 + ima\left(e^{im\bar{z}} - e^{-imz}\right)\right] \qquad \text{on neglecting } a^2$$

$$= c^2\left[1 + ima\left\{e^{im(x-iy)} - e^{-im(x+iy)}\right\}\right]$$

$$= c^2\left[1 + imae^{my}\left(e^{imx} - e^{-imx}\right)\right]$$

$$= c^2\left[1 - 2mae^{my}\sin mx\right]$$

$$q^2 = c^2 + 2mc\,(\psi - cy) \qquad \text{From (5)}$$

Putting the values of q^2 in (6), we get

$$\frac{p}{\rho} + \frac{1}{2}\left[c^2 + 2mc(\psi - cy)\right] + gy = \text{const}$$

$$\Rightarrow \quad \frac{p}{\rho} + y\left(g - mc^2\right) + mc\psi = \text{const} - \frac{1}{2}c^2 = \text{const}$$

On the free surface p = constant and $\psi = 0$

Hence the last equation becomes

$$g - mc^2 = 0$$

$$\Rightarrow \quad c^2 = \frac{g}{m}$$

which is true. Hence pressure is constant along each streamline i.e. free surface and the bottom.

8.18. Waves at the Common Surface of Two Liquids

Two liquids are in motion above the other, find velocity of propagation of oscillatory waves at the common surface.

Let a liquid of density ρ^1 and depth h^1 move with velocity u^1 over another liquid of density ρ and depth h moving in same direction with velocity u. Let the liquids be bounded above and below by two fixed horizontal planes A^1B^1 and AB. If the system is slightly displaced, the waves can be propagated at the common surface of the two liquids. In order to obtain the velocity c of the propagation of the waves at the common surface, we choose the x-axis as shown in the figure 8.3.

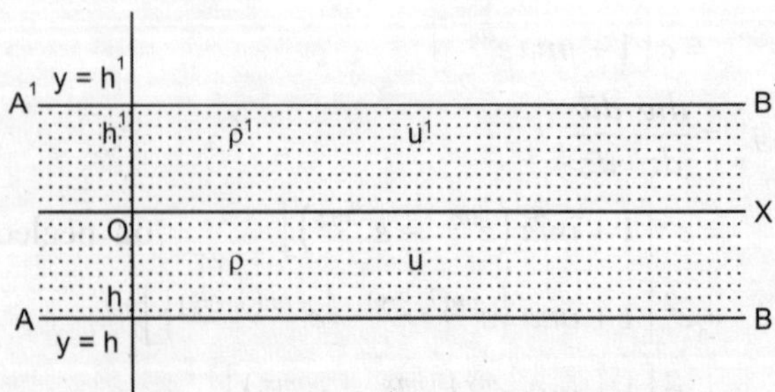

Fig. 8.3

To make the motion steady, we superimpose velocity $-c$ on the whole mass. Then the wave profile $\eta = a \sin (mx - \eta t)$ is reduced to rest so that $\eta = y = a \sin mx$ and the liquids begin to move with velocities $u^1 - c$ and $u - c$ respectively. The lower liquid moves with velocity $u - c$ in positive direction of x-axis so that it will move with velocity $-(u - c)$ in the negative direction of x-axis so that its complex potential is

$$w = -(u - c)z - \frac{a(u - c)}{\sin hmh} \cos m(z + ih) \qquad \text{... (1)}$$

Replacing h by $-h^1$ and u by u^1, we get

$$w^1 = -(u^1 - c)z - \frac{a(u^1 - c)}{\sin h(-mh^1)} \cos m(z - ih^1)$$

for upper liquid

$$\Rightarrow \quad w^1 = -(u^1 - c)z + \frac{a(u^1 - c)}{\sin hmh^1} \cos m(z - ih^1) \qquad \text{... (2)}$$

for upper liquid

It can be easily verified that the wave profile $\eta = a \sin mx$ gives $\psi = 0$, showing thereby the common axis is a streamline.

To determine the velocity of lower liquid

$$q^2 = \frac{dw}{dz}\frac{d\bar{w}}{d\bar{z}}$$

$$= (u - c)^2 \left[-1 + \frac{am}{\sin hmh} \sin m(z + ih)\right]$$

$$\times \left[-1 + \frac{am \sin m(\bar{z} - ih)}{\sin hmh}\right]$$

$$= (u - c)^2 \left[1 - \frac{am}{\sin hmh} \{\sin m(z + ih)\right.$$

$$\left. + \sin m(\bar{z} - ih)\}\right]$$

neglecting a^2

$$= (u - c)^2 \left[1 - \frac{2am}{\sin hmh} \sin mx \cos hm(y + h)\right]$$

If q_0 be the fluid velocity of the lower liquid at the common surface, then

$$q_0 = q \text{ at } y = 0, \text{ so that}$$

$$q_0^2 = (u-c)^2\left[1 - \frac{2am}{\sin hmh}\sin mx \cos hmh\right]$$

$$\Rightarrow \quad q_0^2 = (u-c)^2\left[1 - 2\eta\, m \cot hmh\right] \qquad \dots (3)$$

Similarly velocity q_0^1 of the upper liquid at the common surface is given by

$$q_0^{1^2} = \left(u^1 - c\right)^2\left[1 + 2\eta \cot hmh^1\right] \qquad \dots (4)$$

Also pressure equation for steady motion gives

$$\frac{p}{\rho} + \frac{1}{2}q_0^2 + g\eta = \text{constant for lower liquid}$$

and

$$\frac{p^1}{\rho^1} + \frac{1}{2}q_0^{1^2} + g\eta = \text{constant for upper liquid}$$

But at the common surface $p = p^1$ so that the last two equations gives

$$\left(\rho q_0^2 - \rho^1 q_0^{1^2}\right) + 2g\eta\left(\rho - \rho^1\right) = \text{constant}$$

Using (3) and (4), we get

$$\rho(u-c)^2\left[1 - 2m\eta \cot hmh\right] - \rho^1\left(u^1 - c\right)^2 \left[1 + 2m\eta \cot hmh^1\right] + 2g\eta\left(\rho - \rho^1\right) = \text{constant}$$

$$\Rightarrow \quad 2\eta\left[g\left(\rho - \rho^1\right) - \rho(u-c)^2 m \cot hmh - \rho^1\left(u^1 - c\right)^2 m \cot hmh\right] + \left[\rho(u-c)^2 - \rho^1\left(u^1 - c\right)^2\right] = \text{constant}.$$

This is true for all η and so the coefficient of η must vanish.

$$g\left(\rho - \rho^1\right) = m\left[\rho(u-c)^2 \cot hmh + \rho^1\left(u^1 - c\right)^2 \cot hmh\right] \qquad \dots (5)$$

This is the required expression for determining the velocity of propagation.

Corollary 1. An infinite liquid of density σ lies above an infinite liquid of density ρ, the two liquids being separated by a horizontal plane interface. Show that the velocity v of propagation of waves of length λ along the interface is given by

$$v^2 = \frac{g\lambda}{2\pi}\frac{\rho - \sigma}{\rho + \sigma}$$

Putting $c = v$, $u = 0 = u^1$, $\rho = \rho^1$, $\rho^1 = \sigma$ in (5), we get

$$\sigma(\rho - \sigma) = m\left[\rho \cot hmh + \sigma \cot hmh^1\right]v^2$$

But $m = \dfrac{2\pi}{\lambda}$

Hence

$$v^2 = \frac{g\lambda}{2\pi}\left[\frac{\rho - \sigma}{\rho \cot hmh + \sigma \cot hmh^1}\right] \qquad \dots (6)$$

Again $h \to \infty$, $h^1 \to \infty$ so that $\cot hmh \to 1$ and $mh^1 \to 1$

Thus $v^2 = \dfrac{g\lambda}{2\pi}\left(\dfrac{\rho - \sigma}{\rho + \sigma}\right)$

Corollary 2. If the liquid are at rest then $u^1 = 0 = u$ so that (5) gives.

$$g\left(\rho - \rho^1\right) = mc^2\left[\rho \cot hmh + \rho^1 \cot hmh^1\right]$$

$$\Rightarrow \quad c^2 = \frac{g}{m}\frac{\rho - \rho^1}{\left(\rho \cot hmh + \rho^1 \cot hmh^1\right)}$$

Corollary 3. If the velocities u, u^1 make angles α, α^1 with x-axis, then their components along x-axis are u, cos α, u^1 cos α^1 respectively. The components u sin α, u sin α along y-axis have no effect upon the values of c. Then replacing u, u^1 in (5) by u cos α, u^1 cos α^1, we get

$$g\left(\rho - \rho^1\right) = m\Big[\rho^1\left(u^1 \cos\alpha - c\right)^2 \cot hmh^1 + \rho\left(u \cos\alpha - c\right)^2 \cot hmh\Big]$$

Corollary 4. Let the upper fluid be air of specific gravity $s = \dfrac{\rho^1}{\rho}$ and both the fluids are at rest, i.e. $u^1 = u = 0$. Dividing (5) by ρ, we get

$$g(1-s) = mc^2\left[\cot hmh + s\cot hmh^1\right] \qquad \text{... (7)}$$

If the air is of infinite depth, then $h^1 \to \infty$, so that cot $hmh^1 \to 1$ and $(\cot hmh + s)^{-1}$.

$$= \left[\cot hmh\,(1 + s\tan hmh)\right]^{-1}$$

$$= \tan hmh\,(1 - s\tan hmh) \qquad \text{(Expanding by Binomial Theorem)}$$

Hence (7) becomes

$$c^2 = \frac{g(1-s)}{m}(\cot hmh + s)^{-1}$$

$$= \frac{g(1-s)}{m}\tan hmh(1 - s\tan hmh)$$

$$= \frac{g}{m}[1 - s - s\tan hmh]\tan hmh \qquad \text{neglecting } s^2$$

This shows that the presence of the atmosphere tries to decrease the wave velocity c.

8.19. Waves at the Common Surface of Two Liquids with Upper Surface Free

To determine the velocity of propagation at the common surface of two liquids, bounded below and free at the upper surface, which are at rest except for wave motion.

Let a liquid of density σ^1 and depth h^1 lie over another liquid fo density ρ and depth h and let the liquids be at rest except for the wave motion. If the system is slightly disturbed, then the waves at the common surface and free surface can be propagated. Let us suppose a common velocity c of wave profile at the free surface of the upper liquid and at the common surface. We choose the axis as shown in the figure 8.4.

To make the motion steady, we superimpose a velocity $-c$ on the whole system so that the wave profile is reduced to

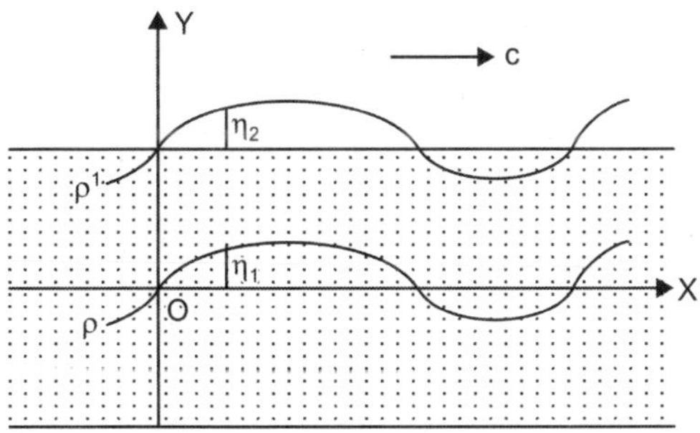

Fig. 8.4

rest and the liquid begins to move with velocity c in negative direction of x-axis.

The complex potential for lower and upper liquids are

$$w = cz + \frac{ac}{\sin hmh} \cos m\left(z + ih\right) \qquad \dots (1)$$

and

$$w^1 = cz + \left[\frac{ac}{\sin hm\left(-h^1\right)}\right] \cos m\left(z - ih^1\right) + \frac{bc \cos mz}{\sin hmh^1}$$

$$\Rightarrow \quad w^1 = cz - \frac{ac}{\sin hmh^1} \cos m\left(z - ih^1\right) + \frac{bc \cos mz}{\sin hmh^1} \qquad \dots (2)$$

In equation (2) the term $\frac{bc \cos mz}{\sin hmh^1}$ represents complex potential of a simple sine wave $\eta_2 = b \sin mx$ at time $t = 0$.

Let q be the fluid velocity in the liquid.

$$q^2 = \frac{dw}{dz}\frac{d\overline{w}}{d\overline{z}}\left[c - \frac{acm}{\sin hmh} \sin m\left(z + ih\right)\right]$$

$$\times \left[c - \frac{acm}{\sin hmh} \sin m\left(\overline{z} - ih\right)\right]$$

$$\Rightarrow \quad q^2 = c^2\left[1 - \frac{am}{\sin hmh}\left\{\sin m\left(z + ih\right) + \sin m\left(\overline{z} - ih\right)\right\}\right]$$

neglecting a^2

$$\Rightarrow \quad q^2 = c^2\left[1 - \frac{2am}{\sin hmh} \sin mx \cos hm\left(y + h\right)\right] \qquad \dots (3)$$

Also

$$q^2 = \frac{dw^1}{dz}\frac{d\bar{w}^1}{d\bar{z}}$$

$$= \left[c + \frac{acm}{\sin hmh^1}\sin m\left(z - ih^1\right) - \frac{bcm}{\sin hmh^1}\sin m\bar{z}\right]$$

$$\times \left[c + \frac{acm}{\sin hmh^1}\sin\left(\bar{z} + ih^1\right) - \frac{bcm}{\sin hmh^1}\sin m\bar{z}\right]$$

Since a, b are small quantities and hence neglecting a^2, ab, b^2 we obtain

$$q^{1^2} = c^2\left[1 + \frac{am}{\sin hmh^1}\left\{\sin\left(z - ih^1\right) + \sin m\left(\bar{z} + ih^1\right)\right\} - \frac{bm}{\sin hmh^1}\left\{\sin m\bar{z} + \sin mz\right\}\right]$$

$$\Rightarrow \quad q^{1^2} = c^2\left[1 + \frac{2am}{\sin hmh^1}\sin mx \cos hm\left(y - h^1\right) - \frac{2bm}{\sin hmh^1}\sin mx \cos hmy\right] \quad \ldots (4)$$

In order to get the values of velocities $q_0{}^1$ and q_0 due to upper and lower liquids at the common surface, we put $\eta_1 = a \sin mx$, $y = 0$ in (3) and (4)

$$q_0^2 = c^2\left(1 - 2m\eta_1 \cot hmh\right)$$

$$\text{and } q_0^1 = c^2\left(1 + 2m\eta_1 \cot hmh^1 - \frac{2mb\eta_1}{a \sin hmh^1}\right)$$

Pressure equation for steady motion gives

$$\frac{p}{\rho} + \frac{1}{2}q_0^2 + g\eta_1 = \text{constant}$$

$$\text{and } \frac{p^1}{\rho^1} + \frac{1}{2}q_0^{1^2} + g\eta_1 = \text{constant}$$

at the common surface. But $p = p^1$ at the common surface.

Hence subtracting these two, we get

$$\frac{1}{2}\left(\rho^1 q_0^{1^2} - \rho q_0^2\right) + g\eta_1\left(\rho^1 - \rho\right) = \text{const.}$$

$$\Rightarrow \quad g\eta_1\left(\rho^1 - \rho\right) + \rho^1 c^2\left(\frac{1}{2} + m\eta_1 \cot hmh^1 - \frac{mb}{a}\frac{\eta_1}{\sin hmh^1}\right)$$

$$- \rho c^2\left(\frac{1}{2} - m\eta_1 \cot hmh\right) = \text{constant.}$$

This equation holds for every value of η_1 and so coefficient of η_1 must vanish.

Hence

$$g\left(\rho^1 - \rho\right) + mc^2\left\{\rho^1 \cot hmh^1 + \rho \cot hmh - \right.$$

$$\left. - \rho^1 \frac{b}{a} \operatorname{cosec} hmh^1\right\} \quad \ldots (5)$$

Since p^1 is constant at the free surface.

$$y = h^1 + b \sin mx = h^1 + \eta_2$$

Then the pressure equation

$$\frac{p^1}{\rho^1} + \frac{1}{2} q^{1^2} + gy = \text{constant}$$

becomes $\frac{1}{2} q^{1^2} \rho^1 + g\rho^1 y = \text{constant}$

$$\Rightarrow \quad g\rho^1\left(h^1 + b \sin mx\right) + \rho^1 c^2\left[\frac{1}{2} + \frac{am \sin mx}{\sin hmh^1}\right.$$

$$\left. - bm \sin mx \cot hmh^1\right] = \text{constant}$$

by putting $y = h^1$ in (4)

As before, coefficient of sin mx must vanish so that

$$bg + c^2 = \left(\frac{am}{\sin hmh^1} - bm \cot hmh^1\right) = 0$$

$$\Rightarrow \quad \frac{b}{a} = \frac{mc^2}{\sin hmh^1\left(c^2 m \cot hmh^1 - g\right)}$$

$$= \frac{c^2 m}{c^2 m \cos mh^1 - g \sin hmh^1}$$

Putting the value of $\frac{b}{a}$ from this equation into the equation (5), we get

$$g(\rho^1 - \rho) + mc^2(\rho^1 \cot hmh^1 + \rho \cot hmh) - \frac{m^2c^4\rho^1(\cot h^2mh^1 - 1)}{c^2m \cot hmh^1 - g} = 0$$

Simplifying, we get

$$c^4m^2(\rho \cot hmh \cot hmh^1 + \rho^1) - c^2mg\rho (\cot hmh + \cot hmh^1) + g^2(\rho - \rho^1) = 0 \quad \text{... (6)}$$

Corollary. If the liquid is deep, then $h \to \infty$ so that cot hmh = 1.

Hence (6) becomes

$$c^4m^2(\rho \cot hmh^1 + \rho^1) - c^2mg\rho(1 + \cot hmh^1) + g^2(\rho - \rho^1) = 0$$

$$\Rightarrow \quad (mc^2 - g)\left[mc^2(\rho \cot hmh^1 + \rho^1) - g(\rho - \rho^1)\right] = 0$$

$$\Rightarrow \quad c^2 = \frac{g}{m} \text{ and } c^2 = \frac{g(\rho - \rho^1)}{m(\rho \cot hmh^1 + \rho^1)}$$

8.20. Motion of a Surface Particle

We assume that the long waves move in x-direction only. Let u be the fluid velocity and η the surface elevation, then

$$\frac{\partial \phi}{\partial t} = g\eta \text{ or } \eta = \frac{1}{g}\frac{\partial}{\partial t}\phi(x - ct) = -\frac{c}{g}\phi^1(x - ct) \quad \text{... (1)}$$

$$u = \frac{-\partial \phi}{\partial x} = -\frac{\partial}{\partial x}\{\phi(x - ct)\} = -\phi^1(x - ct) \quad \text{... (2)}$$

From (1) and (2)

$$\eta = \frac{cu}{g}$$

In case of long waves, $c^2 = gh$

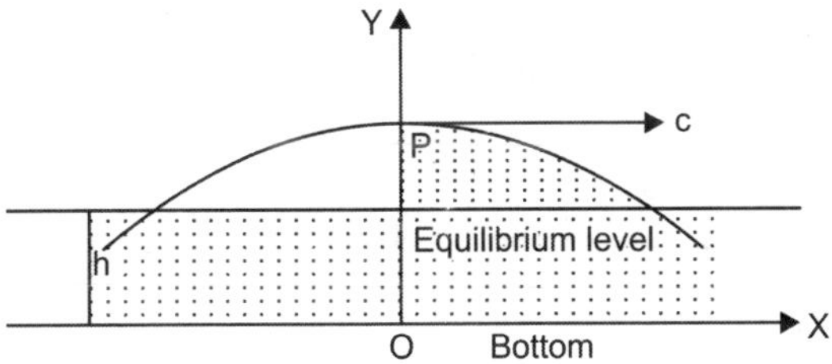

Fig. 8.5

Hence $\eta = \dfrac{cuh}{gh} = \dfrac{cuh}{c^2} = \dfrac{uh}{c}$

$\Rightarrow \quad u = \dfrac{c\eta}{h}$

$$\Rightarrow \quad \int u\,dt = \frac{1}{h}\int c\eta\,dt \qquad \text{... (3)}$$

The integral in R.H.S. of (3) represents shaded area in the figure. We want to determine the motion of a particle initially at P in the undisturbed surface of water in a canal of depth h. We see that

$$\text{displacement of particle } = \frac{1}{h}\int c\eta\,dt$$

$$\Rightarrow \text{ displacement of particle } = \frac{\text{Area of profile through } p}{\text{Depth of undisturbed water}}$$

8.21. Energy of a Long Wave

For a wave in a canal of rectangular section the potential energy V is due to the elevation or depression of the liquid above the mean level and for a unit breadth of wave, we have

$$V = \frac{1}{2}\rho g\int_0^{\lambda}\eta^2 dx \qquad \text{... (1)}$$

Where η is the elevation at x and integration is performed over one wavelength. Again for the same range of integration, the kinetic energy T is given by

$$T = \frac{1}{2}\int_0^{\lambda} u^2 dm$$

$$=\frac{1}{2}\int_0^\lambda u^2\rho h dx$$

$$=\frac{1}{2}\int_0^\lambda\left(\frac{g\eta}{c}\right)^2\rho h dx\left(\because \eta=\frac{cu}{g}\right)$$

But for a long wave $c^2 = gh$

$$T=\frac{1}{2}\int_0^\lambda\frac{g^2\eta^2}{gh}\rho h dx$$

$$=\frac{1}{2}\rho g\int_0^\lambda\eta^2 dx=V \qquad \text{From (1)}$$

Thus $T=V=\frac{1}{2}\rho g\int_0^\lambda\eta^2 dx$

Hence for a long wave travelling in one direction, kinetic energy is equal to potential energy and each equal to half the total energy at any instant.

8.22. Long Waves at the Common Surface of Two Liquids Bounded Above and Below by Two Fixed Horizontal Planes

Let ρ, ρ^1 be the densities of the liquids and A, A^1 the cross sections of lower and upper liquids respectively. Let b be the breadth of the common surface. In order to make wave stationary, we impose a velocity equal and opposite to the velocity u of propagation of the wave. Let u, u^1 be the small additional velocities due to wave motions in the two liquids. If η be elevation of common surface, then the equation of continuity for lower and upper liquids respectively are

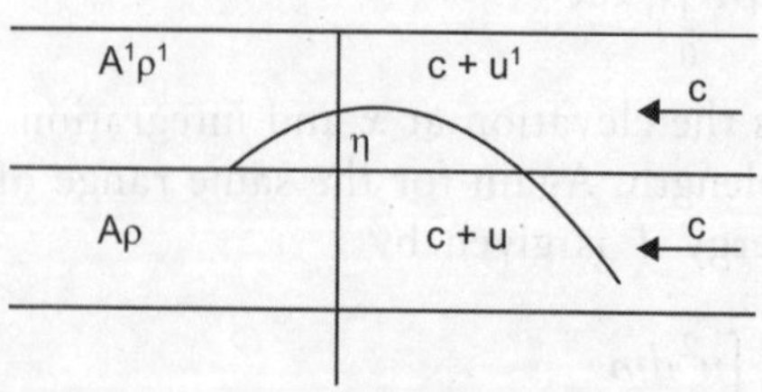

Fig. 8.6

$$Au = (A + b\eta)(U + u) \text{ and } A^1U = (A^1 - b\eta)(U + u^1)$$

Neglecting quantities of second order, we have

$$Au + bU\eta = 0 \quad \text{... (1)}$$

and $A^1U^1 - bU\eta = 0$ ... (2)

If δp and δp^1 be the increments of pressure close to the common surface in the two liquids due to the propagation of waves, then we have

$$\frac{\delta p}{\rho} + g\eta + \frac{1}{2}(U + u)^2 = \frac{1}{2}U^2$$

and $\dfrac{\delta p^1}{\rho^1} + g\eta + \dfrac{1}{2}(U + u^1)^2 = \dfrac{1}{2}U^2$

Also $\phi = 0$ due to steady motion.

$\Rightarrow \quad \delta p + (g\eta + Uu)\rho = 0$ and $\delta p^1 + (g\eta + Uu^1) - \rho^1 = 0$

$\Rightarrow \quad (\delta p - \delta p^1) + g\eta(\rho - \rho^1) + U(u\rho - u^1\rho^1) = 0$

Neglecting surface tension, so that $\delta p = \delta p^1$, we get

$$g\eta(\rho - \rho^1) = U(u^1\rho^1 - u\rho)$$

$$= U\left(\frac{\rho^1 b\, U\eta}{A^1} + \frac{\rho b U\eta}{A}\right) \quad \text{by (1) and (2)}$$

$$\Rightarrow \quad g(\rho - \rho^1) = U^2 b\left(\frac{\rho^1}{A^1} + \frac{\rho}{A}\right)$$

$$\Rightarrow \quad U^2 = \frac{g(\rho - \rho^1)}{b\left(\dfrac{\rho}{A} + \dfrac{\rho^1}{A^1}\right)} \quad \text{... (3)}$$

Deduction. The space between two infinite horizontal planes is filled with two fluids, one of density ρ and depth h and the other of density ρ^1 and depth h^1. Show that the velocity of long wave on the surface of separation is $\left[\dfrac{g(\rho - \rho^1)hh^1}{h^1\rho + \rho^1 h}\right]^{\frac{1}{2}}$.

Proof. Here $A = hb$, $A^1 = h^1b$

So that $\frac{b}{A} = \frac{1}{h}, \frac{b}{A^1} = \frac{1}{h^1}$

Hence from (3)

$$U^2 = \frac{g\left(\rho - \rho^1\right)}{\frac{\rho}{h} + \frac{\rho^1}{h^1}} = \frac{ghh^1\left(\rho - \rho^1\right)}{\rho h^1 + \rho^1 h}$$

8.23. Group Velocity

When a stone is thrown into a pond, a wave packet is formed whose each wave can be analysed into a set of simple harmonic waves of different wavelengths. Since velocity of propagation depends upon wavelength λ, the waves of slightly different wavelengths will form different groups. The velocity with which a group of waves progress is called the group velocity.

Case of Two Simple Harmonic Waves

We consider two simple harmonic waves of the same amplitude such that

$$\eta_1 = a \sin\left(mx - nt\right)$$

$$\eta_2 = a \sin\left[\left(m + \delta m\right) x - \left(n + \delta n\right) t\right]$$

The resulting disturbance is given by

$$\eta = \eta_1 + \eta_2$$

$$\Rightarrow \quad \eta = 2a \sin\left[\left(m + \frac{\delta m}{2}\right) x - \left(n + \frac{\delta \eta}{2}\right) t\right] \cos\left(\frac{x\delta m - t\delta n}{2}\right)$$

Since $\delta m << m$, $\delta n << n$

Hence $\eta = A \sin\,(mx - nt)$ ... (1)

Where $A = 2a \cos\frac{1}{2}\left(x\delta m - t\delta n\right)$ ... (2)

Equation (1) shows that the resulting disturbance is a progressive sine wave whose amplitude A is not constant but is itself varying as a wave of velocity $cg = \frac{\delta n}{\delta m}$. This velocity is known as group velocity.

In differential notation

$$cg = \frac{dn}{dm} \qquad \text{But } c = \frac{n}{m}$$

Hence $$cg = \frac{d}{dm}(mc) = c + m\frac{dc}{dm} \qquad \dots (3)$$

For the waves on the surface of water of depth h, we have the relation

$$c^2 = \frac{n^2}{m^2} = \frac{g}{m}\tan hmh$$

Taking log, we get

$$2\log c = \log g - \log m + \log \tan hmh$$

Differentiating with respect to m, we get

$$\frac{2}{c}\frac{dc}{dm} = -\frac{1}{m} + \frac{h \sec h^2 mh}{\tan hmh}$$

$$\Rightarrow \quad m\frac{dc}{dm} = -\frac{c}{2} + c\frac{mh}{\sin h2mh}$$

Hence (3) becomes

$$cg = \frac{c}{2}\left[1 + \frac{2mh}{\sin h2mh}\right] \qquad \dots (4)$$

$$\Rightarrow \quad \frac{\text{Group velocity}}{\text{Wave velocity}} = \frac{1}{2}\left[1 + \frac{2mh}{\sin h2mh}\right] \qquad \dots (5)$$

with $m = \frac{2\pi}{\lambda}$

Case I. For deep sea, $h \to \infty$ so that $\frac{2mh}{\sin h2mh} \to 0$

Hence $cg = \frac{c}{2}$

It shows that group velocity for deep sea waves is half the wave velocity.

Case II. For shallow water $h \to 0$, so that

$$\frac{2mh}{\sin h2mh} \to 1$$

Hence $cg = \frac{c}{2}(1 + 1) = c$

which shows that the group velocity for shallow water is equal to the wave velocity.

8.24. Transmission of Energy

The rate of transmission of energy is measured by taking a vertical section of the liquid at right angles to the direction of propagation. We shall determine the rate at which the liquid on one side of this section is doing work on the liquid on the other side. We assume that the depth of the liquid be h, we know that

$$\phi = \frac{ga}{n}\frac{\cos hm(y+n)}{\cos hmh}\cos(mx-nt) \qquad \ldots (1)$$

Neglecting squares of small quantities in the Bernoulli's equation the pressure is given by

$$\partial p = \rho\frac{\partial\phi}{\partial t} = \rho ga\,\frac{\cos hm(y+h)}{\cos hmh}\sin(mx-nt) \qquad \ldots (2)$$

Thus the work done W in unit time is given by

$$W = -\int_{-h}^{0}\partial p\left(\frac{\partial\phi}{\partial x}\right)dy$$

$$= \rho ga\int_{-h}^{0}\left\{\frac{\cos hm(y+h)}{\cos hmh}\sin(mx-nt)\right.$$

$$\left.\times\frac{ga}{n}\frac{m\cos hm(y+h)}{\cos hmy}\sin(mx-nt)\right\}dy$$

$$\Rightarrow \quad W = \frac{\rho g^2a^2m}{n}\frac{\sin^2(mx-nt)}{\cos h^2mh}\int_{-h}^{0}\cos h^2m(y+h)\,dy$$

$$\Rightarrow W = \frac{\rho g^2a^2m\sin^2(mx-nt)}{2n\cos h^2mh}\frac{\sin h2mh}{2m}\left(1+\frac{2mh}{\sin h2mh}\right) \qquad \ldots (3)$$

The average value of $\sin^2(mx-nt)$ is $\frac{1}{2}$, so we have

$$W = \frac{\rho g^2a^2m}{4h}\frac{\tan hmh}{m}\left(1+\frac{2mh}{\sin h2mh}\right)$$

$$\Rightarrow \quad W = \frac{1}{4}\rho g a^2 c\left(1 + \frac{2mh}{\sin h2mh}\right)$$

where $c^2 = \frac{g}{m} \tan hmh$ and $c = \frac{n}{m}$

$$\Rightarrow \quad W = \frac{1}{2}\rho g a^2 Cg, \; Cg = \frac{1}{2} C\left(1 + \frac{2mh}{\sin h2mh}\right) \qquad \text{... (4)}$$

Thus the expression for the energy transmitted in unit time is equal to $\frac{1}{2}\rho g a^2 \times \text{group velocity}$. Where $\frac{1}{2}\rho g a^2$ is the whole energy per unit length at any instant. Hence the energy is transmitted at a rate equal to the group velocity which is known as dynamical significance of group velocity.

8.25. Surface Tension and Capillary Waves

The common surface of two fluids (or one fluid and other solid) which do not mix, is found to be in a state of uniform tension called surface tension and is denoted by T. In case of surface tension; p = const. does not hold.

Waves of the kind in which surface tension is taken into account are known as capillary waves.

Surface tension depends on temperature and also on the nature of two fluids. By Hydrostatics we have the relation

$$p_1 - p_2 = T\left(\frac{1}{r_1} + \frac{1}{r_2}\right) \qquad \text{... (1)}$$

where p_1 and p_2 are pressures on the either side of the surface and r_1, r_2 are the principal radii of curvature of the surface.

For a two-dimensional wave, $r_2 \to \infty$ and

$$\frac{1}{r_1} = \frac{1}{r} = -\frac{\partial^2 \eta}{\partial x^2}$$

on neglecting square of small quantities. Here η is the elevation.

In this case (1) becomes

$$p_1 - p_2 = -T\frac{\partial^2 \eta}{\partial x^2} \qquad \text{... (2)}$$

Surface Condition. By pressure equation

$$\frac{p}{\rho}+\frac{1}{2}q^2+g\eta-\frac{\partial\phi}{\partial t}=C$$

for gravity controlled waves. If p_1 and p_2 be the pressures inside and outside of a wave surface, then neglecting q^2

$$p_1=\rho(\dot{\phi}-g\eta)+C$$

so that $p_1-p_2=\rho(\dot{\phi}-g\eta)+C-p_2$

Adjusting ϕ suitably, we can put the above equation as

$$p_1-p_2=\rho(\dot{\phi}-g\eta)$$

Using this in (3), we get

$$\rho(\dot{\phi}-g\eta)=-T\frac{\partial^2\eta}{\partial x^2}$$

Differentiating w.r.t. t

$$\rho(\ddot{\phi}-g\dot{\eta})=-T\frac{\partial}{\partial t}\left(\frac{\partial^2\eta}{\partial x^2}\right)$$

$$=-T\frac{\partial^2}{\partial x^2}\left(\frac{\partial\eta}{\partial t}\right)$$

But $\dot{\eta}=\frac{\partial\eta}{\partial t}=\frac{\partial\psi}{\partial x}$

$$\therefore \quad \rho\left(\ddot{\phi}-g\frac{\partial\psi}{\partial x}\right)=-T\frac{\partial^2}{\partial x^2}\left(\frac{\partial\psi}{\partial x}\right)$$

or $$\frac{\partial^2\phi}{\partial t^2}-g\frac{\partial\psi}{\partial x}+\frac{T}{\rho}\frac{\partial^3\psi}{\partial x^3}=C$$

This is called capillary surface condition.

8.26. Capillary Waves at the Common Surface of Two Liquids

We know that

$$q_0^2=(u-c)^2[1-2m\eta\cot hmh] \qquad \text{... (1)}$$

and

$$q_0^{1^2}=(u^1-c)^2\left[1+2m\eta\cot hmh^1\right] \qquad \text{... (2)}$$

Pressure equation for steady motion gives

$$\frac{p}{\rho}+\frac{1}{2}q_0^2+g\eta=\text{constant}$$

$$\frac{p^1}{\rho^1}+\frac{1}{2}q_0^{1^2}+g\eta=\text{constant}$$

Subtracting these two, we get

$$p-p^1+\frac{1}{2}\left(\rho q_0^2-\rho^1 q_0^{1^2}\right)+g\eta\left(\rho-\rho^1\right)=\text{constant} \quad \ldots (3)$$

But $p-p^1=-T\frac{\partial^2\eta}{\partial r^2}=Tam^2\sin mx=Tm^2\eta$

as $\eta=a\sin mx$

Putting the values of respective terms in (3), we get

$$Tm^2\eta+\frac{1}{2}\rho\,(u-c)^2(1-2m\eta\cot hmh)$$
$$-\frac{1}{2}\rho^1\left(u^1-c\right)^2\left(1+2m\eta\cot hmh^1\right)$$
$$+g\eta.\left(\rho-\rho^1\right)=\text{constant.}$$

Since R.H. is constant. η is only variable on L.H.S. and so coefficient of η must vanish.

Hence

$$Tm^2\rho(u-c)^2 m\cot hmh-\rho^1\left(u^1-c\right)^2 m\cot hmh^1$$
$$+g\left(\rho-\rho^1\right)=0$$

$$\Rightarrow\quad Tm^2+g\left(\rho-\rho^1\right)=m\left[\rho\,(u-c)^2\cot hmh\right.$$
$$\left.+\rho^1\left(u^1-c\right)^2\cot hmh^1\right] \quad \ldots (4)$$

This determines velocity of propagation.

Corollary 1. If the liquids are at rest, then $u^1-u=0$ so that (4) becomes

$$Tm^2+g\left(\rho-\rho^1\right)=mc^2\left[\rho^1\cot hmh^1+\rho\cot hmh\right]$$

2. If the depths of liquids be great compared with wavelength so that $h\to\infty$, $h^1\to\infty$ then (4) becomes

$$Tm^2+g\left(\rho-\rho^1\right)=m\left[(u-c)^2\rho+\left(u^1-c\right)^2\rho^1\right]$$

3. If the liquids are at rest and their depths are great compared with wavelength, then (6) becomes

$$Tm^2 + g(\rho - \rho^1) = mc^2(\rho + \rho^1)$$

$$\Rightarrow \quad c^2 = \frac{Tm}{\rho + \rho^1} + \frac{g(\rho - \rho^1)}{m(\rho + \rho^1)}$$

WORKED OUT EXAMPLES

1. If a canal of rectangular section contains a depth of liquid of density ρ, on which is superposed a depth h^1 of liquid of density ρ^1, the free surface of the latter being exposed to a constant atmospheric pressure. Prove that the velocities of propagation of waves $\frac{2\pi}{m}$ are given by $c^2 = \frac{gu}{m}$ where

$$\rho\,(u \cot hmh - 1)\,(u \cot hmh^1 - 1) = \rho^1(1 - u^2)$$

Solution: We have the relation

$$c^4m^2(\rho \cot hmh \cot hmh^1 + \rho^1)$$
$$- c^2mg\rho(\cot hmh + \cot hmh^1) + g^2(\rho - \rho^1) = 0 \quad \ldots (1)$$

But it is given that $c^2 = \frac{gu}{m}$

So that $c^2m = gu$.

Hence from (1)

$$g^2u^2(\rho \cot hmh \cot hmh^1 + \rho^1)$$
$$- g^2u\rho(\cot hmh + \cot hmh^1) + g^2(\rho - \rho^1) = 0$$

$$\Rightarrow \quad u^2\rho \cot hmh \cot hmh^1 - u\rho(\cot hmh + \cot hmh^1)$$
$$= \rho^1(1 - u^2)$$

$$\Rightarrow \quad \rho(u \cot hmh - 1)(u \cot hmh^1 - 1) = \rho^1(1 - u^2)$$

2. If there be two liquids in a straight canal of uniform section, of densities σ_1, σ_2 and depth l_1 and l_2, show that the velocity c of propagation of long waves is given by the equation

$$\left(\frac{c^2}{l_1 g}-1\right)\left(\frac{c^2}{l_2 g}-1\right)=\frac{\sigma_1}{\sigma_2}$$

where $\sigma_2 > \sigma_1$ and it is assumed that the liquids do not mix.

Solution: Since the two liquids are contained in a straight canal and so the surface of the upper liquid must be free surface. For this we have

$$c^4 m^2 \left(\rho \cot hmh \cot hmh^1 + \rho^1\right)$$
$$- c^2 mg\rho \left(\cot hmh + \cot hmh^1\right) + g^2 \left(\rho - \rho^1\right) = 0 \quad \ldots (1)$$

Putting $\rho = \sigma_2$, $\rho^1 = \sigma_1$, $h = l_2$, $h^1 = l_1$ in (1) as $\rho > \rho^1$ and $\sigma_2 > \sigma_1$, we have

$$c^4 m^2 \left(\sigma_2 \cot hmhl_2 \cot hml_1 + \sigma_1\right)$$
$$- c^2 mg\sigma_2 \left(\cot hml_2 + \cot hml_1\right) + g^2 \left(\sigma_2 - \sigma_1\right) = 0 \quad \ldots (2)$$

Since the waves formed at the common surface are long waves so that $ml_1 \to 0$, $ml_2 \to 0$ since $\frac{h}{\lambda} = \frac{mh}{2\pi} \to 0$

Consequently tan $hml_1 = ml_1$, tan $hml_2 = ml_2$

Hence (2) becomes

$$c^4 m^2 \left(\sigma_2 \frac{1}{ml_1} \cdot \frac{1}{ml_2} + \sigma_1\right) - c^2 mg\sigma_2 \left(\frac{1}{ml_2} + \frac{1}{ml_1}\right)$$
$$+ g^2 \left(\sigma_2 - \sigma_1\right) = 0$$

$$\Rightarrow \frac{c^4}{l_1 l_2} + \frac{\sigma_1}{\sigma_2} c^4 m^2 - c^2 g\left(\frac{1}{l_1} + \frac{1}{l_2}\right) + g^2 \left(1 - \frac{\sigma_1}{\sigma_2}\right) = 0 \quad \ldots (3)$$

But $m^2 = \left(\frac{2\pi}{\lambda}\right)^2$

In case of long waves, λ is very large, so that m^2 is negligible.

Therefore from (3), we have

$$\frac{c^4}{l_1 l_2 g^2} - \frac{c^2}{g}\left(\frac{1}{l_1} + \frac{1}{l_2}\right) + 1 = \frac{\sigma_1}{\cdot\sigma_2}$$

$$\Rightarrow \left(\frac{c^2}{l_1 g} - 1\right)\left(\frac{c^2}{l_2 g} - 1\right) = \frac{\sigma_1}{\sigma_2}$$

3. Let a shallow trough be filled with oil and water, and let the depth of the water be K and its density ρ. Then show that if g be gravity and v the velocity of propagation of long waves,

$$\frac{v^2}{g} = \frac{1}{2}(h+K) + \frac{1}{2}\left[(h-K)^2 + \frac{4hK\rho}{\sigma}\right]^{\frac{1}{2}}$$

assuming that there be slipping between the two fluids.

Solution: Water and liquid both are kept on a trough, i.e. a deep container so that the surface of the upper liquid, i.e. oil will be free surface. For density water > density of oil.

If a liquid of density ρ_2 and depth h_2 lies over a liquid of density ρ_1 and depth h_1, then velocity c of propagation is given by

$$c^4 m^2 (\rho_1 \cot hmh_1 \cot hmh_2 + \rho_2) \\ - c^2 \rho_1 mg(\cot hmh_1 + \cot hmh_2) + g^2(l_1 - l_2) = 0 \qquad \text{... (1)}$$

Where $\rho_1 > \rho_2$

Putting $\rho_1 = \sigma$, $h_1 = K$, $\rho_2 = \rho$, $h_2 = h$, $c = v$ in (1), we get

$$v^4 m^2 (\sigma \cot hmK \cdot \cot hmh + \rho) \\ - v^2 mg\sigma(\cot hmK + \cot hmh) + g^2(\sigma - \rho) = 0$$

For long waves $mK \to 0$, $mh \to 0$

So that tan $hmh = mh$, tan $hmK = mK$

$$\therefore \quad v^4 m^2\left(\frac{\sigma}{mh \cdot mh} + \rho\right) - v^2 mg\sigma\left(\frac{1}{mK} + \frac{1}{mh}\right) \\ + g^2(\sigma - \rho) = 0$$

As m is very small and so neglecting m^2, we get

$$\left(\frac{v^2}{g^2}\right)^2 - \frac{v^2}{g}(h+K) + hK\left(1-\frac{\rho}{\sigma}\right) = 0$$

This is a quadratic equation in $\dfrac{v^2}{g}$

Hence

$$\frac{v^2}{g} = \frac{(h+K) \pm \sqrt{(h+K)^2 - 4\left(1-\frac{\rho}{\sigma}\right)hK}}{2}$$

$$\Rightarrow \quad \frac{v^2}{g} = \frac{1}{2}\left[(h+K) \pm \left\{(h+K)^2 + \frac{4hK\rho}{\sigma}\right\}^{\frac{1}{2}}\right]$$

Since $\dfrac{\rho}{\sigma} = 1$ approximately and velocity v is very large in case of long waves and hence neglecting negative sign

$$\frac{v^2}{g} = \frac{1}{2}\left[(h+K) + \frac{1}{2}\left\{(h-K)^2 + \frac{4hK\rho}{\sigma}\right\}^{\frac{1}{2}}\right]$$

4. An open rectangular box of length a contains two liquids of densities ρ and ρ^1 and depths h, h^1 respectively, that of density ρ being at the bottom. Prove that the periods of oscillation when the liquids are slightly disturbed so that there is no motion perpendicular to the sides of the box, are determined by the type

$$\left[p^2 \cot h\left(\frac{n\pi h}{a}\right) - \frac{gn\pi}{a}\right]\left[p^2 \cot h\left(\frac{n\pi h^1}{a} - \frac{gn\pi}{a}\right)\right] + \frac{\rho^1}{\rho}\left[p^4 - g^2\frac{n^2\pi^2}{a^2}\right] = 0$$

where n is any integer.

Solution: It is given that there is no motion perpendicular to the sides $x = 0$, $x = a$ of the box. Hence stationary waves will be formed at the common surface of the two liquids. We know that

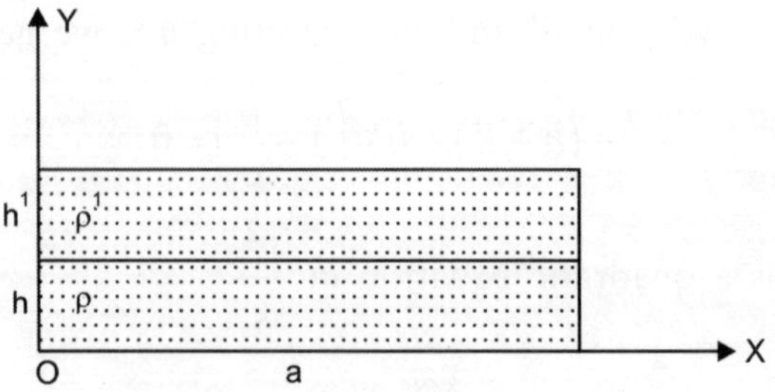

Fig. 8.7

$$c^4 m^2 \left(\rho \cot hmh \cot hmh^1 + \rho^1\right)$$
$$- c^2 mg\rho \left(\cot hmh + \cot hmh^1\right) + g^2 \left(\rho - \rho^1\right) = 0 \quad \dots (1)$$

where $\rho > \rho^1$

When the equation of the wave is

$$\eta = a \sin (mx - nt)$$

In the present case the equation to wave is

$$\eta = a \sin mx \cos pt$$

But a loop will exist at $x = a$ so that $ma = n\pi$, where n is integer.

Putting $m = \frac{n\pi}{a}, c = \frac{p}{m}$ in (1), we get

$$\frac{p^4}{m^2}\left(\rho \cot h \frac{n\pi h}{a} \cot h \frac{n\pi h^1}{a} + \rho^1\right)$$
$$- \frac{p^2}{m} g\rho \left(\cot h \frac{n\pi h}{a} + \cot h \frac{n\pi h^1}{a}\right) + g^2 \left(\rho - \rho^1\right) = 0$$

$$\Rightarrow p^4 \cot h \frac{n\pi h}{a} \cot h \frac{n\pi h^1}{a} - p^2 g \frac{n\pi}{a}\left(\cot h \frac{n\pi h}{a}\right.$$
$$\left. + \cot h \frac{n\pi h^1}{a}\right) + p^4 \frac{\rho^1}{\rho} + g^2 \left(1 - \frac{\rho^1}{\rho}\right)\left(\frac{n\pi}{a}\right)^2 = 0$$

$$\Rightarrow \left(p^2 \cot h \frac{n\pi h}{a} - g \frac{n\pi}{a}\right)\left(p^2 \cot h \frac{n\pi h^1}{a} - g \frac{n\pi}{a}\right)$$
$$+ \frac{\rho^1}{\rho}\left(p^4 - g^2 \frac{n^2\pi^2}{a^2}\right) = 0.$$

5. Prove that in a fluid of depth h, limited by two vertical barriers apart, at right angles to the direction of propagation of straight crested, irrotational waves, the periods of the waves are found by giving r positive integral values in the formula

$$2\left[\left(\frac{\pi d}{\partial g}\right)\cot h\left(\frac{\pi r h}{d}\right)\right]^{\frac{1}{2}}.$$

Solution: Liquid is oscillating in a tank of finite dimensions. Since $x = 0$ has to be a loop, the surface disturbance must be of the form

$$\eta = a \cos mx \cos nt$$

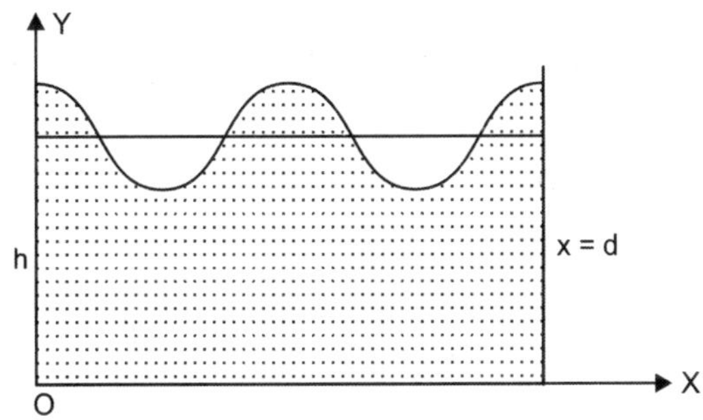

Fig. 8.8

For stationary waves are normal modes of free oscillations of restricted system. Again there must exist a loop at $x = d$ so that $md = r\pi$, where r is an integer. By definition

$$\lambda = \frac{2\pi}{m} = \frac{2\pi}{r\pi/d} = \frac{2d}{r}$$

Hence wavelengths of oscillations are given by $\lambda = \frac{2d}{1}$, $\frac{2d}{2} \ldots\ldots \frac{2d}{r}$ where r is any positive integer.

We know that $\frac{n^2}{m^2} = c^2 = \frac{g}{m} \tan hmh$

$$\Rightarrow \quad n = [gm \tan hmh]^{\frac{1}{2}}$$

$$= \left[\frac{gr\pi}{d} \tan h \left(\frac{4\pi h}{d} \right) \right]^{\frac{1}{2}}$$

Time period T of oscillation is given by

$$T = \frac{2\pi}{n} = 2\pi \sqrt{\frac{d}{gr\pi}} \left[\cot h \left(\frac{\pi rh}{d} \right) \right]^{\frac{1}{2}}$$

$$\Rightarrow \quad T = 2 \left[\left(\frac{\pi d}{rg} \right) \cot h \left(\frac{\pi rh}{d} \right) \right]^{\frac{1}{2}}.$$

6. If a horizontal rectangular canal of great depth has two vertical barriers at a distance l apart, prove that the periods of oscillations of water are $2\left[\frac{\pi l}{sg}\right]^{\frac{1}{2}}$ where l is a positive integer and that the corresponding to any mode, all particles of fluids oscillate in straight lines of length inversely proportional to $\exp\left(\frac{s\pi z}{l}\right)$ where z is the depth.

Solution: We know that stationary waves are normal modes of free oscillation of restricted system. Also $x = 0$ has to be a loop, the surface disturbance must be of the form

$$\eta = a \cos mx \cos nt$$

Then

$$\phi = ace^{my} \cos mx \sin nt \quad \text{... (1)}$$

with

$$c^2 = \frac{n^2}{m^2} = \frac{g}{m} \quad \text{... (2)}$$

On the fixed boundaries

$$x = 0, x = l, -\frac{\partial \phi}{\partial x} = 0 \quad \text{... (3)}$$

$$(1) \Rightarrow -\frac{\partial \phi}{\partial x} = acme^{mx} \sin mx \sin nt \quad \text{... (4)}$$

Using (3) in (4), we get

$$O = 0, 0 = acm\, e^{my} \sin ml \sin nt$$

$$\Rightarrow \quad \sin ml = 0 \Rightarrow ml = s\pi$$

where s is an integer, in particular here, s is a positive integer.

Also from (2) $n^2 = gr\imath = g\dfrac{s\pi}{l}$

Time period T is given by

$$T = \frac{2\pi}{n} = 2\pi\sqrt{\frac{l}{gs\pi}}$$

$$\Rightarrow \quad T = 2\left(\frac{\pi l}{sg}\right)^{\frac{1}{2}}$$

This proves the first part of the problem.

To prove second part, let (X, Y) be the coordinates of a fluid particle relative to the mean position (x, y).

Then

$$\frac{dX}{dt} = acme^{my} \sin mx \sin nt = \frac{-\partial\phi}{\partial x} \qquad \text{... (5)}$$

and

$$\frac{dY}{dt} = -acme^{mx} \cos mx \sin nt = \frac{-\partial\phi}{\partial y} \qquad \text{... (6)}$$

Integrating these two, we have

$$X = \frac{-acm}{n} e^{my} \sin mx \cos nt + A$$

and

$$Y = \frac{acm}{n} e^{my} \cos mx \cos nt + B$$

$$\Rightarrow \quad \frac{Y - B}{X - A} = -a \cot mx = \text{constant}$$

which represents a straight line.

Again $\left[(X - A)^2 + (Y - B)^2\right]^{\frac{1}{2}} = ae^{my} \cos nt$

Its maximum value is ae^{my}.

At a depth z, such that $y = -z$, the maximum value is $ae^{-mz} = \frac{a}{e^{mz}} = \frac{a}{e^{\frac{s\pi z}{l}}}$.

This shows that the length varies inversely as $e^{\frac{s\pi z}{l}}$.

7. Two-dimensional waves of length $\frac{2\pi}{m}$ are produced at the surface of separation of two liquids which are of densities ρ, ρ^1 ($\rho > \rho^1$) and depth h, h^1 confined between two fixed horizontal planes. Prove that if the total potential energy is reckoned zero in the position of equilibrium, the total energy of the lower liquid is to that of the upper in the ratio

$$\rho\left[\left(2\rho-\rho^1\right)\cot hmh+\rho^1\cot hmh^1\right]$$
$$-\rho^1\left[\left(\rho-2\rho^1\right)\cot hmh^1-\rho\cot hmh\right]$$

Solution: Let the equation of wave profile be

$$\eta = a\sin\left(mx-nt\right)$$

Then

$$\phi = \frac{ac}{\sin hmh}\cos hm\left(y+h\right)\cos\left(mx-nt\right)$$

for lower liquid

and

$$\phi^1 = \frac{-ac}{\sin hmh^1}\cos hm\left(y^1-h\right)\cos\left(mx-nt\right)$$

for upper liquid

where $g\left(\rho-\rho^1\right) = c^2 m\left[\rho\cot hmh+\rho^1\cot hmh\right]$

The kinetic energy of lower liquid per wavelength

$$= \frac{1}{2}\rho\int_0^\lambda\left(\phi\frac{\partial\phi}{\partial y}\right)_{y=0} dx$$

$$= \frac{1}{2}\rho\left(\frac{ac}{\sin hmh}\right)^2 m\cos hmh\sin hmh\int_0^\lambda\cos^2\left(mx-nt\right)dx$$

$$= \frac{\lambda}{4}\rho a^2 c^2 m\cot hmh$$

Similarly K.E. for upper liquid

$$= \frac{1}{4} \rho^1 a^2 c^2 \lambda \cot hmh^1$$

Potential energy for lower liquid per wavelength

$$= \frac{1}{2} g\rho \int_0^\lambda \eta^2 dx$$

$$= \frac{1}{4} g\rho a^2 \int_0^\lambda [1 - \cos 2(mx - nt)]\, dx$$

$$= \frac{1}{4} g\rho a^2 \lambda$$

Similarly P.E. for upper liquid $= \frac{1}{2} g\rho^1 a^2 \lambda$

If E and E^1 be total energies for lower and upper liquids, then

$$E = \frac{1}{4} m\rho a^2 c^2 \lambda \cot hmh + \frac{1}{4} g\rho a^2 \lambda$$

$$= \frac{1}{4} \rho a^2 \lambda \left[mc^2 \cot hmh + g \right]$$

$$= \frac{1}{4} \rho a^2 \lambda \left[\frac{g(\rho - \rho^1) \cot hmh}{\rho \cot hmh + \rho^1 \cot hmh^1} + g \right]$$

$$\Rightarrow \quad E = \frac{1}{4} \rho a^2 \lambda g \left[\frac{(2\rho - \rho^1) \cot hmh + \rho^1 \cot hmh^1}{\rho \cot hmh + \rho^1 \cot hmh^1} \right]$$

Also

$$E^1 = \frac{1}{4} \rho^1 a^2 \lambda \left[mc^2 \cot hmh^1 - g \right]$$

$$= \frac{1}{4} \rho^1 a^2 \lambda \left[\frac{(\rho - 2\rho^1) \cot hmh^1 - \rho \cot hmh}{\rho \cot hmh + \rho^1 \cot hmh^1} \right]$$

Hence

$$\frac{E}{E^1} = \frac{\rho\left[(2\rho - \rho^1) \cot hmh + \rho^1 \cot hmh^1\right]}{\rho^1\left[(\rho - 2\rho^1) \cot hmh^1 - \rho \cot hmh\right]}$$

8. Show that if the velocity of the wind is just great enough to prevent the propagation of waves of length λ against it, the velocity of propagation of waves with the wind is $2c\left\{\frac{\sigma}{1+\sigma}\right\}^{\frac{1}{2}}$. Where σ is the specific gravity of air and c is the wave velocity when no air is present. [I.A.S. 87]

Solution: Let a wind of density ρ^1 and depth h^1 move with velocity u^1 over a liquid of density ρ and depth h which is moving with velocity u, then

$$g(\rho-\rho^1)=m\left[\rho(u-v)^2 \cot hmh+\rho^1\left(u^1-v\right)^2 \cot hmh^1\right]$$

where v is the wave velocity. Also $\sigma=\frac{\rho^1}{\rho}$ so that

$$g(1-\sigma)=m\left[(u-v)^2 \cot hmh+\sigma\left(u^1-v\right)^2 \cot hmh^1\right] \quad \ldots (1)$$

Here $h \to \infty$, $h^1 \to \infty$ so that $\cot hmh = 1 = \cot hmh^1$

Also water is assumed to be at rest so that $u = 0$.

Putting these values in (1), we get

$$g(1-\sigma)=m\left[v^2+\sigma\left(u^1-v\right)^2\right] \quad \ldots (2)$$

We are given two boundary conditions.

1. Velocity of wind is just great enough to prevent wave propagation on the surface of water.

2. c is the wave velocity when no air is present.

From (1) $u^1 = u_1$, $v = 0$ ... (3)

From (1) $v = 0$, $u^1 = 0$ ... (4)

We are to determine velocity of wave propagation v_1 with the wind.

Hence

$$u^1 = u_1,\ v = v_1 \quad \ldots (5)$$

Subjecting (2) to the conditions (3), (4) and (5), we get

$$g\,(1-\sigma) = m\sigma u_1^2 \quad \ldots (6)$$

$$\Rightarrow g(1-\sigma) = mc^2(1+\sigma) \qquad \text{... (7)}$$

$$\Rightarrow g(1-\sigma) = m[v_1^2 + (u_1 - v_1)^2 \sigma] \qquad \text{... (8)}$$

From (7) and (8), we have

$$\sigma u_1^2 = v_1^2 + (u_1 - v_1)^2 \sigma$$

$$\Rightarrow \sigma\left[u_1^2 - \left(u_1^2 - 2u_1v_1 + v_1^2\right)\right] = v_1^2$$

$$\Rightarrow \sigma v_1 (2u_1 - v_1) = v_1^2$$

$$\Rightarrow v_1(1+\sigma) = 2\sigma u_1$$

$$\Rightarrow v_1 = \frac{2u_1\sigma}{1+\sigma} \qquad \text{... (9)}$$

Again from (6) and (7)

$$\sigma u_1^2 = c^2(1+\sigma)$$

$$\Rightarrow u_1^2 = \frac{c^2(1+\sigma)}{\sigma}$$

Putting the value of u_1 in (9), we get

$$v_1 = \frac{2\sigma}{1+\sigma} c\left(\frac{1+\sigma}{\sigma}\right)^{\frac{1}{2}}$$

$$= 2c\left(\frac{\sigma}{1+\sigma}\right)^{\frac{1}{2}}.$$

9. When simple harmonic waves of length λ are propagated over the surface of deep water; prove that, at a point whose depth below undisturbed surface is h, the pressure at the instant when the disturbed depth of the point is $h + \eta$ bears to the undisturbed pressure at the same point, the ratio $\left(1 + \frac{\eta}{h} e^{\frac{-2\pi}{h}}\right) : 1$ atmospheric pressure and surface tension being neglected.

Solution: The velocity potential for the wave propagation $\eta = a \sin(mx - nt)$ on the surface of the deep water is given by

$$\phi = ace^{my}\cos(mx - nt) \qquad \text{... (1)}$$

with $c^2 = \frac{n^2}{m^2} = \frac{g}{m}$ so that $n^2 = gm$ ... (2)

By pressure equation

$$\frac{p}{\rho} + \frac{1}{2}q^2 - \frac{\partial\phi}{\partial t} + gy = c \qquad \text{... (3)}$$

For gravity waves, force potential, $\Omega - gy$.

Pressure at any point within water is given by

$$\frac{p}{\rho} - \frac{\partial\phi}{\partial y} + gy = c \qquad \text{as } q^2 \text{ is negligible.}$$

It is given that atmospheric pressure is negligible.

Hence $p = 0$, $\dot{\phi} = 0$ when $y = 0$

Hence C = 0

Thus (3) becomes

$$\frac{p}{\rho} - \frac{\partial\phi}{\partial t} + gy = 0$$

$$\Rightarrow \quad p = \rho(\dot{\phi} - gy) \qquad \text{... (4)}$$

From (1)

$$\dot{\phi} = acn\, e^{my} \sin(mx - nt)$$

$$= \frac{n}{m} \cdot n \cdot \eta e^{my}$$

$$= \frac{gm}{m} \eta e^{my} \quad \text{From (2)}$$

$$= g\eta e^{my}$$

Putting the value of $\dot{\phi}$ in (4) it becomes

$$p = \rho g\left(\eta e^{my} - y\right) \qquad \text{... (5)}$$

Let $p = p_1$ when $y = -h$ i.e. p_1 is the pressure at a depth h while the height of this point from the free surface is $h + \eta$. Putting these values in (5) it becomes

$$p_1 = \rho g\left(\eta e^{-mh} + h\right) = \rho g h\left(1 + \frac{\eta}{h} e^{-mh}\right) \qquad \text{... (6)}$$

Undisturbed pressure p_0 at a depth h is $p_0 = \rho g h$.

Therefore (6) becomes

$$\frac{p}{p_0} = 1 + \frac{n}{h} e^{-\frac{2\pi h}{\lambda}} \qquad \text{since } m = \frac{2\pi}{\lambda}$$

10. Two fluids of densities ρ_1, ρ_2 have horizontal surface of separation but are otherwise unbounded. Show that when waves of small amplitude are propagated at their common surface, the particles of the two fluids describe circles about their mean positions and that at any point of the surface of separation, where the elevation is η, the particles on either side have a relative velocity $\dfrac{4\pi c\eta}{\lambda}$.

Solution: Let the wave profile at the common surface be $\eta = a \sin(mx - nt)$ where $a << 1$, $m = \dfrac{2\pi}{\lambda}$.

Let ϕ, ϕ^1 be velocity potential in lower and upper fluids respectively.

Then $\phi = Ae^{my}\cos(mx - nt)$

and $\phi^1 = Be^{-my}\cos(mx - nt)$

Where ϕ, ϕ^1 are such that

(i) $\vec{\nabla}^2\phi = 0 = \vec{\nabla}^2\phi^1$

(ii) $\displaystyle \mathop{Lt}_{y \to -\infty} \frac{\partial\phi}{\partial y} = 0, \frac{\partial\phi}{\partial y^1} = 0 \text{ as } y \to \infty$

(iii) At the common surface $\eta = \dfrac{-\partial\phi}{\partial y} = -\dfrac{\partial\phi^1}{\partial y}$ at $y = 0$

$$\Rightarrow -an\cos(mx - nt) = Ame^{my}\cos(mx - nt)$$

$$= Bme^{-my}\cos(mx - nt) \text{ at } y = 0$$

$$\Rightarrow -an = -Am = Bm$$

$$\Rightarrow A = ac,\ B = -ac \text{ where } c = \frac{n}{m}$$

$$\therefore \quad \phi = ace^{my}\cos(mx - nt)$$

and

$$\phi^1 = -ace^{-my}\cos(mx - nt)$$

Let (X, Y) be the coordinates of a fluid particle relative to the mean position (x, y) in the lower fluid.

Then $\dfrac{dX}{dt} = -\dfrac{\partial\phi}{\partial x} = acme^{my}\sin(mx - nt)$

and $\dfrac{dY}{dt} = -\dfrac{\partial\phi}{\partial y} = -acme^{my}\cos(mx - nt)$

Integrating these two, we get

$$X = \frac{acm}{n} e^{my}\cos(mx - nt)$$

$$= ae^{my}\cos(mx - nt)$$

and

$$Y = \frac{acm}{n} e^{my}\sin(mx - nt)$$

$$= ae^{my}\sin(mx - nt)$$

Constants of integration vanish as $X = 0 = Y$ at $t = 0$.

Hence $X^2 + Y^2 = a^2e^{2my}$

which represents a circle.

Hence the path of the fluid particle in lower fluid is a circle. Similarly we can prove for the upper fluid also. Relativistic velocity at the common surface

$$= \left[\left(\frac{-\partial\phi}{\partial x}\right) - \left(\frac{-\partial\phi^1}{\partial x}\right)\right]_{y=0}$$

$$= acm\sin(mx - nt) + acm\sin(mx - nt)$$

$$= 2acm\sin(mx - nt)$$

$$= 2\eta\, cm = \frac{4\pi c\eta}{\lambda}$$

Since $\left[\left(\dfrac{-\partial\phi}{\partial y}\right) - \left(\dfrac{-\partial\phi^1}{\partial y}\right)\right]_{y=0} = 0$

11. If water of depth h be flowing with velocity proportional to distance from the bottom, V being the velocity of stream at the surface, prove that the velocity U of propagation of waves in the direction of stream is given by

$$(U-V)^2 + V(U-V)\left(\frac{W^2}{gh}\right) - W^2 = 0.$$

Solution: It is given that

V = velocity of fluid at surface

U = velocity of wave propagation parallel to x-axis

W = velocity of wave propagation in still water so that

$$W^2 = \frac{g}{m}\tan hmh$$

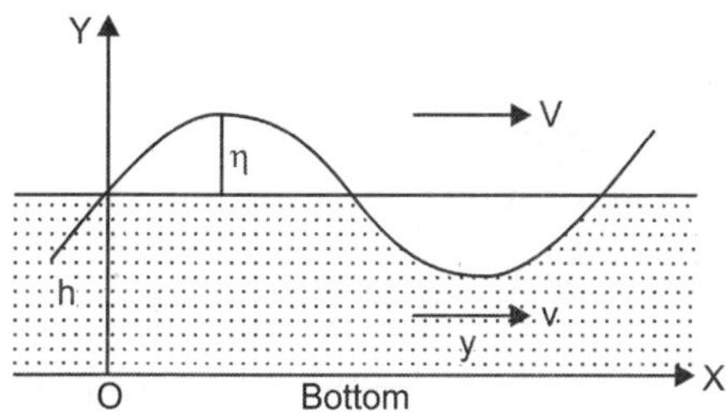

Fig. 8.9

Let us suppose that v is the fluid velocity at any height y from the bottom $y = 0$.

Then given $V = Ky$

But $v = V$ when $y = h$ so that $V = Kh$

$$\Rightarrow \quad K = \frac{V}{h}$$

Hence $v = \dfrac{Vy}{h}$

We impose a velocity $-U$ to the whole system so that the wave profile reduces to rest and water flows with velocity $v - U$ in a steady state.

$$\frac{-\partial\psi}{\partial y} = \frac{-\partial\phi}{\partial x} = -U \rightarrow \psi = Uy$$

Also $\dfrac{-\partial\psi}{\partial y} = \dfrac{-\partial\phi}{\partial x} = v = \dfrac{Vy}{h} \rightarrow \psi = -\dfrac{Vy^2}{h}$

The stream function for the resulting motion is given by

$$\psi = -\left[\frac{Vy^2}{2h} - Uy\right] + (A\cos hmy + B\sin hmy)\times \sin mx \qquad \dots (1)$$

$\therefore \quad \psi = A\sin mx \quad$ at $y = 0$

But the bottom $y = 0$ will be the streamline $\psi = 0$ so that

$$(\psi)_{y=0} = 0$$

$\Rightarrow A \sin mx = 0$

$\Rightarrow A = 0$

$$\therefore \quad \psi = -\left[\frac{Vy^2}{2h} - Uy\right] + B\sin hmy \sin mx \qquad \dots (2)$$

Also the free surface $y = h + \eta = h + a \sin mx$ will be a streamline $\psi = C$ (constant) so that

$$-\left[\frac{V}{2h}(h + a\sin mx)^2 - U(h + a\sin mx)\right] + B\sin hm(h + a\sin mx)\sin mx = C$$

Since $a << 1$ and $m << 1$ and so neglecting m^2, a^2, am. Then

$$-\left[\frac{V}{2h}(h^2 + 2ah\sin mx) - U(h + a\sin mx)\right] + B\sin hmh \sin mx = C$$

Equating the coefficients of sin mx from both sides

$$-(Va - Ua) + B\sin hmh = 0$$

$$\Rightarrow \quad B = a(V - U)/\sin hmh$$

Now (2) becomes

$$\psi = -\left[\frac{Vy^2}{2h} - Uy\right] + \frac{(V - U)}{\sin hmh} a\sin hmy \sin mx \qquad \dots (3)$$

$$q^2 = \left(\frac{\partial\psi}{\partial x}\right)^2 + \left(\frac{\partial\psi}{\partial y}\right)^2$$

$$= \left[\frac{(V-U)}{\sin hmh} am \sin hmy \cos mx\right]^2$$

$$+ \left[\frac{-Vy}{h} + U + \frac{(V-U)}{\sin hmh} am \cos hmy \sin mx\right]^2$$

Neglecting a^2

$$q^2 = \left[-\frac{Vy}{h} + U + \frac{(V-U)}{\sin hmh} am \cos hmy \sin mx\right]^2$$

approximately

Let $q = q_0$ = fluid velocity at free surface, $y = h + a \sin mx$

$$q_0^2 . = \left[-V - \frac{Va}{h} \sin mx + U\right.$$

$$\left. + \frac{(V-U)}{\sin hmh} am \cos hmh \sin mx\right]^2$$

$$= \left[-(V-U) - \frac{Va}{h} \sin mx\right.$$

$$\left. + (V-U)\, am \cot hmh \times \sin mx\right]^2$$

Nelecting a^2

$$q_0^2 = (V-U)^2 + 2(V-U)\frac{Va}{h} \sin mx$$

$$- 2(V-U)^2 am \cot hmh \sin hmx$$

By pressure equation $\frac{p}{\rho} + \frac{1}{2} q^2 + gy = C$

At the free surface this becomes

$$\frac{\pi}{\rho} + \frac{1}{2} q_0^2 + g(h + a \sin mx) = C$$

Putting the value of q_0^2 and then equating the coefficient of sin mx on both sides

$$ga + \frac{1}{2}\left[2(V-U)\frac{Va}{h} - 2(V-U)^2 am \cot hmh\right] = 0$$

$$\Rightarrow \quad (U-V)^2\, m \cot hmh + (U-V)\frac{V}{h} - g = 0$$

But $W^2 = \dfrac{g}{m}\tan hmh$

$$\therefore \quad (U-V)^2\left(\frac{g}{W}\right)^2 + (U-V)\left(\frac{V}{h}\right) - g = 0$$

$$\Rightarrow \quad (U-V)^2 + (U-V)V\left(\frac{W^2}{gh}\right) - W^2 = 0$$

12. A canal of infinite length and rectangular section is of uniform depth h and breadth b in one part but changes gradually to uniform depth h^1 and breadth b^1 in another part. An infinite train of simple harmonic waves travelling in one direction only is propagated along the canal. Prove that if a, a^1 are the heights and $\dfrac{2\pi}{m}, \dfrac{2\pi}{m^1}$ the lengths of the waves in the two uniform portions, then

$$m \tan mh = m^1 \tan hmh^1$$

and $a^2 b \sec h^2 mh\,(\sin h2mh + 2mh)$

$$= a^{1^2} b^1 \sec h^2 m^1 h^1 (\sin h2m^1h^1 + 2m^1h^1)$$

Solution: Let mean profiles of the wave in the two parts be

$$\eta = a\sin(mx - nt),\ \eta^1 = a^1 \sin(m^1 x - n^1 t) \qquad \text{... (1)}$$

Then

$$C^2 = \frac{n^2}{m^2} = \frac{g}{m}\tan hmh \qquad \text{... (2)}$$

and

$$C^{1^2} = \frac{n^{1^2}}{m^{1^2}} = \frac{g}{m^1}\tan hm^1h^1 \qquad \text{... (3)}$$

The time period of simple harmonic wave must remain the same all along the canal so that

$$\frac{2\pi}{n} = \frac{2\pi}{n^1}$$

$$\Rightarrow \quad n^2 = n^{1^2} \qquad \text{... (4)}$$

Dividing (2) by (3), we get

$$\frac{n^2m^{1^2}}{m^2n^{1^2}} = \frac{m^1 \tan hmh}{m \tan hm^1h^1}$$

$$\Rightarrow \quad \frac{m^1}{m} = \frac{\tan hmh}{\tan hm^1h^1} \quad \text{From (4)}$$

$$\Rightarrow \quad m \tan hmh = m^1 \tan hm^1h^1$$

Let E = total energy of liquid per unit length

$$= \frac{1}{2}\rho a^2 g$$

Cg = group velocity

W = rate of transmission of energy

Energy transmitted shall also be the same in either part of the canal.

Hence $W = W^1$ or $b \cdot E \cdot Cg = b^1 \cdot E^1 \cdot Cg^1$ ... (5)

$$\Rightarrow \quad \frac{b}{2}\rho a^2 g \frac{C}{2}\left(1 + \frac{2mh}{\sin h2mh}\right) = \frac{b^1}{2}\rho^1 a^{1^2} g \frac{C^1}{2}\left(1 + \frac{2m^1h^1}{\sin h2m^1h^1}\right)$$

$$\Rightarrow \quad \frac{\rho a^2 Cb(2mh + \sin h2mh)}{\sin h2mh} = \frac{\rho^1 a^{1^2} C^1 b^1 (2m^1h^1 + \sin h2m^1h^1)}{\sin h2m^1h^1} \quad \text{... (6)}$$

But $\frac{C}{C^1} = \frac{n}{m} \cdot \frac{m^1}{n^1} = \frac{m^1}{m}$ From (4)

Also $\rho = \rho^1$

Using (5), we get $\frac{C}{C^1} = \frac{\tan hmh}{\tan hm^1h^1}$

Using this in (6), we obtain

$$\frac{a^2 b(2mh + \sin h2mh)}{\sin h2mh} \frac{\tan hmh}{\tan hm^1h^1} = \frac{a^{1^2} b^1 (2m^1h^1 + \sin h2m^1h^1)}{\sin h2m^1h^1}$$

$$\Rightarrow \quad a^2 b(2mh + \sin h2mh) \sec h^2 mh = a^{1^2} b^1 (2m^1h^1 + \sin h2m^1h^1) \sec h^2 m^1 h^1$$

14. Obtain the conditions to be satisfied for small oscillation at the horizontal interface of two semi-infinite liquids of densities ρ and ρ^1 $(\rho > \rho^1)$ moving with general stream velocities U, U^1 in the same horizontal direction, the surface tension T being taken into account. Show that there are two possible wave velocities for a wavelength λ namely

$$C = \left(\frac{\rho U + \rho^1 U^1}{\rho + \rho^1}\right) \pm \left[\frac{\lambda g}{2\pi}\left(\frac{\rho - \rho^1}{\rho + \rho^1}\right) + \frac{2\pi T}{\lambda(\rho + \rho^1)} - \frac{\rho\rho^1 (U^1 - U)^2}{(\rho + \rho^1)^2}\right]^{\frac{1}{2}}$$

Solution: The required condition is

$$Tm^2 + g(\rho - \rho^1) = m\left[\rho^1 (U^1 - C)^2 \cot hmh^1 + \rho (U - C)^2 \cot hmh\right] \quad \dots (1)$$

Since $h \to \infty$, $h^1 \to \infty$ so that $\cot hmh \to 1$

$\cot hmh^1 \to 1$

Then (1) becomes

$$Tm^2 + g(\rho - \rho^1) = m\left[\rho^1 (U^1 - C)^2 + \rho(U - C)^2\right]$$

$$= m\left[C^2(\rho + \rho^1) - 2mC(\rho^1 U^1 + \rho U) + (\rho^1 U^{1^2} + \rho U^2)\right]$$

$$\Rightarrow \quad mC^2\left(\rho^1+\rho\right)-2mC\left(\rho^1U^1+\rho U\right)$$
$$+\left[\left(\rho^1U^{1^2}+\rho U^2\right)m-Tm^2-g\left(\rho-\rho^1\right)\right]=0$$

This is a quadratic in C

$$C=\frac{2m\left(\rho^1U^1+\rho U\right)\pm\left[4m\left(\rho^1U^1+\rho U\right)^2-4m\left(\rho^1+\rho\right)\left(\rho^1U^{1^2}+\rho U^2\right)m-Tm^2-g\left(\rho-\rho^1\right)\right]^{\frac{1}{2}}}{2m\left(\rho^1+\rho\right)}$$

Taking $A=\dfrac{\rho U+\rho^1U^1}{\rho+\rho^1}$, we get

$$C=A\pm\frac{1}{2m\left(\rho+\rho^1\right)}\left[4m^2\left\{\left(\rho^1U^1+\rho U\right)^2-\left(\rho+\rho^1\right)\left(\rho^{1^2}U^{1^2}+\rho U^2\right)\right\}+4m^2\left(\rho+\rho^1\right)T+4mg\left(\rho^2-\rho^{1^2}\right)\right]^{\frac{1}{2}} \quad \ldots (2)$$

But

$$\left(\rho^1U^1+\rho U\right)^2-\left(\rho+\rho^1\right)\left(\rho^1U^{1^2}+\rho U^2\right)$$
$$=\rho\rho^1\left(2UU^1-U^2-U^{1^2}\right)=-\rho\rho^1\left(U-U^1\right)^2 \quad \ldots (3)$$

Now (3) becomes

$$C=A\pm\left[-\rho\rho^1\left(\frac{U-U^1}{\rho+\rho^1}\right)^2+\frac{Tm}{\rho+\rho^1}+\frac{g}{m}\left(\frac{\rho-\rho^1}{\rho+\rho^1}\right)\right]^{\frac{1}{2}}$$

With $m=\dfrac{2\pi}{\lambda}$, $A=\dfrac{\rho U+\rho^1U^1}{\rho+\rho^1}$.

EXERCISE

1. A fixed buoy in deep water is observed to rise and fall twenty times in a minute. Prove that the velocity of waves is about ten and half miles per hour.

2. The crests of rollers which are directly following a ship 220 ft long are observed to overtake it at intervals of $16\frac{1}{2}$ seconds and it takes a crest of 6 seconds to run along the ship. Find the length of the wave and velocity of the ship

$$\left[\begin{array}{l} C = 37\frac{1}{2} \text{ miles/hour} \\ \text{and } u = 12.5 \text{ miles/hour} \end{array}\right]$$

3. Liquid of density ρ and depth h lies over fixed horizontal bed, above it is a layer of liquid of density $\rho^1(<\rho)$ and thickness h^1 and the upper surface is a fixed horizontal plane. Obtain an equation for the velocity C of waves of length $\frac{2\pi}{m}$ at the common surface, the surface tension between the two liquids being T. Prove that if h, h^1 are both small compared with λ then

$$C^2 = hh^1\left[\frac{\{(\rho-\rho^1)g + m^2T\}}{\rho h^1 + \rho^1 h}\right]$$

$$\left[\begin{array}{l} \textbf{Ans.} \text{ Condition is } Tm^2 + g(\rho-\rho^1) \\ = m\left[\rho^1(U^1 - C)^2 \cot hmh^1 + \rho(U-C)^2 \cot hmh\right] \end{array}\right]$$

LIST OF IMPORTANT FORMULAE

Chapter 1—Kinematics

1. $\vec{q} = -\vec{\nabla}\phi$ where ϕ is a velocity potential.
2. $$\frac{d}{dt} = \frac{\partial}{\partial t} + \vec{q} \cdot \vec{\nabla} = \frac{\partial}{\partial t} + u\frac{\partial}{\partial x} + v\frac{\partial}{\partial y} + w\frac{\partial}{\partial z}.$$
   where $\vec{q} = u\vec{i} + v\vec{j} + w\vec{k}$.
3. Flux = Density · Area of the surface · Normal velocity.
4. Equation of continuity
   (i) In Eulerian form is $\frac{d\rho}{dt} + \rho\vec{\nabla} \cdot \vec{q} = 0$
   (ii) In Lagrangian form is $\rho \dot{J} = \rho_0$
   (iii) In Cartesian form is
   $$\frac{\partial \rho}{\partial t} + \frac{\partial}{\partial x}(\rho u) + \frac{\partial}{\partial y}(\rho v) + \frac{\partial}{\partial z}(\rho w)$$
   (iv) In Orthogonal Curvilinear coordinates is
   $$\frac{\partial}{\partial t}(\rho h_1\, da_1 h_2 da_2 h_3 da_3)$$
   $$= -\left[h_1 da_1 \frac{\partial}{h_1 \partial a_1}(\rho q_1 h_2 da_2 h_3 da_3) + h_2 da_2 \frac{\partial}{h_2 \partial a_2}(\rho q_2 h_1 da_1 h_3 da_3) + h_3 da_3 \frac{\partial}{h_3 \partial a_3}(\rho q_3 h_1 da_1 h_2 da_2)\right]$$

5. The condition for the surface $F(x, y, z, t) = 0$ to be a possible form of a boundary surface is that

$$\left(u\frac{\partial}{\partial x} + v\frac{\partial}{\partial w} + w\frac{\partial}{\partial z} + \frac{\partial}{\partial t}\right)F = 0$$

6. The differential equation of streamlines is $\frac{dx}{u} = \frac{dy}{v} = \frac{dz}{w}$.
7. The condition for irrotational motion is $w = Curl\ \vec{q} = 0$.
8. (i) The spherical line element is

$$ds^2 = (dr)^2 + (rd\theta)^2 + (r\sin\theta dw)^2$$

(ii) The cylindrical line element is

$$ds^2 = (dr)^2 + (rd\theta)^2 + (dz)^2$$

9. $u = \frac{-\partial\phi}{\partial x} = \frac{-\partial\phi}{\partial r}, v = \frac{-\partial\phi}{\partial y} = -\frac{1}{r}\frac{\partial\phi}{\partial\theta}$

$$w = \frac{-\partial\phi}{\partial z} = -\frac{1}{r\sin\theta}\frac{\partial\phi}{\partial w}$$

Chapter 2—Equations of Motion

1. Equation of continuity in different forms

(i) $\frac{\partial u}{\partial x} + \frac{\partial v}{\partial y} + \frac{\partial w}{\partial z} = 0$ for liquid

(ii) $\dot{x}^2 v = F(t)$ when motion has spherical symmetry

(iii) $xv = F(t)$ when motion has cylindrical symmerty

(iv) $\frac{\partial\rho}{\partial t} + \frac{\partial}{\partial x}(\rho v) = 0$ when velocity has one component

2. (i) Euler's equation of motion is

$$\frac{d\vec{q}}{dt} = \vec{F} - \frac{1}{\rho}\vec{\nabla}p$$

(ii) Equation of motion when velocity has one component is

$$\frac{\partial v}{\partial t} + v\frac{\partial v}{\partial x} = F - \frac{1}{\rho}\frac{\partial p}{\partial x}$$

(iii) Equation of motion for symmetrical motion is

$$\frac{\partial v}{\partial t}+v\frac{\partial v}{\partial r}=F-\frac{1}{\rho}\frac{\partial p}{\partial r}$$

(iv) Bernoulli's pressure equations are

$$\int\frac{dp}{\rho}+\frac{1}{2}q^2+\Omega=C \text{ for steady motion}$$

and

$$\int\frac{dp}{\rho}-\frac{\partial\phi}{\partial t}+\frac{1}{2}q^2+\Omega=C \text{ for unsteady motion}$$

3. Equation for impulsive action is

$$\vec{q}_2-\vec{q}_1=I-\frac{1}{\rho}\vec{\nabla}\bar{w}$$

4. Helmholtz vorticity equation is

$$\frac{d}{dt}\left(\frac{\bar{w}}{\rho}\right)=\left(\frac{\vec{w}\cdot\vec{\nabla}}{\rho}\right)\vec{q}$$

Chapter 3—Motion in Two Dimensions

1. Lagrange's function ψ is given by

$$\frac{-\partial\phi}{\partial x}=u=\frac{-\partial\psi}{\partial y},\ \frac{-\partial\phi}{\partial y}=v=\frac{\partial\psi}{\partial x}$$

2. Complex potential w is given by

$$w=\phi+i\psi,\ -\frac{dw}{dz}=u-iv$$

$$\frac{\partial\phi}{\partial r}=\frac{1}{r}\frac{\partial\psi}{\partial\theta},\ \frac{1}{r}\frac{\partial\phi}{\partial\theta}=-\frac{\partial\psi}{\partial r}$$

3. Complex potential due to a source $+m$at $z = a$ is $w=-m\log(z-a)$; in this case $\phi = -m \log r$ $\psi = -m\theta$

4. Complex potential due to a doublet μ at $z = a$, angle α is $w=\frac{\mu e^{i\alpha}}{z-a}$.

5. Image of a source $+m$ with respect to a line is a source $+m$ at an equal distance on the other side of the line.

6. Image of a doublet μ relative to a line is a doublet μ at an equal distance on the other side of the line and its axis is antiparallel to the object doublet.
7. Image of a source $+m$ w.r.t. a circle is a source $+m$ at the inverse point and sin $K - m$ at the centre.
8. Image of a doublet of strength μ with its axis inclined at an angle a w.r.t. circle is a doublet of strength $\mu^1 = \frac{\mu a^2}{f^2}$ with its axis inclined at an angle $\pi - \alpha$ where a is the radius of the circle and f is the distance of doublet from the centre.
9. Blasius Theorem is $X - iY = \frac{i\rho}{2} \int_C \left(\frac{dW}{dz}\right)^2 dz$

Chapter 4—Motion of Cylinders

Part I: Circular Cylinder

1. K.E. of liquid $= -\frac{1}{2}\rho \int \phi \frac{\partial \phi}{\partial \eta} ds$ on $r = a$

$$= -\frac{1}{2}\rho \int_0^{2\pi} \left(\phi \frac{\partial \phi}{\partial r}\right)_{r=a} a d\theta$$

2. General motion of cylinder is

$$\psi = vx - uy + \frac{w}{2}\left(x^2 + y^2\right) + C$$

3. When circular cylinder in motion with velocity u along x-axis, then $w = \frac{ua^2}{z}$.
4. When circular cylinder is streaming past a fixed cylinder, then $w = uz + \frac{ua^2}{z}$ and $w = \frac{iK}{2\pi} \log z$ due to circulation of strength K only.

Part II: Elliptic Cylinders

1. Elliptic coordinates are

$$z = C \cosh \zeta = C \cosh(\xi + i\eta),\ C^2 = a^2 - b^2$$

$$a = C \cosh \alpha,\ b = C \sinh \alpha,\ a + b = Ce^{\alpha}$$

2. $T = K.E.$ of the liquid $= -\frac{1}{2}\int(\phi d\psi)_{\xi=\alpha}$

3. Solution of $\vec{\nabla}^2\psi$ is $\psi = e^{-n\xi}(A\cos n\eta + B\sin n\eta)$

$$\frac{\partial\phi}{\partial\xi} = \frac{\partial\psi}{\partial\eta}$$

4. If an elliptic cylinder is rotating in a liquid, then

$$\phi = \frac{w}{4}(a+b)^2 e^{-2\xi}\sin 2\eta,\ \psi = \frac{w}{4}(a+b)^2 e^{-2\xi}\cos 2\eta$$

$T = K.E.$ of the liquid $= \frac{\pi}{16}\rho w^2 C^4$

5. K.E. of the liquid contained in a rotating elliptic cylinder is $T = \frac{\pi}{8}\rho ab\frac{(a^2-b^2)^2}{(a^2+b^2)}w.$

Chapter 5—Irrotational Motion in Three Dimensions

1. When a sphere is moving with velocity u in a liquid,

$$\phi = \frac{ua^3}{2r^2}\cos\theta$$

2. When liquid is moving with velocity u in presence of a fixed sphere, then $\phi = ur\cos\theta + \frac{ua^3}{2r^2}\cos\theta$

3. K.E. of the liquid $= -\frac{1}{2}\rho\int\phi\frac{\partial\phi}{\partial\eta}ds$

$$= -\frac{1}{2}\rho\int\left(\phi\frac{\partial\phi}{\partial r}\right)_{r=a} 2\pi a\sin\theta\cos\theta$$

Chapter 7—Vortex Motion

1. Complex potential due to a vortex $+K$ at $z = a$ is $w = \frac{iK}{2\pi}\log(z-a)$

2. If there exists two vortices K_1, K_2 at A_1, A_2 then the line A_1, A_2 rotates with angular velocity $w = \frac{K_1+K_2}{2\pi(A_1A_2)^2}$ about O, whereas the points A_1, A_2

move with velocity $\dfrac{K_2}{2\pi A_1A_2}, \dfrac{K_1}{2\pi A_1A_2}$ along a line perpendicular to the line A_1A_2 respectively.

3. The image of a vortex $+K$ inside a circular boundary is a vortex $-K$ at the inverse point.
4. The image of a vortex $+K$ outside a circular boundary is a vortex $-K$ at the inverse point and vortex $+K$ at the centre.
5. The image of a vortex $+K$ w.r.t. a plane is a vortex $-K$ placed at the optical image of the given vortex.

Chapter 8—Waves

1. Equation of simple harmonic progressive wave is $\eta = a\sin(mx - nt)$

   Also $\lambda = \dfrac{2\pi}{m}, T = \dfrac{2\pi}{n}, cT = \lambda, \dfrac{1}{T} =$ Frequency.

2. For total wave $\dfrac{h}{\lambda} \to 0.$
3. Surface condition for small elevation $\eta = a\sin(mx - nt)$ is

   $$\frac{\partial^2\phi}{\partial t^2} + g\frac{\partial\phi}{\partial y} = 0 \text{ at } y = 0 \text{ with } \frac{\partial\eta}{\partial t} = v = \frac{-\partial\phi}{\partial y} \text{ at } y = 0.$$

4. For wave propagation in a canal of uniform depth h, the complex potential w is given by

   $$w = \frac{ac}{\sin hmh}\cos\left[m(z + ih) - nt\right]$$

   with $c^2 = \dfrac{n^2}{m^2} = \dfrac{g}{m}\tan hmh$

   Consequently $\phi = \dfrac{ac}{\sin hmh}\cos hm(y + h)\cos(mx - nt)$

   If $h \to \infty$ then $\phi = ace^{my}\cos(mx - nt)$, $c^2 = \dfrac{g}{m}$

   For long waves $c^2 = gh$.

5. K.E. $= \frac{1}{2}\int_0^\lambda \left(\phi \frac{\partial \phi}{\partial y}\right)_{y=0} dx$, P.E. $= \frac{1}{2}\rho g \int_0^\lambda \eta^2 dx$

6. K.E. = P.E. = $\frac{1}{4}\rho a^2 g\lambda$ per wavelength for the wave $\eta = a \sin(mx - nt)$.

7. For waves at the common surface of two liquids

$$g(\rho - \rho^1) = \rho(u-c)^2 m \cot hmh + \rho^1 (u^1 - c)^2 m \cot hmh^1$$

where $\rho > \rho^1$

8. For waves at the common surface of two liquids with upper surface free

$$c^4 m^2 (\rho \cot hmh \cot hmh^1 + \rho^1) - c^2 mg (\cot hmh + \cot hmh^1) + g^2(\rho - \rho^1) = 0$$

where $\rho > \rho^1$

If $h \to \infty$ then $\cot hmh = 1$.

9. If a liquid of density ρ^1 and depth h^1 lies over a liquid of density ρ and depth h, then

$\phi = \frac{ac}{\sin hmh} \cos hm(y+h) \cos(mx - nt)$ for lower liquid

and

$\phi = \frac{-ac}{\sin hmh^1} \cos hm(y - h^1) \cos(mx - nt)$ for upper liquid

10. Group velocity $= cg = \frac{c}{2}\left[1 + \frac{2mh}{\sin h2mh}\right]$

11. If w be the rate of transmission of energy, then

$w = E \cdot Cg$ where $E = \frac{1}{2}\rho a^2 g$

= total energy per unit length.

12. For capillary waves at the common surface of two liquids

$$Tm^2 + g\left(\rho - \rho^1\right) = m\left[\rho^1\left(u^1 - c\right)^2 \cot hmh^1 + \rho\left(u - c\right)^2 \cot hmh\right]$$

where $p_1 - p_2 = T\left(\dfrac{1}{r_1} + \dfrac{1}{r_2}\right) = -T\dfrac{\partial^2 \eta}{\partial x^2}$

Index of Symbols

Symbol	Meaning
$\forall$	For every
w.r.t.	with respect to
K.E.	Kinetic Energy
P.E.	Potential Energy
$\Rightarrow$	Implies
$\exists$	There exists
$\Leftrightarrow$	Implies and is implied by
iff	if and only if